Immunohistochemical and Physiological Research on Farm Animals

Immunohistochemical and Physiological Research on Farm Animals

Editors

Paola Scocco
Elena De Felice
Alessandro Malfatti

MDPI • Basel • Beijing • Wuhan • Barcelona • Belgrade • Manchester • Tokyo • Cluj • Tianjin

Editors

Paola Scocco
University of Camerino
Italy

Elena De Felice
University of Camerino
Italy

Alessandro Malfatti
University of Camerino
Italy

Editorial Office
MDPI
St. Alban-Anlage 66
4052 Basel, Switzerland

This is a reprint of articles from the Special Issue published online in the open access journal *Animals* (ISSN 2076-2615) (available at: https://www.mdpi.com/journal/animals/special_issues/farm_animals_immunohistochemistry).

For citation purposes, cite each article independently as indicated on the article page online and as indicated below:

LastName, A.A.; LastName, B.B.; LastName, C.C. Article Title. *Journal Name* **Year**, *Volume Number*, Page Range.

ISBN 978-3-0365-6882-9 (Hbk)
ISBN 978-3-0365-6883-6 (PDF)

Cover image courtesy of Dr. Paola Scocco
Flock of sheep on the hill of San Marcello, near the University of Camerino, Italy

Contents

About the Editors ix

Preface to "Immunohistochemical and Physiological Research on Farm Animals" xi

Paola Scocco, Elena De Felice and Alessandro Malfatti
Immunohistochemical and Physiological Research on Farm Animals
Reprinted from: *Animals* **2023**, *13*, 739, doi:10.3390/ani13040739 1

Michele Premi, Matteo Mezzetti, Giulia Ferronato, Mario Barbato, Fiorenzo Piccioli Cappelli, Andrea Minuti, et al.
Changes of Plasma Analytes Reflecting Metabolic Adaptation to the Different Stages of the Lactation Cycle in Healthy Multiparous Holstein Dairy Cows Raised in High-Welfare Conditions
Reprinted from: *Animals* **2021**, *11*, 1714, doi:10.3390/ani11061714 7

Vittoria Lucia Barile, Laura Menchetti, Anna Beatrice Casano, Gabriele Brecchia, Noelita Melo de Sousa, Riccardo Zelli, et al.
Approaches to Identify Pregnancy Failure in Buffalo Cows
Reprinted from: *Animals* **2021**, *11*, 487, doi:10.3390/ani11020487 25

Elena De Felice, Daniela Giaquinto, Sara Damiano, Angela Salzano, Simona Fabroni, Roberto Ciarcia, et al.
Distinct Pattern of NPY in Gastro–Entero–Pancreatic System of Goat Kids Fed with a New Standardized Red Orange and Lemon Extract (RLE)
Reprinted from: *Animals* **2021**, *11*, 449, doi:10.3390/ani11020449 41

Guan Wang, Yuzhu Luo, Jiang Hu, Jiqing Wang, Xiu Liu and Shaobin Li
Effects of Aging on Expression of *Mic60* and *OPA1* and Mitochondrial Morphology in Myocardium of Tibetan Sheep
Reprinted from: *Animals* **2020**, *10*, 2160, doi:10.3390/ani10112160 55

Luca Melotti, Tiziana Martinello, Anna Perazzi, Ilaria Iacopetti, Cinzia Ferrario, Michela Sugni, et al.
A Prototype Skin Substitute, Made of Recycled Marine Collagen, Improves the Skin Regeneration of Sheep
Reprinted from: *Animals* **2021**, *11*, 1219, doi:10.3390/ani11051219 65

Martina De Carolis, Olimpia Barbato, Gabriele Acuti, Massimo Trabalza-Marinucci, Noelita Melo de Sousa, Claudio Canali, et al.
Plasmatic Profile of Pregnancy-Associated Glycoprotein (PAG) during Gestation and Postpartum in Sarda and Lacaune Sheep Determined with Two Radioimmunoassay Systems
Reprinted from: *Animals* **2021**, *10*, 1502, doi:10.3390/ani10091502 85

Olimpia Barbato, Elena De Felice, Luca Todini, Laura Menchetti, Alessandro Malfatti and Paola Scocco
Effects of Feed Supplementation on Nesfatin-1, Insulin, Glucagon, Leptin, T3, Cortisol, and BCS in Milking Ewes Grazing on Semi-Natural Pastures
Reprinted from: *Animals* **2021**, *11*, 682, doi:10.3390/ani11030682 99

Jiangfeng Fan, Xiaohong Han, Honghong He, Yuzhu Luo, Sijiu Yu, Yan Cui, et al.
The Expression of ERK1/2 in Female Yak (*Bos grunniens*) Reproductive Organs
Reprinted from: *Animals* **2020**, *10*, 334, doi:10.3390/ani10020334 113

Cecilia Dall'Aglio, Francesca Mercati, Elena De Felice, Federico Maria Tardella, Josef Kamphues, Maria Grazia Cappai, et al.
Influence of Different Feed Physical Forms on Mandibular Gland in Growing Pigs
Reprinted from: *Animals* **2020**, *10*, 910, doi:10.3390/ani10050910 . . . . **127**

Katarzyna Palus, Michał Bulc and Jarosław Całka
Effect of Acrylamide Supplementation on the CART-, VAChT-, and nNOS-Immunoreactive Nervous Structures in the Porcine Stomach
Reprinted from: *Animals* **2020**, *10*, 555, doi:10.3390/ani10040555 . . . . **139**

Andrea Toschi, Giorgia Galiazzo, Andrea Piva, Claudio Tagliavia, Gemma Mazzuoli-Weber, Roberto Chiocchetti and Ester Grilli
Cannabinoid and Cannabinoid-Related Receptors in the Myenteric Plexus of the Porcine Ileum
Reprinted from: *Animals* **2021**, *11*, 263, doi:10.3390/ani11020263 . . . . **155**

Liang Tian, Jiahe Huang, Aiyou Wen and Peishi Yan
Impaired Mitochondrial Function Results from Oxidative Stress in the Full-Term Placenta of Sows with Excessive Back-Fat
Reprinted from: *Animals* **2020**, *10*, 360, doi:10.3390/ani10020360 . . . . **169**

Barbara Jana and Jarosław Całka
Endometritis Changes the Neurochemical Characteristics of the Caudal Mesenteric Ganglion Neurons Supplying the Gilt Uterus
Reprinted from: *Animals* **2020**, *10*, 891, doi:10.3390/ani10050891 . . . . **189**

Federico Armando, Francesco Godizzi, Elisabetta Razzuoli, Fabio Leonardi, Mario Angelone, Attilio Corradi, et al.
Epithelial to Mesenchymal Transition (EMT) in a Laryngeal Squamous Cell Carcinoma of a Horse: Future Perspectives
Reprinted from: *Animals* **2020**, *10*, 2318, doi:10.3390/ani10122318 . . . . **203**

Ahmed A. A. Abdel-Wareth and Abdallah E. Metwally
Productive and Physiological Response of Male Rabbits to Dietary Supplementation with Thyme Essential Oil
Reprinted from: *Animals* **2020**, *10*, 1844, doi:10.3390/ani10101844 . . . . **217**

Silvia Preziuso
Severe Acute Respiratory Syndrome Coronavirus 2 (SARS-CoV-2) Exhibits High Predicted Binding Affinity to ACE2 from Lagomorphs (Rabbits and Pikas)
Reprinted from: *Animals* **2020**, *10*, 1460, doi:10.3390/ani10091460 . . . . **227**

Magdalena Prusik and Bogdan Lewczuk
Diurnal Rhythm of Plasma Melatonin Concentration in the Domestic Turkey and Its Regulation by Light and Endogenous Oscillators
Reprinted from: *Animals* **2020**, *10*, 678, doi:10.3390/ani10040678 . . . . **239**

Antonio Carminato, Francesco Pascoli, Angela Trocino, Lisa Locatello, Lisa Maccatrozzo, Renato Palazzi, et al.
Productive Results, Oxidative Stress and Contaminant Markers in European Sea Bass: Conventional vs. Organic Feeding
Reprinted from: *Animals* **2020**, *10*, 1226, doi:10.3390/ani10071226 . . . . **249**

Nicole Verdile, Rolando Pasquariello, Marco Scolari, Giulia Scirè, Tiziana A. L. Brevini and Fulvio Gandolfi
A Detailed Study of Rainbow Trout (*Onchorhynchus mykiss*) Intestine Revealed That Digestive and Absorptive Functions Are Not Linearly Distributed along Its Length
Reprinted from: *Animals* **2020**, *10*, 745, doi:10.3390/ani10040745 . . . . **265**

Roberta Imperatore, Lea Tunisi, Isabella Mavaro, Livia D'Angelo, Chiara Attanasio, Omid Safari, et al.
Immunohistochemical Analysis of Intestinal and Central Nervous System Morphology in an Obese Animal Model (*Danio rerio*) Treated with 3,5-T2: A Possible Farm Management Practice?
Reprinted from: *Animals* **2020**, *10*, 1131, doi:10.3390/ani10071131 . . . . **285**

About the Editors

Paola Scocco

Paola Scocco is a Professor of the Anatomy of domestic animals, Histology and Embryology at the University of Camerino, Italy.

She graduated from the University of Camerino in Biological Sciences in 1986 (summa cum laude).

Her main research interests focus on the complex carbohydrates, neuropeptides and adipokines expressed in the digestive and reproductive apparatus of lab and farm animal species. In recent years, she has applied her anatomical competences to the conservative management of pastural ecosystems via the study of grazing animals and the consideration of the wild species.

She was the Scientific Director of the didactic unit "The business management of mountain farms as a tool for the conservation of pastures and development of the livestock farm" and was the Project Leader of the inter-territorial cooperation program "Laboratory of the Environment and Landscape-APE: Apennines Park of Europe". In addition, she was the Scientific Coordinator of the morpho-physiological group that dealt with the "Morphometric analysis of issues related to the state of animal welfare" within the "Zootechnics and fire prevention" project of the Marche Region (Italy).

She is a member of the Editorial Board of Animals and a member of the Topic Board of Veterinary Sciences (MDPI); until 2021, was a member of the Editorial Board of the European Journal of Histochemistry.

Elena De Felice

Elena De Felice is a researcher of the Anatomy of Domestic Animals at the University of Camerino, Italy.

She graduated in Veterinary Medicine in 2008 at the University of Naples Federico II (110/110 with honors). In 2012, she obtained her PhD at the Faculty of Veterinary Medicine of the University of Messina. In 2014, she specialized in technology and the pathology of poultry, rabbit and game species at the Federico II University of Naples. She worked from 2009 to 2014 at the Zoological Station "A. Dohrn" in Naples, from 2015 to 2016 at the Tigem "Telethon Institute of Genetics and Medicine" in Pozzuoli, and in 2017, at the A.O.R.N. Cardarelli of Naples. Since 2017, she has been a researcher of veterinary anatomy s.s.d. VET/01 at the School of Biosciences and Veterinary Medicine of the University of Camerino (2017–2021 type a researcher; 2021–to date type b researcher). Here, since 2017she has taught the course Systematic and Comparative Veterinary Anatomy; since 2021, she has also taught the course Topographic Veterinary Anatomy in the Degree in Veterinary Medicine. She has also been teaching since 2018 at the International School of Advanced Studies; in addition, since 2021, she has taught at the Specialization School of the University of Camerino in Hygiene and the Control of Fishery and Aquaculture Products. She is the author and co-author of several publications in international journals and contributions to national and international conferences. Her main research field is the study of molecules that are involved in the mechanisms that regulate food intake and energy metabolism in animals of veterinary interest.

Alessandro Malfatti

Alessandro Malfatti is a Professor of Animal Physiology, Ethology and Animal Welfare at the University of Camerino, Italy.

He graduated from the University of Perugia (Italy) as a Doctor of Veterinary Medicine in 1982 (summa cum laude) and in 1990, he obtained a PhD in "Physiology of animal reproduction". His main research interests focus on the endocrinology of reproduction and the environmental adaptation of ovine, caprine and buffalo species, on the behaviour and animal welfare of shelter dogs and on free-grazing ruminants.

He is a reviewer for several international journals: Theriogenology, Animal Production Science, Italian Journal of Animal Science, Reproduction in Domestic Animals, Journal of Animal Physiology and Animal Nutrition, Small Ruminant Research, Behavioural Processes.

He has been involved in a number of other academic activities, including as a member of the Academic Senate (Perugia), a member and then Head of the University Evaluation Commission (Camerino), a delegate of the Rector of the Camerino University for the relationship with the University Evaluation Commission, a member of the Quality Assurance Board of the Camerino University and was also the Dean of the Master's degree "Biological Sciences".

Preface to "Immunohistochemical and Physiological Research on Farm Animals"

This book represents a collection of papers that aimed to highlight the importance of immunohistochemical findings and their role in improving the understanding of organ tissue functions. We focused our attention on farm animals because they represent not only a source of food in the supply chain, but also a source of income for breeders, a source of employment, and a primary tool in the quest for sustainable food production. To achieve this last aim, we asked authors for research that could potentially be applied to the purpose of this study, and whose results could be useful in improving farm management practices and the responsible use of natural resources.

Paola Scocco, Elena De Felice, and Alessandro Malfatti
Editors

animals

MDPI

Editorial

Immunohistochemical and Physiological Research on Farm Animals

Paola Scocco, Elena De Felice * and Alessandro Malfatti

School of Biosciences and Veterinary Medicine, University of Camerino (MC), 62032 Camerino, Italy
* Correspondence: elena.defelice@unicam.it

This Special Issue "Immunohistochemical and Physiological Research on Farm Animals" is dedicated to the application of immunohistochemical and physiological studies carried out on farm animals, including traditional (pig, cow, buffalo, horse, sheep, goat, rabbit, turkey and trout) and emerging farm species (yak, sea bass and zebrafish).

Farm species play very important economic and sociocultural roles, such as food supply, source of income, asset saving, source of employment, soil fertility, livelihood, transport, agricultural traction, agricultural diversification and sustainable agricultural production.

For this reason, it is essential to study the anatomy and the physiology of these species in more detail, in order to improve the knowledge about their morpho-physiological features and to guarantee them a high level of welfare and, in turn, better conditions for farmers, including farm income.

In the last few decades, there has been an exponential increase in publications on immunohistochemistry (IHC) and immunocytochemistry techniques, reflecting the current position that IHC holds in most pathological laboratories.

This Special Issue attempts to demonstrate the relevance of immunohistochemical analysis and its relationship with functions in farm animals. However, taking into account the fact that different methods (such as real-time q-PCR, Western blot, FISH, etc.) are necessary to draw more reliable conclusions, the use of these methods is also considered.

In addition to base research contributions, we asked authors for research with potential applied-purpose studies whose results could be useful in improving farm management practices and the responsible use of natural resources, enhancing the nutraceutical properties of animal-derived products and promoting the circular economy and, in turn, the sustainability of livestock.

A total of 20 papers have been contributed for this Special Issue by 118 authors from four countries, comprising 18 research articles and 2 communications.

The work of Premi and colleagues [1] investigates the physiological variations affecting plasma analyte concentrations during the pivotal stages of the lactation cycle in healthy multiparous Holstein dairy cows. They analyzed 34 different analytes, including markers of energy metabolism, protein metabolism and kidney function, mineral metabolism, liver function, inflammation and acute phase proteins, and oxidant status, in four different periods: the dry, the postpartum, the early and the late lactation phases. This study provides a guideline of physiological trends affecting plasma analyte concentrations during the different stages of the lactation cycle.

Barile and coworkers [2] identify the use of pregnancy-associated glycoproteins (PAGs) as the best strategy to diagnose pregnancy failures in buffalo (*Bubalus bubalis*). PAGs constitute a large family of glycoproteins expressed in the outer epithelial layer (chorion/trophectoderm) of the placenta in eutherian species. Through a radioimmunoassay applied on blood samples, PAGs were recognized as the best marker for predicting embryonic mortality between 25 and 40 days of gestation in buffalo.

Citation: Scocco, P.; De Felice, E.; Malfatti, A. Immunohistochemical and Physiological Research on Farm Animals. *Animals* **2023**, *13*, 739. https://doi.org/10.3390/ani13040739

Received: 10 February 2023
Accepted: 15 February 2023
Published: 18 February 2023

De Felice and colleagues [3] evaluate the effect of a diet supplemented with a standardized powder extract, red (blood) orange and lemon extract (RLE), rich in flavanones, anthocyanins and other polyphenols, on the neuropeptide Y (NPY) distribution in the gastro–entero–pancreatic system of goat (*Capra hircus*) kids. Through single and double immunostaining for the first time, the NPY distribution in the abomasum, duodenum, and pancreas and the co-localization with serotonin (5-HT) were documented. Upon RLE feed supplementation, NPY immunoreactive cells increased significantly in abomasal epithelium and pancreatic islets. NPY is known to regulate gastric acid secretion while on the endocrine pancreas to suppress insulin via autocrine and/or paracrine mechanisms, and to stimulate glucagon secretion. These observations represent a baseline for future studies on the interaction between neuropeptides and polyphenols used as feed additive instead of antibiotics.

Wang and coworkers [4] examine the expression of Mic60 and OPA1, two mitochondrial inner membrane proteins, and the morphology of mitochondria in the myocardium of adult and aged Tibetan sheep. With an integrated approach, including real-time q-PCR, ELISA, immunohistochemistry and the ultrastructure morphology of mitochondria with transmission electron microscopy, the authors suggest that the expression of Mic60 and OPA1 genes and OPA1 protein will reduce, while reducing the capacity of myocardial mitochondria with age.

Melotti and coworkers [5] describe the application of a collagen-based skin-like scaffold (CBSS), manufactured with collagen extracted from sea urchin food waste, to treat experimental skin wounds in sheep (*Ovis aries*). Collagen fibrils in their native conformation with endogen fibril-associated glycosaminoglycans can be extracted from sea urchins, obtaining a biomaterial resembling the in vivo structural microenvironment. Clinical, histopathological, immunohistochemical, and molecular experiments assess CBSS' effects on the wound healing process. The authors demonstrate the efficacy of this pioneering skin substitute in an in vivo model. The skin substitute supported and stimulated wound healing throughout the whole process; it controlled inflammation, promoted the deposition and maturation of granulation tissue, enhanced re-epithelialization, and induced the formation of skin appendages. This application explores the possibility of deriving high-value and innovative products for veterinary innovative applications from waste materials.

De Carolis and colleagues [6] study ovine pregnancy-associated glycoprotein (oPAG) levels in the plasma of Sarda and Lacaune ewes throughout gestation and in the first month postpartum, using two heterologous radioimmunoassays (RIA-706 and RIA-srPool). For both breeds, these RIA systems were capable of distinguishing pregnant from non-pregnant ewes starting from day 18 of gestation. The diagnosis of pregnancy at early gestation is fundamental to minimize the costs of unproductive animals.

Barbato and coworkers [7] evaluate the effects of cereal supplementation on body condition score and metabolic hormone profile in milking ewes grazing on semi-natural pastures of the Central Apennines in Italy during the grazing summer period. Sheep are the most-bred species in the Central Apennines, where the natural pastures are used as a trophic resource and grazing activity is fundamental to maintain the grassland biodiversity. Increasing summer aridity decreases the grassland pastoral value, negatively affecting animal morpho-functional features and production with detrimental effects on the sustainability of extensive sheep farming. This work represents a part of a wider study aimed at buffering the negative effects of increasing summer drought stress on farm income and maintaining the grassland biodiversity. Through enzyme immunoassays, radioimmunoassay and ELISA on blood samples, Nesfatin-1, insulin, glucagon, leptin, 3-3′-5-triiodothyronine and cortisol were evaluated. The results of this work indicate that nutritional supplementation has protected ewes from the usual lowering of the body condition linked to lactation, and provides a good maintenance of milk production, also determining a better overall body and metabolic state of the animals, which is important at the beginning of the sexual season.

Fan and colleagues [8], with a multidisciplinary approach including immunohistochemistry, Western blot and real-time q-PCR, investigate the expression and distribution of extracellular signal-regulated kinases1/2 (ERK1/2) in the main reproductive organs of female yak (*Bos grunniens*) during different stages. ERKs are an important subfamily of mitogen-activated protein kinases (MAPKs), which regulate various cellular activities and physiological processes. For example, ERK1/2 has a pivotal role in the ovulatory process or in the mechanism of regression and functional maintenance of corpus luteum. In female yak, the expression of ERK1 and ERK2 proteins and mRNA was most pronounced in the ovary in the luteal phase and gestation period, the oviduct in the luteal phase, and the uterus in the gestation period. The histological appearance and physiological process of the main reproductive organs also varied with the different reproductive stages. These results imply that ERK1/2 plays an important role in the regulation of reproductive functions in different physiological situations.

Dall'Aglio and colleagues [9] test the effects of different feed physical forms (different grinding intensities and compactions of the same diet) on the mandibular gland (MG) of growing pigs (*Sus scrofa domesticus*). Samples were analyzed using conventional histochemistry to identify the glycohistochemical profile and using immunohistochemistry to localize aquaporin 5, apelin and apelin receptor. This study demonstrates that different feed physical forms are capable of inducing morphological and functional modifications of the MG. The intense chewing activity related to the highest feed compaction and hardness promotes an increase in pig MG secretion, and saliva becomes more fluid and richer in acid glycoconjugates in order to better lubricate the bolus and protect the mouth mucosae. It has been hypothesized that the apelinergic system is likely involved in the above modifications, enhancing both the fluidity and the quantity of serous saliva.

Palus and coworkers [10] demonstrate the changes in the population of enteric neurons in the porcine stomach in response to the supplementation of low and high acrylamide doses. Using the double immunofluorescence staining method, it was established that supplementation with both doses resulted in an increased number of the cocaine- and amphetamine-regulated transcript (CART), vesicular acetylcholine transporter (VAChT), and neuronal isoform of nitric oxide synthase (nNOS) immunoreactive neurons. The detected alterations of the porcine stomach neuron phenotype suggest an important role of the enteric nervous system in protecting the gastrointestinal tract during acrylamide intoxication.

Toschi and colleagues [11] immunohistochemically analyze the localization in the myenteric plexus of the porcine ileum of both the cannabinoid receptors, namely CB1R and CB2R, the cannabinoid-related receptors TRP vanilloid 1 (TRPV1) and TRP ankyrin 1 (TRPA1), and 5-HT1, a serotonin receptor (5-HT1aR). In the gastro-intestinal tract, cannabinoid receptors are known to regulate motility, secretion, emesis, food intake, and inflammation. The morphological findings of this article could represent a relevant anatomical basis for future functional, pre-clinical and clinical studies assessing the effects of cannabinoids in the management of the hypermotility associated with gastrointestinal inflammatory diseases and pain in pigs.

Tian and coworkers [12] determine the effect of excessive back fat of sows on placental oxidative stress, ATP generation, mitochondrial alterations in content and structure, and mitochondrial function in isolated trophoblasts. Excessive back fat of sows was associated with increased plasma lipid and leptin levels, which were associated with increased systemic (elevated plasma H2O2 level) and placenta (high levels of placental protein carbonylation and GSSG) oxidative stress, likely because of an increase in placental reactive oxygen species (ROS) production and a reduction in antioxidant defenses. In an in vitro model of pig trophoblast cell culture, cytotrophoblasts from the placenta of sows with excessive back fat reveal the decreased mitochondrial maximum respiration and spare respiratory capacity. The data collected show that excessive back fat exacerbates mitochondrial injury induced by increased oxidative stress in pig term placenta, which may have deleterious repercussions on placental activity and, therefore, impair fetal growth and development.

Jana and Calka [13] analyze the influence of the *E. coli*-induced inflammatory state of the uterus on the neurochemical characteristics of the gilt (crossbred Large White × Landrace pigs) caudal mesenteric ganglion (CaMG) uterus-supplying neurons. After intrauterine bacterial injection, the population of uterine neurons presenting positive staining for dopamine-β-hydroxylase (an enzyme participating in noradrenaline synthesis) and negative staining for galanin, as well as the population of uterine neurons presenting negative staining for dopamine-β-hydroxylase but positive staining for neuropeptide Y, were decreased. Uterine inflammation causes changes in the spatial and neurochemical organization patterns of the CaMG neurons innervating the uterus.

Armando and coworkers [14] examine the so-called epithelial to mesenchymal transition in horse (*Equus caballus*) in a squamous cell carcinoma (SCC) using immunohistochemistry for the first time, thus illustrating an example of tumor cell adaptation during the metastatic process.

Abdel-Wareth and Metwally [15] evaluate the potential effects of thyme essential oil (TEO) as an alternative to dietary antibiotics on the productive performance and serum metabolic profile of male rabbits (*Oryctolagus cuniculus*). TEO levels up to 180 mg/kg can play a major role in improving the productive performance, semen quality, testosterone levels, and the kidney and liver functions, analyzed through blood biochemical assay.

Preziuso [16], in her communications, investigates the interaction between severe acute respiratory syndrome coronavirus 2 (SARS-CoV-2) spikes and ACE2 of lagomorphs, rabbit, and American pika (*Ochotona princeps*), by means of sequence analysis and structure stimulation. Preziuso proved that the simulated complexes ACE2-SARS-CoV-2-RBD showed a high affinity of ACE2 of lagomorphs to the viral spike protein, suggesting that lagomorphs could be susceptible to SARS-CoV-2.

Prusik and Lewczuk [17] typify the diurnal rhythm of plasma melatonin (MLT) concentration and its regulation by light and endogenous oscillators in turkeys (*Meleagris gallopavo domesticus*). Considering the significance of the domestic turkey as a meat-producing animal and the regulation of MLT of feed intake, plasma analysis showed that MLT was secreted in a daily rhythm with a very high amplitude, but also responds quickly and precisely to changes in light conditions.

With regard to teleost fish, Carminato and coworkers [18] investigated the effects of two different diets (organic vs. conventional) on European sea bass (*Dicentrarchus labrax*) in terms of growing performance, oxidative stress, and contaminant markers. The results of this study, which can be considered a pilot, point out a positive trend in the growing performance of both groups but a greater productivity of conventional fish feed compared to the organic ones on one hand, and significant differences among groups in terms of the oxidative stress and contaminant markers on the other.

Verdile and colleagues [19] perform a detailed characterization of the intestinal epithelial cells lining the intestinal tract in rainbow trout (*Oncorhynchus mykiss*) throughout the first year of development. The analysis was performed at typical time points of in vivo feeding trials (50, 150, and 500 g) in order to establish accurate reference values, especially across the productive cycle of animals raised in standardized conditions.

Imperatore and colleagues [20] analyze the effect of 3,5-diiodo-L-thyronine (3,5-T2), an endogenous metabolite of thyroid hormones whose administration to rodents fed a high-fat diet (HFD) prevents body weight increase and reverts the expression pattern of pro-inflammatory factors associated with HFD, in a diet-induced obese (D.I.O.) zebrafish (*Danio rerio*) model. The authors reveal that the effects of 3,5-T2 on fish intestines and brains can deviate from those shown in obese mammals; through the expression of different inflammatory markers, they determined that 3,5-T2 sustained or increased inflammation in the intestine when administered with the obesity-inducing diet.

Acknowledgments: We are grateful to the authors who contributed to this Special Issue for improving the anatomical and physiological knowledge of farm animals.

Conflicts of Interest: The authors declare no conflict of interest.

References

1. Premi, M.; Mezzetti, M.; Ferronato, G.; Barbato, M.; Cappelli Piccioli, F.; Minuti, A.; Trevisi, E. Changes of plasma analytes reflecting metabolic adaptation to the different stages of the lactation cycle in healthy multiparous Holstein dairy cows raised in high-welfare conditions. *Animals* **2021**, *11*, 1714. [CrossRef] [PubMed]
2. Barile, V.L.; Menchetti, L.; Casano, A.B.; Brecchia, G.; Melo de Sousa, N.; Zelli, R.; Canali, C.; Beckers, J.F.; Barbato, O. Approaches to identify pregnancy failure in Buffalo cows. *Animals* **2021**, *11*, 487. [CrossRef] [PubMed]
3. De Felice, E.; Giaquinto, D.; Damiano, S.; Salzano, A.; Fabroni, S.; Ciarcia, R.; Scocco, P.; de Girolamo, P.; D'Angelo, L. Distinct pattern of NPY in gastro–entero–pancreatic system of goat kids fed with a new standardized red orange and lemon extract (RLE). *Animals* **2021**, *11*, 449. [CrossRef] [PubMed]
4. Wang, G.; Luo, Y.; Hu, J.; Wang, J.; Liu, X.; Li, S. Effects of aging on expression of *Mic60* and *OPA1* and mitochondrial morphology in myocardium of Tibetan sheep. *Animals* **2020**, *10*, 2160. [CrossRef] [PubMed]
5. Melotti, L.; Martinello, T.; Perazzi, A.; Iacopetti, I.; Ferrario, C.; Sugni, M.; Sacchetto, R.; Patruno, M. A prototype skin substitute, made of recycled marine collagen, improves the skin regeneration of sheep. *Animals* **2021**, *11*, 1219. [CrossRef] [PubMed]
6. De Carolis, M.; Barbato, O.; Acuti, G.; Trabalza-Marinucci, M.; Melo de Sousa, N.; Canali, C.; Moscati, L. Plasmatic profile of pregnancy-associated glycoprotein (PAG) during gestation and postpartum in Sarda and Lacaune sheep determined with two radioimmunoassay systems. *Animals* **2020**, *10*, 1502. [CrossRef] [PubMed]
7. Barbato, O.; De Felice, E.; Todini, L.; Menchetti, L.; Malfatti, A.; Scocco, P. Effects of feed supplementation on Nesfatin-1, Insulin, Glucagon, Leptin, T3, Cortisol, and BCS in milking ewes grazing on semi-natural pastures. *Animals* **2021**, *11*, 682. [CrossRef] [PubMed]
8. Fan, J.; Han, X.; He, H.; Luo, H.; Yu, S.; Cui, Y.; Xu, G.; Wang, L.; Pan, Y. The expression of ERK1/2 in female yak (*Bos grunniens*) reproductive organs. *Animals* **2020**, *10*, 334. [CrossRef] [PubMed]
9. Dall'Aglio, C.; Mercati, F.; De Felice, E.; Tardella, F.M.; Kamphues, J.; Cappai, M.G.; Scocco, P. Influence of different feed physical forms on mandibular gland in growing pigs. *Animals* **2020**, *10*, 910. [CrossRef] [PubMed]
10. Palus, K.; Bulc, M.; Calka, J. Effect of acrylamide supplementation on the CART-, VAChT-, and nNOS-immunoreactive nervous structures in the porcine stomach. *Animals* **2020**, *10*, 555. [CrossRef] [PubMed]
11. Toschi, A.; Galiazzo, G.; Piva, A.; Tagliava, C.; Mazzuolli-Weber, G.; Chiocchetti, R.; Grilli, E. Cannabinoid and cannabinoid-related receptors in the myenteric plexus of the porcine ileum. *Animals* **2021**, *11*, 263. [CrossRef] [PubMed]
12. Tian, L.; Huang, J.; Wen, A.; Yan, P. Impaired mitochondrial function results from oxidative stress in the full-term placenta of sows with excessive back-fat. *Animals* **2020**, *10*, 360. [CrossRef] [PubMed]
13. Jana, B.; Calka, J. Endometritis changes the neurochemical characteristics of the caudal mesenteric ganglion neurons supplying the gilt uterus. *Animals* **2020**, *10*, 891. [CrossRef] [PubMed]
14. Armando, F.; Godizzi, F.; Razzuoli, E.; Leonardi, F.; Angelone, M.; Corradi, A.; Meloni, D.; Ferrari, L.; Passeri, B. Epithelial to mesenchymal transition (EMT) in a laryngeal squamous cell carcinoma of a horse: Future perspectives. *Animals* **2020**, *10*, 2318. [CrossRef] [PubMed]
15. Abdel-Wareth, A.A.A.; Metwally, A.E. Productive and physiological response of male rabbits to dietary supplementation with thyme essential oil. *Animals* **2020**, *10*, 1844. [CrossRef] [PubMed]
16. Preziuso, S. *Severe Acute Respiratory Syndrome Coronavirus 2* (SARS-CoV-2) exhibits high predicted binding affinity to ace2 from lagomorphs (rabbits and pikas). *Animals* **2020**, *10*, 1460. [CrossRef] [PubMed]
17. Prusik, M.; Lewczuk, B. Diurnal rhythm of plasma melatonin concentration in the domestic turkey and its regulation by light and endogenous oscillators. *Animals* **2020**, *10*, 678. [CrossRef] [PubMed]
18. Carminato, A.; Pascoli, F.; Trocino, A.; Locatello, L.; Maccatrozzo, L.; Palazzi, R.; Radaelli, G.; Ballarin, C.; Bortoletti, M.; Bertotto, D. Productive results, oxidative stress and contaminant markers in European sea bass: Conventional vs. organic feeding. *Animals* **2020**, *10*, 1226. [CrossRef]
19. Verdile, N.; Pasquariello, R.; Scolari, M.; Scirè, G.; Brevini, T.A.L.; Gandolfi, F. A detailed study of rainbow trout (*Onchorhynchus mykiss*) intestine revealed that digestive and absorptive functions are not linearly distributed along its length. *Animals* **2020**, *10*, 745. [CrossRef]
20. Imperatore, R.; Tunisi, L.; Mavaro, I.; D'Angelo, L.; Attanasio, C.; Safari, O.; Motlagh, H.A.; de Girolamo, P.; Cristino, L.; Varricchio, E.; et al. Immunohistochemical analysis of intestinal and central nervous system morphology in an obese animal model (*Danio rerio*) treated with 3,5-t2: A possible farm management practice? *Animals* **2020**, *10*, 1131. [CrossRef] [PubMed]

animals

MDPI

Article

Changes of Plasma Analytes Reflecting Metabolic Adaptation to the Different Stages of the Lactation Cycle in Healthy Multiparous Holstein Dairy Cows Raised in High-Welfare Conditions

Michele Premi †, Matteo Mezzetti †, Giulia Ferronato, Mario Barbato, Fiorenzo Piccioli Cappelli, Andrea Minuti and Erminio Trevisi *

Department of Animal Sciences, Food and Nutrition (DIANA), Facoltà di Scienze Agrarie, Alimentari e Ambientali, Università Cattolica del Sacro Cuore, 29122 Piacenza, Italy; michele.premi@unicatt.it (M.P.); matteo.mezzetti@unicatt.it (M.M.); giulia.ferronato@unicatt.it (G.F.); mario.barbato@unicatt.it (M.B.); Fiorenzo.piccioli@unicatt.it (F.P.C.); andrea.minuti@unicatt.it (A.M.)

* Correspondence: erminio.trevisi@unicatt.it; Tel.: +39-0523-599-278 or +39-328-2828-165; Fax: +39-0523-599-276

† These authors contributed equally.

Citation: Premi, M.; Mezzetti, M.; Ferronato, G.; Barbato, M.; Piccioli Cappelli, F.; Minuti, A.; Trevisi, E. Changes of Plasma Analytes Reflecting Metabolic Adaptation to the Different Stages of the Lactation Cycle in Healthy Multiparous Holstein Dairy Cows Raised in High-Welfare Conditions. *Animals* **2021**, *11*, 1714. https://doi.org/10.3390/ani11061714

Academic Editors: Paola Scocco, Elena De Felice, Alessandro Malfatti and Michael Hässig

Received: 22 March 2021
Accepted: 3 June 2021
Published: 8 June 2021

Publisher's Note: MDPI stays neutral with regard to jurisdictional claims in published maps and institutional affiliations.

Simple Summary: This study investigates the changes occurring in plasma analytes of healthy multiparous Holstein dairy cows during the dry, the postpartum, the early and the late lactation phases. A welfare assessment at the herd level and a retrospective subclinical diseases screening were used as blocking factors for the selection of reference individuals. Thus, this study provides measurements of the physiological variations affecting plasma analytes concentrations during the pivotal stages of the lactation cycle in a healthy, high welfare-raised subset of reference individuals and suggest an explanation for the underlying processes involved. Finally, we propose reference intervals for plasma analytes in the stages investigated.

Abstract: Here, we tested the changes occurring in several plasma analytes during different stages of the lactation cycle of high welfare raised multiparous Holstein cows, and provided reference intervals (RI) for plasma analytes concentrations. Eleven high-welfare farms (HWF) located in Northern Italy were selected and their herds used to recruit 361 clinically healthy cows undergoing the dry (from −30 to −10 days from real calving; DFC), the postpartum (from 3 to 7 DFC), the early lactation (from 28 to 45 DFC) and the late lactation phases (from 160 to 305 DFC). Cows affected by subclinical diseases (SCD) were retrospectively excluded, and a subset of 285 cows was selected. Data of plasma analytes underwent ANOVA testing using physiological phases as predictors. The individual effect of each phase was assessed using a pairwise *t*-test assuming $p \leq 0.05$ as a significance limit. A bootstrap approach was used to define the reference interval (RI) for each blood analyte within physiological phases having a pairwise *t*-test $p \leq 0.05$. The concentration of nonesterified fatty acids, albumin, cholesterol, retinol, paraoxonase and tocopherol changed throughout all the physiological phases, whereas the concentration of K, alkaline phosphatase and thiol groups remained stable. Triglycerides, Zn, and ferric ion reducing antioxidant power in the dry phase and BHB, Ca, myeloperoxidase, haptoglobin, reactive oxygen metabolites and advanced oxidation of protein product in postpartum differed compared with other physiological phases. During the dry phase, Packed cell volume, Cl, and urea concentrations were similar to during the postpartum phase. Similarly, Na, γ-glutamyl transferase and β-carotene concentrations were similar to during the early lactation phase; fructosamine and bilirubin concentrations were similar to during the late lactation phase. During the postpartum phase, fructosamine and P concentrations were similar to during the early lactation phase, and the aspartate transaminase concentration was similar to during the late lactation phase. During the early lactation phase, Mg, creatinine, total protein, globulin and ceruloplasmin concentrations were similar to during the postpartum phase, while the urea concentration was similar to during the late lactation phase. All these plasma analytes differed among the other phases. This study identifies physiological trends affecting plasma analytes concentrations during the different stages of the lactation cycle and provides a guideline for the duration and magnitude of their changes when animals are healthy and raised in optimal welfare conditions.

Keywords: healthy reference individuals; high-welfare farm; metabolic profile; subclinical disorders

1. Introduction

Dairy cows undergoing different phases of the productive cycle face feed, management and environmental changes, as well as alteration in the endocrine, immune and metabolic assets. These changes are reflected by typical trends of plasma analytes. Thus, assessing multiple plasma analytes at the herd level has the potential to provide an overview on the feeding and management strategies adopted in dairy farms, allowing the prompt evaluation of animal responses to the different phases of the lactation cycle [1]. Trends of plasma analytes included in the metabolic profile of dairy cows have already been investigated during the dry period (from 60 to 10 days before expected calving date; [2–4]), the transition to calving (from 7 days before to 7 days after calving [3,5]), the early lactation (from 10 to 100 DIM; [2,5–7]) and the mid-late lactation period (from 90 to 215 DIM; [6]). However, an assessment of the welfare status combined with a subclinical disorders (SCD) screening has never been performed on the animals prior their enrollment in the experiment, despite the known role of these factors in affecting the physiological trends of plasma analytes during the lactation cycle [8–10].

Stressful conditions due to suboptimal welfare are known to induce severe alteration in plasma analytes, both in humans and dairy cows [2,11,12]. Average welfare conditions in dairy herds could be assessed with several models [13–16]. Typically, the available models compare specific management strategies adopted in dairy farms (i.e., facilities, environmental condition, feeding strategies) to guidelines suggested by the current legislation, and they include multiple indicators reflecting the relative animal responses to the management strategies (i.e., health status of the herd, reproductive efficiency, animal behavior and performances).

SCDs commonly affect dairy cows in specific phases of their production cycle [17,18]. Although not inducing any evident clinical sings, the occurrence of a SCD is often paired with severe alterations of several plasma analytes [7,19]. Limiting the concept of "healthy cows" to animals showing no clinical diseases could thus result in the inaccurate prediction of the physiological conditions of a population. This is especially true for dairy cows at their transition to calving, when the incidence of SCDs is known to be the highest across the whole lactation cycle [20].

With this work we aim to identify the physiological processes affecting plasma analytes concentrations during the different stages of the lactation cycle and to provide a guideline for the duration and magnitude of their changes when animals are healthy and raised in optimal field conditions. Hence, we tested changes occurring in several plasma analytes concentrations during different stages of the lactation cycle in high welfare raised multiparous Holstein cows, excluding the SCDs. Additionally, we adopted a 95% confidence interval to define reference intervals (RI) for plasma analytes.

2. Materials and Methods

The experiment was performed between September 2017 and September 2019, in accordance with Italian laws on animal experimentation (D. Lgs. n. 26, 4 March 2014) and ethics. In the present study, plasma samples previously collected for other studies were used (authorization of the Italian Ministry of Health N 451-2017-PR, N 403-2017-PR, N 484-2018-PR, N 851-2018-PR, N 511-2018-PR, N 296-2019-PR, N 464-2019-PR and N 510-2019-PR). Thirty commercial farms located in the provinces of Brescia, Cremona, Mantova and Piacenza (Northern Italy) were enrolled in this study. The weather in the studied areas was a relatively cool, mid-latitude version of the Humid subtropical climate. The chosen farms raised Italian Holstein dairy cows in freestall barns, adopting a total mixed ration feeding system.

2.1. Welfare Status Evaluation and Criteria for Farm Inclusion

The welfare status of the herds was evaluated by skilled personnel according to the computerized Integrated Diagnostic System Welfare (IDSW) model [13]. The IDSW evaluates the welfare condition of the herd adopting a hierarchical structure (Figure 1), and its outcome has been previously submitted to preliminary validation with blood analytes [21]. Scores are expressed as percentages with 0 and 100% representing the worst and best welfare condition (defined according to well established standards), respectively. The total IDSW score is a pooled-mean of the scores attributed to three "clusters". The contribution to the total IDSW score is: 30% from the "environment" cluster, 30% from the "feeding" cluster and 40% from the "animal" cluster. Each "cluster" is scored based on the evaluation of several "components", and each "component" is scored based on the evaluation of several "aspects". Finally, each aspect is scored based on the evaluation of several specific "indicators". Detailed description of the sets of "components", "aspects" and "indicators" considered in the evaluation of each "cluster", as well as their contribution to the total IDSW score have been defined previously [13].

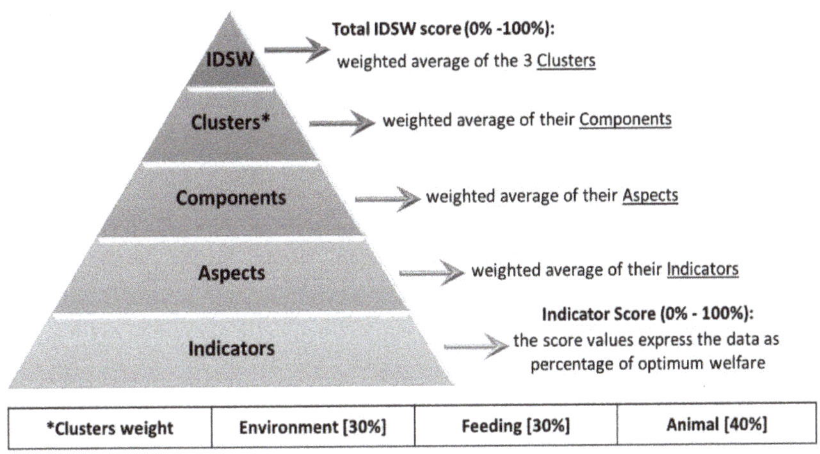

Figure 1. Hierarchical structure and scoring system adopted in the Integrated Diagnostic System Welfare (IDSW) model.

Dairy farms having a total IDSW score >70% and a score >65% for each cluster were considered high-welfare farms (HWF). Eleven conventional HWF ascribed to the Grana Padano Protected Designation of Origin were selected (number of lactating cows: 322.3 ± 201; milking system: 10 milking parlors; 1 automatic units; milking frequency: 2.1 ± 0.3; days open: 115.5 ± 13.8; energy corrected milk: 10,560 ± 8.1 kg per 305 days lactation), while other farms were discharged. Average technical characteristics, cluster scores and overall IDSW scores assigned to the 11 HWF are available in Table 1. Diet and feed provided in the HWF during the dry, the postpartum, the early lactation and the late lactation phases are presented in Tables S1–S3, respectively.

2.2. Criteria for Animal Recruitment

Four physiological phases were defined according to the stage of lactation of the animals (expressed as days from real calving date; DFC): the dry (from −30 to −10 DFC and from 30 to 50 days from dry phase onset), the postpartum (from 3 to 7 DFC), the early lactation (from 28 to 45 DFC), and the late lactation phase (from 160 to 305 DFC). For each HWF, blood samples were collected in a single visit, considering animals undergoing the aforementioned physiological phases at that time. Cows were recruited for blood collection from November until March, to avoid any alteration of the animal physiology due to heat stress. Only cows having parity order $\geq$2 and $\leq$5 were considered for blood collection. Prior to the visit by skilled personnel for blood collection, the health status of the herd was checked by veterinary inspection to exclude any cow affected by clinical diseases. On the day of blood collection, the health status of the cows was checked again by skilled personnel. Rectal temperature was measured with a commercial thermometer (Eco-Fast Digital Thermometer, Agri-Pro Enterprises, Iowa Falls, IA, USA), considering fever rectal temperature values >39.5 °C [22,23]. The animals underwent visual inspection for signals indicating a suboptimal physiological status (e.g., abscesses, trauma, mucopurulent or hemorrhagic nasal discharge, eye discharge, hollow left flank caused by an empty rumen and hoof lesions). For the early and the late lactation phases, the individual milk yield of each cow was recorded during the morning milking and compared to the daily milk yield records from the previous week. Animals showing a drop in milk production >15% between the two records were considered at risk of developing a disease [24]. Additionally, the body condition score (BCS) was evaluated in accordance with the Agricultural Development and Advisory Service (1986) [25]. The BCS value was judged as scarce, optimal or excessive, depending on the physiological phase of the animal [26]. According to these criteria, a minimum of eight cows undergoing each physiological phase were selected for blood collection. Animals having fever, signals indicating a suboptimal physiological status, drop of milk yield or BCS values outside the optimal range were not considered for blood collection. For the late lactation phase, animals that were not pregnant were not considered for blood collection. Based on these criteria, 361 clinically healthy multiparous cows, each in different stages of lactation, were enrolled in the experiment (Table 1).

Table 1. Average characteristics, welfare score assigned, and number of cows enrolled for blood collection in the 11 high welfare farms selected for assessing physiological variations of plasma analytes in high welfare raised multiparous Holstein dairy cows during the pivotal stages of the lactation cycle.

Farm	1	2	3	4	5	6	7	8	9	10	11
Average characteristics, unit											
Milking system [1]	MP	MP	MP	MP	MP	AU	MP	MP	MP	MP	MP
MF [2], number	2	2	2	2	2	2.9	2	2	2	2	2
Days open, d	99	109	93	104	119	120	139	118	129	112	128
Lactating cows, number	184	210	394	864	280	107	311	304	175	378	338
ECM [3], kg	11,291	10,349	12,435	10,841	11,203	10,055	10,007	10,098	10,025	10,147	9695
Welfare score, %											
Environment cluster	75	75	70	80	70	77	75	66	67	70	66
Feeding cluster	88	87	83	76	87	86	82	83	89	76	84
Animal cluster	79	74	80	74	73	67	71	76	70	72	70
Total IDSW score [4]	80	78	78	77	76	76	76	75	75	73	73
Cows enrolled, number											
Dry phase	6	6	6	21	10	6	7	6	10	6	8
Postpartum phase	6	6	4	28	10	5	6	6	7	7	18
Early lactation phase	7	6	8	13	10	6	6	2	6	6	14
Late lactation phase	8	6	8	11	8	6	8	6	8	6	7

[1] MP is milking parlor; AU is automatic unit. [2] Daily milking frequency. [3] Energy corrected milk for a 305-days lactation: ECM = (milk yield × (0.383 × % fat + 0.242 × % protein + 0.7832)/3.1138). [4] Total integrated diagnostic system welfare score = [(Environment cluster score × 0.3) + (Feeding cluster score × 0.3) + (Animal cluster score × 0.4)].

2.3. Blood Sample Collection and Analyses

Blood was harvested by means of jugular venipuncture before the morning feeding. Samples were collected in evacuated heparinized tubes (BD Vacutainer, Becton, Dickinson and Co., Franklin Lakes, NJ, USA) and processed as described by Calamari et al. (2016) [27]. After collection, blood was centrifuged (3500× g, 15 min at 4 °C) and the Packed cell volume

(PCV) was measured through capillary column (ALC Centrifugette 4203) directly on fresh blood after centrifugation. A clinical autoanalyzer (ILAB-650, Instrumentation Laboratory, Bedford, MA, USA) was used to determine the concentration of glucose, nonesterified fatty acids (NEFA), BHB, urea, creatinine, Ca, P, Mg, Na, K, Cl, Zn, aspartate amino transferase–glutamate oxaloacetate transaminase (AST-GOT), gamma glutamyl transferase (GGT), alkaline phosphatase, total protein, haptoglobin, ceruloplasmin, albumin, total bilirubin, cholesterol and globulin in accordance with Calamari et al. (2016) [27]. The reactive oxygen metabolites (ROMt) and ferric ion reducing antioxidant power were determined according to Jacometo et al. (2015) [28], the paraoxonase according to Bionaz et al. (2007) [29], the thiol groups according to Minuti et al. (2014) [30], the myeloperoxidase according to Bradley et al. (1982) [31], the triglycerides according to Bertoni et al. (2008) [32], the fructosamine according to Caré et al. (2018) [33] and the advanced oxidation protein products according to Hanasand et al. (2012) [34]. Retinol, tocopherol and beta-carotene were analyzed by reverse-phase high-performance liquid chromatography (LC-4000, Jasco Europe SRL, Cremella, Italy), as described by Jahan et al. (2015) [35]. Further details on the analytical procedures adopted in blood analysis are reported in Table S4.

2.4. Criteria for Retrospective Exclusion, Statistical Analysis and Reference Intervals Calculation

Statistical analyses were performed using R v3.6.1 [36]. Plasma analytes having values outside 1.5 times the interquartile range (above the third quartile or below the first quartile) were defined as outliers and removed from the statistical analysis, in accordance with the American Society for Veterinary Clinical Pathology (ASVCP) guidelines [37]. In all the phases of the lactation cycle, plasma concentrations of BHB > 1.2 mmol/L and Ca < 2.0 mmol/L were assumed as hyperketonemia and hypocalcemia threshold, respectively [38,39]. Animals affected by hyperketonemia or hypocalcemia were retrospectively excluded from the statistical analysis. Furthermore, as other SCDs might not course with changes in plasma biomarkers mentioned above, animals having more than three outlier plasma analytes were retrospectively excluded from the statistical analysis to minimize the risk of including animals affected by unspecific SCDs (i.e., subclinical displaced abomasum, mild mastitis, and severe inflammatory conditions). A total of 76 animals (16 in dry; 22 in postpartum; 21 in early lactation; 17 in late lactation phase) were retrospectively excluded due to SCDs, while plasma samples from 285 healthy Italian Holstein multiparous cows (76 in dry; 81 in postpartum; 63 in early lactation and 65 in late lactation phase) were included in the final database for the statistical analysis.

Before statistical analysis, the normality of distributions was verified for each analyte, assessing skewness and kurtosis using the 'Skew' and 'Kurt' functions [40], and assuming a ± 1.5 range as skewness and kurtosis limits for normal distribution. Analytes showing deviation from normality were transformed through natural logarithms (BHB, AOPP and Zn). Further details on the distribution of blood analytes, and on the number of samples considered for each phase of the lactating cycle, are reported in Table 2.

In order to understand the association of lactation phases on the variability of plasma analytes, data of plasma analytes underwent ANOVA testing using physiological phases as predictors. The statistical model included the fixed effect of the physiological phase (PHASE) for the dry, the postpartum, the early lactation and the late lactation phases. The individual effect of each phase was assessed using a pairwise *t*-test assuming $p \leq 0.05$ as a significance limit. A bootstrap approach, implemented following Dimauro et al. (2009) [41], and a 95 % confidence interval was used to define the RI for each blood analyte within physiological phases. For each blood analyte, physiological phases having a pairwise t-test $p > 0.05$ were merged in RI calculation.

Table 2. Samples included in the statistical analysis and skewness and kurtosis values for plasma analytes assessed in healthy, high welfare raised multiparous Holstein dairy cows during the pivotal stages of the lactation cycle.

Item, Unit [1]	Physiological Phases (Days from Calving)											
	Dry (−30; −10)			Postpartum (+3; +7)			Early Lactation (+28; +45)			Late Lactation (+160; +305)		
	n	Skew [2]	Kurt [3]	n	Skew [2]	Kurt [3]	n	Skew [2]	Kurt [3]	n	Skew [2]	Kurt [3]
PCV, L/L	75	0.16	−0.69	78	0.36	−0.39	59	−0.13	−0.68	65	0.12	−0.20
Glucose, mmol/L	74	0.10	−0.26	81	0.18	−0.14	60	0.27	−0.40	64	0.26	−0.19
Fructosamine, mmol/L	75	0.46	−0.18	80	0.38	−0.59	62	0.37	−0.34	65	−0.11	−0.35
NEFA, mmol/L	75	0.93	−0.10	80	1.10	0.92	63	1.17	1.33	64	1.16	1.22
BHB ‡, mmol/L	75	0.25	−0.33	80	−0.25	−0.52	63	0.21	0.37	64	0.53	−0.48
Triglycerides, mmol/L	76	0.52	−0.65	75	0.32	−0.36	63	−0.21	−0.21	65	0.06	−0.57
Urea, mmol/L	76	0.31	−0.55	81	0.52	−0.51	63	0.07	−0.49	65	−0.06	−0.41
Creatinine, μmol/L	75	0.29	−0.44	80	0.24	−0.41	63	0.16	−0.45	64	0.06	−0.59
Ca, mmol/L	76	0.30	−0.32	78	−0.08	−0.42	62	−0.05	−0.40	65	−0.33	−0.36
P, mmol/L	65	0.14	−0.20	81	0.45	−0.15	63	−0.42	−0.38	65	−0.09	−0.58
Mg, mmol/L	75	0.19	−0.57	81	0.02	−0.91	63	0.02	−0.91	65	−0.04	−0.68
Na, mmol/L	76	−0.11	−1.09	80	0.05	−0.73	63	0.07	−0.86	65	0.39	−0.69
K, mmol/L	75	0.29	−0.67	79	0.13	−0.26	62	0.07	−0.35	63	−0.26	−0.44
Cl, mmol/L	76	−0.26	−0.73	79	−0.22	−0.89	63	−0.27	−0.38	64	−0.03	−0.68
Zn ‡, mmol/L	76	−0.15	−0.05	81	0.18	−0.13	63	−0.10	0.78	62	1.19	3.84
Total bilirubin, μmol/L	73	0.03	−0.82	75	0.52	−0.74	61	0.28	−0.22	65	−0.05	−0.49
AST, U/L	75	0.62	−0.12	72	0.44	−0.30	56	0.16	−0.73	58	0.40	−0.70
GGT, U/L	75	0.57	−0.49	71	0.30	−0.14	57	0.08	0.04	57	0.25	−0.25
ALP, U/L	70	0.18	−0.56	78	0.33	−0.69	62	0.68	−0.21	65	0.36	−0.26
Myeloperoxidase, U/L	73	0.08	−0.42	77	−0.10	−0.44	57	0.37	−0.61	64	−0.10	−0.29
Total protein, g/L	76	−0.18	−0.26	81	0.01	−0.27	63	0.22	−0.63	62	0.09	−0.22
Globulin, g/L	76	0.28	−0.12	80	0.01	−0.27	63	0.40	−0.51	65	0.70	0.10
Haptoglobin, g/L	71	0.56	0.09	81	0.37	−0.71	53	0.20	−0.07	55	0.61	−0.59
Ceruloplasmin, μmol/L	76	0.26	0.09	81	0.16	−0.41	63	0.15	−0.32	62	0.24	−0.21
Albumin, g/L	75	−0.14	−0.68	77	−0.30	−0.15	63	−0.29	−0.15	63	−0.15	−0.38
Cholesterol, mmol/L	69	0.35	−0.08	80	0.20	−0.22	63	0.15	−0.36	61	−0.28	−0.41
Retinol, μmol/L [4]	73	−0.26	−0.51	81	0.29	−0.73	57	−0.25	−0.54	65	0.32	−0.61
Paraoxonase, U/mL	72	0.48	−0.30	81	0.44	0.03	61	0.10	−0.09	65	0.65	−0.29
Tocopherol, μmol/L	73	0.18	−0.46	80	0.45	−0.40	61	0.39	−0.62	65	0.16	−0.75
β-Carotene, μmol/L	70	0.82	−0.30	69	0.81	0.28	59	0.49	−0.21	59	0.21	−1.01
FRAP, mmol/L	73	0.29	−0.74	78	0.29	−0.97	62	0.58	−0.34	65	−0.47	−0.53
Thiol groups, mmol/L	75	0.40	−0.46	81	0.15	−0.34	62	0.31	−0.50	64	−0.18	−0.84
ROMt, mg H_2O_2/dL	75	0.28	−0.26	77	0.03	−0.34	63	−0.01	−0.82	65	−0.05	−0.81
AOPP ‡, mmol/L	76	0.23	−1.13	81	0.57	−1.05	62	0.04	−1.65	65	−0.14	−1.11

[1] PCV = packed cell volume, NEFA = non-esterified fatty acids, BHB = β-Hydroxybutyrate; AST = aspartate aminotransferase; GGT = γ-glutamyl transferase; ALP = alkaline phosphatase; FRAP = ferric reducing antioxidant power, ROMt = total reactive oxygen metabolites; AOPP = advanced oxidation product of protein. [2] Skew = Skewness. [3] Kurt = Kurtosis. [4] Retinol values could be converted into μg/100 mL as follows: μmol/L × 28.645 = μg/100 mL. ‡ Variable submitted to logarithmic transformation.

3. Results and Discussion

3.1. Effect of the Phase of the Lactation Cycle on the Plasma Analytes Concentrations

Values of plasma analytes during different phases of the lactation cycle are presented in Table 3 as mean ± SD.

3.1.1. Packed Cell Volume

The PCV was higher in the dry and the postpartum than in the late lactation phase, and it was higher in the late lactation than in the early lactation phase ($p < 0.01$). Higher PCV was detected in those phases when milk production was suspended or when dairy cows were at the onset of lactation. This observation, paired with the recovery of PCV we detected once the peak of lactation was surpassed, suggests a modification of plasma volume driven by milk production [42,43].

3.1.2. Energy Metabolism Biomarkers

We recorded the lowest concentration of glucose in the postpartum, and it was lower in the early lactation and the dry than in the late lactation phase ($p < 0.01$). The concentration of fructosamine was lower in postpartum and early lactation than in the late lactation and the dry phases ($p < 0.05$ or less). The concentration of NEFA was the highest in the postpartum phase; it was higher in the early lactation than in the dry phase, and higher in the dry than in the late lactation phase ($p < 0.01$). The highest concentration of triglycerides and BHB was recorded in the dry and the postpartum phase, respectively ($p < 0.01$).

Glucose is the primary energy source for metabolic processes, and fasting plasma glucose concentration is finely reflected by fructosamine within one to three weeks [33,44,45]. Plasma NEFA concentration is proportional to the severity of body fat mobilization; triglycerides reflect the ability of the liver in re-esterifying NEFA and exporting them through very low-density lipoproteins (VLDL), while BHB is released in the bloodstream when a NEFA overload impairs the β-oxidation process at the liver level [38,46,47]. We detected the lowest values of glucose and fructosamine concentration within 3 and 7 DFC, consistently with the hypoglycemic state affecting postparturient cows [48]. This is mainly driven by the synthesis of lactose: glucose uptake by the mammary gland for this purpose is known to more than double during the 2 d before parturition, and an even more substantial increase occurs after parturition [49]. Besides mammary gland uptake, immune cells contribute to reducing circulating glucose concentrations in early lactation [50]. Those two metabolic functions drive the glucose availability of early lactating cows, as they are prioritized by the insulin resistance induced in peripheral tissues by growth hormone trends, and by the activation of leukocytes to cope with calving-related insults [51]. Furthermore, the peak of NEFA and BHB concentration we detected in the postpartum is consistent with the mobilization of body fats occurring in this phase [52,53]. In fact, early lactating cows experience negative energy balance (NEB) due to their limited feed intake and their huge glucose requirements, resulting in the mobilization of body reserves to cover this gap [8,46]. The increase of glucose and the reduction of NEFA concentrations we observed in the early lactation compared to the postpartum phase suggest a partial recovery of the homeostasis condition. Nonetheless, the lower glucose and fructosamine, paired with the higher NEFA concentration detected in the early compared to the late lactation phase, suggests that dairy cows do not fully overcome the NEB condition during their first month of lactation. Despite slight differences, the low concentration of glucose and the high concentration of NEFA we found in the dry compared to the late lactation phase suggest a light energy deficit condition, triggered by the increase of glucose requirements occurring in the last month of gestation. These are likely driven by the gravid uterus paired with the low dietary glucose availability ensured by the dry ration compared to the late lactation one [49,54]. Despite that, the higher concentration of triglycerides in the dry compared to the postpartum phase reflects a greater capability of the liver in re-esterifying circulating NEFA before calving, probably due to a higher VLDL availability. This observation is consistent with the rise of plasma concentrations of triglycerides reported in high yielding cows before parturition [55,56].

Table 3. Coefficient of determination of the statistical model (R^2) and values of plasma analytes (expressed as mean ± standard deviation) assessed in healthy, high welfare raised multiparous Holstein dairy cows during the different stages of the lactation cycle.

Item [1]	R^2	Physiological Phases (Days from Calving)				*p*-Value [2]					
		Dry (−30; −10)	Postpartum (+3; +7)	Early Lactation (+28; +45)	Late Lactation (+160; +305)	DP × PP	DP × EL	DP × LP	PP × EL	PP × LP	EL × LP
PCV, L/L	0.29	0.34 ± 0.03	0.34 ± 0.02	0.30 ± 0.02	0.33 ± 0.03	ns	***	***	***	***	***
Glucose, mmol/L	0.29	4.22 ± 0.23	3.87 ± 0.34	4.15 ± 0.26	4.35 ± 0.28	***	ns	***	***	***	***
Fructosamine, mmol/L	0.11	288 ± 22.3	273 ± 27.2	278 ± 28.8	298 ± 28.2	***	*	ns	ns	***	***
NEFA, mmol/L	0.55	0.18 ± 0.11	0.67 ± 0.33	0.33 ± 0.19	0.11 ± 0.04	***	***	***	***	***	***
BHB ‡, mmol/L	0.26	0.42 ± 0.10	0.67 ± 0.21	0.50 ± 0.19	0.45 ± 0.17	***	ns	ns	***	***	ns
Triglycerides, mmol/L	0.71	0.22 ± 0.05	0.10 ± 0.02	0.10 ± 0.02	0.10 ± 0.02	***	***	***	ns	ns	ns
Urea, mmol/L	0.11	4.22 ± 1.37	4.36 ± 1.20	4.98 ± 1.28	5.22 ± 0.78	ns	***	***	***	***	ns
Creatinine, μmol/L	0.31	97.5 ± 10.5	91.0 ± 8.47	84.3 ± 6.29	84.8 ± 5.50	***	***	***	***	***	ns
Ca, mmol/L	0.23	2.57 ± 0.11	2.40 ± 0.15	2.55 ± 0.13	2.56 ± 0.13	***	ns	ns	***	***	ns
P, mmol/L	0.20	2.02 ± 0.20	1.62 ± 0.45	1.63 ± 0.34	1.86 ± 0.29	***	***	***	ns	***	***
Mg, mmol/L	0.32	0.99 ± 0.08	0.91 ± 0.13	1.09 ± 0.10	1.06 ± 0.10	***	***	***	***	***	ns
Na, mmol/L	0.15	145 ± 4.25	147 ± 3.56	144 ± 4.14	143 ± 3.91	***	ns	***	***	***	***
K, mmol/L	0.01	4.30 ± 0.35	4.23 ± 0.38	4.23 ± 0.48	4.18 ± 0.30	ns	ns	ns	ns	ns	ns
Cl, mmol/L	0.19	106 ± 3.52	106 ± 2.94	104 ± 3.15	103 ± 2.68	ns	***	***	***	***	***
Zn ‡, mmol/L	0.04	14.3 ± 2.54	12.6 ± 3.69	13.2 ± 2.71	13.6 ± 3.74	***	*	*	ns	ns	ns
Total bilirubin, μmol/L	0.54	1.67 ± 0.75	5.65 ± 2.79	2.31 ± 1.00	1.54 ± 0.51	***	***	ns	***	***	***
AST, U/L	0.28	83.8 ± 14.9	109 ± 15.5	95.2 ± 12.7	105 ± 21.4	***	***	***	***	ns	***
GGT, U/L	0.34	24.3 ± 6.81	20.5 ± 4.03	25.4 ± 4.52	31.6 ± 6.21	***	ns	***	***	***	***
ALP, U/L	0.04	48.5 ± 13.9	49.2 ± 15.6	42.8 ± 13.6	49.7 ± 12.8	ns	ns	ns	ns	ns	ns
Myeloperoxidase, U/L	0.17	438 ± 54.4	504 ± 53.3	461 ± 55.5	458 ± 61.0	***	ns	ns	***	***	ns
Total protein, g/L	0.41	80.0 ± 5.23	74.0 ± 4.82	82.3 ± 4.35	83.8 ± 3.84	***	***	***	***	***	ns
Globulin, g/L	0.24	43.4 ± 5.51	38.6 ± 4.14	44.6 ± 5.18	45.7 ± 5.01	***	*	***	***	***	ns
Haptoglobin, g/L	0.55	0.17 ± 0.06	0.84 ± 0.51	0.18 ± 0.07	0.13 ± 0.07	***	ns	ns	***	***	ns
Ceruloplasmin, μmol/L	0.25	2.22 ± 0.39	2.93 ± 0.51	2.46 ± 0.56	2.39 ± 0.45	***	***	*	***	***	ns
Albumin, g/L	0.27	36.6 ± 1.60	35.3 ± 2.17	37.6 ± 2.21	38.5 ± 1.75	***	***	***	***	***	***
Cholesterol, mmol/L	0.73	3.05 ± 0.65	2.06 ± 0.46	4.66 ± 1.34	6.13 ± 1.22	***	***	***	***	***	***
Retinol, μmol/L [3]	0.46	1.04 ± 0.20	0.81 ± 0.27	1.23 ± 0.22	1.41 ± 0.28	***	***	***	***	***	***
Paraoxonase, U/mL	0.27	87.1 ± 16.1	73.2 ± 18.1	101.6 ± 20.3	95.0 ± 16.7	***	***	***	***	***	***
Tocopherol, μmol/L	0.56	5.03 ± 2.03	3.49 ± 1.13	7.26 ± 2.63	10.93 ± 3.80	***	***	***	***	***	***
β-Carotene, μmol/L	0.41	4.02 ± 1.94	2.32 ± 0.98	3.64 ± 1.60	6.74 ± 2.84	***	ns	***	***	***	***
FRAP, mmol/L	0.21	126 ± 13.6	150 ± 25.4	149 ± 21.7	149 ± 18.0	***	***	***	ns	ns	ns
Thiol groups, mmol/L	0.02	338 ± 59.1	351 ± 73.3	354 ± 61.8	361 ± 64.9	ns	ns	ns	ns	ns	ns
ROMt, mg H_2O_2/dL	0.18	13.5 ± 3.10	16.6 ± 3.40	13.5 ± 3.63	12.9 ± 2.71	***	ns	ns	***	***	ns
AOPP ‡, mmol/L	0.11	63.8 ± 23.3	46.6 ± 18.7	59.0 ± 22.5	63.9 ± 19.0	***	ns	ns	***	***	ns

[1] PCV = packed cell volume, NEFA = non-esterified fatty acids, BHB = β-Hydroxybutyrate; AST = aspartate aminotransferase; GGT = γ-glutamyl transferase; ALP = alkaline phosphatase; FRAP = ferric reducing antioxidant power, ROMt = total reactive oxygen metabolites; AOPP = advanced oxidation product of protein. ‡ Variable submitted to logarithmic transformation. [2] DP is dry phase; PP is postpartum phase; EL is early lactation phase; LP is late lactation phase; * is $p < 0.1$; *** is $p < 0.01$. [3] Retinol values could be converted into μg/100 mL as follows: μmol/L × 28.645 = μg/100 mL.

3.1.3. Protein Metabolism and Kidney Function Biomarkers

The urea concentration was lower in the dry and postpartum phases than in the other physiological phases ($p < 0.01$). Plasma urea concentrations closely reflect the trends of rumen fermentations and could be deeply affected by the availability of different protein sources in the ration. In ruminants, urea is synthetized in the liver from two sources: nitrogen deamination from endogenous amino acids and ammonia absorbed by the rumen [57]. Thus, the trend of plasma urea mostly reflects the low ammonia uptake by the rumen fluid, driven both by the low protein content of the dry rations and the low feed volume contained in the rumen of early lactating cows compared to those in mid and late lactation [58,59]. Furthermore, the substantial differences found in the 11 HWF for the diet crude protein contents (Tables S1–S3) and the likely different length of the fasting conditions of the animals before blood sampling, could have contributed to the wide variability recorded in the plasma urea concentration across the physiological phases considered in this study. Feed analyses aimed at identifying the different protein fractions will likely help understanding the source of such variations in future studies.

The concentration of creatinine was higher in the dry than in the postpartum phase, and it was higher in the postpartum than in the early and late lactation phases ($p < 0.01$). Creatinine is related to the used phosphocreatine during normal muscular activity, but the interpretation of this plasma analyte is challenging. On the one hand, plasma creatinine has been reported to reflect the amount of muscle mass in the body [60–62]. On the other hand, the creatinine hematic concentration reflects the kidneys' ability to remove it, thus serving as a reliable marker of glomerular filtration rate efficiency [63]. The highest concentration of creatinine we detected in the dry phase could thus reflect the greater amount of muscle tissue compared with the lactation phase, consistently with the mobilization of 13 to 25% of body proteins occurring in dairy cows at their transition to calving [62]. Furthermore, creatinine trend could suggest a transient impairment of kidney glomerular filtration rate occurring before calving, similarly to those reported in late-pregnant women [64], and likely driven by body fat mobilization and altered blood pressure during the late gestation phase [65]. Both these interpretations are consistent with the decreased creatinine concentration we detected in the lactating phases, suggesting both a reduced amount of muscle tissue and an improved glomerular filtration rate compared with the dry phase.

3.1.4. Mineral Metabolism Biomarkers

The concentration of Ca was lower in the postpartum than in any other physiological phases ($p < 0.01$). The concentration of P was lower in the postpartum and the early lactation than in the late lactation phase, and it was lower in the late lactation than in the dry phase ($p < 0.01$). The concentration of Mg was lower in the postpartum than in the dry phase, and it was lower in the dry than in the early and late lactation phases ($p < 0.01$). Low Ca and P concentrations detected in the postpartum compared with other phases are consistent with the extremely high demand by the mammary gland at the onset of lactation [66,67]. Ca and P homeostasis are finely regulated by parathormone, which promotes Ca reabsorption from urine and bones and stimulates phosphate absorption at gut level when circulating Ca and P decrease [68–70]. Nonetheless, the uptake of circulating Ca and P by the mammary gland at the onset of lactation is too fast to be counterbalanced by these mechanisms, leading to the decrease of their circulating pools [70,71]. Conversely, reduced plasma Mg found in postpartum could be driven by modifications occurring in rumen functions at the onset of lactation [72], as Mg is primely absorbed by rumen epithelium in ruminants.

Among the electrolytes, the concentration of Na was higher in the postpartum than in the dry and the early lactation phases, and it was higher in the dry and the early lactation than in late lactation phase ($p < 0.01$). The concentration of K did not differ between the physiological phases. The concentration of Cl was higher in the dry and the postpartum than in early lactation phase, and it was higher in the early than in the late lactation phase ($p < 0.01$). The higher plasma concentration of Na and Cl detected before calving and at the onset of lactation are consistent with plasma creatinine trends in supporting an impaired glomerular filtration rate occurring in late pregnancy. In fact, an increased plasma electrolyte concentration has been reported in humans affected by kidney dysfunctions, and also in cattle [73,74]. Despite that, increased plasma concentration of electrolytes (hypernatremia) in postpartum cows has also been related to the stressful conditions driven by adrenocorticotropin hormone release and inflammatory conditions occurring around calving [75].

The Zn concentration was higher in the dry phase than in the other physiological phases ($p < 0.05$), as previously observed by Anon (1973) and Pryor (1976) [76,77]. Despite several factors (including inflammatory condition) that are known to contribute to reducing plasma Zn availability in the postpartum phase, a clear explanation behind the reduced concentration of this mineral during lactation phases is still lacking. Moreover, our findings are in contrast with those of previous studies performed on Hereford cattle, reporting the lowest plasma Zn concentration during the dry period and relating these changes to the constriction of the posterior genital tract in late pregnancy [78].

3.1.5. Liver Function Biomarkers

Bilirubin concentration was higher in postpartum than in the early lactation phase, and it was higher in the early lactation than in the dry and the late lactation phases ($p < 0.01$). Among liver damage indicators, the AST-GOT concentration was higher in the postpartum and the late lactation than in the early lactation phase, and it was higher in the early lactation than in the dry phase ($p < 0.01$). The GGT concentration was higher in the late lactation than in the dry and the early lactation phases, and it was higher in the dry and the early lactation than in the postpartum phase ($p < 0.01$). The concentration of ALP did not differ among the physiological phases.

Bilirubin results from degradation of red blood cells. Its plasma concentration could be assumed as a cholestasis index, as it reflects the efficiency of liver enzymes in removing it [79]. Both AST-GOT and GGT are involved in amino acids metabolism, while ALP exerts a role in dephosphorylating compounds. Increased plasma concentration of these enzymes serves as a cytolysis and liver damage biomarker [79]. The higher plasma bilirubin and AST-GOT concentrations found in the postpartum phase suggests transient impairment of liver function, paired with liver cell damages occurring at the onset of lactation. These are probably related to the massive liver activities occurring in this phase, combined with the detrimental effect sorted by mobilized NEFA on the liver functions [80]. The decreased concentration of AST-GOT and bilirubin found in the early lactation phase suggests a gradual recovery of liver function [32]. Both bilirubin and AST-GOT reached their minimum during the dry phase, indicating that this phase displays the best liver status and probably the lower liver activity. Surprisingly, AST-GOT and bilirubin do not overlap with GGT trends in the postpartum phase: the GGT had higher plasma concentrations at the end of lactation and during the dry phase than in early lactation. We hypothesize that this could be due to the concentration of GGT in the plasma fraction at the end of lactation, because the plasma half-life of GGT is most likely longer than that of the AST-GOT [81].

3.1.6. Inflammation and Acute Phase Proteins Biomarkers

The concentration of myeloperoxidase was higher in the postpartum than in the other physiological phases ($p < 0.01$). These trends suggest the activation of leukocytes occurring after calving, as documented by others [82]. In fact, myeloperoxidase is involved in the generation of reactive oxygen species in activated neutrophils, thus serving as a reliable marker of inflammation [83].

Among the positive acute phase proteins (APP), haptoglobin was higher in the postpartum than in the other physiological phases ($p < 0.01$). Ceruloplasmin was higher in the postpartum than in the early and the late lactation phases ($p < 0.01$), and it was higher in the early and late lactation than in the dry phase ($p < 0.05$ or less). Among the negative APP biomarkers, albumin, total cholesterol, and retinol were lower in the postpartum than in the dry phase, in the dry than in the early lactation phase and in the early than in the late lactation phase ($p < 0.01$). The concentration of paraoxonase was lower in the postpartum than in the dry phase, in the dry than in the late lactation phase and in the late lactation than in the early lactation phase ($p < 0.01$). The concentration of total protein and globulin was higher in the early and the late lactation than in the dry phase, and it was higher in the dry than in the postpartum phase ($p < 0.05$ or less).

During an acute phase response, the plasma concentration of positive APP increases as the liver upregulates the synthesis of alpha globulins [84,85]. Conversely, the liver reduces the synthesis of albumin, paraoxonase, retinol-binding protein, and lipoproteins (reflected by plasma concentration of retinol and cholesterol respectively). The drop of their plasma concentration is proportional to the severity of the acute phase response [84–86]. Total protein includes serum albumin and globulin; thus, trends of total protein and globulin could serve as reliable markers of changes in plasma protein fractions during the acute phase response. The higher concentration of positive APP, paired with the lower concentration of negative APP and total protein, reflects that a marked acute phase response occurred after calving [32]. The similar trends of myeloperoxidase and acute phase biomarkers also suggest a contribution of activated leukocytes in boosting the severity of the acute phase response in this phase [10]. The reduction of haptoglobin supports the mitigation of the acute phase response condition to occur since the early lactation phase. In fact, haptoglobin is released into the blood at the onset of the acute phase, and its concentration then decreases within 60–96 h [87]. Such a mitigation of the acute phase response in the early lactation phase is further supported by the increase of albumin, total cholesterol, and retinol concentrations. We detected that ceruloplasmin had a delayed recovery of homeostasis condition compared to haptoglobin, as it reached its minimum in the dry phase. Further, ceruloplasmin is known to serve as a longer-lasting plasma biomarker relative to acute phase onset than haptoglobin [88,89].

3.1.7. Oxidant Status Biomarkers

Among the antioxidant systems, tocopherol was lower in the postpartum than in the dry phase, in the dry than in the early lactation phase and in the early lactation than in the late lactation phase ($p < 0.01$). β-carotene was lower in the postpartum than in the early lactation and the dry phases, and it was lower in the early lactation and the dry than in the late lactation phase ($p < 0.01$). Concentration of ferric ion reducing antioxidant power (FRAP) was lowest in the dry phase ($p < 0.01$). Concentration of thiol groups was comparable across physiological phases. Among oxidant species and oxidative stress biomarkers, the highest concentration of ROMt and the lowest concentration of advanced oxidation of protein products (AOPP) were recorded in the postpartum phase ($p < 0.01$).

Tocopherol is known to act as a secondary antioxidant by reducing the chain propagation and amplification of lipid peroxidation. β-carotene is known to exert an indirect antioxidant action by maintaining other antioxidant molecules in their reduced form [90]. FRAP reflects the antioxidant power of bilirubin, uric-acid, proteins and vitamins C and E [91]. Thiols could be used as reliable marker for glutathione availability [92]. ROMt include a wide range of oxidant molecules. AOPP classically reflects protein oxidation driven by hypochlorous acid, thus representing a synthetic marker of oxidative stress caused by activated leukocytes through the respiratory burst [93]. Lower concentrations of tocopherol and β-carotene, paired with higher concentration of ROMt detected in the postpartum phase, are consistent with the depletion of circulating antioxidant systems occurring in early lactating cows as a consequence of the increased oxidative metabolism, mainly driven by the milk synthesis and the activation of the immune system [93,94].

The trends of FRAP and AOPP we detected in postpartum are harder to interpret. We hypothesize that the high FRAP concentration we found in the postpartum compared with the dry phase could be due to the upregulation of body antioxidants synthesis occurring to cope with high ROMt concentrations [92]. This is supported by the high bilirubin concentration found in this phase compared to the dry phase. However, a rise of plasma AOPP was expected in the postpartum phase, when leukocytes were at their maximum activated status. We can speculate that the lower AOPP concentration we found in the postpartum phase could be caused by the primary contribution of albumin in facing oxidative damage, as supported by the detection of nadir for plasma albumin during the postpartum phase. The low ROMt concentration and the high availability of antioxidants (tocopherol and β-carotene) in every other physiological phase compared with the postpartum suggests a recovery of the oxidant/antioxidant balance, likely driven by a mitigation of the oxidative metabolism.

3.2. Reference Intervals Calculation

Reference intervals for plasma analytes during the different stages of the lactation cycle are presented in Table 4 as lower limit and upper limit (confidence intervals for 95% observations referred to the reference interval extreme values). We found several differences between the RI defined in this study and those reported in previous works, and the largest discrepancy was detected during the postpartum phase. The minor values of our RI in postpartum were higher for plasma glucose, Ca and P and lower for lipid mobilization-related analytes (NEFA and BHB) than those proposed in previous studies [3,5]. We infer that excluding animals affected by hyperketonemia and hypocalcemia from our reference dataset mitigated the NEB condition and the mineral metabolism dysfunctions faced by the cows at the onset of lactation. Furthermore, the minor values of the RI we identified in postpartum were lower for plasma positive APP (haptoglobin and ceruloplasmin), and higher for plasma albumin and total protein than those proposed by others [3,5]. Such outcomes suggest that the cows in our experiment underwent a milder acute phase response during postpartum phase as compared to previous works [3,5]. Excluding animals having hyperketonemia from our reference dataset could partially account for such a difference, as animals undergoing subclinical ketosis are prone to develop severe acute phase responses at the onset of lactation [19].

Despite positive effects sorted by the rigorous procedures adopted in the recruitment of reference individuals on RIs defined in this study, we caution that the adoption of BHB and Ca as subclinical ketosis and subclinical hypocalcemia markers prevented the exclusion of those animals undergoing other less-specific SCDs (i.e., mild mastitis, subclinical displaced abomasum and mild liver lipidosis). Importantly, the lack of any veterinary inspection concurrent with the blood collection further accrued this limitation in our study, as dedicated veterinary inspections would have likely improved the identification and selection of healthy individuals. Most of the unspecific SCDs are often paired with inflammatory conditions in postpartum cows, leading to the simultaneous variation of multiple plasma analytes concentrations [95]. Consequently, the presence of animals with undetected SCDs within the reference database has the potential to decrease the reliability of RIs. For example, the reduced RI for plasma Ca, P and Zn concentration during the postpartum phase could be expected if animals affected by severe acute-phase responses are included in the reference population. Under this scenario, the conversion of vitamin D to 2,5-hydroxyvitamin D by parathormone is impaired during the acute phase [96], thus accounting for a reduced Ca and P uptake from bone and gut cells in animals having severe acute phase responses after calving, while the liver sequesters circulating Zn from the bloodstream during the acute phase [97]. Further, trends of liver enzymes, liver function indicators and oxidant status biomarkers are deeply affected by the severity of the inflammatory conditions during the postpartum phase [86,94].

Table 4. Reference intervals (expressed as lower and upper limits (confidence intervals for 95% observations referred to the reference interval extreme values)) for plasma analytes assessed in healthy, high welfare raised multiparous Holstein dairy cows during the pivotal stages of the lactation cycle.

Item, Unit [1]	Physiological Phases (Days from Calving)							
	Dry (−30; −10)		Postpartum (+3; +7)		Early Lactation (+28; +45)		Late (+160; +305)	
	Lower Limit	Upper Limit	Lower Limit	Upper Limit	Lower Limit	Upper Limit	Lower Limit	Upper Limit
PCV, L/L	0.28 (0.28–0.29)	0.39 (0.38–0.40)	0.28 (0.28–0.29)	0.39 (0.38–0.40)	0.25 (0.24–0.26)	0.34 (0.33–0.35)	0.27 (0.26–0.29)	0.37 (0.36–0.39)
Glucose, mmol/L	3.70 (3.63–3.77)	4.68 (4.60–4.75)	3.19 (3.06–3.32)	4.55 (4.43–4.68)	3.70 (3.63–3.77)	4.68 (4.60–4.75)	3.80 (3.69–3.92)	4.90 (4.78–5.02)
Fructosamine, mmol/L	242 (235–250)	343 (336–351)	220 (212–228)	331 (323–339)	220 (212–228)	331 (323–339)	242 (235–250)	343 (336–351)
NEFA, mmol/L	0.00 (0.00–0.01)	0.39 (0.34–0.43)	0.01 (0.00–0.14)	1.33 (1.21–1.46)	0.00 (0.00–0.03)	0.70 (0.62–0.78)	0.02 (0.01–0.04)	0.18 (0.17–0.20)
BHB, mmol/L	0.13 (0.10–0.17)	0.77 (0.74–0.81)	0.26 (0.19–0.34)	1.08 (1.00–1.16)	0.13 (0.10–0.17)	0.77 (0.74–0.81)	0.13 (0.10–0.17)	0.77 (0.74–0.81)
Triglycerides, mmol/L	0.11 (0.09–0.13)	0.32 (0.30–0.34)	0.06 (0.06–0.06)	0.14 (0.13–0.14)	0.06 (0.06–0.06)	0.14 (0.13–0.14)	0.06 (0.06–0.06)	0.14 (0.13–0.14)
Urea, mmol/L	1.70 (1.35–2.05)	6.88 (6.53–7.23)	1.70 (1.35–2.05)	6.88 (6.53–7.23)	3.04 (2.73–3.35)	7.17 (6.86–7.48)	3.04 (2.73–3.35)	7.17 (6.86–7.48)
Creatinine, µmol/L	76.6 (72.5–80.7)	118 (114–122)	74.2 (71.0–77.4)	108 (105–111)	72.9 (71.1–74.7)	96.2 (94.4–97.9)	72.9 (71.1–74.7)	96.2 (94.4–97.9)
Ca, mmol/L	2.32 (2.29–2.35)	2.81 (2.78–2.84)	2.11 (2.05–2.16)	2.7 (2.64–2.76)	2.32 (2.29–2.35)	2.81 (2.78–2.84)	2.32 (2.29–2.35)	2.81 (2.78–2.84)
P, mmol/L	1.62 (1.54–1.70)	2.42 (2.33–2.50)	0.80 (0.69–0.92)	2.44 (2.32–2.55)	0.80 (0.69–0.92)	2.44 (2.32–2.55)	1.30 (1.18–1.42)	2.43 (2.31–2.55)
Mg, mmol/L	0.8 (0.81–0.87)	1.14 (1.11–1.17)	0.67 (0.62–0.71)	1.16 (1.11–1.21)	0.88 (0.85–0.91)	1.27 (1.24–1.30)	0.88 (0.85–0.91)	1.27 (1.24–1.30)
Na, mmol/L	2.32 (2.29–2.35)	2.81 (2.78–2.84)	140 (139–141)	154 (153–156)	2.32 (2.29–2.35)	2.81 (2.78–2.84)	135 (133–136)	150 (149–152)
K, mmol/L	3.47 (3.40–3.55)	5.00 (4.92–5.08)	3.47 (3.40–3.55)	5.00 (4.92–5.08)	3.47 (3.40–3.55)	5.00 (4.92–5.08)	3.47 (3.40–3.55)	5.00 (4.92–5.08)
Cl, mmol/L	99.6 (98.7–101)	113 (112–114)	99.6 (98.7–101)	113 (112–114)	97.7 (96.3–99.0)	110 (109–112)	97.3 (96.2–98.5)	108 (107–109)
Zn, mmol/L	9.26 (8.28–10.3)	19.4 (18.4–20.3)	6.15 (5.34–6.97)	20.0 (19.2–20.8)	6.15 (5.34–6.97)	20.0 (19.2–20.8)	6.15 (5.34–6.97)	20.0 (19.2–20.8)
Total bilirubin, µmol/L	0.33 (0.14–0.51)	2.89 (2.71–3.08)	0.21 (0.00–1.28)	11.1 (10.0–12.2)	0.33 (0.00–0.76)	4.29 (3.86–4.72)	0.33 (0.14–0.51)	2.89 (2.71–3.08)
AST, U/L	54.2 (48.3–60.0)	113 (108–119)	70.8 (65.4–76.2)	143 (138–149)	69.8 (64.1–75.6)	120 (115–126)	70.8 (65.4–76.2)	143 (138–149)
GGT, U/L	12.9 (11.1–14.6)	36.7 (34.9–38.4)	12.5 (10.8–14.1)	28.5 (26.8–30.1)	12.9 (11.1–14.6)	36.7 (34.9–38.4)	19.3 (16.5–22.1)	44.0 (41.2–46.7)
ALP, U/L	19.5 (16.6–22.4)	75.9 (73.0–78.8)	19.5 (16.6–22.4)	75.9 (73.0–78.8)	19.5 (16.6–22.4)	75.9 (73.0–78.8)	19.5 (16.6–22.4)	75.9 (73.0–78.8)
Myeloperoxidase, U/L	336 (322–350)	566 (552–580)	398 (378–419)	609 (589–630)	336 (322–350)	566 (552–580)	336 (322–350)	566 (552–580)
Total protein, g/L	69.5 (67.4–71.6)	90.5 (88.4–92.5)	64.3 (62.5–66.1)	83.6 (81.8–85.4)	74.7 (73.4–75.9)	91.4 (90.1–92.7)	74.7 (73.4–75.9)	91.4 (90.1–92.7)
Globulin, g/L	32.5 (30.4–34.7)	54.4 (52.2–56.5)	30.4 (28.9–32.0)	46.8 (45.3–48.4)	35.1 (33.6–36.6)	55.2 (53.7–56.7)	35.1 (33.6–36.6)	55.2 (53.7–56.7)
Haptoglobin, g/L	0.02 (0.00–0.03)	0.30 (0.28–0.32)	0.00 (0.00–0.02)	1.86 (1.66–2.05)	0.02 (0.00–0.03)	0.30 (0.28–0.32)	0.02 (0.00–0.03)	0.30 (0.28–0.32)
Ceruloplasmin, µmol/L	1.44 (1.29–1.59)	3.00 (2.84–3.15)	1.92 (1.73–2.11)	3.94 (3.75–4.13)	1.42 (1.27–1.58)	3.43 (3.28–3.59)	1.42 (1.27–1.58)	3.43 (3.28–3.59)
Albumin, g/L	33.4 (32.8–34.1)	39.9 (39.2–40.5)	31.0 (30.2–31.9)	39.6 (38.8–40.5)	33.2 (32.3–34.2)	42.0 (41.1–43.0)	35.0 (34.2–35.7)	42.0 (41.2–42.7)
Cholesterol, mmol/L	1.75 (1.49–2.02)	4.34 (4.07–4.60)	1.14 (0.97–1.32)	2.97 (2.80–3.14)	1.98 (1.40–2.56)	7.35 (6.77–7.93)	3.68 (3.15–4.22)	8.57 (8.03–9.10)
Retinol, µmol/L [2]	0.65 (0.57–0.73)	1.43 (1.36–1.51)	0.29 (0.19–0.39)	1.34 (1.24–1.44)	0.78 (0.68–0.88)	1.67 (1.57–1.77)	0.86 (0.75–0.98)	1.96 (1.84–2.07)
Paraoxonase, U/mL	55.1 (48.6–61.5)	119 (113–126)	37.1 (30.2–43.9)	109 (102–116)	61.7 (52.9–70.4)	142 (133–150)	61.9 (54.9–68.9)	128 (121–135)
Tocopherol, µmol/L	1.04 (0.24–1.84)	9.03 (8.23–9.83)	1.26 (0.83–1.68)	5.72 (5.30–6.15)	2.09 (0.96–3.22)	12.4 (11.3–13.6)	3.38 (1.78–4.98)	18.5 (16.9–20.1)
β-Carotene, µmol/L	0.25 (0.00–0.79)	7.43 (6.89–7.97)	0.38 (0.00–0.77)	4.25 (3.86–4.65)	0.25 (0.00–0.79)	7.43 (6.89–7.97)	1.10 (0.15–2.36)	12.4 (11.1–13.6)
FRAP, mmol/L	98.6 (93.2–104)	153 (147–158)	105 (99.9–110)	193 (188–198)	105 (99.9–110)	193 (188–198)	105 (99.9–110)	193 (188–198)
Thiol groups, mmol/L	221 (208–234)	480 (467–493)	221 (208–234)	480 (467–493)	221 (208–234)	480 (467–493)	221 (208–234)	480 (467–493)
ROMt, mg H_2O_2/dL	7.00 (6.25–7.75)	19.6 (18.8–20.3)	9.78 (8.47–11.1)	23.3 (22.0–24.7)	7.00 (6.25–7.75)	19.6 (18.8–20.3)	7.00 (6.25–7.75)	19.6 (18.8–20.3)
AOPP, mmol/L	19.0 (13.8–24.2)	106 (101–111)	9.20 (2.10–16.3)	83.8 (76.8–90.9)	19.0 (13.8–24.2)	106 (101–111)	19.0 (13.8–24.2)	106 (101–111)

[1] PCV = packed cell volume, NEFA = non-esterified fatty acids, BHB = β-Hydroxybutyrate; AST = aspartate aminotransferase; GGT = γ-glutamyl transferase; ALP = alkaline phosphatase; FRAP = ferric reducing antioxidant power, ROMt = total reactive oxygen metabolites; AOPP = advanced oxidation product of protein. [2] Retinol values could be converted into µg/100 mL as follows: µmol/L × 28.645 = µg/100 mL.

Likely, the retrospective exclusion from the reference database of animals with more than three plasma analyte outliers allowed us to remove some of the bias driven by unspecific SCDs from RI calculated in this study, and this could further account for the milder inflammatory state affecting our reference individuals in postpartum as compared with previous studies [3,5]. Despite that, we caution the use of the RIs obtained in this study as "golden standard" for plasma analytes of dairy cows, especially for the postpartum phase. To counteract the possible inclusion in the reference dataset of cows affected by unspecific SCDs, we can speculate that performing a veterinary inspection concurrent with the blood collection and adopting a more stringent confidence interval threshold (with respect to the widely used 95% threshold; [98]) should be considered in future studies to develop reliable RIs for the postpartum phase of dairy cows.

4. Conclusions

Our results show that, with the exclusion of K, alkaline phosphatase and thiol groups, trends of all the other investigated plasma analytes differed among the physiological phases considered. We show that assessing welfare condition at the herd level and performing a SCD screening likely reduced the variability among reference individuals used in this study, especially in postpartum phase. The inclusion of BHB as a biomarker for subclinical ketosis accounts for the higher minor value for glucose and the lower minor value for NEFA and BHB, while the inclusion of Ca as a biomarker for subclinical hypocalcemia accounts for the higher minor value for Ca and P we proposed in postpartum as compared with previous works. The inclusion of these specific biomarkers for subclinical disorders and the high welfare standard of the selected herds likely contributed to excluding animals affected by severe acute phase responses in postpartum, as suggested by the lower minor value for haptoglobin and ceruloplasmin, and for the higher minor value for albumin and total protein we proposed as compared with previous works. However, we could not prevent the exclusion of animals affected by unspecific SCDs other than hypocalcemia and hyperketonemia, which likely affected the RIs for postpartum phase. Here, we provided a field-measurement of the physiological variability affecting plasma analytes and a possible explanation for the processes behind the occurrence of such variations in healthy, high-welfare raised multiparous Holstein dairy cows at different stages of the lactation cycle. Although the sets of reference intervals defined in the current study is tightly linked to the population of cows included in the experiment, similar variations of plasma analytes among the four phases of the lactation cycle investigated in this research are likely to be expected in any high-welfare herd.

Supplementary Materials: The following are available online at https://www.mdpi.com/article/10.3390/ani11061714/s1, Table S1: Diet used during the dry phase in the 11 high welfare farms included in the study, Table S2: Diet used during the postpartum and early lactation phases in the 11 high welfare farms included in the study, Table S3: Diet used during the late lactation phase in the 11 high welfare farms included in the study, Table S4: Intra- and inter-assay coefficient of variations, limit of quantification (LOQ), codes of commercial kits used, references for their validation in the bovine plasma, calibrators and quality controls used for plasma parameters included in the study. References [99–101] are cited in the supplementary materials.

Author Contributions: Conceptualization, E.T.; methodology, M.P. and E.T.; formal analysis M.P., M.B.; investigation, M.P., G.F., F.P.C. and M.M.; resources, E.T.; data curation, M.P., M.B., M.M. and A.M.; writing—original draft preparation, M.M. and M.P.; writing—review and editing, A.M., M.B., G.F., F.P.C. and E.T.; supervision, E.T.; project administration, E.T.; funding acquisition, E.T. All authors have read and agreed to the published version of the manuscript.

Funding: This work was conducted in the framework of projects supported by CREI (Romeo and Enrica Invernizzi Research Center of the Università Cattolica del S. Cuore) funded by the "Fondazione Romeo ed Enrica Invernizzi", Milan, Italy. The research was also partly supported by the BENELAT and RABOLA projects, funded by the Lombardy Region under the Rural Development Program 2014–2020.

Institutional Review Board Statement: The study did not directly require ethical approval, as blood samples collected for other research trials were used (authorization of the Italian Ministry of Health N 451-2017-PR, N 403-2017-PR, N 484-2018-PR, N 851-2018-PR, N 511-2018-PR, N 296-2019-PR, N 464-2019-PR and N 510-2019-PR).

Data Availability Statement: The data presented in this study are available on request from the corresponding author.

Conflicts of Interest: The authors declare no conflict of interest. The funders had no role in the design of the study; in the collection, analyses, or interpretation of data; in the writing of the manuscript, or in the decision to publish the results.

References

1. Herdt, T.H. Variability characteristics and test selection in herdlevel nutritional and metabolic profile testing. *Vet. Clin. N. Am. Food Anim. Pract.* **2000**, *16*, 387–403. [CrossRef]
2. Bertoni, G. Welfare, health and management of dairy cows. In *Recent Progress in Animal Production Science*; Piva, G., Bertoni, G., Masoero, F., Bani, P., Calamari, L., Eds.; FrancoAngeli: Milan, Italy, 1999; pp. 59–78.
3. Quiroz-Rocha, G.F.; LeBlanc, S.J.; Duffield, T.F.; Wood, D.; Leslie, K.E.; Jacobs, R.M. Reference limits for biochemical and hematological analytes of dairy cows one week before and one week after parturition. *Can. Vet. J. La Rev. Vet. Can.* **2009**, *50*, 383–388.
4. Brscic, M.; Cozzi, G.; Lora, I.; Stefani, A.L.; Contiero, B.; Ravarotto, L.; Gottardo, F. Short communication: Reference limits for blood analytes in Holstein late-pregnant heifers and dry cows: Effects of parity, days relative to calving, and season. *J. Dairy Sci.* **2015**, *98*, 7886–7892. [CrossRef]
5. Moretti, P.; Paltrinieri, S.; Trevisi, E.; Probo, M.; Ferrari, A.; Minuti, A.; Giordano, A. Reference intervals for hematological and biochemical parameters, acute phase proteins and markers of oxidation in Holstein dairy cows around 3 and 30 days after calving. *Res. Vet. Sci.* **2017**, *114*, 322–331. [CrossRef]
6. Cozzi, G.; Ravarotto, L.; Gottardo, F.; Stefani, A.L.; Contiero, B.; Moro, L.; Brscic, M.; Dalvit, P. Short communication: Reference values for blood parameters in Holstein dairy cows: Effects of parity, stage of lactation, and season of production. *J. Dairy Sci.* **2011**, *94*, 3895–3901. [CrossRef] [PubMed]
7. Bertoni, G.; Trevisi, E. Use of the liver activity index and other metabolic variables in the assessment of metabolic health in dairy herds. *Vet. Clin. N. Am. Food Anim. Pract.* **2013**, *29*, 413–431. [CrossRef] [PubMed]
8. Drackley, J.K. Biology of dairy cows during the transition period: The final frontier? *J. Dairy Sci.* **1999**, *82*, 2259–2273. [CrossRef]
9. Drackley, J.K.; Dann, H.M.; Douglas, N.; Guretzky, N.A.J.; Litherland, N.B.; Underwood, J.P.; Loor, J.J. Physiological and pathological adaptations in dairy cows that may increase susceptibility to periparturient diseases and disorders. *Ital. J. Anim. Sci.* **2005**, *4*, 323–344. [CrossRef]
10. Mezzetti, M.; Bionaz, M.; Trevisi, E. Interaction between inflammation and metabolism in periparturient dairy cows. *J. Anim. Sci.* **2020**, *98*, S155–S174. [CrossRef]
11. Ametaj, B.N.; Bradford, B.J.; Bobe, G.; Nafikov, R.A.; Lu, Y.; Young, J.W.; Beitz, D.C. Strong relationships between mediators of the acute phase response and fatty liver in dairy cows. *Can. J. Anim. Sci.* **2005**, *85*, 165–175. [CrossRef]
12. Huzzey, J.M.; Nydam, D.V.; Grant, R.J.; Overton, T.R. Associations of prepartum plasma cortisol, haptoglobin, fecal cortisol metabolites, and nonesterified fatty acids with postpartum health status in Holstein dairy cows. *J. Dairy Sci.* **2011**, *94*, 5878–5889. [CrossRef] [PubMed]
13. Calamari, L.; Bertoni, G. Model to evaluate welfare in dairy cow farms. *Ital. J. Anim. Sci.* **2009**, *8*, 301–323. [CrossRef]
14. Welfare Quality®Consortium. *Welfare Quality®Assessment Protocol for Cattle*; Welfare Quality: Lelystad, The Netherlands, 2009.
15. Jordan, K.; Cook, N.; Darr, D.; DeCoite, C.; Doak, R.; Endres, M.; Humphrey, P.; Keyserlingk, N.V.; Maddox, S.; Mahoney, J.; et al. *Farmers Assuring Responsible Management: Animal Care Reference Manual*; FARM: Arlington, VA, USA, 2016.
16. Krueger, A.; Cruickshank, J.; Trevisi, E.; Bionaz, M. Systems for evaluation of welfare on dairy farms. *J. Dairy Res.* **2020**, *87*, 13–19. [CrossRef] [PubMed]
17. Duffield, T.F. Subclinical ketosis in lactating dairy cattle. *Vet. Clin. N. Am. Food Anim. Pract.* **2000**, *16*, 231–253. [CrossRef]
18. Goff, J.P. The monitoring, prevention, and treatment of milk fever and subclinical hypocalcemia in dairy cows. *Vet. J.* **2008**, *176*, 50–57. [CrossRef] [PubMed]
19. Mezzetti, M.; Minuti, A.; Piccioli-Cappelli, F.; Amadori, M.; Bionaz, M.; Trevisi, E. The role of altered immune function during the dry period in promoting the development of subclinical ketosis in early lactation. *J. Dairy Sci.* **2019**, *102*, 9241–9258. [CrossRef] [PubMed]
20. Ingvartsen, K.L.; Moyes, K. Nutrition, immune function and health of dairy cattle. *Animal* **2013**, *7*, 112–122. [CrossRef]
21. Calamari, L.; Bionaz, M.; Trevisi, E.; Bertoni, G. Preliminary study to validate a model of animal welfare assessment in dairy farms. In Proceedings of the Fifth Congress of the European Society for Agricultural and Food Ethics (EURSAFE), Leuven, Belgium, 2–4 September 2004; pp. 38–42.
22. Wagner, S.A.; Schimek, D.E.; Cheng, F.C. Body temperature and white blood cell count in postpartum dairy cows. *Bov. Pract.* **2008**, *42*, 18–26.

23. Smith, B.I.; Risco, C.A. Management of periparturient disorders in dairy cattle. *Vet. Clin. N. Am. Food Anim. Pract.* **2005**, *21*, 503–521. [CrossRef] [PubMed]
24. Edwards, J.L.; Tozer, P.R. Using activity and milk yield as predictors of fresh cow disorders. *J. Dairy Sci.* **2004**, *87*, 524–531. [CrossRef]
25. Agricultural Development and Advisory Service Condition scoring of dairy cows. In *Proceedings of the Publication 612*; Ministry of Agriculture, Fisheries Food: Alhwick, Northumberland, UK, 1986.
26. Roche, J.R.; Friggens, N.C.; Kay, J.K.; Fisher, M.W.; Stafford, K.J.; Berry, D.P. Invited review: Body condition score and its association with dairy cow productivity, health, and welfare. *J. Dairy Sci.* **2009**, *92*, 5769–5801. [CrossRef]
27. Calamari, L.; Ferrari, A.; Minuti, A.; Trevisi, E. Assessment of the main plasma parameters included in a metabolic profile of dairy cow based on Fourier Transform mid-infrared spectroscopy: Preliminary results. *BMC Vet. Res.* **2016**, *12*, 4. [CrossRef]
28. Jacometo, C.B.; Osorio, J.S.; Socha, M.; Corrêa, M.N.; Piccioli-Capelli, F.; Trevisi, E.; Loor, J.J. Maternal consumption of organic trace minerals (4-Plex) alters calf systemic and neutrophil mRNA and microRNA indicators of inflammation and oxidative stress. *J. Dairy Sci.* **2015**, *98*, 7717–7729. [CrossRef] [PubMed]
29. Bionaz, M.; Trevisi, E.; Calamari, L.; Librandi, F.; Ferrari, A.; Bertoni, G. Plasma paraoxonase, health, inflammatory conditions, and liver function in transition dairy cows. *J. Dairy Sci.* **2007**, *90*, 1740–1750. [CrossRef]
30. Minuti, A.; Ahmed, S.; Trevisi, E.; Piccioli-Cappelli, F.; Bertoni, G.; Jahan, N.; Bani, P. Experimental acute rumen acidosis in sheep: Consequences on clinical, rumen, and gastrointestinal permeability conditions and blood chemistry. *J. Anim. Sci.* **2014**, *92*, 3966–3977. [CrossRef] [PubMed]
31. Bradley, P.P.; Priebat, D.A.; Christensen, R.D.; Rothsein, G. Measurement of cutaneous inflammation: Estimation of neutrophil content with an enzyme marker. *J. Investig. Dermatol.* **1982**, *78*, 206–209. [CrossRef] [PubMed]
32. Bertoni, G.; Trevisi, E.; Han, X.; Bionaz, M. Effects of inflammatory conditions on liver activity in puerperium period and consequences for performance in dairy cows. *J. Dairy Sci.* **2008**, *91*, 3300–3310. [CrossRef]
33. Caré, S.; Trevisi, E.; Minuti, A.; Ferrari, A.; Loor, J.J.; Calamari, L. Plasma fructosamine during the transition period and its relationship with energy metabolism and inflammation biomarkers in dairy cows. *Livest. Sci.* **2018**, *216*, 138–147. [CrossRef]
34. Hanasand, M.; Omdal, R.; Norheim, K.B.; Gøransson, L.G.; Brede, C.; Jonsson, G. Improved detection of advanced oxidation protein products in plasma. *Clin. Chim. Acta* **2012**, *413*, 901–906. [CrossRef]
35. Jahan, N.; Minuti, A.; Trevisi, E. Assessment of immune response in periparturient dairy cows using ex vivo whole blood stimulation assay with lipopolysaccharides and carrageenan skin test. *Vet. Immunol. Immunopathol.* **2015**, *165*, 119–126. [CrossRef]
36. The R Core Team. *R: A Language and Environment for Statistical Computing*; R Project for Statistical Computing: Vienna, Austria, 2012.
37. Friedrichs, K.R.; Harr, K.E.; Freeman, K.P.; Szladovits, B.; Walton, R.M.; Barnhart, K.F.; Blanco-Chavez, J. ASVCP reference interval guidelines: Determination of de novo reference intervals in veterinary species and other related topics. *Vet. Clin. Pathol.* **2012**, *41*, 441–453. [CrossRef] [PubMed]
38. Herdt, T.H.; Stevens, J.B.; Olson, W.G.; Larson, V. Blood concentrations of beta hydroxybutyrate in clinically normal Holstein-Friesian herds and in those with a high prevalence of clinical ketosis. *Am. J. Vet. Res.* **1981**, *42*, 503–506. [PubMed]
39. Reinhardt, T.A.; Lippolis, J.D.; McCluskey, B.J.; Goff, J.P.; Horst, R.L. Prevalence of subclinical hypocalcemia in dairy herds. *Vet. J.* **2011**, *188*, 122–124. [CrossRef] [PubMed]
40. Signorell, A.; Aho, K.; Alfons, A.; Anderegg, N.; Aragon, T.; Arachchige, C.; Arppe, A.; Baddeley, A.; Barton, K.; Bolker, B.; et al. DescTools: Tools for Descriptive Statistics. R Package version 0.99.41. 2019. Available online: https://cran.r-project.org/package=DescTools (accessed on 7 June 2021).
41. Dimauro, C.; Macciotta, N.P.P.; Rassu, S.P.G.; Patta, C.; Pulina, G. A bootstrap approach to estimate reference intervals of biochemical variables in sheep using reduced sample sizes. *Small Rumin. Res.* **2009**, *83*, 34–41. [CrossRef]
42. Wohlt, J.E.; Evans, J.L.; Trout, J.R. Blood constituents in lactating holstein cows influenced by hematocrit, sampling site, and diet protein and calcium. *J. Dairy Sci.* **1984**, *67*, 2236–2246. [CrossRef]
43. Shalit, U.; Maltz, E.; Silanikove, N.; Berman, A. Water, sodium, potassium, and chlorine metabolism of dairy cows at the onset of lactation in hot weather. *J. Dairy Sci.* **1991**, *74*, 1874–1883. [CrossRef]
44. Baker, J.R.; O'Connor, J.P.; Metcalf, P.A.; Lawson, M.R.; Johnson, R.N. Clinical usefulness of estimation of serum fructosamine concentration as a screening test for diabetes mellitus. *BMJ* **1983**, *287*, 863–867. [CrossRef]
45. Armbruster, D.A. Fructosamine: Structure, analysis, and clinical usefulness. *Clin. Chem.* **1987**, *33*, 2153–2163. [CrossRef] [PubMed]
46. Herdt, T.H. Ruminant adaptation to negative energy balance. Influences on the etiology of ketosis and fatty liver. *Vet. Clin. N. Am. Food Anim. Pract.* **2000**, *16*, 215–230. [CrossRef]
47. Vazquez-Añon, M.; Bertics, S.; Luck, M.; Grummer, R.R.; Pinheiro, J. Peripartum liver triglyceride and plasma metabolites in dairy cows. *J. Dairy Sci.* **1994**, *77*, 6. [CrossRef]
48. Butler, W.R. Nutrition, negative energy balance and fertility in the postpartum dairy cow. *Cattle Pract.* **2005**, *13*, 13–18.
49. Bell, A.W.; Bauman, D.E. Adaptations of glucose metabolism during pregnancy and lactation. *J. Mammary Gland Biol. Neoplasia* **1997**, *2*, 265–278. [CrossRef] [PubMed]
50. Kvidera, S.K.; Horst, E.A.; Abuajamieh, M.; Mayorga, E.J.; Fernandez, M.V.S.; Baumgard, L.H. Glucose requirements of an activated immune system in lactating Holstein cows. *J. Dairy Sci.* **2017**, *100*, 2360–2374. [CrossRef]
51. Youssef, M.; El-Ashker, M. Significance of insulin resistance and oxidative stress in dairy cattle with subclinical ketosis during the transition period. *Trop. Anim. Health Prod.* **2017**, *49*, 239–244. [CrossRef] [PubMed]

52. Grum, D.E.; Drackley, J.K.; Younker, R.S.; LaCount, D.W.; Veenhuizen, J.J. Nutrition during the dry period and hepatic lipid metabolism of periparturient dairy cows. *J. Dairy Sci.* **1996**, *13*, 13–18. [CrossRef]
53. Reist, M.; Erdin, D.; Von Euw, D.; Tschuemperlin, K.; Leuenberger, H.; Chilliard, Y.; Hammon, H.M.; Morel, C.; Philipona, C.; Zbinden, Y.; et al. Estimation of energy balance at the individual and herd level using blood and milk traits in high-yielding dairy cows. *J. Dairy Sci.* **2002**, *85*, 3314–3327. [CrossRef]
54. Janovick, N.A.; Boisclair, Y.R.; Drackley, J.K. Prepartum dietary energy intake affects metabolism and health during the periparturient period in primiparous and multiparous Holstein cows. *J. Dairy Sci.* **2011**, *94*, 1385–1400. [CrossRef] [PubMed]
55. Bertics, S.J.; Grummer, R.R.; Cadorniga-Valino, C.; Stoddard, E.E. Effect of prepartum dry matter intake on liver triglyceride concentration and early lactation. *J. Dairy Sci.* **1992**, *75*, 1914–1922. [CrossRef]
56. Skaar, T.C.; Grummer, R.R.; Dentine, M.R.; Stauffacher, R.H. Seasonal effects of prepartum and postpartum fat and niacin feeding on lactation performance and lipid metabolism. *J. Dairy Sci.* **1989**, *72*, 2028–2038. [CrossRef]
57. Herdt, T.H. Gastrointestinal physiology and metabolism. In *Textbook of Veterinary Physiology*; Saunders, W.B., Ed.; Elsevier: Amsterdam, The Netherlands, 2002; pp. 222–224.
58. Marini, J.C.; Van Amburgh, M.E. Nitrogen metabolism and recycling in Holstein heifers. *J. Anim. Sci.* **2003**, *81*, 545–552. [CrossRef]
59. Odensten, M.O.; Chilliard, Y.; Holtenius, K. Effects of two different feeding strategies during dry-off on metabolism in high-yielding dairy cows. *J. Dairy Sci.* **2005**, *88*, 2072–2082. [CrossRef]
60. Osorio, J.S.; Trevisi, E.; Ji, P.; Drackley, J.K.; Luchini, D.; Bertoni, G.; Loor, J.J. Biomarkers of inflammation, metabolism, and oxidative stress in blood, liver, and milk reveal a better immunometabolic status in peripartal cows supplemented with Smartamine M or MetaSmart. *J. Dairy Sci.* **2014**, *97*, 7437–7450. [CrossRef] [PubMed]
61. Pires, J.A.A.; Delavaud, C.; Faulconnier, Y.; Pomiès, D.; Chilliard, Y. Effects of body condition score at calving on indicators of fat and protein mobilization of periparturient Holstein-Friesian cows. *J. Dairy Sci.* **2013**, *96*, 6423–6439. [CrossRef] [PubMed]
62. McCabe, C.J.; Boerman, J.P. Invited review: Quantifying protein mobilization in dairy cows during the transition period. *Appl. Anim. Sci.* **2020**, *36*, 389–396. [CrossRef]
63. Zoccali, C.; Maio, R.; Tripepi, G.; Mallamaci, F.; Perticone, F. Inflammation as a mediator of the link between mild to moderate renal insufficiency and endothelial dysfunction in essential hypertension. *J. Am. Soc. Nephrol.* **2006**, *4*, S64–S68. [CrossRef] [PubMed]
64. Davison, J.M.; Hytten, F.E. Glomerular filtration during and after pregnancy. *BJOG An. Int. J. Obstet. Gynaecol.* **1974**, *81*, 588–595. [CrossRef]
65. Thompson, M.; Ray, U.; Yu, R.; Hudspeth, A.; Smillie, M.; Jordan, N.; Bartle, J. Kidney function as a determinant of HDL and triglyceride concentrations in the australian population. *J. Clin. Med.* **2016**, *5*, 35. [CrossRef] [PubMed]
66. Horst, R.L.; Goff, J.P.; Reinhardt, T.A. Calcium and vitamin D metabolism during lactation. *J. Mammary Gland Biol. Neoplasia* **1997**, *2*, 253–263. [CrossRef]
67. Ramberg, C.F.; Mayer, G.P.; Kronfeld, D.S.; Phang, J.M.; Berman, M. Calcium kinetics in cows during late pregnancy, parturition, and early lactation. *Am. J. Physiol.* **1970**, *219*, 1166–1177. [CrossRef] [PubMed]
68. Cavestany, D.; Blanc, J.E.; Kulcsar, M.; Uriarte, G.; Chilibroste, P.; Meikle, A.; Febel, H.; Ferraris, A.; Krall, E. Studies of the transition cow under a pasture-based milk production system: Metabolic profiles. *J. Vet. Med. Ser. A Physiol. Pathol. Clin. Med.* **2005**, *52*, 1–7. [CrossRef]
69. Peterson, A.B.; Orth, M.W.; Goff, J.P.; Beede, D.K. Periparturient responses of multiparous holstein cows fed different dietary phosphorus concentrations prepartum. *J. Dairy Sci.* **2005**, *88*, 3582–3594. [CrossRef]
70. Goff, J.P.; Liesegang, A.; Horst, R.L. Diet-induced pseudohypoparathyroidism: A hypocalcemia and milk fever risk factor. *J. Dairy Sci.* **2014**, *97*, 1520–1528. [CrossRef] [PubMed]
71. Goff, J.P. Treatment of calcium, phosphorus, and magnesium balance disorders. *Vet. Clin. N. Am. Food Anim. Pract.* **1999**, *15*, 619–639. [CrossRef]
72. Calamari, L.; Soriani, N.; Panella, G.; Petrera, F.; Minuti, A.; Trevisi, E. Rumination time around calving: An early signal to detect cows at greater risk of disease. *J. Dairy Sci.* **2014**, *97*, 3635–3647. [CrossRef] [PubMed]
73. Moyer, J.H.; Mills, L.C.; Smith, C.P. Hexamethonium—Its effect on glomerular filtration rate, maximal tubular function, and renal excretion of electrolytes. *J. Clin. Investig.* **1953**, *32*, 172–184. [CrossRef]
74. Weeth, H.J.; Lesperance, A.L. Renal function of cattle under various water and salt loads. *J. Anim. Sci.* **1965**, *24*, 441–447. [CrossRef]
75. Harshfield, G.A.; Dong, Y.; Kapuku, G.K.; Zhu, H.; Hanevold, C.D. Stress-induced sodium retention and hypertension: A review and hypothesis. *Curr. Hypertens. Rep.* **2009**, *11*, 29–34. [CrossRef]
76. Committee on Mineral Nutrition. *Tracing and Treating Mineral. Disorders in Dairy Cattle*; Centre for Agricultural Publishing and Documentation: Wageningen, The Netherlands, 1973.
77. Pryor, W.J. Plasma zinc status of dairy cattle in the periparturient period. *N. Z. Vet. J.* **1976**, *24*, 57–58. [CrossRef]
78. Dufty, J.H.; Bingley, J.B.; Cove, L.Y. The plasma zinc concentration of nonpregnant, pregnant and parturient hereford cattle. *Aust. Vet. J.* **1977**, *53*, 519–522. [CrossRef]
79. Rodriguez-Jimenez, S.; Haerr, K.J.; Trevisi, E.; Loor, J.J.; Cardoso, F.C.; Osorio, J.S. Prepartal standing behavior as a parameter for early detection of postpartal subclinical ketosis associated with inflammation and liver function biomarkers in peripartal dairy cows. *J. Dairy Sci.* **2018**, *101*, 8224–8235. [CrossRef]

80. Graugnard, D.E.; Moyes, K.M.; Trevisi, E.; Khan, M.J.; Keisler, D.; Drackley, J.K.; Bertoni, G.; Loor, J.J. Liver lipid content and inflammometabolic indices in peripartal dairy cows are altered in response to prepartal energy intake and postpartal intramammary inflammatory challenge. *J. Dairy Sci.* **2013**, *96*, 918–935. [CrossRef] [PubMed]
81. Lumeij, J.T.; De Bruijne, J.J.; Slob, A.; Wolfswinkel, J.; Rothuizen, J. Enzyme activities in tissues and elimination half-lives of homologous muscle and liver enzymes in the racing pigeon (Columba Livia domestica). *Avian Pathol.* **1988**, *17*, 851–864. [CrossRef]
82. Sordillo, L.M.; Raphael, W. Significance of metabolic stress, lipid mobilization, and inflammation on transition cow disorders. *Vet. Clin. N. Am. Food Anim. Pract.* **2013**, *29*, 267–278. [CrossRef] [PubMed]
83. Faith, M.; Sukumaran, A.; Pulimood, A.B.; Jacob, M. How reliable an indicator of inflammation is myeloperoxidase activity? *Clin. Chim. Acta* **2008**, *396*, 23–25. [CrossRef] [PubMed]
84. Schreiber, G.; Howlettf, G.; Nagashima, M.; Millership, A.; Martin, H.; Urban, J.; Kotlerg, L. The acute phase response of plasma protein synthesis during experimental inflammation. *J. Biol. Chem.* **1982**, *257*, 10271–10277. [CrossRef]
85. Fleck, A. Clinical and nutritional aspects of changes in acute-phase proteins during inflammation. *Proc. Nutr. Soc.* **1989**, *48*, 347–354. [CrossRef] [PubMed]
86. Ceciliani, F.; Ceron, J.J.; Eckersall, P.D.; Sauerwein, H. Acute phase proteins in ruminants. *J. Proteomics* **2012**, *75*, 4207–4231. [CrossRef] [PubMed]
87. Giblett, E.R. Haptoglobin: A review. *Int. J. Transfus. Med.* **1961**, *6*, 513–524. [CrossRef] [PubMed]
88. Humblet, M.-F.; Guyot, H.; Boudry, B.; Mbayahi, F.; Hanzen, C.; Rollin, F.; Godeau, J.-M. Relationship between haptoglobin, serum amyloid A, and clinical status in a survey of dairy herds during a 6-month period. *Vet. Clin. Pathol.* **2006**, *35*, 188–193. [CrossRef]
89. Trevisi, E.; Bertoni, G. Some physiological and biochemical methods for acute and chronic stress evaluationin dairy cows. *Ital. J. Anim. Sci.* **2009**, *8*, 265–286. [CrossRef]
90. Ghiselli, A.; Serafini, M.; Maiani, G.; Azzini, E.; Ferro-Luzzi, A. A fluorescence-based method for measuring total plasma antioxidant capability. *Free Radic. Biol. Med.* **1995**, *18*, 29–36. [CrossRef]
91. Benzie, I.F.; Strain, J.J. The ferric reducing ability of plasma (FRAP) as a measure of "antioxidant power": The FRAP assay. *Anal. Biochem.* **1996**, *239*, 70–76. [CrossRef] [PubMed]
92. Sordillo, L.M.; Aitken, S.L. Impact of oxidative stress on the health and immune function of dairy cattle. *Vet. Immunol. Immunopathol.* **2009**, *128*, 104–109. [CrossRef] [PubMed]
93. Celi, P.; Gabai, G. Oxidant/antioxidant balance in animal nutrition and health: The role of protein oxidation. *Front. Vet. Sci.* **2015**, *2*, 48. [CrossRef] [PubMed]
94. Celi, P. Oxidative stress in ruminants. In *Studies on Veterinary Medicine. Oxidative Stress in Applied Basic Research and Clinical Practice*; Mandelker, L., Vajdovich, P., Eds.; Humana Press: Totowa, NJ, USA, 2011; pp. 191–231. ISBN 978-1-61779-070-6.
95. Sordillo, L.M.; Contreras, G.A.; Aitken, S.L. Metabolic factors affecting the inflammatory response of periparturient dairy cows. *Anim. Health Res. Rev.* **2009**, *10*, 53–63. [CrossRef] [PubMed]
96. Reid, D.; Toole, B.J.; Knox, S.; Talwar, D.; Harten, J.; O'Reilly, D.S.J.; Blackwell, S.; Kinsella, J.; McMillan, D.C.; Wallace, A.M. The relation between acute changes in the systemic inflammatory response and plasma 25-hydroxyvitamin D concentrations after elective knee arthroplasty. *Am. J. Clin. Nutr.* **2011**, *93*, 1006–1011. [CrossRef]
97. Bertoni, G.; Grossi, P.; Trevisi, E. Use of nutraceuticals for improving animal health during the transition period of dairy cows. In *Enhancing Animal Welfare and Farmer Income through Strategic Animal Feeding. Some Case Studies*; Harinder, P.S., Makkar, Eds.; Food and Agriculture Organization of the United Nations (FAO): Rome, Italy, 2013; pp. 79–83. ISBN 9789251074527.
98. Jones, M.L.; Allison, R.W. Evaluation of the ruminant complete blood cell count. *Vet. Clin. N. Am. Food Anim. Pract.* **2007**, *23*, 377–402. [CrossRef] [PubMed]
99. Skinner, J.G.; Brown, R.A.; Roberts, L. Bovine haptoglobin response in clinically defined field conditions. *Vet. Rec.* **1991**, *128*, 147–149. [CrossRef] [PubMed]
100. Sunderman, F.W.; Nomoto, S. Measurement of human serum ceruloplasmin by its p-phenylenediamine oxidase activity. *Clin. Chem.* **1970**, *16*, 903–910. [CrossRef]
101. Ferré, N.; Camps, J.; Prats, E.; Vilella, E.; Paul, A.; Figuera, L.; Joven, J. Serum paraoxonase activity: A new additional test for the improved evaluation of chronic liver damage. *Clin. Chem.* **2002**, *48*, 261–268. [CrossRef]

animals

MDPI

Article

Approaches to Identify Pregnancy Failure in Buffalo Cows

Vittoria Lucia Barile [1], Laura Menchetti [2], Anna Beatrice Casano [3], Gabriele Brecchia [4], Noelita Melo de Sousa [5], Riccardo Zelli [3], Claudio Canali [3], Jean François Beckers [5] and Olimpia Barbato [3,*]

1 Council for Agricultural Research and Analysis of Agricultural Economics (CREA), Research Centre for Animal Production and Aquaculture, Monterotondo, 00015 Rome, Italy; vittorialucia.barile@crea.gov.it
2 Department of Agricultural and Food Sciences, University of Bologna, Viale G. Fanin 4, 40137 Bologna, Italy; laura.menchetti7@gmail.com
3 Department of Veterinary Medicine, University of Perugia, via San Costanzo, 06126 Perugia, Italy; annabeatrice.casano@libero.it (A.B.C.); riccardo.zelli@unipg.it (R.Z.); claudio.canali@unipg.it (C.C.)
4 Department of Veterinary Medicine, University of Milano, via dell'Università, 26900 Lodi, Italy; gabriele.brecchia@unimi.it
5 Laboratory of Animal Endocrinology and Reproduction, Faculty of Veterinary Medicine, University of Liege, 4000 Liege, Belgium; noelitamelo@gmail.com (N.M.d.S.); jfbeckers@ulg.ac.be (J.F.B.)
* Correspondence: olimpia.barbato@unipg.it

Simple Summary: Embryonic mortality and pregnancy failures still represent a major issue in domestic livestock production, particularly in dairy cattle. Despite the presence of extensive work in this research area, there is still no effective, accurate and practical method able to determine timing and viability of embryo specifically during early gestation. Indeed, technologies and techniques for predicting pregnancy success must continue to be developed. The aim of this work was to find the best strategy to diagnose pregnancy failures in buffalo cows in order to improve farm reproductive management. Among the methods compared in this study (ultrasonography, progesterone, PAGs), pregnancy-associated glycoproteins (PAGs) seem to be the best marker for predicting embryonic mortality between 25 and 40 days of gestation to be utilized as a diagnostic tool to improve reproductive management in buffalo farms.

Abstract: The aim of this work was to find the best strategy to diagnose pregnancy failures in buffalo. A total of 109 animals belonging to a buffalo herd subjected to a synchronization and artificial insemination (AI) program were enrolled in this study. Blood samples were collected at days 0, 14, 25, 28 and 40 after AI for the determination of progesterone (P4) and pregnancy-associated glycoproteins (PAGs) by the radioimmunoassay (RIA) method. Transrectal ultrasonography was performed on day 25, 28 and 40 after AI to monitor pregnancy. The animals included in the data analysis were assigned ex post in pregnant (n = 50) and mortality (n = 12) groups. By ultrasonography, the predictive sign of mortality was the heartbeat. At day 25, the PAGs concentration was significant in predicting embryonic mortality with respect to ultrasonography and P4, at the cut-off of 1.1 ng/mL. At day 28, either PAGs, at a cut-off of 2.2 ng/mL, or ultrasonography, with no detection of heartbeat, were highly predictive of embryonic mortality. PAGs were the best marker ($p < 0.05$) for predicting embryonic mortality between 25 and 40 days of gestation in buffalo. Its utilization as a diagnostic tool can influence management decisions in order to improve farm reproductive management.

Keywords: embryonic mortality; pregnancy-associated glycoproteins; progesterone; ultrasonography; buffalo

Citation: Barile, V.L.; Menchetti, L.; Casano, A.B.; Brecchia, G.; Melo de Sousa, N.; Zelli, R.; Canali, C.; Beckers, J.F.; Barbato, O. Approaches to Identify Pregnancy Failure in Buffalo Cows. *Animals* **2021**, *11*, 487. https://doi.org/10.3390/ani11020487

Academic Editor: Paola Scocco

Received: 18 December 2020
Accepted: 5 February 2021
Published: 12 February 2021

1. Introduction

Animal reproductive biotechnology is continually evolving. Advances have been made in our understanding of embryo development and early embryonic mortality in domestic animals, which has improved the selection and success of in vivo and in vitro technologies. Declining fertility is a globally recognized problem that represents a major source of economic loss and culling in livestock species [1,2]. The low reproductive efficiency of cattle has had a negative financial impact on the dairy industry because the higher yields of milk cannot compensate for dwindling herd sizes. The demand for assisted reproductive techniques is therefore increasing, by creating an additional cost for dairy farmers [3]. Many factors contribute to the decline in reproductive fitness. The major cause of poor reproductive success is early embryonic mortality [4], which is defined as pregnancy failure during the period between fertilization and day 42 of gestation [5]. This is true especially in the animals that are not mated during their reproductive period.

Although buffaloes are polyestrus, their reproductive efficiency changes throughout the year, showing a distinct seasonal pattern [6]. This reproductive seasonality affects the efficiency of breeding programs, particularly during the spring–summer season with the daylight lengthening period, which corresponds to the low breeding period for buffalo [7]. It was observed a higher incidence of embryo loss (20 to 40%) in buffaloes that conceive during the daylight lengthening period, whereas a lower incidence (7%) was observed during decreasing daylight length [8–11]. Such high rates of early embryonic mortality lead to substantial losses in time and money spent rebreeding cows, slower genetic progress, as well as a significant loss in the number of potential calves [12,13]. One strategy for improving reproductive performance aims to shorten the calving-conception interval by rapidly identify embryo losses and rebreeding non-pregnant cows [14]. Pregnancy can be monitored using a variety of methods, including direct methods such as ultrasonography, or indirect methods like progesterone (P4) or pregnancy-associated glycoproteins (PAGs) measurement in maternal blood [15,16].

In buffalo, applications of transrectal ultrasonography to monitor early pregnancy and embryonic development have been described by different authors [17–21]. The sensitivity (true positive) of transrectal ultrasonography between days 19 and 24 is reported to be 44.4%, reaching 100% from day 31 after mating, while the specificity (true negative) ranged between 92.5 and 100% from days 19 to 55 after mating [22]. The time of heartbeat detection can be considered as the moment at which the sensitivity of the gestational diagnosis is 100% [20]. In buffalo, embryo proper with a heartbeat can be visualized between 23 and 28 days after gestation [20]. No detectable heartbeat together with membrane detachment/disorganization, reduced amount of intrauterine fluid or echoic floating structures, including remnants of the conceptus, are ultrasound findings of an embryo death [23]. Therefore, using ultrasonography, early pregnancy losses can be clearly diagnosed. The disadvantage of this method is that accuracy is limited before 28 to 30 days of gestation and pregnancy status is only guaranteed at the time of diagnosis; moreover, when the fetal heartbeat of an embryo is viewed on an ultrasound, there is no indication of whether or not embryo mortality will occur [16].

Chemical-based methods have been developed to identify pregnancy in buffalo, as an alternative to the ultrasound technique [24].

Progesterone (P4) is the most biologically active progestagen and is primarily produced and secreted by the corpus luteum (CL) during the estrous cycle and by the placenta during pregnancy. Quantification of P4 in blood or milk at days 20, 22 and 24 post-breeding has been utilized for early pregnancy diagnosis in buffaloes [25,26]. Campanile et al. [8] found higher P4 plasma levels in pregnant buffaloes than in buffaloes that showed embryonic mortality since day 10 after AI. Although P4 concentration helps in detecting early pregnancy, a single analysis does not provide sufficient information to evaluate the pregnancy status accurately [27,28]. Indeed, the concentration of P4 reflects the function of the CL and not the presence or the vitality of an embryo or foetus.

Characterized for the first time in the early eighties, the pregnancy-associated glycoproteins (PAGs) constitute a large family of glycoproteins expressed in the outer epithelial layer (chorion/trophectoderm) of the placenta in eutherian species [29–31]. They are synthesized by the mononucleate and binucleate trophoblastic cells, some of them being secreted into the maternal blood from the moment when the conceptus becomes more closely attached to the uterine wall and placentome formation begins [32,33]. Using different chromatographic procedures, some members of the PAGs family have been isolated from the cotyledons of different species of Cetartiodactyla order [29,34–43], included buffalo species [44,45]. The accumulation of PAGs in the maternal blood of ruminant ungulates has become a useful tool for monitoring pregnancy, thanks to the development of homologous [46] and heterologous radioimmunoassay (RIA). In our previous works, our research group [45,47–49] described the use of antisera raised against buffalo PAGs for RIA development and pregnancy detection in buffalo cows. Previously, other authors [22,50,51] used the RIA-706 system to measure PAGs in this species. Recently, the molecular biology technique has been utilized to detect the mRNA expression of PAGs in the maternal blood of pregnant buffaloes [52–54], improving knowledge regarding the peri-implantation period and the earliest time in which PAGs could be detected in this species.

The aim of this work was to find the best strategy to diagnose pregnancy failures in buffalo comparing ultrasonography, P4 and PAGs in order to improve farm reproductive management. Moreover, the work highlights whether PAGs determination could support the diagnosis of early pregnancy failures in buffalo.

2. Materials and Methods

2.1. Animals and Experimental Design

The trial was carried out at the CREA Animal Production and Aquaculture experimental farm of Monterotondo (Rome, Italy, 42° N parallel). A total of 109 animals belonging to the Italian Mediterranean buffalo herd subjected to a synchronization and artificial AI program were enrolled in this study and grouped as described below. Buffaloes were kept on a loose-housing system, fed ad libitum once a day with a total mixed ration based on sorghum silage, hay and concentrate and milked twice a day in a milking parlor. Before estrous synchronization, regular clinical examination excluded diseases such as endometritis, mastitis and metabolic disorders.

Buffaloes were synchronized with a progesterone releasing intravaginal device (PRID; Sanofi, Paris, France), containing 1.55 g natural progesterone inserted in situ for 10 days and an i.m. injection of 1000 IU of Pregnant Mare Serum Gonadotrophin (PMSG; Ciclogonina, Fort Dodge, Bologna, Italy) and 0.15 mg of cloprostenol (PGF2α analogue; Dalmazin, Ozzano Emilia, Fatro, Italy) on day seven. On day 10, the PRID was removed and cows were artificially inseminated at 72 and 96 h from device withdrawal.

Blood samples were taken from the jugular vein in 10 mL EDTA tubes at days 0 (0d), 25 (25d), 28 (28d) and 40 (40d) after AI for P4 and PAGs analysis. On day 14 (14d), a further sample was taken for P4 analysis. The day of the second AI was considered as day zero. Plasma was immediately separated by centrifugation (1200× *g* for 15 min at 5 °C) and stored at −20 °C until assayed.

The animals were classified ex post as pregnant (n = 50) and mortality (n = 12) groups, as determined by ultrasonography at day 40 based on diagnostic criteria as reported below in paragraph 2.2. A total of 47 buffaloes were diagnosed as non-pregnant. Only animals related to pregnant and mortality groups were included in the data analysis.

The animals involved in this experiment were treated in compliance with the animal testing regulations established under Italian law. The experimental design was carried out according to good veterinary practices under farm conditions. The CREA Research Centre for Animal Production and Aquaculture is authorized to use farm animals for experimental design (as stated in DM 26/96-4 of the Italian Welfare Ministry).

2.2. Pregnancy Diagnosis

Transrectal ultrasonography (Aloka SSD Prosound 2 scanner, Hitachi Medical System, Buccinasco, Italy, equipped with a 7.5 MHz linear-array transducer) was performed by the same operator on days 25 (25d), 28 (28d) and 40 (40d) after AI in correspondence with blood sampling for PAGs. On the day of scan, the ultrasound observations were classified into four categories: no vesicle, vesicle, vesicle + embryo, and vesicle + embryo + beat.

Recognition of the embryonic vesicle and embryo proper with a heartbeat was used as the criterion for positive diagnosis. Embryo mortality was diagnosed when a first embryo vesicle and/or embryo proper with a heartbeat was no longer visible in later ultrasound scans. Non-pregnant buffaloes had no embryonic vesicle detected at any time point. Pregnancy status was confirmed on day 60 post-AI.

2.3. Progesterone Radioimmunoassay

Samples were assayed for P4 using an extraction step as described previously [47,55,56]. P4 was extracted from plasma using ethyl ether. Each sample was assayed in duplicate. The efficiency of the extraction of a tracer amount of [3H]-progesterone ranged from 82 to 95%. Extraction was conducted using 0.2 mL plasma. Briefly, 0.8 mL distilled water and 3 mL diethyl ether were added to each sample. After stirring, samples were centrifuged (1000× *g*, 10 min), frozen, and the supernatant was discarded. Then, 1 mL borate buffer containing 10% ethanol was added to each sample. A 100-mL volume of increasing concentrations of P4 (0.1, 0.25, 0.5, 2, 5, 10, 20, and 40 ng/mL) constituted the standard curve. Tubes corresponding with the total count (Tc), nonspecific binding (NSB), and reference samples (0.5 and 10 ng/mL) were also prepared. Volumes of 100 mL [3H]-progesterone and 100 mL of diluted antiserum were added to the tubes containing the extracted samples and standard curve. Incubation was performed for at least 4 h at 4 °C. Bound and free P4 were separated by centrifugation after dextran-charcoal adsorption. Tubes were counted in a beta-counter (Tri-carb 2100 TR; Packard; Milan, Italy). The minimum detection limit (MDL) was 0.08 ng/mL. Intra-assay and interassay coefficients of variation (CV) were 7% and 11%, respectively.

Based on P4 assay, buffaloes were considered pregnant when the plasma P4 concentrations on days 14, 25, 28 and 40 were over 1 ng/mL.

2.4. PAGs Radioimmunoassay

For PAGs concentration, RIA-860 previously described by Barbato et al. [45] was used. Pure boPAG$_{67kDa}$ preparation was used as the standard and tracer. Iodination (Na-I^{125}, Amersham Pharmacia Biotech, Uppsala, Sweden) was carried out according to the chloramine-T method previously described by Greenwood et al. [57]. The samples were assayed in a preincubated system in which the standard curve ranged from 0.2 to 25 ng/mL.

The minimum detection limit (MDL), calculated as the mean concentration minus twice the standard deviation (mean–2 SD) of 20 duplicates of the zero (B_0) standard [58], was 0.4 ng/mL. The intra- and inter-assay coefficients were 2.8% and 7.1%, respectively.

Based on PAGs assay, (cut-off value: ≥1 ng/mL) buffaloes were considered non-pregnant when concentrations remained very close to zero at all time points and pregnant when concentrations were higher than 1 ng/mL at days 25, 28 and 40. When PAGs concentrations were ≥1 ng/mL at day 25 and dropped under 0.2 ng/mL by day 40, embryo mortality was considered to have occurred.

2.5. Statistical Analysis

The association between the outcome (2 levels: Pregnant and Mortality) and results of the ultrasound analysis (4 levels: No vesicle, Vesicle, Vesicle + embryo, Vesicle + embryo + beat) was analyzed by Chi-square or Fisher's exact test, stratifying for day of observation (time). Z-tests were used to compare column proportions. This association was not evaluated at time 0 as all the animals were in the no vesicle group.

The changes of P4 and PAGs concentrations with time in the two groups were instead analyzed using linear mixed models (LMMs). In LMMs, animals and days were included as subjects and repeated factors, respectively. The LMMs evaluated the main effects of time (4 levels: 0, 25, 28, and 40 days post-AI), outcome (2 levels: pregnant and mortality), and the interaction between the outcome and time. Sidak adjustment was used for carrying out multiple comparisons. Diagnostic graphics were used for testing assumptions and outliers.

Because non-normality of the data was detected for P4 and PAGs concentrations, log and log(x + 1) transformation, respectively, were used for analysis. Back-transformed estimated marginal means were presented as results while the row data were presented as figures. Moreover, the association between P4 and PAGs concentrations was analysed using the Spearman's correlation coefficient (ρ).

In order to test the ability of PAGs and P4 concentrations to discriminate between mortality and pregnant outcomes at each time (days 25, 28, and 40 post-AI), the receiver operating characteristic (ROC) analysis was also performed and the optimal cut-off was determined by Youden index [59]. Finally, univariate models were built using the generalized linear models procedure with a binomial distribution and the logit link function to evaluate the accuracy of the parameters in predicting mortality at day 25 and 28. PAGs and P4 concentrations were categorized according to the cut-off for each day while ultrasound outcome was categorized according to the identification of the embryo heartbeat. Odds ratios (ORs) with the corresponding 95% confidence interval (CI) and p-values were calculated.

Statistical analyses were performed with SPSS Statistics version 25 (IBM, SPSS Inc., Chicago, IL, USA). Statistical significance occurred when $p \leq 0.05$.

3. Results

A total of 50 out of 109 buffalo cows enrolled in this study became pregnant (pregnant group) while 12 had embryo mortality (mortality group) and 47 remained non-pregnant as determined by ultrasonography at day 40. All buffaloes diagnosed as pregnant were confirmed at day 60.

3.1. Ultrasound Observations and Embryo Mortality

Results of ultrasound observations according to the outcomes (pregnant vs. mortality) are summarized in Table 1. At day 25, the proportion of animals in which the vesicle was not identified was greater in the mortality group, while the proportion of animals in which the vesicle, vesicle + embryo or vesicle + embryo + beat was identified were higher in the pregnant group ($p < 0.0001$). At day 28, the proportion of animals in which the vesicle + embryo was identified was higher in the mortality group while the animals in which vesicle + embryo + beat were identified was higher in pregnant buffaloes than those in the mortality group ($p < 0.0001$). A significant association between the ultrasound observations and the outcome was also found at day 40; vesicle + embryo + beat was identified in all the buffaloes in the pregnant group; in the mortality group, 11 buffalo cows (91.7%) showed only the vesicle and 1 cow (8.3%) the vesicle + embryo ($p < 0.0001$).

Table 1. Results of ultrasound observations according to number of days post-artificial insemination (AI) and outcome.

Day Post-AI	Ultrasound Observation	Outcome		Significance
		Mortality	Pregnant	
25	No vesicle	8 ^{a} (66.7%)	4 ^{b} (8.0%)	0.0001
	Vesicle	1 ^{a} (8.3%)	22 ^{b} (44.0%)	
	Vesicle + embryo	1 ^{a} (8.3%)	3 ^{a} (6.0%)	
	Vesicle + embryo + beat	2 ^{a} (16.7%)	21 ^{b} (42.0%)	
28	No vesicle	1 ^{a} (8.3%)	0 ^{a} (0.0%)	0.0001
	Vesicle	2 ^{a} (16.7%)	3 ^{a} (6.0%)	
	Vesicle + embryo	3 ^{b} (25.0%)	0 ^{a} (0.0%)	
	Vesicle + embryo + beat	6 ^{a} (50.0%)	47 ^{b} (94.0%)	
40	No vesicle	11 ^{b} (91.7%)	0 ^{a} (0.0%)	0.0001
	Vesicle + embryo	1 ^{a} (8.3%)	0 ^{a} (0.0%)	
	Vesicle + embryo + beat	0 ^{a} (0.0%)	50 ^{b} (100.0%)	

Values in the same row not sharing the same superscript (a, b) are significantly different at $p < 0.05$ (z-test).

3.2. P4 Concentrations and Mortality

P4 concentrations were affected by time ($p < 0.001$), outcome ($p < 0.001$), and interaction ($p < 0.001$; Figure 1). Significant differences in P4 concentrations between the mortality and pregnant groups were found at days 25 (mean difference: 1.8 ± 0.1 ng/mL, $p < 0.001$), 28 (mean difference: 1.4 ± 0.1 ng/mL, $p < 0.05$) and 40 (mean difference: 3.1 ± 0.1 ng/mL, $p < 0.001$) post-AI.

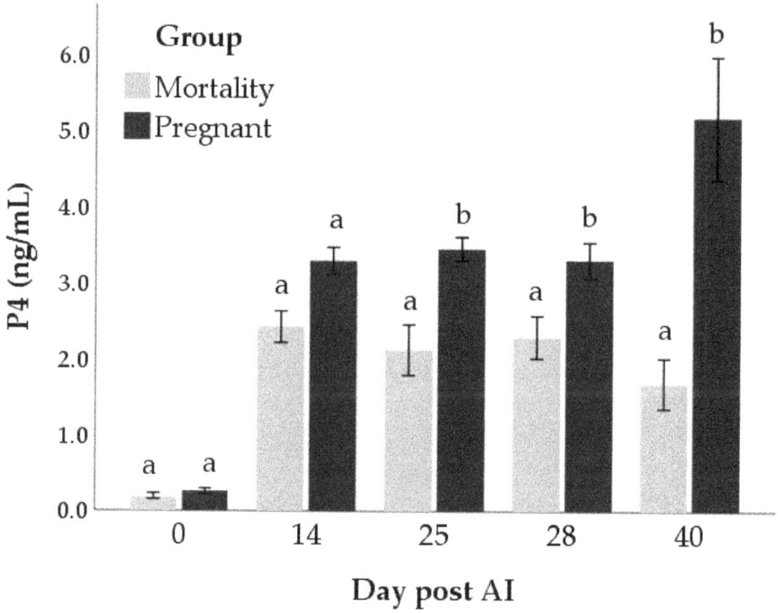

Figure 1. Concentrations of progesterone (P4) in mortality and pregnant groups at day 0, 14, 25, 28 and 40 post-AI. Bars not sharing the same superscript within each day are significantly different at $p < 0.05$.

The ROC curves for detection of embryo mortality by P4 and the optimal cut-off for predicting mortality are reported in Table 2 and Figure 2.

Table 2. Results of Receiver Operator Characteristic (ROC) analysis including the area under the ROC curve (AUC) with corresponding 95% confidence intervals (CI) and *p*-values, optimal cut-off with associated sensitivity, specificity and accuracy values.

Day	Parameter	AUC	95% CI	*p*-Value	Cut-off (ng/mL)	Sensitivity (%)	Specificity (%)	Accuracy (%)
25	PAG	0.837	0.706–0.967	<0.001	1.1	75	74	74
	P4	0.793	0.639–0.946	0.002	2.6	67	76	74
28	PAG	0.700	0.516–0.884	0.033	2.2	67	76	74
	P4	0.722	0.537–0.906	0.018	2.6	75	66	68
40	PAG	1.000	1.000–1.000	<0.001	2.7	100	100	100
	P4	0.883	0.758–1000	<0.001	2.4	83	86	85

PAG = Pregnancy-Associated Glycoproteins. P4 = Progesterone.

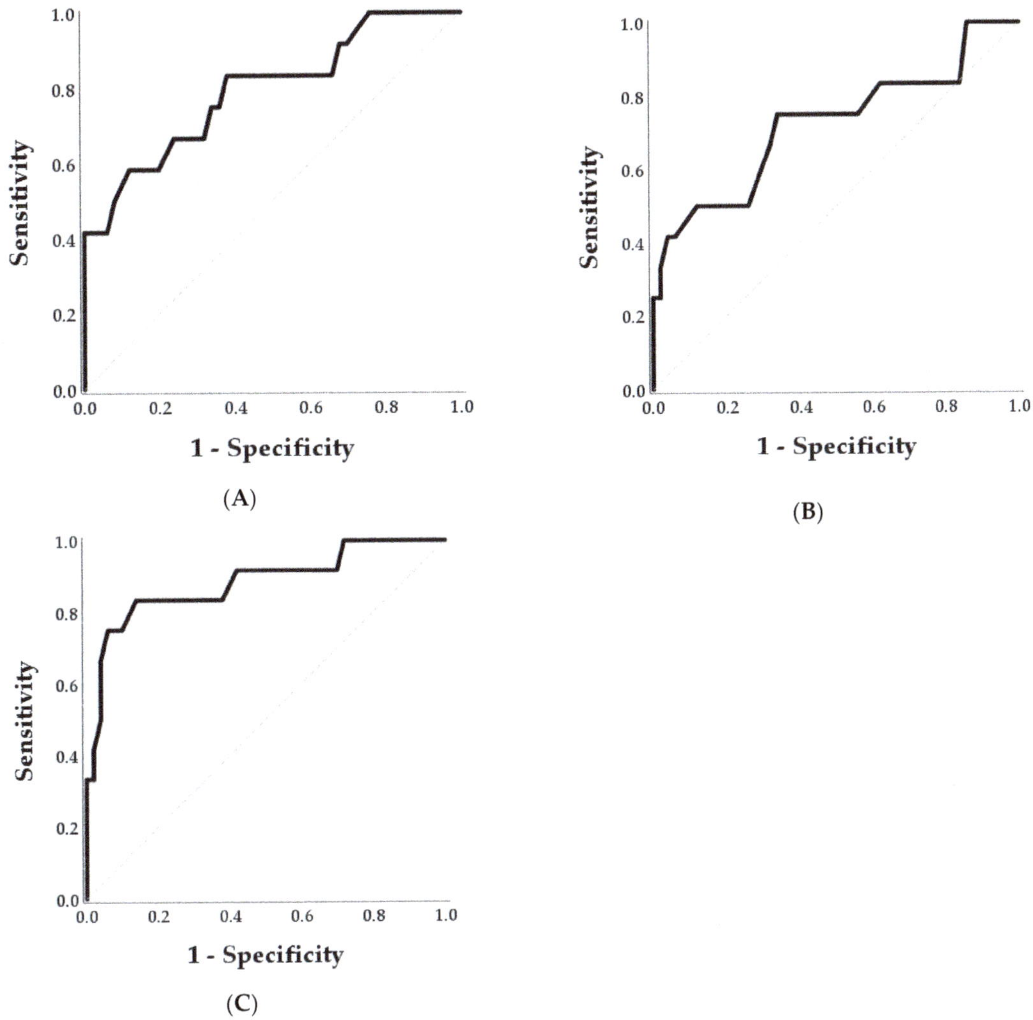

Figure 2. Receiver operating characteristics curves for the detection of embryonic mortality by P4 at days 25 (Panel **A**), 28 (Panel **B**), and 40 (Panel **C**) post-AI. Optimal cut-offs for predicting mortality were 2.6 ng/mL, 2.6 ng/mL, and 2.4 ng/mL at days 25, 28 and 40 post-AI, respectively.

ROC analysis for P4 concentrations at day 25 showed an area of 0.793 (Figure 2, Panel A) and the optimal cut-off for predicting mortality of 2.6 ng/mL. A sensitivity of 100% was achieved for P4 concentrations lower than 4.1 ng/mL. At day 28, the AUC was 0.722 (Figure 2, Panel B) and the optimal cut-off was 2.6 ng/mL. At day 40, the AUC was 0.883 (Figure 2, Panel C) and the optimal cut-off was 2.4 ng/mL. A sensitivity of 100% at day 40 post-AI was achieved for P4 concentrations lower than 4.2 ng/mL.

3.3. PAGs Concentrations and Mortality

Significant effects of time ($p < 0.001$), outcome ($p < 0.001$) and interaction were observed in the log-PAG-1 concentrations ($p < 0.001$). Differences in PAGs levels between the two groups started from day 25 post-AI (mean difference: 0.3 ± 0.1 ng/mL, 0.3 ± 0.1 ng/mL, and 11.6 ± 0.1 ng/mL at days 25, 28, and 40 post-AI, respectively; $p < 0.01$; Figure 3).

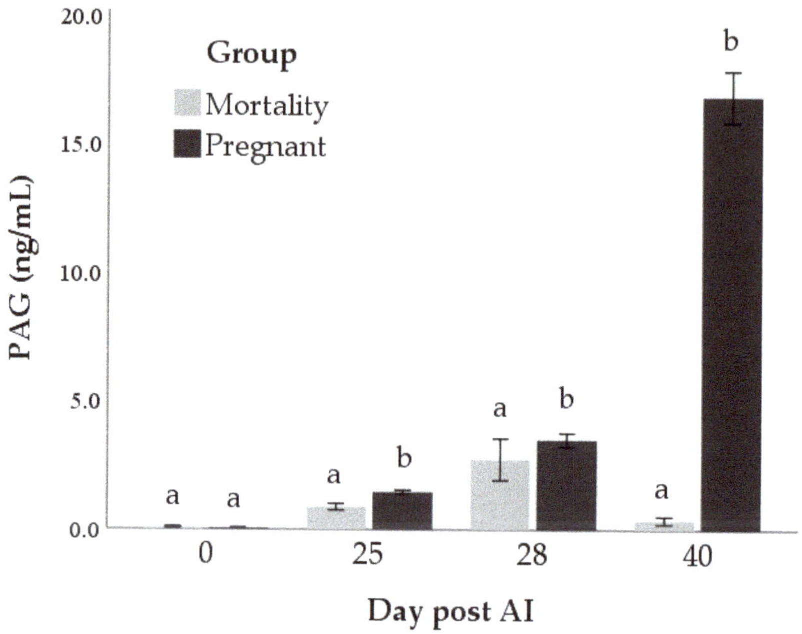

Figure 3. Concentrations of pregnancy-associated glycoproteins (PAG) in mortality and pregnant groups at days 0, 25, 28, and 40 post-AI. Bars not sharing the same superscript within each day are significantly different at $p < 0.05$.

The results of the ROC analyses performed at each day are reported in Table 2 and Figure 4.

At day 25, the area under ROC was 0.837 and the Youden Index analysis revealed that the optimal cut-off for predicting mortality was 1.1 ng/mL (Figure 4, Panel A). A sensitivity of 100% was achieved for PAGs concentrations lower than 1.6 ng/mL. At day 28, the AUC was lower (0.700) and increased the optimal cut-off value (2.2 ng/mL; Figure 4, Panel B). At day 40, PAGs concentration perfectly discriminated between the mortality and pregnant groups (AUC = 1.000; Figure 4, Panel C) and the cut-off of 2.7 ng/mL identified cases of mortality with a sensitivity of 100% and a specificity of 100%.

3.4. Association between Ultrasound Outcome and PAGs Concentrations

In the model including days 25 and 28 as time and the classification based on the ultrasound observations as fixed effect, we found a significant effect of time ($p < 0.05$), group ($p < 0.001$), and interaction ($p < 0.05$) on PAGs concentrations.

There were no differences on PAGs concentrations at day 25 according to ultrasound observations. At day 28, Vesicle (1.5 ± 0.1 ng/mL; $p < 0.01$) and Vesicle + embryo (1.5 ± 0.1 ng/mL; $p < 0.05$) had lower PAGs concentrations compared with Vesicle + embryo + beat (3.3 ± 0.1 ng/mL).

3.5. Univariate Models Determining Predictors of Mortality at Days 25 and 28 Post-AI

The lack of embryo heartbeat detected by ultrasound at day 25 was not a predictor of embryonic mortality ($p < 0.119$) while at day 28 the odds of mortality increased by about 15 times if the heartbeat was not detected (Table 3).

Conversely, P4 and PAGs were predictors of mortality both at day 25 and 28 post-AI (Table 3). Both at days 25 and 28, the odds of mortality were six times higher when PAGs concentrations were lower than or equal to the cut-off (1.1 ng/mL and 2.2 at days 25 and 28, respectively; $p < 0.05$) and five times higher if P4 was lower than or equal to 2.6 ng/mL ($p < 0.05$).

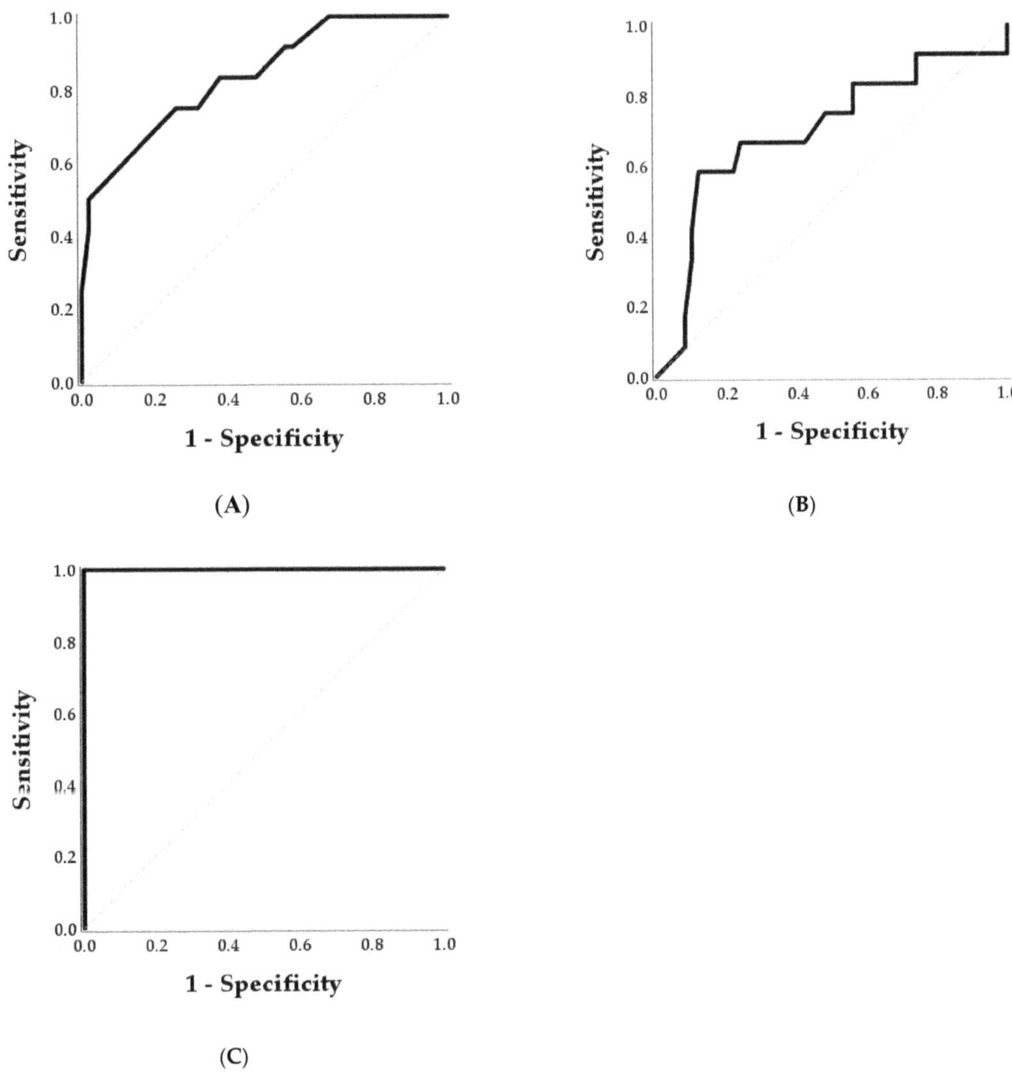

Figure 4. Receiver operating characteristic (ROC) curves for the detection of embryonic mortality by PAG at days 25 (Panel **A**), 28 (Panel **B**), and 40 (Panel **C**) post-AI. Optimal cut-offs for predicting mortality were 1.1 ng/mL, 2.2 ng/mL, and 2.7 ng/mL at days 25, 28 and 40 post-AI, respectively.

Table 3. Univariate models to evaluate the predictors of mortality in buffalo cows at days 25 and 28 post-AI.

Day Post-AI	Predictor	OR	95% CI		p-Value
			Lower	Upper	
25	**Ultrasound outcome**				
	No embryo + beat vs. Embryo + beat	3.621	0.717	18.272	0.119
	PAG concentrations ≤1.1 ng/mL vs. >1.1 ng/mL	6.375	1.517	26.784	0.011
	P4 concentration ≤2.6 ng/mL vs. >2.6 ng/mL	4.667	1.217	17.894	0.025
28	**Ultrasound outcome**				
	No Embryo + beat vs. Embryo + beat	15.667	3.083	79.613	0.001
	PAG concentrations ≤2.2 ng/mL vs. >2.2 ng/mL	6.333	1.618	24.786	0.008
	P4 concentrations ≤2.6 ng/mL vs. >2.6 ng/mL	4.895	1.176	20.372	0.029

Dependent variable: mortality. OR = odds ratio. CI = Wald confidence interval.

4. Discussions

The aim of this work was to find the best strategy to diagnose pregnancy failures in buffalo cows in order to improve farm reproductive management through intervention strategies for animals identified to be at risk for embryo loss. From the results of ultrasound at day 25, it seems that the animals that experienced embryonic mortality had a delayed growth of the vesicle and embryo since the vesicle was not observed in more than half of the buffalo cows in the mortality group, while those that maintained pregnancy already showed vesicle + embryo + beat in a higher percentage at the same time. Confirming this hypothesis, the higher rate of vesicle + embryo + beat in the animals that later experienced embryonic mortality was found at day 28.

Uterine environment is critical as the embryo transcends from the oviduct to the site of implantation. Asynchrony between the uterus and the embryo can be problematic as the uterine environment will not wait for an embryo, although an embryo can accelerate or decelerate its development to some degree [60,61]. At day 40, all the animals in which there was a recognized embryonic vesicle and the embryo proper with heartbeat maintained pregnancy until the end of the observation period (60 days post-AI), while no buffaloes in the mortality group showed vesicle + embryo + heartbeat. Pregnancy diagnosis is considered positive only when gestational vesicle and embryo, together with a heartbeat, can be detected [62]. In bovine pregnancy, based on the absence of a heartbeat, placental detachment, or reduced placental fluid volume, pregnancy losses have been diagnosed and reported to occur between 24 and 40 days of gestation [63–65]. In the present study, using the ultrasound observations, most of embryonic mortality was diagnosed as occurring between 28 and 40 days after insemination as reported by other authors in this species [8,66,67].

Embryo death can be clearly diagnosed by an undetectable heartbeat. In fact, at day 28 post-AI, the odds of mortality increased by about 15 times if the heartbeat was not detected. The disadvantage of this method is that before 28 days of gestation it is not predictive of embryonic mortality. This is in agreement with what has been reported by other authors of bovine pregnancy studies [16,60,68].

Regarding the biological markers of pregnancy, buffaloes that maintained pregnancy had significantly higher circulating concentrations either of PAGs or P4 starting from day 25 after AI until the end of the observation period (40 days after AI).

Progesterone is primarily produced and secreted by the CL and subsequently by the placenta during pregnancy [69]; thus, the concentration of progesterone in the first weeks of pregnancy reflects the function of the CL more than the presence of an embryo. The persistence of the CL after an embryonic death or the start of a new oestrus cycle after embryo resorption leads to a positive value of P4 in the bloodstream [24]; therefore, the presence of P4 that we found in the mortality group at 40 days post-AI, when embryo resorptions were recorded by ultrasound observation, could be due to this reason.

Starting from day 25 post-AI, significant differences in the P4 value were found between animals that experienced embryo mortality and those that maintained pregnancy. The lower P4 value in the mortality group could indicate a lower functionality of the CL that could have contributed to the embryo resorption [69,70]. In a previous work, the diameter and echogenicity of the CL seemed to affect its functionality as there was a positive correlation between plasma P4 concentration and CL diameter that was found to be significantly larger in pregnant buffalo in contrast to non-pregnant buffalo [71]. Campanile et al. [72] reported that buffaloes that underwent late embryonic development had relatively low concentrations of P4 in their blood; this finding was interpreted to indicate that reduced P4 concentrations were inadequate to induce changes in the uterus required for attachment of the conceptus.

Although P4 concentration helps in detecting early pregnancy, it is not highly specific and has been the major problem for the prevalence of false-positives (identifying a non-pregnant animal as pregnant) [16,73]; moreover, P4 is not useful for verifying the presence of a viable embryo in the uterus [74]. Therefore, the non-specificity of P4 and the presence of possible false positives has limited the use of this method as a marker of embryonic viability [28,75,76].

Different from P4, PAGs indicate the presence of an embryo in the uterus of eutherian species. Several authors showed the relationship between PAGs and fetal wellbeing [77–81]. These glycoproteins are synthesized by the giant trophoblast cells (TGCs) which migrate from the fetal to the uterine epithelium and release PAGs into the maternal blood [32,33,82]. This process presumes the presence of a healthy trophoblastic tissue and thus of a healthy embryo. If this condition fails, the source of production of the proteins is missing. Furthermore, thanks to this "active migration", PAGs play an important role in the remodeling of fetal membranes and in the formation of placentome during pregnancy [81,83,84], and is a possible factor controlling maternal immune modulation [85,86]. Therefore, in addition to serving as an accurate tool for diagnosing pregnancy in ruminants, PAGs may also serve as a marker for monitoring embryonic/fetal viability and placental function [16,60,87–89]. In buffalo, this protein can be detected in the blood of pregnant animals starting from day 25 post-AI, with an accuracy of 99% on day 28 of gestation, thus providing an accurate test to follow up pregnancy [45,49,90].

In this work, buffaloes that experienced embryo mortality had a lower concentration of PAGs starting from day 25 post-AI, suggesting a possible embryo suffering in these animals at the early stage of gestation. In cows, physiologically, pregnancy loss after day 28 of gestation coincides with a period of active placentation marked by extensive endometrium remodeling, binucleate cell migration and changes in PAGs expression production [68].

On the basis of the ROC curves and positive and negative analysis, an optimal PAGs cut value has been established that is 74% accurate in predicting early embryonic mortality for a concentration less than 1.1 and 2.2 ng/mL at day 25 and 28 post-AI, respectively. To our knowledge, no data are reported in the literature regarding a cut-off predicting embryonic mortality in buffalo. In bovine pregnancy, Polher et al. [16] suggest that PAGs are predictive of embryonic mortality between day 28 and 45 of gestation. The same authors in a later work [91] have shown that at day 28 of gestation a circulating concentration of PAGs greater than 7.9 ng/mL was 95% accurate in predicting embryonic maintenance (to day 100), and a concentration of glycoproteins less than 0.72 ng/mL was 95% accurate in predicting embryonic mortality by day 100. Gatea et al., [92] reported that circulating concentrations of PAGs on days 28 to 31 have been shown to be measurably lower in cows that experienced late embryonic mortality.

Based on odds ratio analysis, our data showed that for buffalo cows with PAGs values $\leq$1.1 and $\leq$28 ng/mL at day 25 and 28 post-AI, respectively, there was a 6.3 times greater chance to undergo pregnancy loss within the first 40 days of gestation compared to those that had higher values. Higher PAGs concentrations were correlated to a decreased incidence of embryonic loss as reported also in other bovine studies [93,94].

In a study on the accuracy of ultrasonography and PAGs for pregnancy diagnosis in buffaloes, Karen et al. [22] reported no significant differences between the sensitivity and specificity of the two tests in the examined period (19 and 55 days after mating), and concluded that both are highly accurate tests for detecting pregnant buffaloes from day 31 onwards after mating. In this study, the ultrasound allowed us to significantly detect the presence of a viable embryo (vesicle + embryo + heartbeat) at day 28 post-AI, predicting the animals that would maintain pregnancy compared to those that would experience embryonic mortality, while PAGs concentrations permitted the discrimination between buffalo that experienced embryonic mortality and those that maintained pregnancy starting from 25 days of gestation.

5. Conclusions

Among the methods investigated in this study, PAGs were the best marker for predicting embryonic mortality in buffalo between 25 and 40 days of gestation. Although low values of P4 were associated with pregnancy failures, as a predictor for pregnancy loss, P4 is less reliable compared to PAGs and ultrasonography. The disadvantage of ultrasonography is that the pregnancy status is only guaranteed at the time of diagnosis. Differently, PAGs reflect embryo wellbeing and therefore the reduction of its circulating concentrations is a prognostic sign of pregnancy failure. Notwithstanding that the assay refinement and studies with a large sample size are needed to improve the predictive value to an acceptable point for use in applied reproductive management, our data have shown that PAGs could be utilized as a diagnostic tool in order to improve farm reproductive management through intervention strategies for animals identified to be at risk for embryo loss.

Author Contributions: Conceptualization, writing and editing, V.L.B.; formal analysis and data curation, L.M.; samples analysis, A.B.C.; visualization and review, G.B.; supervision, N.M.d.S.; visualization, R.Z., laboratory analysis, C.C.; supervision, J.F.B.; conceptualization, writing and draft preparation, O.B. All authors have read and agreed to the published version of the manuscript.

Funding: This research was founded by Scientific and Technological Research Program founded by Fondazione Cassa di Risparmio of Perugia and Sterling SPA.

Institutional Review Board Statement: The trial was carried out in the period January-June 2012. The animals involved in this experiment were treated in compliance with the animal testing regulations established under Italian law in force at that time (DL 27 January 1992, N 116). The experimental design was carried out according to good veterinary practices under farm conditions. The CREA Research Centre for Animal Production and Aquaculture is authorized to use farm animals for experimental design (as stated in DM 26/96-4 of Italian Welfare Ministry). The animals were supevised by the responsable of animal welfare and the designated veterinarian as required by the law in force at the time of the trial.

Data Availability Statement: No new data were created or analyzed in this study. Data sharing is not applicable to this article.

Conflicts of Interest: The authors declare no conflict of interest.

References

1. Khatib, H.; Huang, W.; Mikheil, D.; Schutzkus, V.; Monson, R. Effects of signal transducer and activator of transcription (STAT) genes STAT1 and STAT3 genotypic combinations on fertilization and embryonic survival rates in Holstein cattle. *J. Dairy Sci.* **2009**, *92*, 6186–6191. [CrossRef] [PubMed]
2. Wiltbank, M.C.; Baez, G.M.; Garcia-Guerra, A.; Toledo, M.Z.; Monteiro, P.L.; Melo, L.F.; Ochoa, J.C.; Santos, J.E.; Sartori, R. Pivotal periods for pregnancy loss during the first trimester of gestation in lactating dairy cows. *Theriogenology* **2016**, *86*, 239–253. [CrossRef]
3. Fricke, P.M.; Ricci, A.; Giordano, J.O.; Carvalho, P.D. Methods for and implementation of pregnancy diagnosis in dairy cows. *Vet. Clin. Food Anim. Pract.* **2016**, *32*, 165–180. [CrossRef] [PubMed]

4. Santos, J.; Thatcher, W.; Chebel, R.; Cerri, R.; Galvao, K. The effect of embryonic death rates in cattle on the efficacy of estrus synchronization programs. *Anim. Reprod. Sci.* **2004**, *82*, 513–535. [CrossRef] [PubMed]
5. Walsh, S.; Williams, E.; Evans, A. A review of the causes of poor fertility in high milk producing dairy cows. *Anim. Reprod. Sci.* **2011**, *123*, 127–138. [CrossRef] [PubMed]
6. Barile, V.L. Reproductive efficiency in female buffaloes. In *Buffalo Production and Research*; Borghese, A., Ed.; REU Technical Series 67; FAO: Roma, Italy, 2005; pp. 77–108. Available online: http://ftp.fao.org/docrep/fao/010/ah847e/ah847e.pdf (accessed on 16 December 2020).
7. Barile, V.; Terzano, G.; Pacelli, C.; Todini, L.; Malfatti, A.; Barbato, O. LH peak and ovulation after two different estrus synchronization treatments in buffalo cows in the daylight-lengthening period. *Theriogenology* **2015**, *84*, 286–293. [CrossRef]
8. Campanile, G.; Neglia, G.; Gasparrini, B.; Galiero, G.; Prandi, A.; Di Palo, R.; Michael, J.; Zicarelli, L. Embryonic mortality in buffaloes synchronized and mated by AI during the seasonal decline in reproductive function. *Theriogenology* **2005**, *63*, 2334–2340. [CrossRef]
9. Campanile, G.; Neglia, G. Embryonic mortality in buffalo cows. *Ital. J. Anim. Sci.* **2007**, *6*, 119–129. [CrossRef]
10. Baruselli, P.; Visintin, J.; Barnabe, V.; Barnabe, R.; Amaral, R.; Souza, A. Early pregnancy ultrasonography and embryonic mortality occurence in buffalo. In Proceedings of the V World Buffalo Congress, Caserta, Italy, 13–18 October 1997; pp. 13–16.
11. Vale, W.; Ohashi, O.; Sousa, J.; Ribeiro, H.; Silva, A.; Nanba, S. Morte embrionaria e fetal em bufalos, *Bubalus bubalis* Lin. *Rev. Bras. Reprod. Anim.* **1989**, *13*, 157–165.
12. Hansel, W.; Spalding, R.W.; Larson, L.L.; Laster, D.B.; Wagner, J.F.; Braun, R.K. Influence of human chorionic gonadotropin on pregnancy rates in lactating dairy and beef cows. *J. Dairy Sci.* **1976**, *59*, 751–754. [CrossRef]
13. Morris, D.G.; Grealy, M.; Leese, H.; Diskin, M.G.; Sreenan, J. *Cattle Embryo Growth Development and Viabilty*; Teagasc: Dublin, Ireland, 2001; ISBN 1-84170-224-2.
14. Fricke, P. Scanning the future—Ultrasonography as a reproductive management tool for dairy cattle. *J. Dairy Sci.* **2002**, *85*, 1918–1926. [CrossRef]
15. Balhara, A.K.; Gupta, M.; Singh, S.; Mohanty, A.K.; Singh, I. Early Pregnancy Diagnosis in Bovines: Current Status and Future Directions. *Sci. World J.* **2013**, *2013*. [CrossRef]
16. Pohler, K.G.; Pereira, M.H.C.; Lopes, F.R.; Lawrence, J.C.; Keisler, D.H.; Smith, M.F.; Vasconcelos, J.L.M.; Green, J.A. Circulating concentrations of bovine pregnancy-associated glycoproteins and late embryonic mortality in lactating dairy herds. *J. Dairy Sci.* **2016**, *99*, 1584–1594. [CrossRef] [PubMed]
17. Pawshe, C.H.; Appa Rao, K.B.C.; Totey, S.M. Ultrasonographic imaging to monitor early pregnancy and embryonic development in the buffalo (*Bubalus bubalis*). *Theriogenology* **1994**, *41*, 697–709. [CrossRef]
18. Ali, A.; Fahmy, S. Ultrasonographic fetometry and determination of fetal sex in buffaloes (*Bubalus bubalis*). *Anim. Reprod. Sci.* **2008**, *106*, 90–99. [CrossRef] [PubMed]
19. Bhosreker, M.R.; Hangarge, I.M. Ultra sonography for early pregnancy diagnosis in buffaloes. *Indian J. Anim. Reprod.* **2000**, *21*, 143–144.
20. Pinheiro Ferreira, J.C.P.; Martin, I.; Irikura, C.R.; Gimenes, L.U.; Fujihara, C.J.; Jorge, A.M.; Oba, E. Ultrasonographic monitoring of early pregnancy development in Murrah buffalo heifers (*Bubalus bubalis*). *Livest. Sci.* **2011**, *138*, 174–179. [CrossRef]
21. Sharma, R.; Singh, J.; Khanna, S.; Phulia, S.; Sarkar, S.K.; Singh, I. Fetal age determination in Murrah buffaloes from days 22 through 60 with ultrasonography. *Indian J. Anim. Sci.* **2012**, *82*, 374–376.
22. Karen, A.; Darwish, S.; Ramoun, A.; Tawfeek, K.; Van Hanh, N.; De Sousa, N.; Sulon, J.; Szenci, O.; Beckers, J.-F. Accuracy of ultrasonography and pregnancy-associated glycoprotein test for pregnancy diagnosis in buffaloes. *Theriogenology* **2007**, *68*, 1150–1155. [CrossRef]
23. López-Gatius, F.; García-Ispierto, I. Ultrasound and Endocrine Findings that Help to Assess the Risk of Late Embryo/Early Foetal Loss by Non-Infectious Cause in Dairy Cattle. *Reprod. Domest. Anim.* **2010**, *45*, 15–24. [CrossRef]
24. Barbato, O.; Barile, V.L. The Pregnancy Diagnosis in Buffalo Species: Laboratory Methods. *J. Buffalo Sci.* **2012**, *1*, 157–162. [CrossRef]
25. Arora, R.; Pandey, R. Changes in peripheral plasma concentrations of progesterone, estradiol-17β, and luteinizing hormone during pregnancy and around parturition in the buffalo (*Bubalus bubalis*). *Gen. Comp. Endocrinol.* **1982**, *48*, 403–410. [CrossRef]
26. Kaul, V.; Prakash, B.S. Accuracy of pregnancy/non pregnancy diagnosis in zebu and crossbred cattle and Murrah buffaloes by milk progesterone determination post insemination. *Trop. Anim. Health Prod.* **1994**, *26*, 187–192. [CrossRef]
27. Batra, S.K.; Prakash, B.S.; Madan, M.L. Relationship of progesterone. *Trop. Anim. Health Prod.* **1993**, *25*, 185–192.
28. Starbuck, M.J.; Dailey, R.A.; Inskeep, E.K. Factors affecting retention of early pregnancy in dairy cattle. *Anim. Reprod. Sci.* **2004**, *84*, 27–39. [CrossRef]
29. Zoli, A.P.; Beckers, J.F.; Wouters-Ballman, P.; Closset, J.; Falmagne, P.; Ectors, F. Purification and Characterization of a Bovine Pregnancy-Associated Glycoprotein. *Biol. Reprod.* **1991**, *45*, 1–10. [CrossRef] [PubMed]
30. Garbayo, J.M.; Serrano, B.; Lopez-Gatius, F. Identification of novel pregnancy-associated glycoproteins (PAG) expressed by the peri-implantation conceptus of domestic ruminants. *Anim. Reprod. Sci.* **2008**, *103*, 120–134. [CrossRef]
31. Wallace, R.M.; Pohler, K.G.; Smith, M.F.; Green, J.A. Placental PAGs: Gene origins, expression patterns, and use as markers of pregnancy. *Reproduction* **2015**, *149*, R115–R126. [CrossRef]
32. Zoli, A.P.; Guilbault, L.A.; Delahaut, P.; Ortiz, W.B.; Beckers, J.-F. Radioimmunoassay of a Bovine Pregnancy-Associated Glycoprotein in Serum: Its Application for Pregnancy Diagnosis. *Biol. Reprod.* **1992**, *46*, 83–92. [CrossRef]

33. Touzard, E.; Reinaud, P.; Dubois, O.; Guyader-Joly, C.; Humblot, P.; Ponsart, C.; Charpigny, G. Specific expression patterns and cell distribution of ancient and modern PAG in bovine placenta during pregnancy. *Reproduction* **2013**, *146*, 347–362. [CrossRef]
34. Szafranska, B.; Xie, S.; Green, J.; Roberts, R.M. Porcine pregnancy-associated glycoproteins: New members of the aspartic proteinase gene family expressed in trophectoderm. *Biol. Reprod.* **1995**, *53*, 21–28. [CrossRef]
35. Xie, S.; Green, J.; Bixby, J.B.; Szafranska, B.; DeMartini, J.C.; Hecht, S.; Roberts, R.M. The diversity and evolutionary relationships of the pregnancy-associated glycoproteins, an aspartic proteinase subfamily consisting of many trophoblast-expressed genes. *Proc. Natl. Acad. Sci. USA* **1997**, *94*, 12809–12816. [CrossRef] [PubMed]
36. Garbayo, J.M.; Remy, B.; Alabart, J.L.; Folch, J.; Wattiez, R.; Falmagne, P.; Beckers, J.F. Isolation and Partial Characterization of a Pregnancy-Associated Glycoprotein Family from the Goat Placenta. *Biol. Reprod.* **1998**, *58*, 109–115. [CrossRef]
37. Green, J.A.; Xie, S.; Quan, X.; Bao, B.; Gan, X.; Mathialagan, N.; Beckers, J.-F.; Roberts, R.M. Pregnancy-Associated Bovine and Ovine Glycoproteins Exhibit Spatially and Temporally Distinct Expression Patterns During Pregnancy1. *Biol. Reprod.* **2000**, *62*, 1624–1631. [CrossRef] [PubMed]
38. El Amiri, B.; Sousa, N.M.; Alvarez Oxiley, A.; Hadarbach, D.; Beckers, J.-F. Pregnancy-associated glycoprotein (PAG) concentration in plasma and milk samples for early pregnancy diagnosis in Lacaune dairy sheep. *Res. Vet. Sci.* **2015**, *99*, 30–36. [CrossRef]
39. Sousa, N.M.; Zongo, M.; Pitala, W.; Boly, H.; Sawadogo, L.; Sanon, M.; de Figueiredo, J.R.; Gonçalves, P.B.D.; El Amiri, B.; Perènyi, Z.; et al. Pregnancy-associated glycoprotein concentrations during pregnancy and the postpartum period in Azawak Zebu cattle. *Theriogenology* **2003**, *59*, 1131–1142. [CrossRef]
40. Majewska, M.; Panasiewicz, G.; Majewski, M.; Szafranska, B. Localization of chorionic pregnancy-associated glycoprotein family in the pig. *Reprod. Biol.* **2006**, *6*, 205–230.
41. Brandt, G.A.; Parks, T.E.; Killian, G.; Ealy, A.D.; Green, J.A. A cloning and expression analysis of pregnancy-associated glycoproteins expressed in trophoblasts of the white-tail deer placenta. *Mol. Reprod. Dev.* **2007**, *74*, 1355–1362. [CrossRef]
42. Bériot, M.; Tchimbou, A.F.; Barbato, O.; Beckers, J.-F.; de Sousa, N.M. Identification of pregnancy-associated glycoproteins and alpha-fetoprotein in fallow deer (Dama dama) placenta. *Acta Vet. Scand.* **2014**, *56*, 4. [CrossRef]
43. De Carolis, M.; Barbato, O.; Acuti, G.; Trabalza Marinucci, M.; de Sousa, N.M.; Canali, C.; Moscati, L. Plasmatic Profile of Pregnancy-Associated Glycoprotein (PAG) during Gestation and Postpartum in Sarda and Lacaune Sheep Determined with Two Radioimmunoassay Systems. *Animals* **2020**, *10*, 1502. [CrossRef] [PubMed]
44. Barbato, O.; Sousa, N.M.; Klisch, K.; Clerget, E.; Debenedetti, A.; Barile, V.L.; Malfatti, A.; Beckers, J.F. Isolation of new pregnancy-associated glycoproteins from water buffalo (*Bubalus bubalis*) placenta by Vicia villosa affinity chromatography. *Res. Vet. Sci.* **2008**, *85*, 457–466. [CrossRef]
45. Barbato, O.; Melo de Sousa, N.; Barile, V.L.; Canali, C.; Beckers, J.-F. Purification of pregnancy-associated glycoproteins from late-pregnancy *Bubalus bubalis* placentas and development of a radioimmunoassay for pregnancy diagnosis in water buffalo females. *BMC Vet. Res.* **2013**, *9*, 89. [CrossRef]
46. Zoli, A.P.; Demez, P.; Beckers, J.-F.; Reznik, M.; Beckers, A. Light and Electron Microscopic Immunolocalization of Bovine Pregnancy-Associated Glycoprotein in the Bovine Placentome. *Biol. Reprod.* **1992**, *46*, 623–629. [CrossRef]
47. Barbato, O.; Sousa, N.M.; Debenedetti, A.; Canali, C.; Todini, L.; Beckers, J.F. Validation of a new pregnancy-associated glycoprotein radioimmunoassay method for the detection of early pregnancy in ewes. *Theriogenology* **2009**, *72*, 993–1000. [CrossRef]
48. Barbato, O.; Chiaradia, E.; Barile, V.L.; Pierri, F.; de Sousa, N.M.; Terracina, L.; Canali, C.; Avellini, L. Investigation into omocysteine, vitamin E and malondialdehyde as indicators of successful artificial insemination in synchronized buffalo cows (*Bubalus bubalis*). *Res. Vet. Sci.* **2016**, *104*, 100–105. [CrossRef]
49. Barbato, O.; Menchetti, L.; Sousa, N.M.; Malfatti, A.; Brecchia, G.; Canali, C.; Beckers, J.F.; Barile, V.L. Pregnancy-associated glycoproteins (PAGs) concentrations in water buffaloes (*Bubalus bubalis*) during gestation and the postpartum period. *Theriogenology* **2017**, *97*, 73–77. [CrossRef]
50. El-Battawy, K.A.; Sousa, N.M.; Szenci, O.; Beckers, J.F. Pregnancy-associated glycoprotein profile during the first trimester of pregnancy in Egyptian buffalo cows. *Reprod. Domest. Anim. Zuchthyg.* **2009**, *44*, 161–166. [CrossRef] [PubMed]
51. Nguyen, V.H.; Barbato, O.; Bui, X.N.; Beckers, J.-F.; de Sousa, N.M. Assessment of pregnancy-associated glycoprotein (PAG) concentrations in swamp buffalo samples from fetal and maternal origins by using interspecies antisera. *Anim. Sci. J.* **2012**, *83*, 683–689. [CrossRef] [PubMed]
52. Guelfi, G.; Stefanetti, V.; De Luca, S.; Giontella, A.; Barile, V.L.; Barbato, O. Serum microRNAs in buffalo cows: Potential biomarkers of pregnancy. *Res. Vet. Sci.* **2017**, *115*, 294–300. [CrossRef]
53. Barbato, O.; Guelfi, G.; Barile, V.; Menchetti, L.; Tortiello, C.; Canali, C.; Brecchia, G.; Traina, G.; Beckers, J.-F.; de Sousa, N.M. Using real-time PCR to identify pregnancy-associated glycoprotein 2 (PAG-2) in water buffalo (*Bubalus bubalis*) blood in early pregnancy. *Theriogenology* **2017**, *89*, 106–113. [CrossRef]
54. Barbato, O.; Guelfi, G.; Menchetti, L.; Brecchia, G.; Melo de Sousa, N.; Canali, C.; Grandoni, F.; Scatà, M.C.; De Matteis, G.; Casano, A.B.; et al. Investigation of PAG2 mRNA Expression in Water Buffalo Peripheral Blood Mononuclear Cells and Polymorphonuclear Leukocytes from Maternal Blood at the Peri-Implantation Period. *Vet. Sci.* **2019**, *6*, 8. [CrossRef]
55. Boiti, C.; Ceccarelli, P.; Beghelli, V.; Daniotti, P.; Pennisi, F. Messa a punto di un metodo per la determinazione radioimmunologica del progesterone plasmatico (A RIA method for plasma progesterone determination). *Proc. Soc. Ital. Sci. Vet.* **1974**, *26*, 366–371.
56. Todini, L.; Malfatti, A.; Barbato, O.; Costarelli, S.; Debenedetti, A. Progesterone Plus PMSG Priming in Seasonally Anovulatory Lactating Sarda Ewes Exposed to the Ram Effect. *J. Reprod. Dev.* **2007**, *53*, 437–441. [CrossRef]

57. Greenwood, F.C.; Hunter, W.M.; Glover, J.S. The preparation of 131I-labelled human growth hormone of high specific radioactivity. *Biochem. J.* **1963**, *89*, 114–123. [CrossRef]
58. Skelley, D.S.; Brown, L.P.; Besch, P.K. Radioimmunoassay. *Clin. Chem.* **1973**, *19*, 146–186. [CrossRef] [PubMed]
59. Greiner, M.; Pfeiffer, D.; Smith, R.D. Principles and practical application of the receiver-operating characteristic analysis for diagnostic tests. *Prev. Vet. Med.* **2000**, *45*, 23–41. [CrossRef]
60. Pohler, K.G.; Reese, S.T.; Franco, G.A.; Oliveira, R.V.; Paiva, R.; Fernandez, L.; de Melo, G.; Vasconcelos, J.L.M.; Cooke, R.; Poole, R.K. New approaches to diagnose and target reproductive failure in cattle. *Anim. Reprod.* **2020**, *17*. [CrossRef]
61. Pope, W.F. Uterine Asynchrony: A Cause of Embryonic Loss12. *Biol. Reprod.* **1988**, *39*, 999–1003. [CrossRef]
62. Kastelic, J.P.; Curran, S.; Ginther, O.J. Accuracy of ultrasonography for pregnancy diagnosis on days 10 to 22 in heifers. *Theriogenology* **1989**, *31*, 813–820. [CrossRef]
63. Pierson, R.A.; Ginther, O.J. Ultrasonography for detection of pregnancy and study of embryonic development in heifers. *Theriogenology* **1984**, *22*, 225–233. [CrossRef]
64. Curran, S.; Pierson, R.; Ginther, O. Ultrasonographic appearance of the bovine conceptus from days 10 through 20. *J. Am. Vet. Med. Assoc.* **1986**, *189*, 1295–1302.
65. Kastelic, J.; Curran, S.; Pierson, R.; Ginther, O. Ultrasonic evaluation of the bovine conceptus. *Theriogenology* **1988**, *29*, 39–54. [CrossRef]
66. Vecchio, D.; Di Palo, R.; Zicarelli, L.; Grassi, C.; Cammarano, A.; D'Occhio, M.; Campanile, G. Embryonic mortality in buffalo naturally mated. *Ital. J. Anim. Sci.* **2007**, *6*, 677–679. [CrossRef]
67. Naikoo, M.; Patel, D.; Derashri, H. Early pregnancy diagnosis by transrectal ultrasonography in Mehsana buffaloes (*Bubalus bubalis*). *Buffalo Bull.* **2013**, *32*, 120–125.
68. Reese, S.T.; Geary, T.W.; Franco, G.A.; Moraes, J.G.N.; Spencer, T.E.; Pohler, K.G. Pregnancy associated glycoproteins (PAGs) and pregnancy loss in high vs sub fertility heifers. *Theriogenology* **2019**, *135*, 7–12. [CrossRef] [PubMed]
69. Niswender, G.D.; Juengel, J.L.; Silva, P.J.; Rollyson, M.K.; McIntush, E.W. Mechanisms Controlling the Function and Life Span of the Corpus Luteum. *Physiol. Rev.* **2000**, *80*, 1–29. [CrossRef]
70. Inskeep, E.K. Preovulatory, postovulatory, and postmaternal recognition effects of concentrations of progesterone on embryonic survival in the cow1,2. *J. Anim. Sci.* **2004**, *82*, E24–E39. [CrossRef] [PubMed]
71. Barile, V.; Terzano, G.; Allegrini, S.; Maschio, M.; Razzano, M.; Neglia, G.; Pacelli, C. Relationship among preovulatory follicle, corpus luteum and progesterone in oestrus synchronized buffaloes. *Ital. J. Anim. Sci.* **2007**, *6*, 663–666. [CrossRef]
72. Campanile, G.; Neglia, G.; Michael, J. Embryonic and fetal mortality in river buffalo (*Bubalus bubalis*). *Theriogenology* **2016**, *86*, 207–213. [CrossRef] [PubMed]
73. Pieterse, M.C.; Szenci, O.; Willemse, A.H.; Bajcsy, C.S.A.; Dieleman, S.J.; Taverne, M.A.M. Early pregnancy diagnosis in cattle by means of linear-array real-time ultrasound scanning of the uterus and a qualitative and quantitative milk progesterone test. *Theriogenology* **1990**, *33*, 697–707. [CrossRef]
74. Humblot, F.; Camous, S.; Martal, J.; Charlery, J.; Jeanguyot, N.; Thibier, M.; Sasser, R.G. Pregnancy-specific protein B, progesterone concentrations and embryonic mortality during early pregnancy in dairy cows. *J. Reprod. Fertil.* **1988**, *83*, 215–223. [CrossRef]
75. Pohler, K.G.; Geary, T.W.; Johnson, C.L.; Atkins, J.A.; Jinks, E.M.; Busch, D.C.; Green, J.A.; MacNeil, M.D.; Smith, M.F. Circulating bovine pregnancy associated glycoproteins are associated with late embryonic/fetal survival but not ovulatory follicle size in suckled beef cows. *J. Anim. Sci.* **2013**, *91*, 4158–4167. [CrossRef]
76. Pohler, K.G.; Green, J.A.; Geary, T.W.; Peres, R.F.G.; Pereira, M.H.C.; Vasconcelos, J.L.M.; Smith, M.F. Predicting Embryo Presence and Viability. In *Regulation of Implantation and Establishment of Pregnancy in Mammals: Tribute to 45 Year Anniversary of Roger V. Short's "Maternal Recognition of Pregnancy"*; Geisert, R.D., Bazer, F.W., Eds.; Advances in Anatomy, Embryology and Cell Biology; Springer International Publishing: Cham, Switzerland, 2015; pp. 253–270. ISBN 978-3-319-15856-3.
77. Kindahl, H.; Kornmatitsuk, B.; Königsson, K.; Gustafsson, H. Endocrine changes in late bovine pregnancy with special emphasis on fetal well-being. *Domest. Anim. Endocrinol.* **2002**, *23*, 321–328. [CrossRef]
78. Kornmatitsuk, B.; Veronesi, M.C.; Madej, A.; Dahl, E.; Ropstad, E.; Beckers, J.F.; Forsberg, M.; Gustafsson, H.; Kindahl, H. Hormonal measurements in late pregnancy and parturition in dairy cows—Possible tools to monitor foetal well being. *Anim. Reprod. Sci.* **2002**, *72*, 153–164. [CrossRef]
79. Dobson, H.; Rowan, T.G.; Kippax, I.S.; Humblot, P. Assessment of fetal number, and fetal and placental viability throughout pregnancy in cattle. *Theriogenology* **1993**, *40*, 411–425. [CrossRef]
80. Patel, O.V.; Sulon, J.; Beckers, J.F.; Takahashi, T.; Hirako, M.; Sasaki, N.; Domeki, I. Plasma bovine pregnancy-associated glycoprotein concentrations throughout gestation in relationship to fetal number in the cow. *Eur. J. Endocrinol.* **1997**, *137*, 423–428. [CrossRef]
81. Beckers, J.F.; Drion, P.V.; Garbayo, J.M.; Perény, Z.S.; Zarrouk, A.; Sulon, J.; Remy, B.; Szenci, O. Pregnancy Associated Glycoproteins in ruminants: Inctive members of the aspartic proteinase family. *Acta Vet. Hung.* **1999**, *47*, 461–469. [PubMed]
82. Wooding, F. Structure and function of placental binucleate (giant) cells. *Bibl. Anat.* **1982**, *22*, 134–139.
83. Klisch, K.; Jeanrond, E.; Pang, P.-C.; Pich, A.; Schuler, G.; Dantzer, V.; Kowalewski, M.P.; Dell, A. A Tetraantennary Glycan with Bisecting N-Acetylglucosamine and the Sda Antigen is the Predominant N-Glycan on Bovine Pregnancy-Associated Glycoproteins. *Glycobiology* **2007**, *18*, 42–52. [CrossRef] [PubMed]

84. Carvalho, A.F.; Klisch, K.; Miglino, M.A.; Pereira, F.T.V.; Bevilacqua, E. Binucleate trophoblast giant cells in the water buffalo (*Bubalus bubalis*) placenta. *J. Morphol.* **2006**, *267*, 50–56. [CrossRef]
85. Austin, K.J.; King, C.P.; Vierk, J.E.; Sasser, R.G.; Hansen, T.R. Pregnancy-Specific Protein B Induces Release of an Alpha Chemokine in Bovine Endometrium. *Endocrinology* **1999**, *140*, 542–545. [CrossRef] [PubMed]
86. Hoeben, D.; Monfardini, E.; Opsomer, G.; Burvenich, C.; Dosogne, H.; de Kruif, A.; Beckers, J.-F. Chemiluminescence of bovine polymorphonuclear leucocytes during the periparturient period and relation with metabolic markers and bovine pregnancy-associated glycoprotein. *J. Dairy Res.* **2000**, *67*, 249–259. [CrossRef]
87. Silva, E.; Sterry, R.A.; Kolb, D.; Mathialagan, N.; McGrath, M.F.; Ballam, J.M.; Fricke, P.M. Accuracy of a Pregnancy-Associated Glycoprotein ELISA to Determine Pregnancy Status of Lactating Dairy Cows Twenty-Seven Days After Timed Artificial Insemination. *J. Dairy Sci.* **2007**, *90*, 4612–4622. [CrossRef]
88. Romano, J.E.; Larson, J.E. Accuracy of pregnancy specific protein-B test for early pregnancy diagnosis in dairy cattle. *Theriogenology* **2010**, *74*, 932–939. [CrossRef]
89. Oliveira Filho, R.; Franco, G.; Reese, S.; Dantas, F.; Fontes, P.; Cooke, R.; Rhinehart, J.; Thompson, K.; Pohler, K. Using pregnancy associated glycoproteins (PAG) for pregnancy detection at day 24 of gestation in beef cattle. *Theriogenology* **2020**, *141*, 128–133. [CrossRef]
90. Barbato, O.; Menchetti, L.; Sousa, N.M.; Brecchia, G.; Malfatti, A.; Canali, C.; Beckers, J.-F.; Barile, V.L. Correlation of two radioimmunoassay systems for measuring plasma pregnancy-associated glycoproteins concentrations during early pregnancy and postpartum periods in water buffalo. *Reprod. Domest. Anim.* **2018**, *53*, 1483–1490. [CrossRef]
91. Pohler, K.G.; Peres, R.F.G.; Green, J.A.; Graff, H.; Martins, T.; Vasconcelos, J.L.M.; Smith, M.F. Use of bovine pregnancy-associated glycoproteins to predict late embryonic mortality in postpartum Nelore beef cows. *Theriogenology* **2016**, *85*, 1652–1659. [CrossRef]
92. Gatea, A.O.; Smith, M.F.; Pohler, K.G.; Egen, T.; Pereira, M.H.C.; Vasconselos, J.L.M.; Lawrence, J.C.; Green, J.A. The ability to predict pregnancy loss in cattle with ELISAs that detect pregnancy associated glycoproteins is antibody dependent. *Theriogenology* **2018**, *108*, 269–276. [CrossRef] [PubMed]
93. Szenci, O.; Humblot, P.; Beckers, J.F.; Sasser, G.; Sulon, J.; Baltusen, R.; Varga, J.; Bajcsy, C.S.A.; Taverne, M.A.M. Plasma Profiles of Progesterone and Conceptus Proteins in Cows with Spontaneous Embryonic/Fetal Mortality as Diagnosed by Ultrasonography. *Vet. J.* **2000**, *159*, 287–290. [CrossRef]
94. Thompson, I.M.; Cerri, R.L.A.; Kim, I.H.; Green, J.A.; Santos, J.E.P.; Thatcher, W.W. Effects of resynchronization programs on pregnancy per artificial insemination, progesterone, and pregnancy-associated glycoproteins in plasma of lactating dairy cows. *J. Dairy Sci.* **2010**, *93*, 4006–4018. [CrossRef]

animals

MDPI

Article

Distinct Pattern of NPY in Gastro–Entero–Pancreatic System of Goat Kids Fed with a New Standardized Red Orange and Lemon Extract (RLE)

Elena De Felice [1], Daniela Giaquinto [1], Sara Damiano [2], Angela Salzano [2], Simona Fabroni [3], Roberto Ciarcia [2], Paola Scocco [1], Paolo de Girolamo [2,*] and Livia D'Angelo [2]

1 School of Biosciences and Veterinary Medicine, University of Camerino, Via Pontoni 5, 62032 Camerino, Italy; elena.defelice@unicam.it (E.D.F.); daniela.giaquinto@unicam.it (D.G.); paola.scocco@unicam.it (P.S.)
2 Department of Veterinary Medicine and Animal Productions, University of Naples Federico II, 80137 Naples, Italy; sara.damiano@unina.it (S.D.); angela.salzano@unina.it (A.S.); roberto.ciarcia@unina.it (R.C.); livia.dangelo@unina.it (L.D.)
3 Research Centre for Olive, Fruit and Citrus Crops, Council for Agricultural Research and Economics (CREA), 95024 Acireale, Italy; simona.fabroni@crea.gov.it
* Correspondence: degirola@unina.it

Simple Summary: In the last decades the European ban towards antibiotics resulted in an increase of the number of studies on the effects of natural feed additives, that can enhance the health of farm animals intended for human consumption. Polyphenols such as flavanones and anthocyanins (responsible of the red, purple or blue colors) are bioactive compounds found in fruits and vegetables. Polyphenols possess multiple pharmacological characteristics, like antioxidant, anti-inflammatory and immunostimulant properties. Although many of the biological effects of polyphenols are known, only a limited number of studies has been focused on the effects of their supplementation in ruminant diet. Therefore, we evaluated the effect of a diet supplemented with a standardized powder extract, red (blood) orange and lemon extract (RLE), rich in flavanones, anthocyanins and other polyphenols on the neuropeptide Y (NPY) distribution in the gastro–entero–pancreatic system of goat kids. In mammals, NPY occurs in both the central and peripheral nervous systems and it is involved in the control of different physiological processes, including food intake regulation. For the first time, we document that NPY is widely distributed in the abomasum, duodenum and pancreas of goat kids and that significantly increases in the abomasum and pancreas of RLE supplemented feed animals.

Abstract: The use of natural compounds as feed additive is also increasing in farm animals, thanks to the beneficial effect on both animals and consumers health. Here, we questioned whether natural extracts, such as red orange and lemon extract (RLE) rich in flavanones, anthocyanins, and other polyphenols, used as feed additives could display an effect on the neuropeptide Y (NPY) in the gastro–entero–pancreatic tract of goat kids. NPY is one of the most abundant neuropeptides in mammals, known for its orexigenic role although it is involved in many central and peripheral functions. We carried out immunohistochemical analyses on samples of abomasum, duodenum and pancreas collected from two experimental groups: one fed with standard diet and one with standard diet + RLE. For the first time we document NPY distribution in the abomasum, duodenum and pancreas of goats and observe the highest number of NPY positive cells in neuroendocrine cells of duodenum. Remarkably, upon RLE feed supplementation, NPY immunoreactive cells increased significantly in abomasal epithelium and pancreatic islets but not in duodenum, likely due to pH variation of abomasum and duodenum. Our observations represent a baseline for future studies on the interaction between neuropeptides and polyphenols, used as feed additive.

Keywords: gastro-intestinal apparatus; feed additive; small ruminants; goats; neuropeptides; natural compound; polyphenols

Citation: De Felice, E.; Giaquinto, D.; Damiano, S.; Salzano, A.; Fabroni, S.; Ciarcia, R.; Scocco, P.; de Girolamo, P.; D'Angelo, L. Distinct Pattern of NPY in Gastro–Entero–Pancreatic System of Goat Kids Fed with a New Standardized Red Orange and Lemon Extract (RLE). *Animals* **2021**, *11*, 449. https://doi.org/10.3390/ani11020449

Academic Editor: María Teresa Paramio

Received: 11 January 2021
Accepted: 5 February 2021
Published: 9 February 2021

1. Introduction

The use of natural compounds as feed supplementation has been facing a new era, especially since the European ban on ionophore antibiotics (European Commission, Directive 1831/2003/CEE, 2003). For many years antibiotics' employ has been a usual praxis in livestock farming, to augment growth, enhance feed conversion efficiency and prevent and treat diseases [1]. However, the utilization of antibiotics has amplified the extent of antimicrobial resistance, posing severe warnings to both animal and human health and food security [2].

Recently, among natural compounds there has been a growing interest in the consumption of anthocyanin-rich food. Anthocyanins (ANTs), which belong to the flavonoid family, are water soluble polyphenolic pigments widespread in the plant kingdom, responsible of the blue, purple, red and orange pigmentation of various vegetables and fruits [3]. ANTs have been demonstrated to have numerous benefits, such as anti-inflammatory, antioxidant, anti-obesity, anti-angiogenesis, anti-cancer, anti-diabetes, anti-microbial, anti-proliferative, neuroprotection and immunomodulation properties [4]. The anti-inflammatory and antioxidant properties have been demonstrated by inducing the downregulation of cyclooxygenases-2 and the inhibition of prostaglandin E2 production [5] and by decreasing the activation of NF-kB transcription factor [6]. ANTs have been shown to contrast oxidative stress directly, donating or transferring electrons from hydrogen atoms [7] and indirectly through the activation of specific detoxification enzymes [8]. Recent studies documented a beneficial effect of ANTs also on gut microbiota populations, acting as prebiotics. In particular the consumption of food rich in ANTs is associated with the increase in the large intestine of *Bifidobacterium* spp. and *Lactobacillus* spp., two Gram-positive species beneficial in the treatment of diarrhoea and other intestinal diseases [4,9,10]. Very recently, the value of ANTs and proanthocyanidin as feed supplementation has been investigated and proposed as ruminant feedstuffs because of the positive effects on ruminal degradation, as well as antioxidant contents and activities [11,12]. Anthocyanin-rich purple corn (*Zea mays* L.) showed a lowering effect on aminotransferase activity with concomitant enhancement of superoxide dismutase (SOD) activity in the plasma of lactating dairy cows [11], while in dairy goats it improved SOD activity in plasma and up-regulated nuclear factor (erythroid-derive 2)-like 2 gene expression in mammary gland [13]. However, ANTs were found to have poor palatability owing to their bitter taste [14], with consequences on food consumption and food intake behaviour.

Despite the great number of papers devoted to characterize the disparate biological effects of ANTs and flavanones (FLAVs), only a limited number of studies have so far focused on the potential effect of natural compound rich in ANTs and FLAVs for the diet supplementation in farm animals. FLAVs and hydroxycinnamic acids are currently used for the treatment of capillary fragility and for other diseases whose aetiology is linked to the disruptive action of free radicals [15–19]. Recently, it has been demonstrated that a lemon fruit extract rich in eriocitrin and other FLAVs can positively act in the regulation of adipocyte differentiation and lipid accumulation [20].

Neuropeptide Y (NPY) is a 36-amino acid peptide widely distributed in both the central and peripheral nervous systems and it has been functionally related to regulation of feeding behaviour and gastrointestinal tract motility [21]. It belongs to the neuroendocrine peptide NPY family, which also includes peptide YY (PYY) and pancreatic polypeptide, mainly occurring in the gastro-enteric tract of all mammals. In goats, NPY is localized at 75, 95 and 113 days of gestation in the lamina propria-submucosa, tunica muscularis and myenteric of the three prestomachs [22]. In the abomasum, NPY was detected at 75 days of gestation [23]. Remarkably, compared to sheep and cow, NPY is detected earlier in goats [22,23]. However, there are no more data on the occurrence and regulation of NPY in the postnatal glandular stomach of goats neither in the duodenum nor in the pancreas.

Recently, it has been demonstrated that ANTs have the capability to reduce body weight and food intake through their modulatory effect on NPY and $GABAB_1R$ in rat hypothalamus [24]. In the rat hypothalamus, the consumption of fruits containing ANTs induced an increase of NPY mRNA expression in non-obese animals [25]. Knowledge on the peripheral regulation of NPY upon ANTs and FLAVs supplemented feeding is still lacking, although it has well established that ANTs are bioavailable and metabolized at gastro-intestinal level, as demonstrated in weanling pigs [26]. Therefore, we propose in this study to investigate whether RLE may regulate the occurrence of NPY protein in the abomasum, duodenum and pancreas of goat kids.

We therefore undertook this study to investigate the effect of a red orange and lemon extract (RLE) rich in FLAVs, ANTs and other polyphenols, on the gastro-enteric regulation of NPY in goat kids from the birth until weaning.

2. Materials and Methods

2.1. Experimental Design and Animals

The research was approved with Protocol PG/2019/0028161 of 19 March 2019 by the Animal Welfare Body of the University of Naples Federico II. The experimental procedures were carried out in the experimental farm of the Council for Agricultural Research and Economics, Research Center for Animal Production and Aquaculture (CREA-ZA, Potenza, Italy).

The standardized powder phytoextract rich in flavanones, anthocyanins, and other polyphenols was obtained by a patented extraction process (Italian Patent No. 102017000057761) from blood orange and lemon processing wastes (red orange and lemon extract, here named RLE). The standardized extract was realized in the laboratories of CREA-Research Centre for Olive, Fruit and Citrus Crops (CREA-OFA, Acireale, Italy) for research purposes only. Information concerning RLE chemical composition are shown in Table 1. Identification and relative concentrations of individual flavanones and anthocyanins are reported in previous works [19–27].

Table 1. Chemical composition of red orange and lemon extract (RLE) used as food supplement for the study.

Class of Compounds	Relative Composition (%)
Total flavanones	15.91 ± 0.01
Total Anthocyanins (as cyanidin 3-glucoside equivalents)	2.66 ± 0.01
Total Hydroxycinnamic acids	1.77 ± 0.02
Ascorbic acid	2.40 ± 0.01

Sixty kids of Saanen bred, males and females, after colostrum administration, were randomly divided into two homogenous (homogeneous for body weight, sex, age and physiological condition) groups (named Control and Treated respectively) of 30 kids each and housed in single boxes. The numerousness of animal for each group was calculated and considered optimal for a significance level of 0.05, a test power of 0.9 and an effect size of 0.85.

The two experimental groups were fed for the whole experimental period (40 days) with: (1) standard diet made of hay (100 g) and kids starter (150 g) (CTRL group); (2) standard diet supplemented with RLE (90 mg/kg) (TRT group). The dose of RLE extract (90 mg/kg of live weight) was defined according to a previous work [27]. Kids were weighed daily in order to record the average weight gain and RLE extract was mixed with water to obtain a cream [28], which was then administered through a syringe directly in the mouth.

All kids underwent natural suckling throughout the experimental period. Water was provided *ad libitum* and the diets were wet (water to feed ratio 3:1). Data about daily average weight gain were collected. At the end of the experimental period animals were slaughtered at a public abattoir in accordance with the Art. 29 of the Council Regulation (EC) No. 1099/2009 on the protection of animals at the time of killing.

2.2. Tissue Sampling

Six samples of abomasum, duodenum (removed 4 cm from the pyloric sphincter) and pancreas were collected from subjects of each group. The samples were promptly immersed in 10% neutral-buffered formalin solution for 24 h and so carefully fixed [29]. All specimens were then dehydrated with a graded ethanol series, cleared in xylene, embedded in paraffin wax and cut into 7-µm-thick serial sections. The sections were stained with hematoxylin and eosin for the morphological analysis.

2.3. Single Immunostaining

After deparaffinization, the sections were incubated with 3% hydrogen peroxide for 30 min at room temperature (RT) to block endogenous peroxidase activity, and then were rinsed in 0.1 M phosphate-buffered saline (PBS), pH 7.4 for 15 min, subsequently pre-incubated for 1 h at RT with the blocking solution (cat# n191356, MP biomedical LLC, Solon, OH, USA) (1:5 in 0.01 M PBS). The sections were then incubated with polyclonal antibody raised in rabbit against NPY (1:500, cat# ab30914, Abcam, Cambridge, UK) and polyclonal antibody raised in rabbit against 5-HT (1:5000, cat# 20080, Immunostar, Hudson, Wisconsin, WI, USA), overnight at 4 °C in humid chamber. The day after, the sections were rinsed in PBS for 15 min and incubated for 30 min at RT with ultrapolymer cocktail (cat# UNIHRP-015, ImmunoReagents, Inc., Raleigh, NC, USA). Immunoreactive sites were visualized using a fresh solution of 10 mg of 3–3′ diaminobenzidine tetrahydrocloride (DAB, cat# D5905, Sigma-Aldrich, Darmstadt, Germany) in 15 mL of 0.5 M Tris buffer, pH 7.6, containing 0.03% hydrogen peroxide.

Double Immunostaining

Double immunostaining of NPY and 5-HT was performed according to Wessel & McClay [30]. After dewaxing, the sections were rinsed in 0.1 M PBS for 10 min and pre-incubated for 1 h at RT with the blocking solution, and then incubated with the first primary antiserum 5-HT (1:500) for 36 h at 4 °C in humid chamber. Then the sections were washed in PBS and incubated with goat anti-rabbit Fab fragment conjugated to tetramethylrhodamine-5-(and 6) isothiocyanate fluorochrome (1:30, cat# 111-297-003, Jackson ImmunoResearch, Cambridge, UK) for 2 h at 37 °C. Thereafter, the sections were rinsed in PBS and incubated with NPY (1:50) over night at 4 °C in humid chamber. After rinsing in PBS, the sections were treated with affinity-pure goat anti-rabbit IgG conjugated to fluoroscein isothiocyanate fluorochrome (1:50, cat# 111-095-006, Jackson ImmunoResearch, Cambridge, UK) for 2 h at 37 °C. Finally, the sections were washed with PBS and mounted.

2.4. Controls of Specificity

Positive controls were made by sections of mouse duodenum (Supplementary Figure S1). Internal reaction controls were carried out by substituting primary antisera or secondary antisera with PBS or normal serum in the specific step.

2.5. Image Acquisition

Fluorescent and light images were observed and analyzed by Nikon Eclipse 90i. The digital raw images were optimized for image resolution, contrast, evenness of illumination, and background by using Adobe Photoshop CS5 (Adobe Systems, San Jose, CA, USA). For cell counting, micrographs were saved in TIFF format and adjusted for light and contrast.

2.6. Cell Counting

In order to evaluate the number of NPY containing cells, six random sections of abomasum, intestine and pancreas for each group, composed of an equal number of males and females, were selected. Positive cells were counted in ten randomly selected observation fields per section (observation field was equal to 26.67 × 20.00 inches; 1920 × 1440 pixels; RGB 11 MB; microscopy magnification 20×. Therefore, 60 observation fields were evaluated for each sample from each animal. The obtained data were pooled and analysed as comparison of mean value [31]. Statistical analysis was performed using Student's t-test (GraphPad Prism v. 3.0, GraphPad Software Inc., San Diego, CA, USA). The differences were considered statistically significant at $p < 0.05$.

3. Results

Any difference was recorded in terms of daily average weight gain that was 128 ± 0.01 vs. 113 ± 0.01 g, respectively for group TRT and CTRL.

3.1. Distribution of NPY and 5-HT in the Abomasum, Duodenum and Pancreas of Control and Treated Animals

Immunoreactivity to NPY was observed in all analyzed tracts of the digestive apparatus of kids, in both experimental groups. In the abomasum, NPY was seen in scattered cells of the epithelium (Figure 1a), which appeared more abundantly distributed, also at the basis of epithelium, in treated animals (Figure 1b). Immunoreactivity to NPY was also seen in varicose fibers in the muscular layer of the abomasum in both groups (Figure 1a,b).

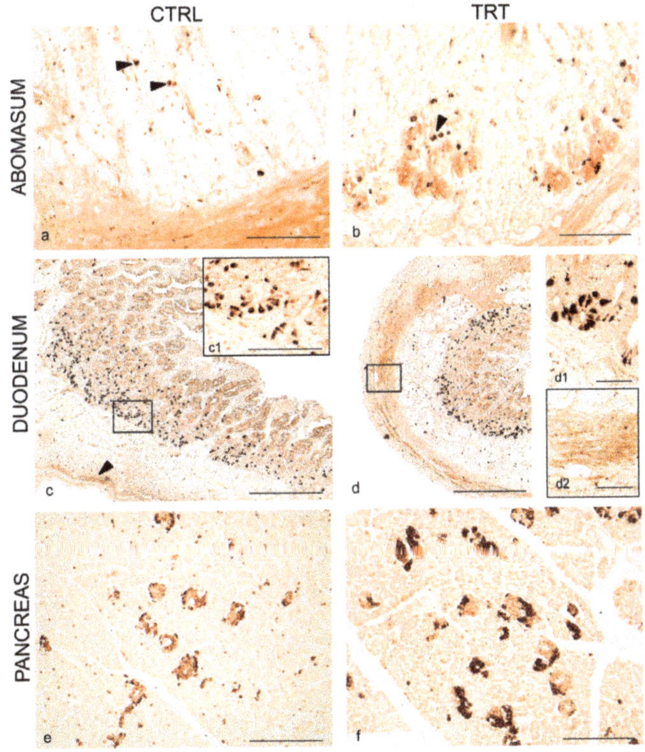

Figure 1. Neuropeptide Y (NPY) immunoreactivity in the abomasum, duodenum and pancreas of control and treated goat kids. Scattered positive cells (arrowheads) in the epithelium of gastric mucosa of control (**a**) and anthocyanins (ANTs) fed animals (**b**). Overview of NPY distribution in neuroendocrine cells (**c**) in the crypt of Lieberkühn and varicose positive fibers (arrowhead) in the muscular layer of duodenum of control (**a**) and ANTs fed animals (**d**). Higher magnification of neuroendocrine cells in the crypt of Lieberkühn (**c1**,**d1**) and fibers in the muscular layer (**d2**). Overview of NPY distribution in the pancreatic islets of control (**e**) and ANTs fed animals (**f**). Scale bar: (a,b,e,f) = 50 µm; (c,d) = 100 µm; (c1,d1,d2) = 25 µm.

In the duodenum, NPY immunostaining was visualized in numerous cells in the crypt of Lieberkhün in animals of both groups (Figure 1c,c1,d,d1). NPY immunoreactivity was seen in varicose fibers of muscular layer of the duodenum and appeared weaker in control animals (Figure 1c) compared to treated animals (Figure 1d,d2).

In the pancreas, numerous NPY immunoreactive cells were seen in the pancreatic islets of control and treated animals (Figure 1e,f), displaced at the margin of the islet. Some scattered positive NPY cells were also seen over the pancreatic parenchyma.

To better characterize NPY immunoreactive cells we conducted immunohistochemical experiments by using 5-HT as classical neuroendocrine marker. In single staining, immunoreactivity to 5-HT was clearly seen in all tracts, although to a less extent in the pancreas. In the abomasum immunoreactivity to 5-HT was seen in cells of the abomasal glands displaced at the basis of the epithelium of control and treated animals (Figure 2a,b).

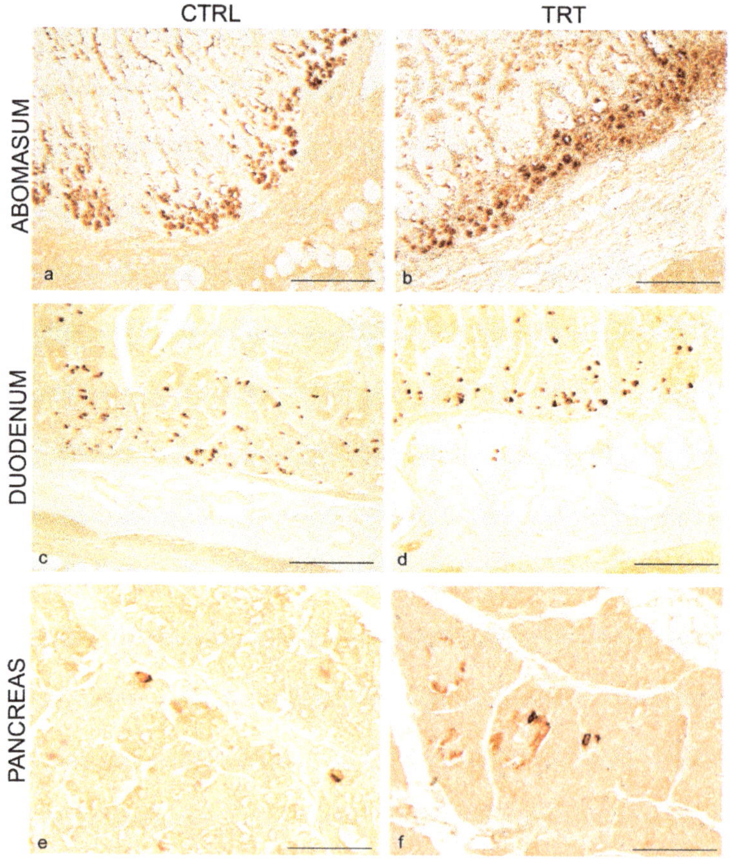

Figure 2. 5-HT immunoreactivity in the abomasum, duodenum and pancreas of control and treated goat kids. Numerous positive cells in the glands at the basis of gastric epithelium of control (**a**) and ANTs fed animals (**b**). Overview of 5-HT distribution in neuroendocrine cells (**c**) in the crypt of Lieberkühn of duodenum of control (**c**) and ANTs fed animals (**d**). Overview of 5-HT distribution in scattered cells in the pancreatic parenchyma of control (**e**) and ANTs fed animals (**f**). Scale bar: (a,b,f) = 50 μm; (c,d) = 100 μm; (e) = 25 μm.

In the duodenum, 5-HT immunostaining was observed in cells in the crypt of Lieberkhün in animals of both groups (Figure 2c,d).

In the pancreas, immunoreactivity to 5-HT was seen in scattered and isolated cells over the pancreatic parenchyma in the two experimental groups (Figure 2e,f).

3.2. Co-Localization of NPY and 5-HT in the Abomasum, Duodenum and Pancreas

We then performed experiment of immunofluorescence on sections of treated animals to evaluate whether NPY was co-localized with 5-HT.

In the abomasum, NPY and 5-HT displayed a clear different distribution, with NPY distributed in scattered cells of the epithelium and at the basis of epithelium while 5-HT abundantly distributed in cells of glands (Figure 3a). In the duodenum all cells immunopositive to NPY were co-localized with 5-HT (Figure 3b). Remarkably, NPY and 5-HT in the same cells displayed a different localization: the former in the cytoplasm and the latter on the cytoplasmic membrane (Figure 3b). In the pancreas, any co-distribution was seen between NPY and 5-HT, because NPY was distributed in the pancreatic islets while 5-HT was scattered along the pancreatic parenchyma (Figure 3c).

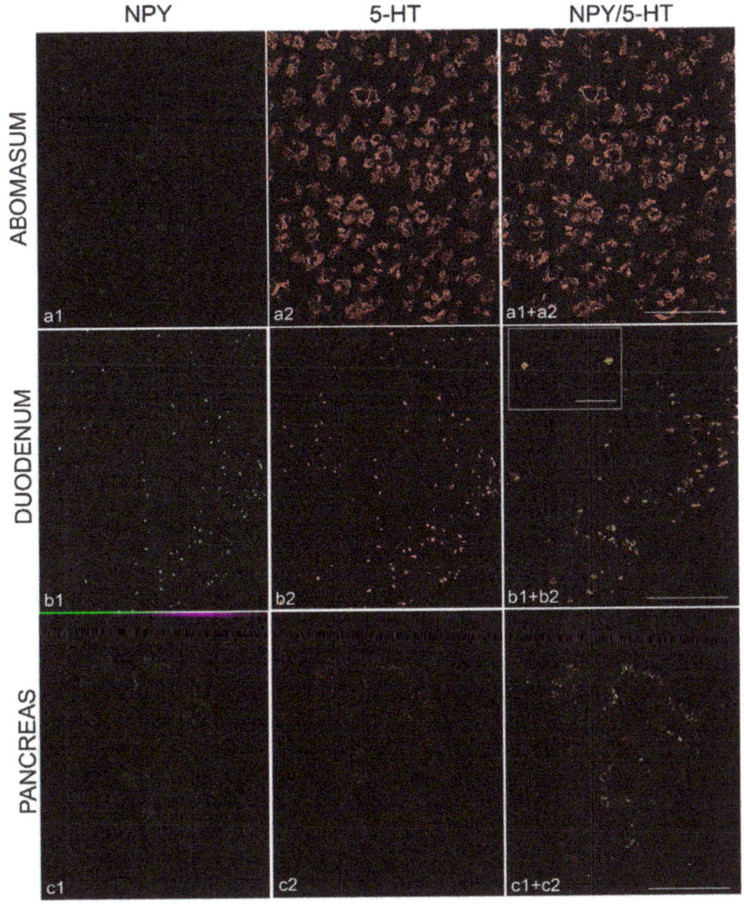

Figure 3. Immunofluorescence of NPY (green) and 5-HT (red) in the abomasum, duodenum and pancreas of goat kids fed with red orange and lemon extract (RLE) supplementation. NPY distributed in sparse cells of abomasal epithelium (**a1**) and 5-HT in glands at the basis of gastric epithelium (**a2**). Absence of NPY/5-HT co-localization (**a1+a2**). NPY (**b1**) and 5-HT (**b2**) co-distributed in the neuroendocrine cells of duodenum, with NPY mainly localized in the cytoplasm and 5-HT on the cellular membrane (**b1+b2**). NPY (**c1**) and 5-HT (**c2**) in the pancreatic islet, localized in different cells (**c1+c2**). Scale bar: (a) = 25 μm; (b,c) = 50 μm.

3.3. NPY Cell Distribution

We counted the number of immunoreactive cells to NPY in abomasum, duodenum and pancreas to quantify the relative distribution in each tract. Statistically significant number of NPY positive cells was counted in the duodenum ($p > 0.0001$) compared to abomasum and pancreas ($p < 0.05$) (Figure 4a).

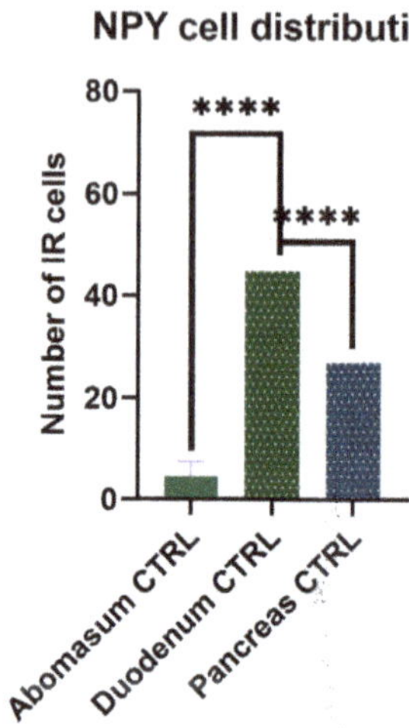

(a)

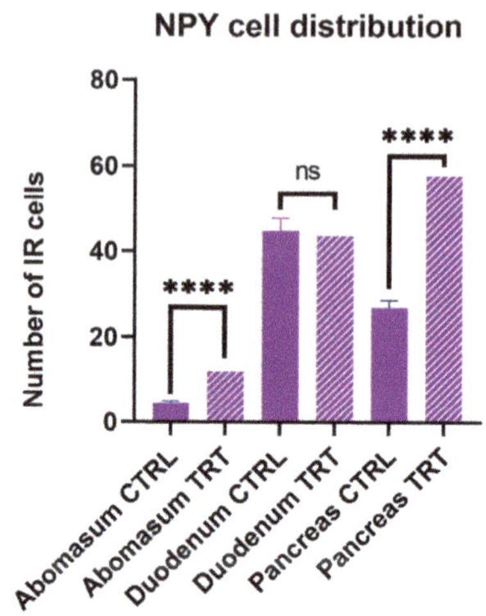

(b)

Figure 4. Number of immunoreactive cells in the gastro–entero–pancreatic system of goat kids. (**a**) Higher distribution in the duodenum and pancreas in control animals. (**b**) Statistically increased number of NPY immunoreactive cells in the abomasum and pancreas but not in the duodenum following RLE feed supplementation. **** $p > 0.0001$; ns non-significant.

To assess if ANTs feed supplementation could have an effect of the regulation of NPY protein, we quantified the number of immunoreactive cells in the three different selected segments. Notably, we observed a strong significantly increase of NPY immunolabeled cells in the abomasum and pancreas ($p > 0.0001$) of ANTs supplemented feed animals while a non-significant increase was observed in the duodenum ($p < 0.05$) of the same animals (Figure 4).

4. Discussion

Polyphenols and ANTs have aroused considerable interest because of their high antioxidant, antimicrobial, anti-inflammatory activities in human and animal health [27,32,33]. RLE extracted from citrus and lemon fruits are a rich source of health-promoting compounds. Indeed, pigmented oranges (*Citrus sinensis* L. Osbeck) are particularly rich in anthocyanins (95% of these being represented by cyanidin-3-glucoside and cyanidin-3-6′′-malonyl-glucoside), flavanones (hesperidin, narirutin and didymin) and hydroxycinnamic acids (caffeic, coumaric, sinapic and ferulic acids). Lemon fruit (*Citrus limon* L. Burm) are also rich in flavanones (eriocitrin, hesperidin and diosmin) and other polyphenols. All these compounds exert their biological activity due to their capacity to scavenge free radicals. Increasing demand of food supplements containing relevant amounts of these compounds have been recently recorded for both human and animal healthy diets [34,35]. Dietary consumption of ANTs has revealed benefits in animal performance [36]. The preventive potential of ANTs depends on their absorption and metabolism. Several studies conducted on animal models indicated that the absorption likely starts in the stomach [37] and in the small intestine, mainly in the duodenum and jejunum [38]. Here we investigated whether feeding supplementation of RLE rich in FLAVs, ANTs and other polyphenols had local effect on gastro–entero–pancreatic cells containing NPY.

For the first time, we document that NPY is widely distributed in the gastro–entero–pancreatic tract of goat kids and its occurrence is regulated by RLE supplemented feed, both in male and female subjects. NPY was observed in scattered cells of the epithelium and in typical varicosity in the muscular layer of the abomasum; in neuroendocrine cells and varicose fibers of the muscular layer in the duodenum and in pancreatic islet cells. The wider distribution was seen in the duodenum, and to a less extent in the pancreas and abomasum. NPY in the abomasum appears already during prenatal development [23] and it seems to contribute to the earlier secretory capacity development compared to sheep and cattle [22]. NPY is known to regulate gastric acid secretion [39] and inhibit water and electrolyte secretion, in particular Cl^- secretion, from the small intestine [40]. In addition, NPY serves many other functions in the gastrointestinal system, such as inhibition of motility [41], mediating action between the nervous and immune system [42], regulation of inflammation by recruiting immature dendritic cells and promoting helper T-cell polarization [43]. As expected, we observed NPY in neuroendocrine cells of duodenum epithelium and in the varicose fibers in the muscle layer, accordingly to the vast literature available in mammals [44]. To better characterize NPY containing cells, we co-stained with 5-HT, which is a multifunctional bioamine, derived from the amino acid tryptophan, playing a role in several complex biological functions, such as suppression of appetite; influence of learning, memory and happiness; regulation of sleep and behaviour. Although serotonin has long been recognized for its critical functions in central nervous system development and function, the majority of the body's serotonin, however, is synthesized in specialized enteroendocrine cells within the gastrointestinal mucosa called enterochromaffin cells, where it plays key roles in enteric nervous system development and function [45]. The main function of 5-HT in the gastrointestinal tract concerns motility and secretion, regulating muscular peristaltic activity [46] and being directly implicated in the pathways mediating the bicarbonate secretion in response to luminal acidification [47] and enhancing water secretion [48]. Interestingly, although NPY was co-localized with 5-HT in neuroendocrine cells of duodenum, the cellular localization appeared different, with NPY mainly localized in the cytoplasm and 5-HT on the cell membrane. This pattern was also confirmed in the positive control. At intestinal levels, NPY acts in autocrine or paracrine manner to regulate all aspects of nutrient homeostasis including satiety, mechanical and chemical digestion, nutrient absorption, storage and utilization [49]. In addition, we observed NPY immunoreactivity in glucagon cells of pancreatic islets, further confirmed by immunofluorescence experiments which excluded co-presence of NPY with 5-HT, known to be marker of insulin cells [50,51]. 5-HT is synthesized by β-cells and promotes their functions [52], regulating insulin secretion as local autocrine/paracrine signal [53] and inhibiting glucan secretion in paracrine manner [50]. The localization of NPY in the pancreas is still matter of debate in mammalian pancreas. In very close species to goat, immunoreactivity to NPY was appreciated in nerve fibers in the endocrine pancreas of calf and cow [54], and in neurons of endocrine pancreas of sheep [55]. Further studies from animal models have concluded that NPY is primarily expressed in β-cells of murine models [56–58] while others have reported that it is found in the glucagon secreting alpha cells [59] or in the somatostatin-containing delta cells [60]. A most recent study suggests that NPY-immunoreactivity reported in alpha and delta cells from other studies was likely due to the presence of NPY-related [61]. The antibody we employed here is raised against NPY, and we thus exclude cross-reaction with other peptides of the NPY family. The direct effect of NPY on the endocrine pancreas is to suppress insulin via autocrine and/or paracrine mechanisms [62] and stimulate glucagon secretion [63]. Very interestingly, the pattern of NPY distribution in all analysed gastro–entero–pancreatic tracts changed drastically in RLE fed goats, differently from data reported in the hypothalamus of rat fed with ANTs, where levels of NPY were decreased with a consequent effect on lipogenesis [24]. These differences may be due to the percentage of ANTs contents in the feed supplementation of the two experimental design. In our study, a statistically significant higher number of NPY immunoreactive cells was observed in the abomasal epithelium and in the pancreatic islets but not in the duodenum. The remarkable difference observed between abomasum and duodenum could be ascribed to the pH variation of the two environments, which affect the chemistry and the availability of compounds rich of FLAVs, ANTs and other polyphenols [64]. ANTs are indeed pH sensitive [38]. ANTs are reported also to influence and increase insulin secretion in pancreatic β-cells, as demonstrated in vitro studies [65]. In goat kids, we observed an increase of NPY pancreatic α-cells and we suspect a different metabolic regulation of ANTs and NPY, likely due to evolutionary adaptation and feeding habits of the species. Future experiments, also in closed related species, may be helpful to clarify this aspect.

5. Conclusions

The effect of feeding supplementation of RLE extract rich in FLAVs, ANTs and other polyphenols on the increase of NPY at gastro-intestinal levels opens new avenues for future studies on the metabolic effect of ANTs and could support the use of natural compounds as alternative feeding resources to ameliorate animal and human health. Furthermore, the peripheral increase of NPY associated to RLE supplementation represents the baseline for developing more accurate experiments to investigate the interactions between NPY and ANTs, in addition to provide a complete characterization of meat quality and animal performance of goat kids fed with RLE used as feed supplementation.

Supplementary Materials: The following are available online at https://www.mdpi.com/2076-2615/11/2/449/s1, Figure S1: NPY and 5-HT immunoreactivity in the duodenum of mouse used as positive control. (a) NPY mainly localized in cells of basal epithelial lamina. (b) 5-HT localized in epithelial cells of intestinal villi. (c) Co-staining of NPY/5-HT in enteroendocrine cells of duodenum. Scale bar: a = 100 μm; b = 50 μm; b_1-c = 25 μm.

Author Contributions: Conceptualization, L.D. and P.d.G.; methodology and validation, S.F., R.C. and S.D.; formal analysis, P.S. and A.S.; investigation, E.D.F. and D.G.; data curation, E.D.F. and D.G.; writing—original draft preparation, E.D.F. and L.D.; writing—review and editing, P.d.G., P.S. and S.D.; supervision, L.D. and P.d.G. All authors have read and agreed to the published version of the manuscript.

Funding: This research received no external funding.

Institutional Review Board Statement: The research was approved with Protocol PG/2019/0028161 of 19 March 2019 by the Animal Welfare Body of the University of Naples Federico II.

Informed Consent Statement: Not applicable.

Data Availability Statement: Datasets used in the analyses are stored at the authors' home institution and will be provided upon request.

Conflicts of Interest: The authors declare no conflict of interest.

References

1. Landers, T.F.; Cohen, B.; Wittum, T.E.; Larson, E.L. A review of antibiotic use in food animals: Perspective, policy, and potential. *Publ. Health Rep.* **2012**, *127*, 4–22. [CrossRef]
2. Ma, Z.; Lee, S.; Jeong, K.C. Mitigating antibiotic resistance at the livestock-environment Interface: A review. *J. Microbiol. Biotechnol.* **2019**, *29*, 1683–1692. [CrossRef]
3. Cisowska, A.; Wojnicz, D.; Hendrich, A.B. Anthocyanins as antimicrobial agents of natural plant origin. *Nat. Prod. Commun.* **2011**, *6*, 149–156. [CrossRef] [PubMed]
4. Li, S.; Wu, B.; Fu, W.; Reddivari, L. The anti-inflammatory effects of dietary anthocyanins against ulcerative colitis. *Int. J. Mol. Sci.* **2019**, *20*, 2588. [CrossRef]
5. Hassimotto, N.M.; Moreira, V.; do Nascimento, N.G.; Souto, P.C.; Teixeira, C.; Lajolo, F.M. Inhibition of carrageenan-induced acute inflammation in mice by oral administration of anthocyanin mixture from wild mulberry and cyanidin-3-glucoside. *Biomed. Res. Int.* **2013**, *2013*, 146716. [CrossRef] [PubMed]
6. Taverniti, V.; Fracassetti, D.; Del Bo', C.; Lanti, C.; Minuzzo, M.; Klimis-Zacas, D.; Riso, P.; Guglielmetti, S. Immunomodulatory effect of a wild blueberry anthocyanin-rich extract in human Caco-2 intestinal cells. *J. Agric. Food Chem.* **2014**, *62*, 8346–8351. [CrossRef] [PubMed]

7. Bicudo, M.O.; Ribani, R.H.; Beta, T. Anthocyanins, phenolic acids and antioxidant properties of Juçara fruits (*Euterpe edulis* M.) along the on-tree ripening process. *Plant Foods Hum. Nutr.* **2014**, *69*, 142–147. [CrossRef] [PubMed]
8. Shih, P.H.; Yeh, C.T.; Yen, G.C. Anthocyanins induce the activation of phase II enzymes through the antioxidant response element pathway against oxidative stress-induced apoptosis. *J. Agric. Food Chem.* **2007**, *55*, 9427–9435. [CrossRef]
9. Morais, C.A.; Vera de Rosso, V.; Estadella, D.; Pellegrini Pisani, L. Anthocyanins as inflammatory modulators and the role of the gut microbiota. *J. Nutr. Biochem.* **2016**, *33*, 1–7. [CrossRef] [PubMed]
10. Igwe, E.O.; Charlton, K.E.; Probst, Y.C.; Kent, K.; Netzel, M.E. A systematic literature review of the effect of anthocyanins on gut microbiota populations. *J. Hum. Nutr. Diet.* **2018**, *32*, 53–62. [CrossRef]
11. Hosoda, K.; Eruden, B.; Matsuyama, H.; Shioya, S. Effect of anthocyanin-rich corn silage on digestibility, milk production and plasma enzyme activities in lactating dairy cows. *Anim. Sci. J.* **2012**, *83*, 453–459. [CrossRef]
12. Hosoda, K.; Sasahara, H.; Matsushita, K.; Tamura, Y.; Miyaji, M.; Matsuyama, H. Anthocyanin and proanthocyanidin contents, antioxidant activity and in situ degradability of black and red rice grains. *Asian Australas J. Anim. Sci.* **2018**, *31*, 1213–1220. [CrossRef]
13. Tian, X.; Xin, H.; Paengkoum, P.; Paengkoum, S.; Ban, C.; Sorasak, T. Effects of anthocyanin-rich purple corn (*Zea mays* L.) stover silage on nutrient utilization, rumen fermentation, plasma antioxidant capacity, and mammary gland gene expression in dairy goats. *J. Anim. Sci.* **2019**, *97*, 1384–1397. [CrossRef]
14. Jobstl, E.; O'Connell, J.; Fairclough, J.P.; Williamson, M.P. Molecular model for astringency produced by polyphenol/protein interactions. *Biomacromolecules* **2004**, *5*, 942–949. [CrossRef]
15. González-Molina, E.; Domínguez-Perles, R.; Moreno, D.A.; García-Viguera, C. Natural bioactive compounds of citrus limon for food and health. *J. Pharm. Biomed. Anal.* **2010**, *51*, 327–345. [CrossRef] [PubMed]
16. Miyake, Y.; Yamamoto, K.; Morimitsu, Y.; Osawa, T. Isolation of C-glucosylflavone from lemon peel and antioxidative activity of flavonoid compounds in lemon fruit. *J. Agr. Food Chem.* **1997**, *45*, 4619–4623. [CrossRef]
17. Del Río, J.A.; Fuster, M.D.; Gómez, P.; Porras, I.; García-Lidón, A.; Ortuño, A. Citrus limon: A source of flavonoids of pharmaceutical interest. *Food Chem.* **2004**, *84*, 457–461. [CrossRef]
18. Amenta, M.; Ballistreri, G.; Fabroni, S.; Romeo, F.V.; Spina, A.; Rapisarda, P. Qualitative and nutraceutical aspects of lemon fruits grown on the mountainsides of the Mount Etna: A first step for a protected designation of origin or protected geographical indication application of the brand name 'Limone dell'Etna'. *Food Res. Int.* **2015**, *74*, 250–259. [CrossRef] [PubMed]
19. Caruso, M.; Fabroni, S.; Emma, R.; Ballistreri, G.; Amenta, M.; Currenti, W.; Rinzivillo, C.; Rapisarda, P. A new standardized phytoextract from red orange and lemon wastes (red orange and lemon extract) reduces basophil degranulation and activation. *Nat. Prod. Res.* **2020**, 1–6. [CrossRef]
20. Carota, G.; Raffaele, M.; Amenta, M.; Ballistreri, G.; Fabroni, S.; Rapisarda, P.; Vanella, L.; Sorrenti, V. In vitro effects of bioflavonoids rich lemon extract on pre-adipocyte differentiation. *Nat. Prod. Res.* **2020**, 1–5. [CrossRef] [PubMed]
21. Kageyama, H.; Takenoya, F.; Hirako, S.; Wada, N.; Kintaka, Y.; Inoue, S.; Ota, E.; Ogawa, T.; Shioda, S. Neuronal circuits involving neuropeptide Y in hypothalamic arcuate nucleus-mediated feeding regulation. *Neuropeptides* **2012**, *46*, 285–289. [CrossRef]
22. Garcia, A.; Masot, J.; Franco, A.; Gazquez, A.; Redondo, E. Immunohistochemical evaluation of the goat forestomach during prenatal development. *J. Vet Sci.* **2014**, *15*, 35–43. [CrossRef]
23. Garcia, A.; Masot, J.; Franco, A.; Gazquez, A.; Redondo, E. Histomorphometric and immunohistochemical study of the goat abomasum during prenatal development. *Histol. Histopathol.* **2013**, *28*, 1639–1649. [PubMed]
24. Badshah, H.; Ullah, I.; Kim, S.E.; Kim, T.; Lee, H.Y.; Kim, M.O. Anthocyanins attenuate body weight gain via modulating neuropeptide Y and GABAB1 receptor in rats hypothalamus. *Neuropeptides* **2013**, *47*, 347–353. [CrossRef] [PubMed]
25. Ibars, M.; Aragonès, G.; Ardid-Ruiz, A.; Gibert-Ramos, A.; Arola-Arnal, A.; Suárez, M.; Bladé, C. Seasonal consumption of polyphenol-rich fruits affects the hypothalamic leptin signaling system in a photoperiod-depend mode. *Sci. Rep.* **2018**, *8*, 3572. [CrossRef]
26. Wu, X.; Pittman 3rd, H.E.; Prior, R.L. Fate of anthocyanins and antioxidant capacity in contents of the gastrointestinal tract of weanling pigs following black raspberry consumption. *J. Agric. Food Chem.* **2006**, *54*, 583–589. [CrossRef]
27. Damiano, S.; Lombari, P.; Salvi, E.; Papale, M.; Giordano, A.; Amenta, M.; Ballistreri, G.; Fabroni, S.; Rapisarda, P.; Capasso, G.; et al. A red orange and lemon by-products extract rich in anthocyanins inhibits the progression of diabetic nephropathy. *J. Cell. Physiol.* **2019**, *234*, 23268–23278. [CrossRef]
28. Maggiolino, A.; Lorenzo, J.M.; Quiñones, J.; Latorre, M.A.; Blando, F.; Centoducati, G.; Dahl, E.; De Palo, P. Effects of dietary supplementation with Pinus taeda hydrolyzed lignin on in vivo performances, in vitro nutrient apparent digestibility, and gas emission in beef steers. *Anim. Feed Sci. Technol.* **2019**, *255*, 114217. [CrossRef]
29. Dall'Aglio, C.; Scocco, P.; Maranesi, M.; Petrucci, L.; Acuti, G.; De Felice, E.; Mercati, F. Immunohistochemical identification of resistin in the uterus of ewes subjected to different diets: Preliminary results. *Eur. J. Histochem.* **2019**, *63*, 3020. [CrossRef]
30. Wessel, G.M.; McClay, D.R. Two embryonic, tissue-specific molecules identified by a double-label immunofluorescence technique for monoclonal antibodies. *J. Histochem. Cytochem.* **1986**, *34*, 703–706. [CrossRef] [PubMed]
31. Gonkowski, S. Bisphenol A (BPA)-induced changes in the number of serotonin-positive cells in the mucosal layer of porcine small intestine-the preliminary studies. *Int. J. Mol. Sci.* **2020**, *21*, 1079. [CrossRef]

32. Piovezana Gomes, J.V.; Buttow Rigolon, T.C.; Sampaio da Silveira Souza, M.; Alvarez-Leite, J.I.; Della Lucia, C.M.; Stampini Duarte Martino, H.; de Oliveira Barbosa Rosa, C. Antiobesity effects of anthocyanins on mitochondrial biogenesis, inflammation, and oxidative stress: A systematic review. *Nutrition* **2019**, *66*, 192–202. [CrossRef]
33. Damiano, S.; Iovane, V.; Squillacioti, C.; Amenta, M.; Florio, S.; Ciarcia, R. Red orange and lemon extract prevents the renal toxicity induced by ochratoxin A in rats. *J. Cell. Physiol.* **2020**, *235*, 5386–5393. [CrossRef]
34. Mora-Sánchez, B.; Fuertes, H.; Balcázar, J.L.; Perez-Sanchez, T. Effect of a multi-citrus extract-based feed additive on the survival of rainbow trout (*Oncorhynchus mykiss*) following challenge with *Lactococcus garvieae*. *Acta Vet. Scand.* **2020**, *62*, 38. [CrossRef]
35. Lauro, M.R.; Crascì, L.; Giannone, V.; Ballistreri, G.; Fabroni, S.; Sansone, F.; Rapisarda, P.; Panico, A.M.; Puglisi, G. An alginate/cyclodextrin spray drying matrix to improve shelf life and antioxidant efficiency of a blood orange by-product extract rich in polyphenols: MMPs inhibition and antiglycation activity in dysmetabolic diseases. *Oxid. Med. Cell. Long.* **2017**, *2017*, 2867630. [CrossRef] [PubMed]
36. Changxing, L.; Chenling, M.; Alagawany, M.; Jianhua, L.; Dongfang, D.; Gaichao, W.; Wenyin, Z.; Syed, S.F.; Arain, M.A.; Saeed, M.; et al. Health benefits and potential applications of anthocyanins in poultry feed industry. *World's Poult. Sci. J.* **2018**, *74*, 251–264. [CrossRef]
37. Passamonti, S.; Vrhovsek, U.; Vanzo, A.; Mattivi, F. The stomach as a site for anthocyanins absorption from food. *FEBS Lett.* **2003**, *544*, 210–213. [CrossRef]
38. Matuschek, M.C.; Hendriks, W.H.; McGhie, T.K.; Reynolds, G.W. The jejunum is the main site of absorption for anthocyanins in mice. *J. Nutr. Biochem.* **2006**, *17*, 31–36. [CrossRef]
39. Vona-Davis, L.C.; McFadden, D.W. NPY family of hormones: Clinical relevance and potential use in gastrointestinal disease. *Curr. Top. Med. Chem.* **2007**, *7*, 1710–1720. [CrossRef]
40. Cox, H.M. Neuropeptide Y receptors; antisecretory control of intestinal epithelial function. *Auton. Neurosci.* **2007**, *133*, 76–85. [CrossRef]
41. Sheikh, S.P. Neuropeptide Y and peptide YY: Major modulators modulators of gastrointestinal blood flow and function. *Am. J. Physiol.* **1991**, *261*, G701–G715. [CrossRef] [PubMed]
42. Sipos, G.; Altdorfer, K.; Pongor, E.; Chen, L.P.; Fehér, E. Neuroimmune link in the mucosa of chronic gastritis with Helicobacter pylori infection. *Dig. Dis. Sci.* **2006**, *51*, 1810–1817. [CrossRef]
43. Palus, K.; Calka, J. Neurochemical plasticity of the coeliac-superior mesenteric ganglion complex neurons projecting to the prepyloric area of the porcine stomach following hyperacidity. *Neural Plast.* **2016**, *2016*, 8596214. [CrossRef] [PubMed]
44. McCauley, H. Enteroendocrine regulation of nutrient absorption. *J. Nutr.* **2020**, *150*, 10–21. [CrossRef]
45. Martin, A.M.; Young, R.L.; Leong, L.; Rogers, G.B.; Spencer, N.J.; Jessup, C.F.; Keating, D.J. The diverse metabolic roles of peripheral serotonin. *Endocrinology* **2017**, *158*, 1049–1063. [CrossRef]
46. Yabut, J.M.; Crane, J.D.; Green, A.E.; Keating, D.J.; Khan, W.I.; Steinberg, G.R. Emerging roles for serotonin in regulating metabolism: New implications for an ancient molecule. *Endocr. Rev.* **2019**, *40*, 1092–1107. [CrossRef] [PubMed]
47. Smith, A.J.; Chappell, A.E.; Buret, A.G.; Barrett, K.E.; Dong, H. 5-Hydroxytryptamine contributes significantly to a reflex pathway by which the duodenal mucosa protects itself from gastric acid injury. *FASEB J.* **2006**, *20*, 2486–2495. [CrossRef] [PubMed]
48. Diwakarla, S.; Fothergill, L.J.; Fakhry, J.; Callaghan, B.; Furness, J.B. Heterogeneity of enterochromaffin cells within the gastrointestinal tract. *Neurogastroenterol. Motil.* **2017**, 29. [CrossRef] [PubMed]
49. Gribble, F.M.; Reimann, F. Enteroendocrine cells: Chemosensors in the intestinal epithelium. *Ann. Rev. Physiol.* **2016**, *78*, 277–299. [CrossRef] [PubMed]
50. Almaça, J.; Molina, J.; Menegaz, D.; Pronin, A.N.; Tamayo, A.; Slepak, V.; Berggren, P.O.; Caicedo, A. Human beta cells produce and release serotonin to inhibit glucagon secretion from alpha cells. *Cell. Rep.* **2016**, *17*, 3281–3291. [CrossRef]
51. Moon, J.H.; Kim, Y.G.; Kim, K.; Osonoi, S.; Wang, S.; Saunders, D.C.; Wang, J.; Yang, K.; Kim, H.; Lee, J.; et al. Serotonin regulates adult β-cell mass by stimulating perinatal β-cell proliferation. *Diabetes* **2020**, *69*, 205–214. [CrossRef]
52. El-Merahbi, R.; Löffler, M.; Mayer, A.; Sumara, G. The roles of peripheral serotonin in metabolic homeostasis. *FEBS Lett.* **2015**, *589*, 1728–1734. [CrossRef]
53. Wyler, S.C.; Lord, C.C.; Lee, S.; Elmquist, J.K.; Liu, C. Serotonergic control of metabolic homeostasis. *Front. Cell. Neurosci.* **2017**, *11*, 277. [CrossRef]
54. Myojin, T.; Kitamura, N.; Hondo, E.; Baltazar, E.T.; Pearson, G.T.; Yamada, J. Immunohistochemical localization of neuropeptides in bovine pancreas. *Anat. Histol. Embryol.* **2000**, *29*, 167–172. [CrossRef] [PubMed]
55. Arciszewski, M.B.; Zacharko-Siembida, A. A co-localization study on the ovine pancreas innervation. *Ann. Anat.* **2007**, *189*, 157–167. [CrossRef] [PubMed]
56. Myrsen-Axcrona, U.; Ekblad, E.; Sundler, F. Developmental expression of NPY, PYY and PP in the rat pancreas and their coexistence with islet hormones. *Regul. Pept.* **1997**, *68*, 165–175. [CrossRef]
57. Jackerott, M.; Oster, A.; Larsson, L. PYY in developing murine islet cells: Comparisons to development of islet hormones, NPY, and BrdU incorporation. *J. Histochem. Cytochem.* **1996**, *44*, 809–817. [CrossRef] [PubMed]
58. Gu, C.; Stein, G.H.; Pan, N.; Goebbels, S.; Hornberg, H.; Nave, K.-A.; Herrera, P.; White, P.; Kaestner, K.H.; Sussel, L.; et al. Pancreatic beta cells require NeuroD to achieve and maintain functional maturity. *Cell. Metab.* **2010**, *11*, 298–310.

59. Teitelman, G.; Alpert, S.; Polak, J.M.; Martinez, A.; Hanahan, D. Precursor cells of mouse endocrine pancreas coexpress insulin, glucagon and the neuronal proteins tyrosine hydroxylase and neuropeptide Y, but not pancreatic polypeptide. *Development* **1993**, *118*, 1031–1039.
60. Myrsen, U.; Sundler, F. Neuropeptide Y is expressed in islet somatostatin cells of the hamster pancreas: A combined immunocytochemical and in situ hybridization study. *Regul. Pept.* **1995**, *57*, 65–76. [CrossRef]
61. Whim, M.D. Pancreatic beta cells synthesize neuropeptide Y and can rapidly release peptide co-transmitters. *PLoS ONE* **2011**, *6*, e19478. [CrossRef] [PubMed]
62. Machida, Y.; Bruinsma, C.; Hallinger, D.R.; Roper, S.M.; Garcia, E.; Trevino, M.B.; Nadler, J.; Ahima, R.; Imai, Y. Pancreatic islet Neuropeptide Y overexpression has minimal effect on islet morphology and β-cell adaptation to high-fat diet. *Endocrinolgy* **2014**, *155*, 4634–4640. [CrossRef] [PubMed]
63. Opara, E.C.; Burch, W.M.; Taylor, I.L.; Akawi, O.E. Pancreatic hormone response to Neuropeptide Y (NPY) perifusion in vitro. *Regul. Peptides* **1991**, *34*, 225–233. [CrossRef]
64. McGhie, T.K.; Ainge, G.D.; Barnett, L.E.; Cooney, J.M.; Jensen, D.J. Anthocyanin glycosides from berry fruit are absorbed and excreted unmetabolized by both humans and rats. *J. Agric. Food Chem.* **2003**, *51*, 4539–4548. [CrossRef] [PubMed]
65. Jayaprakasam, B.; Vareed, S.K.; Olson, L.K.; Nair, M.G. Insulin secretion by bioactive anthocyanins and anthocyanidins present in fruits. *J. Agric. Food Chem.* **2005**, *53*, 28–31. [CrossRef]

animals

MDPI

Article

Effects of Aging on Expression of *Mic60* and *OPA1* and Mitochondrial Morphology in Myocardium of Tibetan Sheep

Guan Wang [1,2], Yuzhu Luo [1,*], Jiang Hu [1], Jiqing Wang [1], Xiu Liu [1] and Shaobin Li [1,*]

1 Gansu Key Laboratory of Herbivorous Animal Biotechnology, Faculty of Animal Science and Technology, Gansu Agricultural University, Lanzhou 730070, China; 13680740a@sina.com (G.W.); huj@gsau.edu.cn (J.H.); wangjq@gsau.edu.cn (J.W.); liuxiu@gsau.edu.cn (X.L.)

2 School of Biotechnology and Food, Shangqiu Normal University, Shangqiu 476000, China

* Correspondence: luoyz@gsau.edu.cn (Y.L.); lisb@gsau.edu.cn (S.L.); Tel./Fax: +86-931-7603991 (Y.L.); +86-931-7631870 (S.L.)

Received: 11 November 2020; Accepted: 18 November 2020; Published: 20 November 2020

Simple Summary: Mitochondria play a crucial role in the adaptation to high altitude hypoxia environment in Tibetan sheep, and the changes of its morphology and structure directly affect its function. *OPA1* and *Mic60* are important mitochondria-shaping proteins that work together to regulate the morphology of mitochondrial inner membrane and cristae. It has been shown that aging affects the expression of *OPA1* and *Mic60* in mice, but it has not been investigated in sheep and hence it is not known whether it might affect the ultrastructure of mitochondria. In this context, reverse transcription-quantitative PCR (RT-qPCR), enzyme-linked immunosorbent assay (ELISA) and immunohistochemistry method were used to measure the expression of *Mic60* and *OPA1* genes and proteins in myocardium of adult and aged Tibetan sheep, and the ultrastructure of mitochondria were compared by transmission electron microscope. The results suggest that aging can reduces the expression of *Mic60* and *OPA1* genes and *OPA1* protein, which can affect the mitochondrial function.

Abstract: In order to investigate the effects of aging on the expression of *Mic60* and *OPA1* and mitochondrial morphology in plateau animals, the expression of *Mic60* and *OPA1* genes and proteins, and the morphology of mitochondria in the myocardium of adult and aged Tibetan sheep were investigated. The expression of *Mic60* and *OPA1* genes and *OPA1* protein were higher ($p < 0.05$) in the myocardium of adult Tibetan sheep than in those of the aged ones. The number of mitochondrial cristae in the myocardium of adult was higher than that in aged ($p < 0.05$). The density of mitochondria in the myocardium of adult was higher than that in aged ($p < 0.01$). Compared with the adult Tibetan sheep, the mitochondrial crista of aged were relatively sparse, the crista membrane was wide, and the mitochondria were not closely linked, showing fragmentation. These results suggest that the myocardial mitochondria of the adult have better energy supply ability, indicating that aging can lead to the weakening of oxygen supply in the myocardial mitochondria of Tibetan sheep.

Keywords: aging; *Mic60*; *OPA1*; myocardial mitochondria; Tibetan sheep

1. Introduction

Tibetan sheep live in the Qinghai-Tibet plateau and its adjacent areas at an altitude of 3000 to 5000 m above sea level. Generation reproduction under the altitude environment makes Tibetan sheep well adapted to the low oxygen concentration. It also makes it a good model experimental animal for cardiovascular research. The heart is one of the most important organs in animals, which is the power center for the supply of oxygen in whole body tissues. Under increased metabolism and high oxygen consumption, the heart can provide pressure to increase the blood flow into all parts of the body. As an aerobic tissue, the energy of myocardial metabolism is almost entirely dependent on the oxidative phosphorylation of mitochondria. Therefore, the quantity, function and structure of mitochondria in cardiomyocytes determine the energy supply of the heart.

Mitochondria are the main sites for aerobic respiration of eukaryotic cells and play an important role in the maintenance of calcium homeostasis and the regulation of apoptosis [1,2]. The morphology of mitochondria is regulated by fusion and fission [3,4]. The complete morphological structure is the basis for the function of mitochondria and the integrity of the crista membrane is particularly important [5]. Mitochondrial crista membrane is the main site of oxidative phosphorylation. More than 90% of mitochondrial oxidative phosphorylation complex III, ATP synthase and many proteins and enzymes associated with the oxidative phosphorylation are located in crista [6–8].

Mitochondria-shaping proteins *Mic60* (Mitofilin) and *OPA1* (Dominant Optic Atrophy), which are located in the mitochondrial membrane space, play an important role in the maintenance of mitochondrial morphology and crista membrane structure [9,10]. The nuclear gene *OPA1* is widely expressed in many organs, but the expression level is different. The main functions of *OPA1* is to promote mitochondrial fusion [11,12]. Inhibition of *OPA1* gene expression or inactivation of *OPA1* protein will lead to the disruption of mitochondrial network fusion [13]. The loss of *OPA1* subtype will cause damage to the crista membrane structure and seriously affects cell proliferation [14,15]. Mic60, as the core component of MINOS (Mitochondrial Contact Site) complex [16], plays an important role in maintaining the morphology and structure of mitochondrial crista [17]. Interfering with the expression of *Mic60* not only leads to mitochondrial malformation, but also inhibits the mitochondrial oxidative phosphorylation activity and the normal function of mitochondria [18]. Recent studies showed that *Mic60* was crucial for the formation of mitochondrial crista junctions [19,20].

The change of heart function is closely related to age [21]. The age-related changes of cardiac myocardium focus on three main aspects: myocardial cells, mitochondria related proteins and enzymes, and collagen fibers. In this study, the effects of aging on the expression of *Mic60* and *OPA1* genes and proteins and the mitochondrial morphology in myocardial of Tibetan sheep were studied in combination with the observation of molecular regulation mechanism, protein expression, and mitochondrial ultrastructure.

2. Materials and Methods

All animal work in this study was conducted according to the guidelines of the care and use of experimental animals established by the Ministry of Science and Technology of the People's Republic of China (Approval number 2006-398), and was approved by the Animal Care Committee of Gansu Agricultural University.

2.1. Materials

Six ewes of Tibetan sheep in Gannan Luqu (the average altitude is 3500 m) were selected randomly, three adult (2 years old) and three aged (6 years old) each. All ewes were slaughtered according to the conventional methods, and the same site of myocardial tissue was immediately collected. Of these, 200 mg was immediately used for the isolation of the mitochondria. The sample for light microscopy was cut into pieces of 1 × 1 × 1 cm, washed with PBS and fixed in a 4% paraformaldehyde fixative solution for later use. Transmission electron microscope samples were cut into strips of 1 × 1 × 2 mm, washed with PBS and fixed in 3% glutaraldehyde phosphate buffer (pH 7.2–7.4, 4 °C). The rest of the samples was rapidly frozen by liquid nitrogen and then stored in the refrigerator at −80 °C for the determination of *Mic60* and *OPA1* genes and proteins expression.

2.2. Isolation of the Myocardial Mitochondria

Mitochondria in myocardial tissue were isolated according to the Biovision mitochondrial isolation kit (Mammalian Mitochondria Isolation Kit for Tissue and Cultured Cells, K288). The whole process of isolation was carried out at low temperature to ensure the complete structure and function of mitochondria. The isolated mitochondria were fixed with 3% glutaraldehyde phosphate buffer (pH 7.2–7.4, 4 °C).

2.3. Extraction of Tissue RNA

The total RNA in myocardium was extracted according to the RNA extraction kit (RNA isolator Total RNA Extraction Reagent, Vazyme, Nanjing, China). The quality of RNA integrity was tested by 1.5% agarose gel electrophoresis. Next, the OD260/OD280 value and its concentration were measured by ultramicro spectrophotometer, and the qualified RNA was preserved in the freezer at −80 °C. Reverse transcription of RNA was completed by using reverse transcription kit (HiScript IIO RT SuperMix for qPCR, +gDNA wiper, Vazyme, Nanjing, China). The extracted cDNA was diluted into 1:9 solution and stored at −20 °C.

2.4. Design and Synthesis of Primers

The complete sequence of sheep *Mic60* and *OPA1* genes published by GenBank were used to design PCR primers by Primer Premier 5.0 software (PREMIER Biosoft, San Francisco, CA, USA). The β-actin gene was used as the reference gene, and the designed primers were tested for its specificity by Blast. Primers were synthesized by BGI Liuhe Company (Beijing, China). The detail information of primers is shown in Table 1.

Table 1. Primer information for Tibetan sheep *MIC60*, *OPA1* and β-actin genes.

Gene	GenBank Accession No.	Primer Sequence (5′–3′)	Primer Length	Product Length
Mic60	XM_012169573	F:TTGAGATGGTCCTTGGTT R:TTGTTTCTGAGGTGGTGAG	18 19	136
OPA1	XM_012140446	F:ATGAAATAGAACTCCGAATG R:GTCAACAAGCACCATCCT	20 18	112
β-actin	NM_001009784	F: AGCCTTCCTTCCTGGGCATGGA R:GGACAGCACCGTGTTGGCGTAGA	22 23	113

2.5. RT-qPCR Reaction

The cDNA working fluid was carried out by SYBR Green I chimeric fluorescence method using ChamQ Universal SYBR qPCR Master Mix (Q711-02/03, Vazyme, Nanjing, China) for RT-qPCR reaction. The 20 μL reaction containing 2 μL of cDNA working fluid, 0.4 μL of upstream and downstream primers (10 μM), respectively, 2× ChamQ Universal SYBR qPCR Master Mix (10 μL) and RNase-free dd H_2O (7.2 μL). The optimal conditions of qPCR amplification procedure were as follows: Pre-denaturation at 94 °C for 4 min; Cyclic reaction, 95 °C 10 s, 60 °C 30 s, 40 cycles; Dissolving curve, 95 °C 15 s, 60 °C 60 s, 95 °C 15 s. The specificity of the PCR product and the presence of primer dimers were determined according to the resulting dissolution curve. Data were collected and gene expression was calculated [22] and analyzed for variance using SPSS software (SPSS, Chicago, IL, USA).

2.6. Immunohistochemistry and ELISA

The samples of cut sections were fixed with 4% paraformaldehyde fixative (4 °C, 24 h) and then cut into 5 μm thick sections by routine paraffin embedding. Cut sections were deparaffinized to water for antigen retrieval (Citrate Buffer-Microwave Heat Recovery Method). After rinsed in PBS, the cut sections were blocked (37 °C, 10 min) by adding 3% H_2O_2 (SP KIT), rinsed in PBS, placed in wet box and added blocking solution (SP KIT A) on it (at room temperature, 10 min), and then incubated overnight at room temperature with primary antibody (Rabbit Anti-Mitofilin antibody and Rabbit Anti-*OPA1* antibody, 1:200, Bioss, Beijing, bs-11764R and bs-1824R). The sections were rinsed by PBS, and then added the secondary antibody (SP KIT B) on it (37 °C, 15 min). The sections were rinsed by PBS and the third antibody (SP KIT C) was added (37 °C, 15 min). At last, the sections were stained by DAB, slightly stained by hematoxylin and fixed by neutral balata. An Olympus BX61 positive universal microscope (Olympus, Tokyo, Japan) was used to observe the sections. The *Mic60* and *OPA1* proteins in the myocardium were detected by ELISA. To measure this, 100 mg of fresh tissue was triturated using a homogenizer (Biospec, Bartlesville, OK, USA) and mixed with 500 μL PBS. The solution was then centrifuged for 10 min at 3000× *g* at 4 °C and the supernatant collected. The protein was measured and a solution containing 100 μg proteins was analyzed according to the manufacturer's instructions (Mic60 and *OPA1* ELISA Kit, Bio-Swamp, Wuhan Beinglay Biotech Co., LTD, Wuhan, China). The PBS solution used for grinding tissue was analyzed as blank control and the absorbance was measured at 450 nm. The concentrations of *Mic60* and *OPA1* were calculated from corresponding with the *Mic60* and *OPA1* standard curves.

2.7. Transmission Electron Microscopy Preparation

The tissue and isolated mitochondria for cut sections were fixed with 3% glutaraldehyde, rinsed twice with 0.2 mmol/L PB (10 min/time), fixed with 1% osmium tetroxide (4 °C, 2 h), and rinsed three times with PB (10 min/time). Then, the samples were dehydrated by gradient alcohol (50, 60, 70, 80, and 90% ethanol for 10 min and 100% acetone dehydration twice for 10 min each). The samples were soaked in the mixture of epoxy resin (SPION-PON812) with acetone (1:1) for 5 h (room temperature), embedded with epoxy resin and dried (45 °C, 12 h; 65 °C, 48 h). Semi-thin sections were cut before making ultrathin sections (70 nm). The ultrathin sections were stained by uranium and lead (Uranium acetate stained for 30 min and lead citrate stained for 10 min). The ultrathin sections were observed using a JEM-1230 transmission electron microscope (Jeol, Nihon Denshi, Japan).

2.8. Statistical Analyses

To measure the number of mitochondrial crista and density of mitochondria, 50 microscopic fields (10 cut sections were made for each animal and 5 microscopic fields were randomly selected from each cut section, 20,000×) were analyzed from each animal included in the study. All data in this study were analyzed by Independent Samples t-test using SPSS 19.0 (SPSS, Chicago, IL, USA) and were expressed as the mean ± SD. All p-values were considered statistically significant when $p < 0.05$.

3. Results

3.1. Analysis of the Result of RT-qPCR

The expression of *Mic60* and *OPA1* genes in myocardium of adult and aged Tibetan sheep was quantitatively analyzed by RT-qPCR. The results showed that both the *Mic60* and *OPA1* genes were expressed in the myocardium of adult and aged Tibetan sheep, but the expression levels were different. The Tm values of the two gene amplification products were uniform, and the single peak melt curve indicated that both primer sets had good specificity in the RT-qPCR reaction, and no non-specific amplification and primer dimers were observed (Figure 1a,b). The relative expression levels of *Mic60* and *OPA1* genes in the myocardium of adult Tibetan sheep were 0.7471 ± 0.3105 and 1.3922 ± 0.43547, respectively, which were significantly higher than those in aged Tibetan sheep (0.2366 ± 0.02534 and 0.307 ± 0.08628, respectively) ($p < 0.01$) (Figure 1c,d).

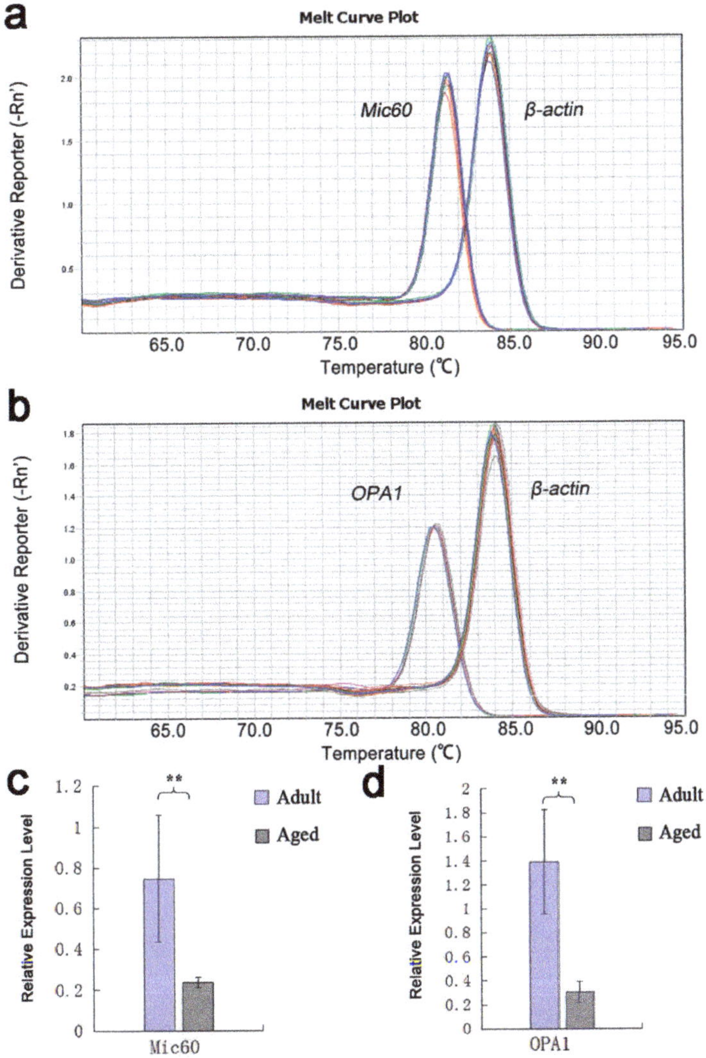

Figure 1. Melting curves of *Mic60*, *OPA1* and β-actin genes and their relative expression in adult and aged Tibetan sheep cardiomyocytes. Note: (**a**,**b**): Melting curve of *Mic60*, *OPA1* and β-actin genes. (**c**,**d**): Relative expression of *Mic60* and *OPA1* genes. All data showed as means ± SD ($n = 3$). ** $p < 0.01$.

3.2. Expression and Analysis of Mic60 and OPA1 Proteins

ELISA method was used to locate and quantify *Mic60* and *OPA1* proteins. The results from the immunohistochemical photographs showed that *Mic60* protein expressed in the myocardium of both adult and aged Tibetan sheep. *OPA1* protein expressed in the myocardium of adult Tibetan sheep, but hardly expressed in the aged sheep (Figure 2a,b,d,e). ELISA results showed that the concentrations of *Mic60* protein in the myocardium had no significant difference between adult (153.27 ± 26.28 ng/mL) and aged (150.24 ± 25.98 ng/mL) Tibetan sheep ($p > 0.05$) (Figure 2g). The concentrations of *OPA1* protein in the myocardium had significant difference between adult (49.77 ± 16.45 ng/mL) and aged (34.06 ± 12.29 ng/mL) Tibetan sheep ($p < 0.05$) (Figure 2g).

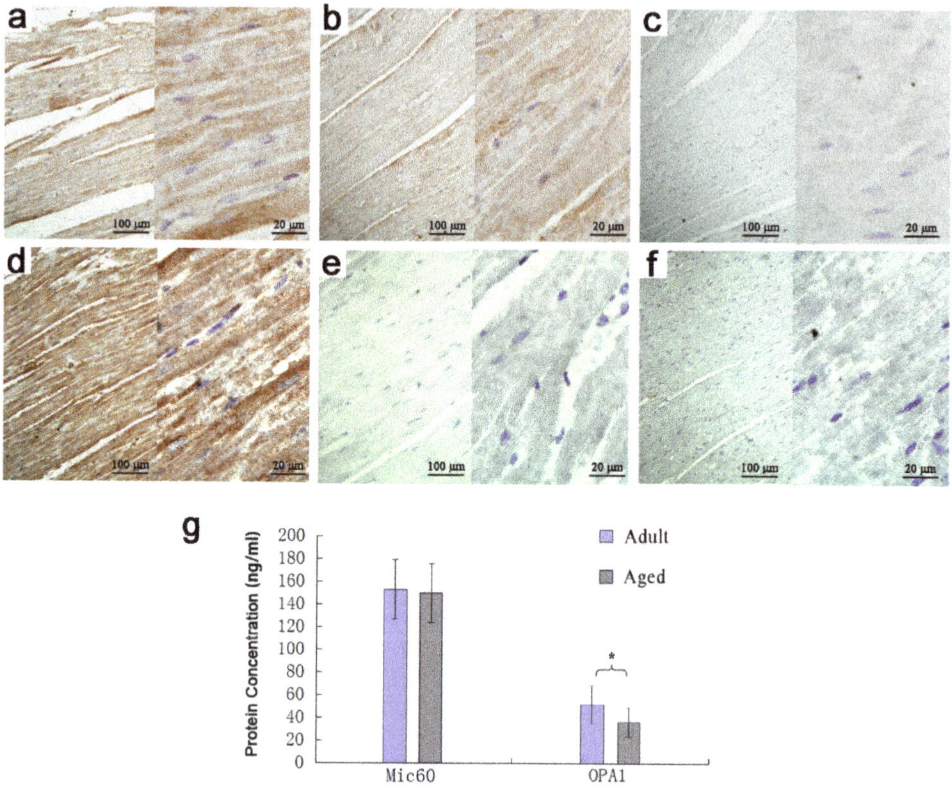

Figure 2. Expression of *Mic60* and *OPA1* proteins in cardiomyocytes. Note: (**a–c**): Expression of *Mic60* protein in adult and aged Tibetan sheep myocardium and blank control. (**d–f**): Expression of *OPA1* protein in adult and aged Tibetan sheep myocardium and blank control. (**g**): Protein concentration of *Mic60* and *OPA1*. Data show means ± SD ($n = 3$). * $p < 0.05$.

3.3. Mitochondrial Morphological Structure Comparison

The mitochondrial ultrastructure of myocardium was observed by transmission electron microscopy, and the morphological and the difference of myocardial mitochondria between adult and aged Tibetan sheep were compared. The average number of mitochondrial crista and the density of mitochondria isolated were calculated. The mitochondria in the myocardium of both adult and aged Tibetan sheep had clear double membranes and tubular cristae, and their structures were complete (Figure 3a,b). The mitochondrial density in the myocardium of adult Tibetan sheep was 57.4 ± 6.07/100 μm^2, which was significantly higher than that of the older age group (38.6 ± 4.51/100 μm^2) ($p < 0.01$) (Figure 4b). The number of mitochondria cristae in the myocardium of adult Tibetan sheep was 21.5 ± 2.65, which was denser than that of 16.5 ± 3.11 in aged Tibetan sheep ($p < 0.05$) (Figure 4a). The mitochondria of adult were closely related but were not closely related in the aged Tibetan sheep, being slightly fragmented, and the cristae membrane were slightly swelled (Figure 3a,b).

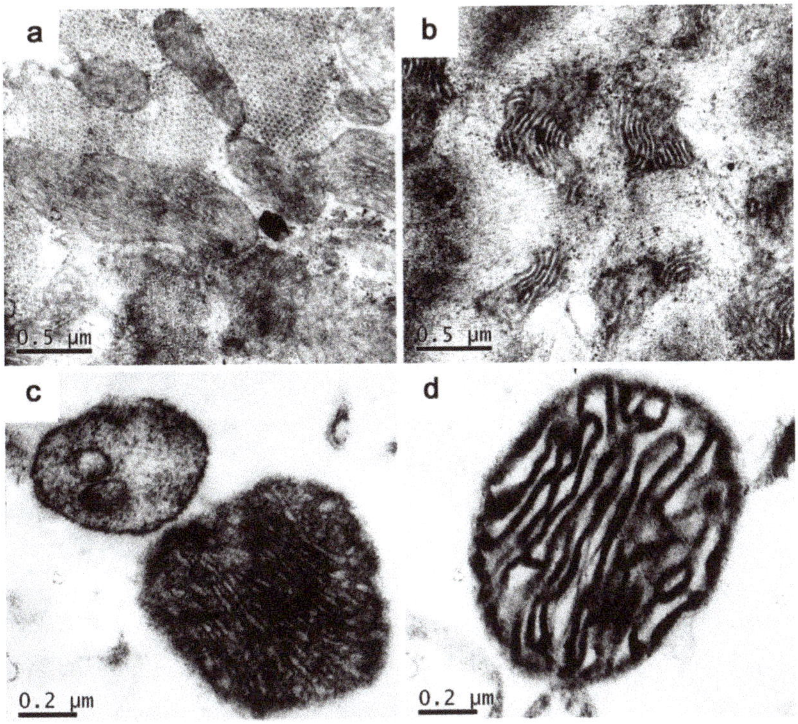

Figure 3. Ultrastructure observation of myocardial mitochondria in adult and aged Tibetan sheep. Note: (**a**,**b**): Adult and aged Tibetan sheep myocardial mitochondria, bar = 0.5 μm. (**c**,**d**): Mitochondria isolated from adult and aged Tibetan Sheep myocardium, bar = 0.2 μm.

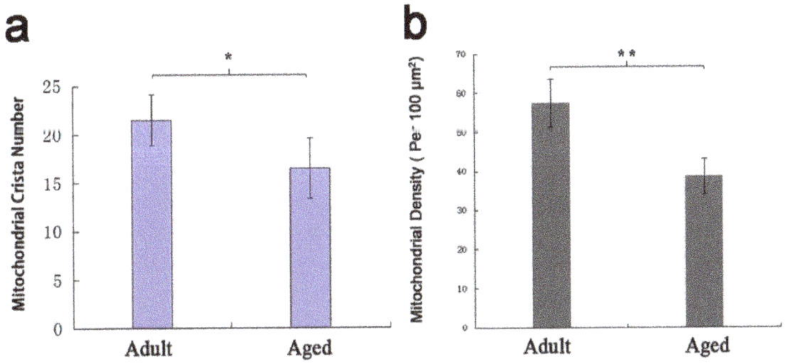

Figure 4. The number of mitochondrial crista (**a**) and density of mitochondria (**b**) in the myocardium of adult and aged Tibetan Sheep. Data show means ± SD. * $p < 0.05$, ** $p < 0.01$.

4. Discussion

The expression of mitochondrial inner membrane proteins *Mic60* and *OPA1* may be related to aging. Studies found that with the *Mic60* heterozygous deletion, the expression of senescence-associated protein p21 in the myocardium of mice was significantly increased, indicated that the mitochondrial inner membrane protein *Mic60* may be related to cardiac senescence in mice [23]. Eckert et al. found that the expression of *OPA1* in mice was decreased with age [24]. A similar result was obtained by

Xu et al., the expression of *OPA1* in Drosophila was significantly decreased with age [25]. In this study, the expression levels of *Mic60* and *OPA1* genes in the myocardium of aged Tibetan sheep were significantly lower than that of adult, which were 0.32 and 0.22 times of that of adult Tibetan sheep, respectively, which indicated that the expression of *Mic60* and *OPA1* genes in the myocardium of Tibetan sheep would reduce with the age. The expression of *OPA1* protein in the aged Tibetan sheep was significantly lower than that of adult. In spite of this, the expression of *Mic60* protein in the myocardium of aged Tibetan sheep was similar to that of adult, indicating that the expression of *Mic60* protein may affect by other factors, which deserved further investigation.

In this study, the density of myocardial mitochondria in aged Tibetan sheep was significantly lower than that of adult, which indicated that myocardial mitochondrial density was age-dependent. Preston et al. reported that the adult rat heart mitochondrial density was higher than that of older rat [26]. The similar phenomenon was found in the studies of yak heart, which showed that mitochondrial density of heart gradually decreased with age, and the concentration of mitochondrial protein gradually increased [27]. Studies on the heart of yak from six months to two years old showed that the density of myocardial mitochondria gradually decreased with age [28], which was consistent with this research.

The expression of *Mic60*, *OPA1* genes and *OPA1* protein were significantly lower in aged Tibetan sheep myocardium than that of adult in this study. Besides, the number of mitochondrial crista was significantly lower, the crista membrane was wide and the mitochondria were not closely linked, showing fragmentation in aged Tibetan sheep. Von et al. reported that absence of fcj1 (Mic60 in mammals) would lead to the mitochondrial crista membrane becoming hypertrophic and stacked, the crista junction being lost and the crista separating from the inner boundary membrane in the yeast cells [16]. Studies found that the oligomerization of *OPA1* in starvation cells was significantly increased, while the crista width of the mitochondria in the starving cells were narrowed obviously and *OPA1* had a regulatory effect on mitochondrial crista independent of its fusion effect [29]. *OPA1* protein was also found to be related to the tubular network structure of mitochondria, and inactivation of *OPA1* would make the network cracked and aggravate mitochondrial fragmentation [30]. Some studies have shown that the expression of *OPA1* was positively correlated with the number of mitochondria crista. Researchers found that a large number of crista structures disappeared in mitochondria when the *OPA1* protein was inhibited in Hela cells [15]. Studies also showed that the interaction between *OPA1* and *Mic60* regulated the morphology of mitochondrial [31]. Therefore, the mitochondrial morphology of aged Tibetan sheep is different from the adult Tibetan ones, which may be influenced by the regulation of *Mic60* and *OPA1* genes and proteins expression, with *OPA1* possibly playing a more important role.

Mitochondria cristae can greatly increase mitochondrial inner membrane surface area and provide more attachment sites for oxidative phosphorylation-related proteins and enzymes. The more mitochondria cristae there are, the stronger oxidative phosphorylation ability they have, and the higher metabolic activity and energy consumption. The mitochondrial of the adult Tibetan sheep has a larger crista membrane area compared with the aged sheep. This structural feature indicates that the mitochondria of adult Tibetan sheep have stronger oxidative phosphorylation ability.

5. Conclusions

In conclusion, our study investigated the expression of *Mic60* and *OPA1* genes and proteins and the morphology of mitochondria in the myocardium of adult and aged Tibetan sheep. Our data showed that the expression of *Mic60*, *OPA1* genes and *OPA1* protein were significantly lower in aged Tibetan sheep myocardium than that of adult in this study. Besides, the number of mitochondrial crista was significantly lower, the crista membrane was wide and the mitochondria were not closely linked, showing fragmentation in aged Tibetan sheep. These results suggest the expression of *Mic60* and *OPA1* genes and *OPA1* protein will reduce, while reducing the capacity of myocardial mitochondria in Tibetan sheep with age.

Author Contributions: G.W. and Y.L. conceived and designed the experiments; G.W., J.H., J.W., X.L. and S.L. performed the experiments; G.W. analyzed the data; G.W. and Y.L. wrote the manuscript. All authors reviewed and commented on the manuscript. All authors have read and agreed to the published version of the manuscript.

Funding: This work was financially supported by Discipline Construction Fund Project of Gansu Agricultural University (GAU-XKJS-2018-029).

Conflicts of Interest: The authors declare no conflict of interest.

References

1. Dimmer, K.S.; Scorrano, L. (De) constructing mitochondria: What for? *Physiology* **2006**, *21*, 233–241. [CrossRef] [PubMed]
2. Frezza, C.; Cipolat, S.; Brito, O.M.D.; Micaroni, M.; Beznoussenko, G.V.; Rudka, R.T.; Bartoli, D.; Polishuck, R.S.; Danial, N.N.; Strooper, B.D.; et al. OPA1 Controls Apoptotic Cristae Remodeling Independently from Mitochondrial Fusion. *Cell* **2006**, *126*, 177–189. [CrossRef] [PubMed]
3. Okamoto, K.; Shaw, J.M. Mitochondrial Morphology and Dynamics in Yeast and Multicellular Eukaryotes. *Annu. Rev. Genet.* **2005**, *39*, 503–536. [CrossRef] [PubMed]
4. Chan, D.C. Mitochondrial Fusion and Fission in Mammals. *Annu. Rev. Cell. Dev. Bi.* **2006**, *22*, 79–99. [CrossRef]
5. Cogliati, S.; Frezza, C.; Soriano, M.E.; Varanita, T.; Quintana-Cabrera, R.; Corrado, M.; Cipolat, S.; Costa, V.; Casarin, A.; Gomes, L.C.; et al. Mitochondrial Cristae Shape Determines Respiratory Chain Supercomplexes Assembly and Respiratory Efficiency. *Cell* **2013**, *155*, 160–171. [CrossRef]
6. Gilkerson, R.W.; Selker, J.M.L.; Capaldi, R.A. The cristal membrane of mitochondria is the principal site of oxidative phosphorylation. *FEBS Lett.* **2003**, *546*, 355–358. [CrossRef]
7. Chaban, Y.; Boekema, E.J.; Dudkina, N.V. Structures of mitochondrial oxidative phosphorylation supercomplexes and mechanisms for their stabilisation. *BBA-Bioenerg.* **2014**, *1837*, 418–426. [CrossRef]
8. Vogel, F.; Bornhovd, C.; Neupert, W.; Reicher, A.S. Dynamic subcompartmentalization of the mitochondrial inner membrane. *J. Cell Biol.* **2006**, *175*, 237–247. [CrossRef]
9. Carelli, V.; Morgia, C.L.; Iommarini, L.; Carroccia, R.; Mattiazzi, M.; Sangiorgi, S.; Farne, S.; Maresca, A.; Foscarini, B.; Lanzi, L.; et al. Mitochondrial Optic Neuropathies: How Two Genomes may Kill the Same Cell Type? *Biosci. Rep.* **2007**, *27*, 173–184. [CrossRef]
10. Gieffers, C.; Korioth, F.; Heimann, P.; Ungermann, C.; Frey, J. Mitofilin is a transmembrane protein of the inner mitochondrial membrane expressed as two isoforms. *Exp. Cell. Res.* **1997**, *232*, 395–399. [CrossRef]
11. Olichon, A.; Elachouri, G.; Baricault, L.; Delettre, C.; Belenguer, P.; Lenaers, G. OPA1 alternate splicing uncouples an evolutionary conserved function in mitochondrial fusion from a vertebrate restricted function in apoptosis. *Cell Death Differ.* **2007**, *14*, 682–692. [CrossRef] [PubMed]
12. Youle, R.J.; van der Bliek, A.M. Mitochondrial Fission, Fusion, and Stress. *Science* **2012**, *337*, 1062–1065. [CrossRef] [PubMed]
13. Spinazzi, M.; Cazzola, S.; Bortolozzi, M.; Baracca, A.; Loro, E.; Casarin, A.; Solaini, G.; Sgarbi, G.; Casalena, G.; Cenacchi, G.; et al. A novel deletion in the gtpase domain of opa1 causes defects in mitochondrial morphology and distribution, but not in function. *Hum. Mol. Genet.* **2008**, *17*, 3291–3302. [CrossRef] [PubMed]
14. Griparic, L.; Kanazawa, T.; Van Der Bliek, A.M. Regulation of the mitochondrial dynamin-like protein Opa1 by proteolytic cleavage. *J. Cell Biol.* **2007**, *178*, 757–764. [CrossRef]
15. Griparic, L.; Van Der Wel, N.N.; Orozco, I.J.; Peters, P.J.; Van Der Bliek, A.M. Loss of the intermembrane space protein Mgm1/OPA1 induces swelling and localized constrictions along the lengths of mitochondria. *J. Biol. Chem.* **2004**, *279*, 18792–18798. [CrossRef]
16. Von der Malsburg, K.; Mueller, J.M.; Bohnert, M.; Oeljeklaus, S.; Kwiatkowska, P.; Becker, T.; Loniewska-Lwowska, A.; Wiese, S.; Rao, S.; Milenkovic, D.; et al. Dual Role of Mitofilin in Mitochondrial Membrane Organization and Protein Biogenesis. *Dev. Cell* **2011**, *21*, 694–707. [CrossRef]
17. Head, B.P.; Zulaika, M.; Ryazantsev, S.; van der Bliek, A.M. A novel mitochondrial outer membrane protein, MOMA-1, that affects cristae morphology in Caenorhabditis elegans. *Mol. Biol. Cell* **2011**, *22*, 831–841. [CrossRef]

18. Yang, R.F.; Sun, L.H.; Zhang, R.; Zhang, Y.; Luo, Y.X.; Zheng, W.; Zhang, Z.Q.; Chen, H.Z.; Liu, D.P. Suppression of Mic60 compromises mitochondrial transcription and oxidative phosphorylation. *Sci. Rep.* **2015**, *5*, 7990. [CrossRef]
19. Koerner, C.; Barrera, M.; Dukanovic, J.; Eydt, K.; Harner, M.; Rabl, R.; Vogel, F.; Rapaport, D.; Neupert, W.; Reichert, A.S. The C-terminal domain of Fcj1 is required for formation of crista junctions and interacts with the TOB/SAM complex in mitochondria. *Mol. Biol. Cell* **2012**, *23*, 2143–2155. [CrossRef]
20. John, G.B.; Shang, Y.; Li, L.; Renken, C.; Mannella, C.A.; Selker, J.M.L.; Rangell, L.; Bennett, M.J.; Zha, J. The mitochondrial inner membrane protein mitofilin controls cristae morphology. *Mol. Biol. Cell* **2005**, *16*, 1543–1554. [CrossRef]
21. Chantler, P.D.; Goldspink, D.F.; Clements, R.E.; Sharp, L.; Schlosshan, D.; Tan, L.B. Congestive heart failure: Extent of cardiac functional changes caused by aging and organ dysfunction. *Heart* **2006**, *92*, 686–688. [CrossRef] [PubMed]
22. Livak, K.; Schmittgen, T. Analysis of Relative Gene Expression Data Using Real-Time Quantitative PCR and the 2-△△Ct Method. *Methods* **2000**, *25*, 402–408. [CrossRef] [PubMed]
23. Wang, C.L.; Sun, L.H.; Yue, Y.S. Correlation between Mic60 haploid insufficiency and cardiac aging in mouse. *Chin. J. Pathol.* **2017**, *46*, 406–410.
24. Eckert, G.P.; Schiborr, C.; Hagl, S.; Abdel-Kader, R.; Muller, W.E.; Rimbach, G.; Frank, J. Curcumin prevents mitochondrial dysfunction in the brain of the senescence-accelerated mouse-prone 8. *Neurochem. Int.* **2013**, *62*, 595–602. [CrossRef] [PubMed]
25. Xu, C.; Wang, H.; Wang, Z.; Wang, Q.; Zhang, S.; Deng, Y.; Fang, Y. In vivo imaging reveals mitophagy independence in the maintenance of axonal mitochondria during normal aging. *Aging Cell* **2017**, *16*, 1180–1190.
26. Preston, C.C.; Oberlin, A.S.; Holmuhamedov, E.L.; Gupta, A.; Sagar, S.; Syed, R.H.K.; Siddiqui, S.A.; Raghavakaimal, S.; Terzic, A.; Jahangir, A. Aging-induced alterations in gene transcripts and functional activity of mitochondrial oxidative phosphorylation complexes in the heart. *Mech. Ageing Dev.* **2008**, *129*, 304. [CrossRef]
27. Simonson, T.S.; Yang, Y.; Huff, C.D.; Yun, H.; Qin, G.; Witherspoon, D.J.; Bai, Z.; Lorenzo, F.R.; Xing, J.; Jorde, L.B.; et al. Genetic evidence for high-altitude adaptation in Tibet. *Science* **2010**, *329*, 72. [CrossRef]
28. He, Y.; Yu, S.; Hu, J.; Cui, Y.; Liu, P. Changes in the Anatomic and Microscopic Structure and the Expression of HIF-1α and VEGF of the Yak Heart with Aging and Hypoxia. *PLoS ONE* **2016**, *11*, e0149947. [CrossRef]
29. Patten, D.A.; Wong, J.; Khacho, M.; Soubannier, V.; Mailloux, R.J.; Pilon-Larose, K.; MacLaurin, J.G.; Park, D.S.; McBride, H.M.; Trinkle-Mulcahy, L.; et al. OPA1-dependent cristae modulation is essential for cellular adaptation to metabolic demand. *Embo J.* **2014**, *33*, 2676–2691. [CrossRef]
30. Meeusen, S.; DeVay, R.; Block, J.; Cassidy-Stone, A.; Wayson, S.; McCaffery, J.M.; Nunnari, J. Mitochondrial inner-membrane fusion and crista maintenance requires the dynamin-related GTPase Mgm1. *Cell* **2006**, *127*, 383–395. [CrossRef]
31. Darshi, M.; Mendiola, V.L.; Mackey, M.R.; Murphy, A.N.; Koller, A.; Perkins, G.A.; Ellisman, M.H.; Taylor, S.S. ChChd3, an Inner Mitochondrial Membrane Protein, Is Essential for Maintaining Crista Integrity and Mitochondrial Function. *J. Biol. Chem.* **2011**, *286*, 2918–2932. [CrossRef] [PubMed]

animals

MDPI

Article

A Prototype Skin Substitute, Made of Recycled Marine Collagen, Improves the Skin Regeneration of Sheep

Luca Melotti [1], Tiziana Martinello [2], Anna Perazzi [3], Ilaria Iacopetti [3,*], Cinzia Ferrario [4,5], Michela Sugni [4,5,*], Roberta Sacchetto [1] and Marco Patruno [1,*]

1 Department of Comparative Biomedicine and Food Science, University of Padova, Viale dell'Università 16, Legnaro, 35020 Padova, Italy; luca.melotti@unipd.it (L.M.); roberta.sacchetto@unipd.it (R.S.)
2 Department of Veterinary Medicine, University of Bari, SP. Casamassima Km.3, Valenzano, 70010 Bari, Italy; tiziana.martinello@uniba.it
3 Department of Animal Medicine, Production and Health, University of Padova, Viale dell'Università 16, Legnaro, 35020 Padova, Italy; anna.perazzi@unipd.it
4 Department of Environmental Science and Policy, University of Milan, Via Celoria, 2, 20133 Milan, Italy; cinzia89.ferrario@gmail.com
5 Center for Complexity and Biosystems, Department of Physics, University of Milan, Via Celoria, 16, 20133 Milan, Italy
* Correspondence: ilaria.iacopetti@unipd.it (I.I.); michela.sugni@unimi.it (M.S.); marco.pat@unipd.it (M.P.)

Simple Summary: Marine ecosystems are a huge source of unexplored "blue" materials for different applications. The edible part of sea urchin is limited, and the vast majority of the product ends up as waste. Our studies intend to fully recycle wastes from the food industry and reconvert them in high added-value products, as innovative biocompatible skin substitutes for tissue regeneration. The aim of the present work is to apply the pioneering skin substitute in in vivo experimental wounds to test its regenerative potential and compare it, in a future study, to the available commercial membranes produced with collagen of bovine, porcine, and equine origin. Results are encouraging since the skin substitute made with marine collagen reduced inflammation, promoted the deposition of granulation tissue, and enhanced a proper re-epithelialization with the adequate development of skin appendages. In summary, our findings might be of great interest for processing industries and biotech companies which transform waste materials in high-valuable and innovative products for Veterinary advanced applications.

Abstract: Skin wound healing is a complex and dynamic process that aims to restore lesioned tissues. Collagen-based skin substitutes are a promising treatment to promote wound healing by mimicking the native skin structure. Recently, collagen from marine organisms has gained interest as a source for producing biomaterials for skin regenerative strategies. This preliminary study aimed to describe the application of a collagen-based skin-like scaffold (CBSS), manufactured with collagen extracted from sea urchin food waste, to treat experimental skin wounds in a large animal. The wound-healing process was assessed over different time points by the means of clinical, histopathological, and molecular analysis. The CBSS treatment improved wound re-epithelialization along with cell proliferation, gene expression of growth factors (VEGF-A), and development of skin adnexa throughout the healing process. Furthermore, it regulated the gene expression of collagen type I and III, thus enhancing the maturation of the granulation tissue into a mature dermis without any signs of scarring as observed in untreated wounds. The observed results (reduced inflammation, better re-epithelialization, proper development of mature dermis and skin adnexa) suggest that sea urchin-derived CBSS is a promising biomaterial for skin wound healing in a "blue biotechnologies" perspective for animals of Veterinary interest.

Keywords: marine collagen; wound healing; regenerative medicine; innovative therapies; skin substitute; sea urchin; biomaterials; tissue engineering; circular economy; 3D scaffolds

Citation: Melotti, L.; Martinello, T.; Perazzi, A.; Iacopetti, I.; Ferrario, C.; Sugni, M.; Sacchetto, R.; Patruno, M. A Prototype Skin Substitute, Made of Recycled Marine Collagen, Improves the Skin Regeneration of Sheep. *Animals* **2021**, *11*, 1219. https://doi.org/10.3390/ani11051219

Academic Editor: Paola Scocco

Received: 26 February 2021
Accepted: 21 April 2021
Published: 23 April 2021

1. Introduction

Skin lesions are a common event in the veterinary practice, regardless of sex or age, and they may be of different etiology such as trauma, burns, fights, predation, etc. In addition, these injuries can often result in loss of skin function due to extensive scarring or inability of the skin to close the wound (e.g., excessive loss of tissue after tumor resection or bite wounds). Indeed, the treatment of these kind of wounds, which might evolve into chronic or non-healing wounds, is a challenging and expensive issue to manage in the veterinary clinical field of domestic animals [1–3]. Wound healing is a complex multi-phase (hemostasis, inflammation, proliferation, and remodeling) process, which is regulated and led by the dynamic interplay of cells, growth factors, and the extracellular matrix (ECM) [4]. Large skin wounds are commonly left open to heal by second intention in veterinary patients like horses or dogs [5,6]. Nonetheless, this common practice has disadvantages such as delayed healing and excessive wound contracture with consequent tissue dysfunction (e.g., a wound on the limb) [7,8]. Nowadays, the application of skin autografts to treat severe non-healing wounds is considered the gold standard [9]. However, this practice has many limitations such as the availability of healthy donor tissue or high pain level for the patients. In addition, graft failure is frequently observed as a consequence of excessive inflammation, infection, or animal locomotion [9,10]. Tissue engineering might offer a solution to this health issue by creating a scaffold that can accelerate wound closure and support the complete regeneration of the skin in terms of function and anatomy, namely skin appendages and anatomical structure.

According to Ferreira and colleagues [11], "skin substitutes are a heterogeneous group of biological and/or synthetic elements that enable the temporary or permanent occlusion of wounds". To date, many tissue-engineered products have been used to treat skin defects. The ideal skin substitute should resemble the skin structure and architecture in order to stimulate and support wound healing. In general, the aim of researchers is to create an easy-to-handle and resistant skin substitute. Different materials of different origin, such as bovine collagen, polyglycolic acid, or acellular cadaver dermis [12–15], have been exploited to obtain skin substitutes, which could also contain lipids, fibrin, glycosaminoglycans, and proteoglycans in addition to collagen as its main substituent. Among them, the use of collagen-based biomaterials, being the main protein of the ECM, has been widely investigated for tissue engineering studies and showed promising results. As a matter of fact, the application of these biomaterials for the treatment of skin wounds has been reinforced for their characteristics such as biocompatibility [16], low immunogenicity [17], and bioactive properties [18].

Since, a one-step skin grafting and a split-thickness skin graft approach to cover the wound at the same time is desirable, numerous bi-layer skin substitutes have been developed for the management of full-thickness skin defects, such as Alloderm® [19,20], Integra® [21], Pelnac® [22], and Matriderm® [23,24]. These commercial matrices are all produced with collagen of bovine, porcine, and equine origin. In the last years, alternative collagen sources have been investigated to overcome ethical and economical problems associated with collagen of the aforementioned origin because of risk of diseases transmission (e.g., BSE, bovine spongiform encephalopathy); among these, marine organisms have been proposed as promising sources of collagen [25–27]. Nevertheless, the majority of marine-derived collagens (sponges, jellyfish, mollusks, and fish), as well as bovine or swine collagens, are used in their hydrolyzed form, since hydrolysis is a step necessary for an efficient extraction. However, this chemical extraction presents two main limitations. First of all, the collagen-associated molecules, i.e., GAGs, are generally lost during the hydrolysis and they need to be artificially added to reproduce the natural characteristics of the ECM, such as hydration [28]. Moreover, when collagen is re-assembled in vitro to form fibrils [29], it often fails to fully reconstitute the original structure and, therefore, its functional efficiency, providing sub-optimal mechanical features [30]. Echinoderms, in particular sea urchins, are one of the most promising source of collagen among marine invertebrates [31,32]. In fact, they offer the possibility to extract collagen fibrils in their native conformation [31,32] with endogen fibril-associated GAGs [31,33], obtaining a biomaterial resembling the in vivo structural microenvironment. In addition, sea urchin collagen has the remarkable advantage to be extracted from wastes originating from the food industry (restaurants and seafood enterprises), thus following circular economy principles and sustainable approaches [31].

In a previous study [27], we described the production and the characterization of sea urchin (*Paracentrosus lividus*)-derived collagen-based skin-like scaffolds (CBSS) composed by an "epidermis-like" layer and a "dermis-like layer," i.e., a thin 2D collagen membrane and a sponge-like 3D collagen scaffold, respectively. In that work, we demonstrated that the sea urchin-derived CBSS is a promising biomaterial for biomedical applications because of its in vitro non-cytotoxicity and barrier function against bacteria, dehydration, and protein loss. Moreover, it possesses the advantage to be biodegradable, thus in an in vivo application no further removal operation of the synthetic upper layer will be necessary as observed for example with Integra®.

The aim of this current work was to apply this innovative skin substitute in in vivo experimental wounds to test its regenerative potential using a sheep surgical wound model. Wounds were treated with the CBSS and clinical, histopathological, and molecular studies were carried out to assess its effects on the wound healing process. The obtained results provide evidences that CBSS could lead to a reduction of inflammation, promote granulation tissue maturation, and enhance re-epithelialization along with the development of skin appendages.

2. Materials and Methods

2.1. Animal Model and Ethical Statement

Three female 2-year-old Bergamasca sheep were included in this experimental study. Sheep were allocated in an experimental barn (MAPS Department, University of Padova, Legnaro, Italy) at least two weeks prior to the start of the experimental study for acclimating themselves. In addition, the animals were subjected to clinical, hematological, and parasitological examinations in order to assess their health status before beginning the study. The animals had ad libitum access to water and feed.

The experiment was approved by the Italian Ministry of Health (n°51/2015-PR), in accordance with the Body for the Protection of Animals (OPBA). The number of animals included in the present study was based on a sample size calculation, taking into account the "3Rs" principle [34]. At the end of the experimentation, sheep were transferred to a teaching farm.

2.2. Sea Urchin Collagen Extraction, Production of 2D Membranes, and 3D Scaffolds

Fibrillar collagen in its native conformation, i.e., with its typical decoration of glycosaminoglycans (GAGs), was extracted from the sea urchin *Paracentrotus lividus*, specifically from the peristomial membranes obtained from restaurant wastes, as previously described [31,32].

The so obtained suspension of fibrillar collagen in autoclaved filtered distilled water (dH_2O) was stored at −80 °C until use. After thawing at room temperature (RT), it was used to produce both 2D thin membranes and 3D sponge-like scaffolds following two different procedures, as described in [27,32]. Briefly, 2D membranes (about 4 cm × 4 cm × 10–15 μm) were produced by adding a proper amount of collagen suspension (2 mg/mL in 0.01% TritonX-100 in autoclaved filtered dH_2O) to rubber silicone molds of the desired size and left to dry at 37 °C overnight. 3D scaffolds (about 4 cm × 4 cm × 1.5 cm) were produced by adding a proper amount of collagen suspension (6 mg/mL) in 6% ethanol in autoclaved filtered dH_2O to rubber silicone molds, frozen overnight at −80 °C, and lyophilized (Edwards Pirani 1001) overnight. Finally, both 2D membranes and 3D scaffolds were cross-linked and sterilized under a 15 W UV lamp at RT overnight (Figure 1). The so obtained sea urchin-derived collagen biomaterials (CBSS) were stored at −20 °C until use. Immediately before the surgical operation, they were sterilized for 1 h under an UV lamp.

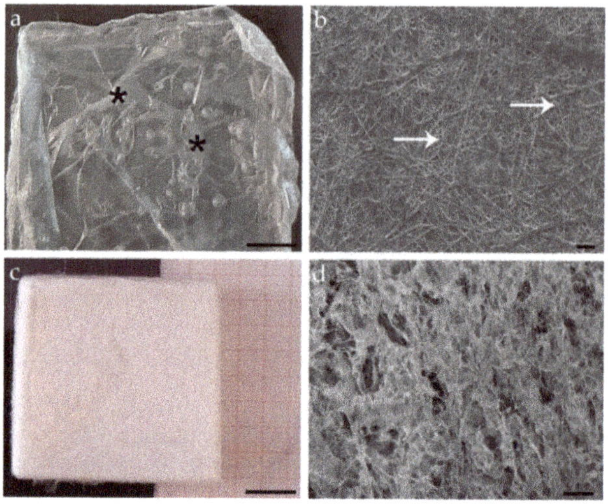

Figure 1. Examples of 2D membranes and 3D scaffolds of sea urchin-derived collagen. (**a**) Top view of a 2D membrane (light microscopy). Asterisks mark macroscopic folds of the thin 2D membrane. (**b**) Micrograph of a 2D membrane where the random distribution of the single collagen fibrils (arrows) in the two-dimensional network is visible (scanning electron microscopy). (**c**) Top view of a 3D scaffold (light microscopy). (**d**) Micrograph of a 3D scaffold where the porous microstructure of the biomaterial is detectable (scanning electron microscopy). Scalebar: (**a**) = 500 μm; (**b**) = 2 μm; (**c**) = 1 cm; (**d**) = 200 μm.

2.3. Surgical Procedure and Animal Model

A total of six full-thickness skin defects (16 cm^2, 4 cm × 4 cm) were created on the back of each sheep under general anesthesia ensuring complete analgesia. The wounds were equidistant (6 cm ipsilteral and 10 cm contralateral) and the distance between them did not affect the healing process or the outcomes of the experiment. The loss of skin tissue was considered significant and the wound edges could not be approximated: wounds will heal by second intention [35]. Five wounds were treated with five different therapies while one was left untreated, and considered as the control wound. The application of the five different therapies and the untreated (control) wound were randomized for each sheep. In the current study, we consider the application of the previously described CBSS comparing it to control wounds. Other treatments, not related to the application of biomaterials, are reported in other original research manuscripts [36].

Before surgery, the animals were treated with an antibiotic (amoxicilline, 15 mg/Kg) and an analgesic therapy (tramadol, 4 mg/Kg) via intramuscular injection. Then, they were premedicated and sedated with medetomidine (0.01 mg/Kg) via intravenous administration. After a proper level of sedation, the animal was placed in sternal decubitus position and the trichotomy of the *thoracolumbar* area was performed for the preparation of the surgical field. Anesthesia was induced by administering propofol (4 mg/Kg). Afterwards to maintain anesthesia, the animal was orotracheally intubated and administered with isoflurane combined with a mixture of oxygen and medical air.

The skin on the back of each sheep was marked using as a guide a sterilized square model (as a template) in order to designate the wound area where to perform the skin surgery. At the same time, this procedure was performed to take into account the skin retraction during wound healing in order to have a representative collection of skin biopsies at different time points. After scrubbing the surgical field with iodine-povidone (10%), six square full-thickness skin wounds (16 cm^2, 4 × 4 cm) were created on the back of each sheep using the previously marked point as a guide. After skin incision to define the wound borders with a sterile surgical blade, the skin was removed using dissection scissors

(Figure 2a,b). Then, one of the wounds was covered with the CBSS: first, the dermal layer (3D porous layer) (Figure 2c) was placed on the wound bed (Figure 2d) followed by the placement of the epidermis-like layer (2D thin membrane) on the top of the dermal one. Afterwards, each wound was medicated and covered with a sterile non-woven gauze. The sterile gauzes were covered with a protective covering bandage and a tubular mesh gauze, which was fixed to the peripheral wool to keep bandages on site (Figure 2e). The covering bandages were changed every three days until day 42. In order to change them, non-woven gauzes were wet with a sterile saline physiological solution to avoid damage to the healing surface.

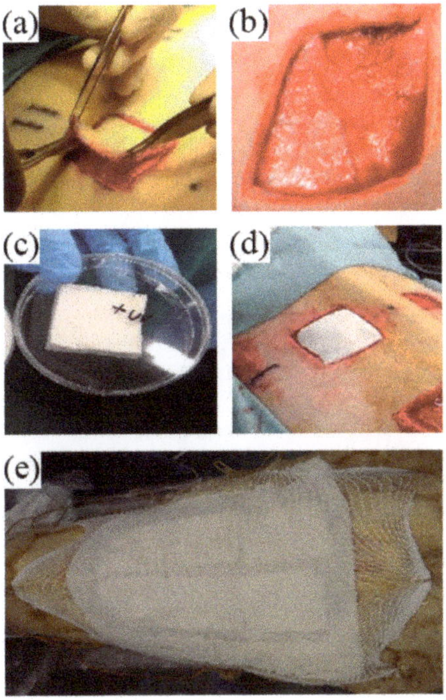

Figure 2. Representative images of the surgery procedure and biomaterial implantation. (**a**) Skin full-thickness removal by using surgical scissors for detaching the dermis from the subcutis; (**b**) macroscopic appearance of the wound after surgery; (**c**) 3D scaffold after UV sterilization and before implantation; (**d**) 3D scaffold implantation in the wound, the biomaterial was directly placed onto the wound bed; (**e**) representative figure of the bandages applied to the back of each sheep after the surgery.

After surgery, the animals were all housed in an adequate barn to ensure their well-being as gregarious animals. The covering bandage was enough to prevent allogrooming between animals. They received antibiotic therapy (amoxicilline, 150 mg/Kg) for five days and anti-inflammatory therapy (carprofen, 3.5 mg/Kg) for two days. Both therapies were performed by subcutaneous injection.

At 7, 14, 21, and 42 days after surgery skin biopsy were collected for assessing the skin wound healing process and comparing the efficacy of the treatment to control wounds (i.e., physiologically healing wounds). Biopsies collection was performed with the animals properly sedated and treated with medetomidine (0.01 mg/Kg, IV injection) and tramadol (4 mg/Kg, intramuscular injection) for analgesia and pain management. At each time point, two biopsies were collected by means of a 6-mm punch for all lesions. Skin samples were collected in two different and precise position of the wound, opposite to each other and equidistant from the original wound margins (day 0), previously marked on the healthy skin, after shearing the wool. Biopsies were then processed as follows: one for histological analysis (histopathology and immunohistochemistry) and one for gene expression analysis (real time PCR).

2.4. Clinical Follow-Up

The macroscopic appearance of each lesion was evaluated at 3, 7, 14, 21, and 42 days after surgery by taking pictures. This operation was performed by the same blinded operator at each time point. Moreover, the percentages of re-epithelialization and wound contraction were measured and calculated using an image processing program (ImageJ®).

2.5. Histopathological and Immunohistochemical Analysis

A total of 24 biopsies, 6 for each time point (3 for treated and 3 for untreated lesions), were processed for histological analysis. Skin samples were fixed in 10% neutral-buffered formalin (4% formaldehyde) for 24 h. They were then dehydrated by using a gradual dilution of ethanol and embedded in paraffin following standard procedures. After embedding, samples were cut with a microtome (Leica—RM2035, Leica Microsystems, Wetzlar, Germany) into 4-μm thick slices.

For histopathological evaluation, sections were stained with Harris hematoxylin and eosin (H&E) following the standard protocol. All sections were examined by two blinded independent operators and assessed for different histological parameters: superficial and deep inflammation, immature granulation tissue, and development of skin adnexa. The parameters were scored with a semi-quantitative scale from 0 to 3 (0 absence, 1 mild, 2 moderate, 3 abundant). Data were calculated for each animal and parameter, and presented as relative frequencies. The histopathological evaluation was performed by observation under a light microscope (Olympus Vanox AHBT3, Olympus, Tokyo, Japan).

In addition, the epidermal thickness index (ETI) was calculated to define the epidermal hypertrophy degree. To do so, the average thickness of the epidermis of the skin samples at 0, 21, and 42 days after surgery was measured in ten randomly selected fields of view for each slide (400× magnification). The ETI was calculated based on an Equation (1) described by Rahmani-Neishaboor and colleagues [37]:

$$\mathrm{ETI} = \frac{\text{height of epidermis at day } 21/42}{\text{height of epidermis at day } 0} \tag{1}$$

ETI values equivalent to 1 indicate a full healing of the wound without scar formation while values >1 represent a newly formed hypertrophic epidermis.

For immunohistochemistry, samples were immunostained with Ki67 (1:50, Dako, Santa Clara, CA, USA), a nuclear marker for cell proliferation, and alpha smooth muscle actin (α-SMA, 1:1000, Dako, Santa Clara, CA, USA), a marker to detect the presence of myofibroblasts (Table 1), following the manufacturer's protocol. Briefly, sections were deparaffinized and rehydrated by immerging them in graded ethanol series. In order to avoid false positive signals, endogenous peroxidase activity was blocked by immerging the tissue sections in a 0.3% H2O2 solution for 20 min at room temperature. Only sections for Ki67 immunostaining were exposed to antigen retrieval (heat induced epitope retrieval) by using a 10 mM sodium citrate buffer (pH 6) at 95 °C for 20 min. Moreover, nonspecific binding sites were blocked by using 2.5% normal goat serum for 1 h at room temperature. Then, sections were incubated with the primary antibody for a different amount of time: Ki67 overnight at 4 °C and α-SMA for 1 h at room temperature. After washing three times with PBS, sections were incubated with the secondary antibody (goat anti-mouse biotin-conjugated IgG, 1:200, Dako, Santa Clara, CA, USA) for 30 min at room temperature. Thereafter, an avidin–biotin complex (ABC Reagent, VECTASTAIN® ABC Kit, PK-4000, Vector Laboratories, Burlingame, CA, USA) and 3,3′-diamobenzidine (DAB) system (ImmPACT® DAB Substrate, SK-4105, Vector Laboratories, Burlingame, CA, USA) were used to achieve the immunolabeling. All sections were counterstained with Mayer's hematoxylin solution. Positive and negative sections controls were always performed in parallel with experimental sections in order to assess the specificity of immunostaining reaction. Immunostained sections were observed under a light microscope provided with a camera (Olympus Vanox AHBT3, Olympus, Tokyo, Japan).

Table 1. List and details of antibodies used in this study for immunohistochemistry (HIER, heat-induced epitope retrieval; α-SMA, alpha smooth muscle actin).

Antibody	Clonality	Host	Clone	Antigen Retrieval	Dilution	Catalog#
Ki67	Monoclonal	Mouse	MIB-1	HIER, sodium citrate 10 mM pH 6.0	1:50	M7240
α-SMA	Monoclonal	Mouse	1A4	-	1:1000	M0851

The positive area of Ki67 was quantitatively analyzed. The percentage of positive area was measured in ten randomly selected (400× magnification) fields of view for each slide. The positive signal was measured using an image processing program (ImageJ®) and expressed as a percentage of positive area.

α-SMA immunoreactivity was analyzed in a semi-quantitative fashion by scoring the abundancy and orientation of myofibroblasts using a histological score as follows: 0 for absence of immunostaining, 1 for mild presence of immunostaining and irregular pattern, 2 as moderate immunostaining and well-oriented myofibroblasts, and 3 as abundant immunostaining and compact organized myofibroblasts.

2.6. Gene Expression Analysis

A total of 24 biopsies, 6 for each time point (3 for treated and 3 for untreated lesions) were used for gene expression analysis by real time PCR (RT-PCR). Total RNA was isolated from skin samples by using TRIzol reagent (Life Technologies, Carlsbad, CA, USA). Then, the RNA extracted was assessed for its quality (260/280 nm wavelengths ratio) and quantified using a Nanodrop spectrophotometer (Thermo Scientific, Waltham, MA, USA). A total amount of 2 μg of RNA was retrotranscribed with Superscript™ II Reverse Transcriptase (Invitrogen, Carlsbad, CA, USA) to obtain complementary DNA (cDNA). Thus, the obtained cDNA was used as template for the RT-PCR gene expression analysis using the ABI 7500 Real-Time PCR system (Applied Biosystems, Foster City, CA, USA).

The relative expression of genes involved in the skin wound-healing process was evaluated: Collagen 1α1 (Collagen type I, Col1α1), Collagen 3α1 (Collagen type III, Col3α1), vascular endothelial growth factor A (VEGF-A), and hair-Keratin (hKER). The 18S ribosomal RNA (18S) and Ribosomal Protein S24 (RPS24) were used as reference genes in order to normalize the obtained data. Specific pairs of primers for each gene were designed using the Primer Express 3.0 software (Applied Biosystems, Foster City, CA, USA) based on the sheep annotated genome sequence on the GenBank database (sheep genome assembly: GCA_000298735.1) (Supplementary Table S1). Validation and efficiency of the designed primer were assessed by using the standard curve method. All pairs of primers presented an acceptable slope (between −3.8 and −3.3) with a corresponding efficiency of 90–100%. To calculate the efficiency, the ABI 7500 System SDS Software (Applied Biosystems, Foster City, CA, USA) was used.

All experiments were run in triplicate to study the relative gene expression of each gene of interest. A melting curve analysis (dissociation curve) was performed as well to detect the non-specific amplification. The relative expression was calculated by using the $2^{-\Delta\Delta Ct}$ method to normalize the cDNA level of expression of the gene of interest to the reference genes. The uninjured skin was used as the calibrator sample.

2.7. Statistical Analysis

All obtained data are expressed as the mean ± standard error of the mean (SEM). All statistical differences were assessed by means of the two-way ANOVA test and statistical differences ($p \leq 0.05$) were further detected by Dunnett "*post-hoc*" test. Differences in the ETI score were assessed by a Student's t distribution test and a p value less than 0.05 was considered statistically significant. All statistical analyses were performed using the GraphPad Prism 7.0 software (San Diego, CA, USA).

3. Results

3.1. Clinical Follow-Up

Wounds treated with the CBSS showed an improved macroscopic quality of the regenerated skin throughout the wound healing process in comparison to untreated (control) wounds. However, neither group reached a complete closure of the wound at 42 days. The presence of the CBSS was observed in treated wounds until day 14 (Figure 3).

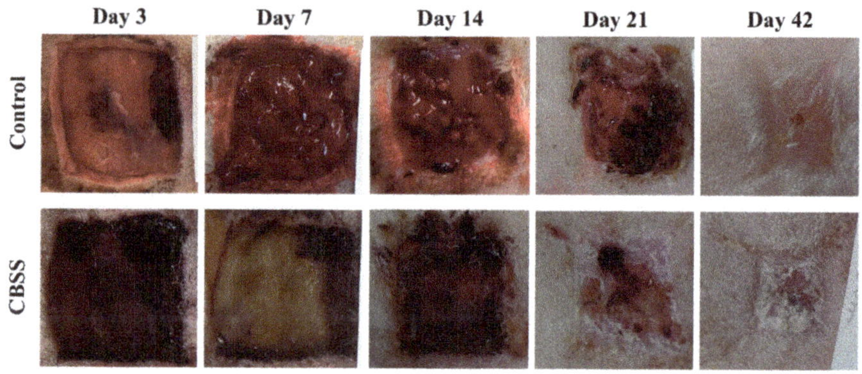

Figure 3. Representative images of the skin ulcers during wound healing.

One week after wound creation, untreated and CBSS-treated wounds presented a similar percentage of wound contraction (16.250 ± 0.695 vs. 15.313 ± 1.782). At two weeks, CBSS treated wounds showed a slightly higher contraction than the untreated ones (39.188 ± 0.937 vs. 44.708 ± 1.553). Between 21 and 42 days, untreated wounds had a higher percentage of contraction with respect to the CBSS-treated ones (70.917 ± 8.178 vs. 63.583 ± 11.510 at day 21 and 89.333 ± 1.752 vs. 86.896 ± 3.545 at day 42). No significant statistical differences were observed at any time point.

At day 7, all wounds showed a similar percentage of re-epithelialization (12.779 ± 1.363 vs. 13.273 ± 0.04). Between 14 and 21 days, CBSS-treated wounds showed a significantly higher percentage of re-epithelialization compared to the untreated wounds: 29.061 ± 5.100 vs. 37.709 ± 0.052 at day 14 (p = 0.034) and 67.109 ± 3.667 vs. 79.052 ± 7.398 at day 21 (p = 0.0031). At 42 days, all animals presented a similar percentage of re-epithelialization (99.316 ± 1.184 vs. 99.037 ± 1.668) (Figure 4).

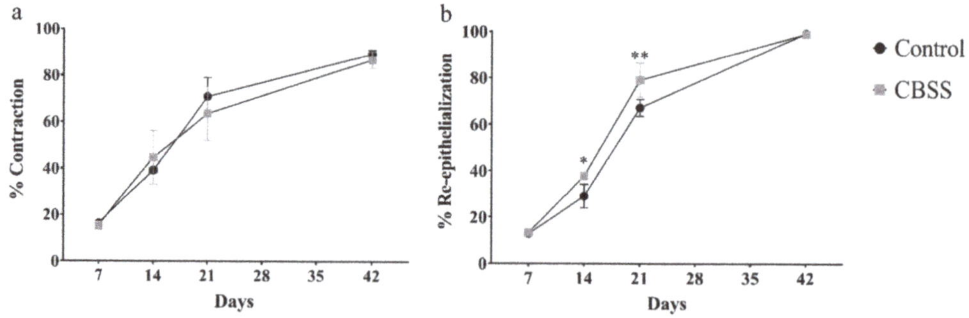

Figure 4. (**a**) Percentage of contraction. (**b**) Percentage of re-epithelialization. Data are shown as mean ± SEM. * $p < 0.05$; ** $p < 0.01$.

3.2. Histopathological Analysis

3.2.1. Superficial Inflammation

At 7 days, the 67% of CBSS-treated wounds showed an abundant inflammatory infiltrate in the superficial region of the dermis while the 33% presented a mild inflammation (Figure 5b). On the contrary, all untreated (control) wounds presented a mild superficial inflammation (Figure 5a). After two weeks, all wounds showed the same amount of inflammation with 67% of the wounds presenting a moderate presence and 33% mild one (Figure 5c,d). At 21 days, CBSS-treated wounds showed a reduction of inflammation with 67% of them presenting a mild inflammation and 33% with no inflammation (Figure 5f). On the contrary, in untreated wounds 67% still showed a moderate amount of inflammation while 33% displayed a mild inflammation (Figure 5e). At 42 days, all wounds showed no evidences of superficial inflammation (100% absent) (Figure 5g,h).

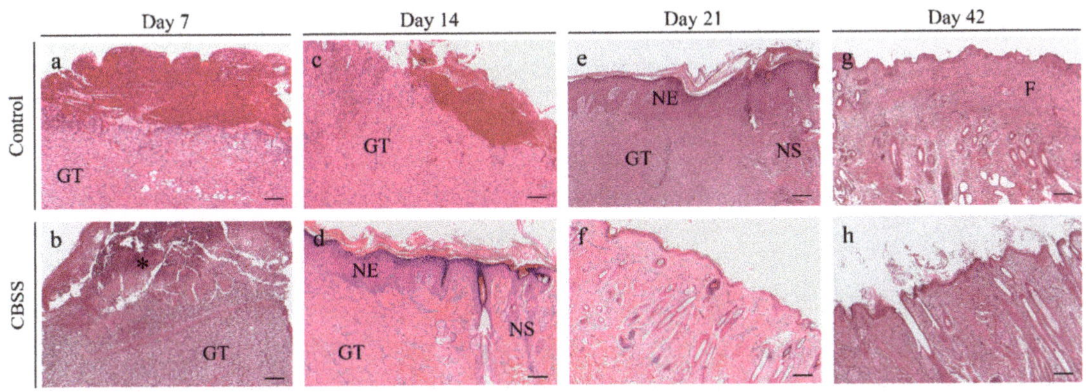

Figure 5. Histopathological microphotographs of skin biopsies at different time points after wounding: control and CBSS-treated wounds comparison. (**a**,**b**) Skin wounds at 7 days, in the CBSS-treated wounds (b) is possible to appreciate the presence of the 3D scaffold; (**c**,**d**) wounds at 14 days, treated wounds started to show a neoepidermis (NE, characterized by an hyperplastic appearance) and skin adnexa; (**e**,**f**) wounds at 21 days after wounding; (**g**,**h**) wounds at 42 days. GT = granulation tissue; NE = neoepidermis; NS = neoskin; F = fibrosis; asterisk = 3D sponge-like scaffold. Scalebar = 200 μm.

3.2.2. Deep Inflammation

As observed for the superficial inflammation, the deep inflammation parameter showed a decreasing trend throughout the experimentation. At 7 days, 67% of CBSS-treated wounds presented with a moderate inflammation while 33% showed mild inflammation (Figure 5b). At 14 days, in the same group an opposite tendency was observed with 67% of the wounds showing a mild inflammation and 33% a moderate one (Figure 5d). At 21 days, all wounds (100%) showed a mild inflammation in the deeper layer of the skin biopsy (Figure 5f) while at 42 days only 33% presented a mild inflammation, with the remnant 67% without any sign of inflammation (Figure 5h). Untreated wounds showed a mild inflammation (100%) at every time point (Figure 5a–d).

3.2.3. Immature Granulation Tissue

The treatment with the biomaterial led to a higher deposition of granulation tissue than in untreated wounds at 7 days (33% abundant and 67% mild in treated wounds vs. 67% mild and 33% absent in untreated wounds) (Figure 5a,b). At 14 days, a similar amount, but in a different ratio, of immature granulation tissue was observed in treated wounds (67% moderate and 33% mild presence) while all control wounds showed a moderate presence of granulation tissue (100%) (Figure 5c,d). Three weeks post-wounding, 33% of treated wounds showed a moderate amount of immature granulation tissue and 67% a mild one (Figure 5f). Only 33% of untreated wounds showed a mild amount of this parameter and 67% a moderate one (Figure 5e). At 42 days, all wounds did not show any sign of the presence of immature granulation tissue (100% absent) (Figure 5g,h).

3.2.4. Skin Adnexa

Skin adnexa appeared for the first time at 14 days only in CBSS-treated wounds (33% of wounds) (Figure 5d). At 21 days, all treated wounds (100%) showed a moderate amount of skin adnexa while in control wounds only the 33% showed a moderate amount (Figure 5e,f). At 42 days, all treated wounds presented an abundant presence of skin adnexa (Figure 5 h); on the contrary, in untreated wounds 67% presented a moderate amount and 33% an abundant amount (Figure 5g).

3.2.5. Epidermal Thickness Index (ETI)

The control group showed a higher ETI at 21 and 42 days than wounds treated with the CBSS (7.487 ± 0.558 vs. 5.599 ± 0.412 at 21 days and 3.657 ± 0.891 vs. 1.543 ± 0.436 at 42 days). The application of the CBSS led to a reduction of the epidermal thickness at 42 days in a significant way as compared to the control group, more similar to the one of unwounded skin (Figure 6).

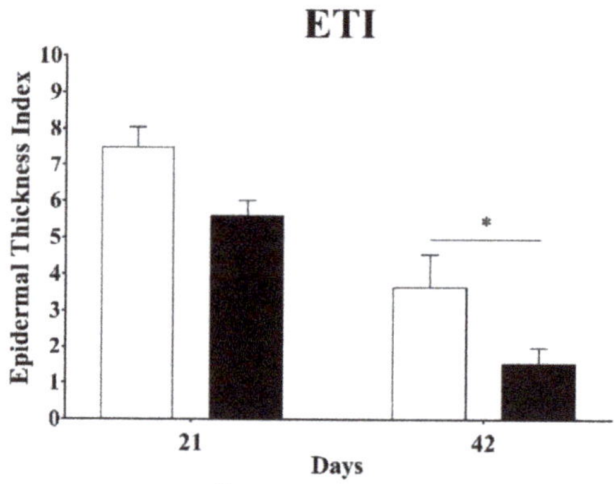

Figure 6. Epidermal thickness index (ETI) at 21 and 42 days respect to unwounded skin. Data are shown as mean ± SEM. * $p < 0.05$.

3.2.6. Ki67 Immunohistochemistry

The immunolabeling for the nuclear protein Ki67, a biological marker for cell proliferation, showed a higher amount of positivity at all time-points in CBSS-treated wounds compared to the untreated wounds, with a statistically significant difference at 7 and 14 days: 4,0080 ± 0.304 vs. 6.902 ± 0.628 at day 7 ($p < 0.0001$), and 0.545 ± 0.072 vs. 3.811 ± 0.390 at day 14 ($p < 0.0001$) (Figure 7a–d,i). The higher amount of Ki67 expression at 14 in CBSS-treated wounds might be attributed to the abundant presence of positivity in the basal layer of the neoepidermis (Figure 7d).

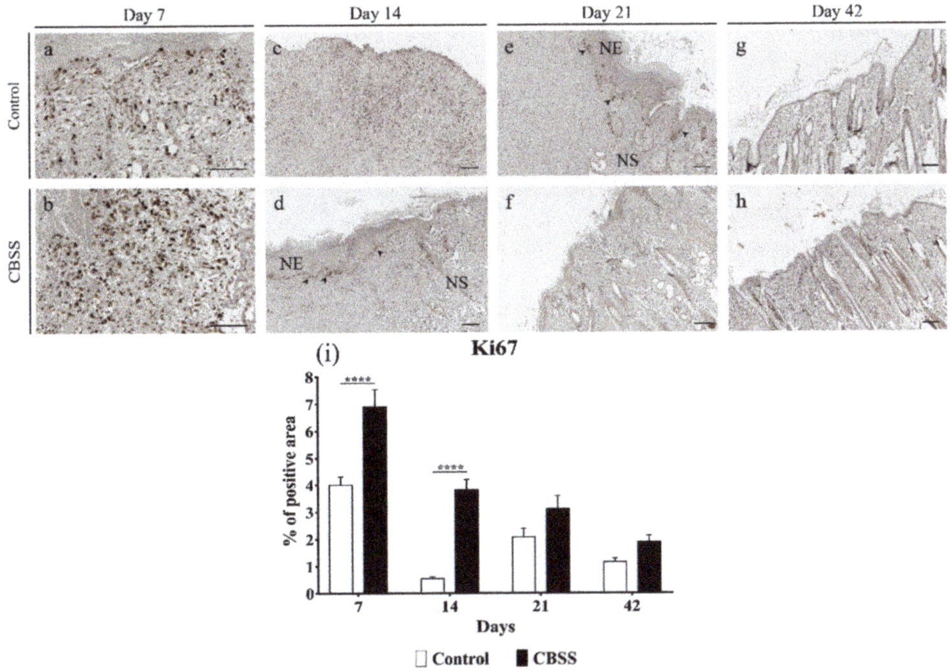

Figure 7. Immunohistochemistry microphotographs for Ki67 immunolabeling. Wounds are showed at (**a**,**b**) 7 days, (**c**,**d**) 14 days, (**e**,**f**) 21 days, and (**g**,**h**) 42 days. (**i**) Quantitative analysis of the percentage of positive area of each sample at 7, 14, 21, and 42 days. Arrowhead = active proliferating keratinocytes in the epidermal basal layer; NE = neoepidermis; NS = neoskin. Scalebar = 200 µm. (**i**) Data are expressed as mean ± SEM. Statistical differences were measured between the two experimental groups at the same time-point. **** $p < 0.0001$.

3.2.7. α-SMA Immunohistochemistry

The action of myofibroblasts during skin wound healing is fundamental for wound contraction and deposition of new extracellular matrix components [38]. α-SMA is a reliable myofibroblast marker and was detected since day 7 in all wounds. At 7 and 14 days post-wounding, both treated and untreated wounds showed a similar immunopositive pattern for α-SMA immunostaining. In particular, at 7 days CBSS-treated wounds showed a slightly lower presence of myofibroblasts but better organized (i.e., cells were parallel to each other and to the wound surface) (Figure 8b). At 21 days, the amount of myofibroblasts decreased in wounds treated with the biomaterial while in control wounds the amount of positive cells was similar to day 14; this difference between groups was significant ($p = 0.0263$) (Figure 8e,f). At 42 days, myofibroblasts were still present in untreated wounds (Figure 8g) in the upper layer of the dermis, oriented in parallel to the newly formed epidermis. On the contrary, no myofibroblasts were observed in treated wounds (Figure 8g). It is important to mention that the positivity observed in CBSS-treated wounds at 21 (mainly) and at 42 days has to be ascribed to the myoepithelial cells of apocrine glands and to arrector pili muscles (Figure 8f,h). As a matter of fact, α-SMA is also a marker for smooth muscle cells.

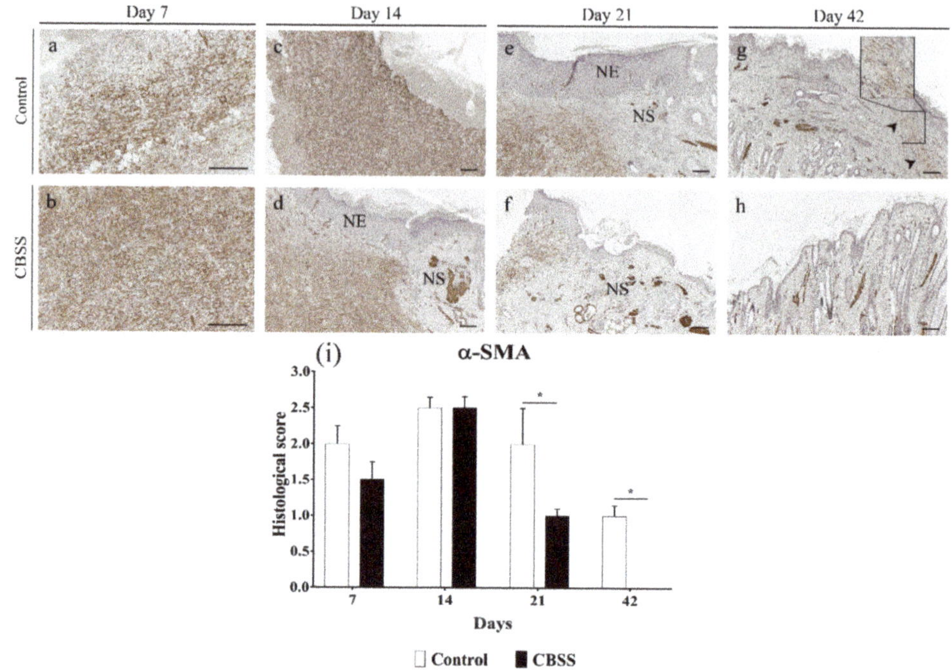

Figure 8. Immunohistochemistry microphotographs for α-SMA immunostaining. Wounds are showed at (**a**,**b**) 7 days, (**c**,**d**) 14 days, (**e**,**f**) 21 days, and (**g**,**h**) 42 days. (**i**) Semi-quantitative analysis based on the score for presence and orientation of myofibroblasts in wounds at 7, 14, 21, and 42 days. Arrowhead = myofibroblasts in the mature dermis; NE = neoepidermis; NS = neoskin. Scalebar = 200 μm; inset = higher magnification of the dermal fibrosis. (**i**) Data are expressed as mean ± SEM. Statistical differences were measured between the two experimental groups at the same time point * $p < 0.05$.

3.3. Gene Expression Analysis

RT-PCR analysis for the gene expression of the two principal collagen types of the skin (I and III) involved in the wound-healing process (in particular, granulation tissue formation and dermis remodeling) revealed an up-regulation of the expression of both genes during the first two weeks after wounding in both groups. At day 7, the gene expression of either type I or III was slightly different between the two groups with a statistically significant difference for collagen type I expression ($p < 0.0001$). However, after two weeks an opposite trend of expression was observed for collagen type I and type III. Indeed, collagen type I gene expression was upregulated in CBSS-treated wounds, while in control group it decreased as compared to day 7, and the difference was statistically different at this time point ($p < 0.0001$). At day 14, even if collagen type III gene expression was lower than day 7 in both groups, it was significantly higher in the control group with respect to the CBSS-treated wounds ($p < 0.0001$). At 21 days, the collagen type I gene expression was reduced for both groups, while the gene expression of collagen type III was strongly upregulated in the control group and decreased in the CBSS-treated group with a statically significant difference ($p < 0.0001$). At 42 days, the gene expression of both genes was similar in the two groups without showing any statistical difference (Figure 9).

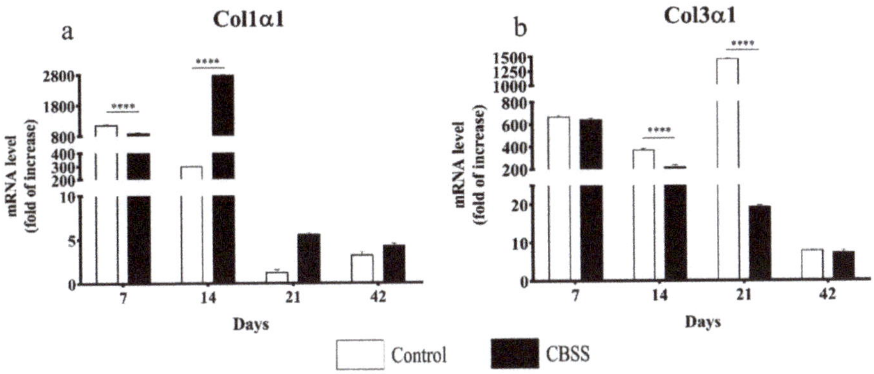

Figure 9. Gene expression analysis for collagen genes involved in skin wound healing. (**a**) Relative expression of the collagen type I (Col1α1) gene and (**b**) collagen type III (Col3α1) gene at 7, 14, 21, and 42 days after wounding in control and CBSS-treated wounds. Data are shown as mean ± SEM. Unwounded skin was used as the calibrator sample. Statistical differences were measured between the two experimental groups at the same time point. **** $p < 0.0001$.

The gene expression of VEGF-A, an essential growth factor involved in the process of neoangiogenesis, was always higher in the treated wounds in comparison to the untreated ones. In particular, the observed difference of expression was statistically significant at 7 ($p < 0.05$) and 14 days ($p < 0.0001$) (Figure 10a).

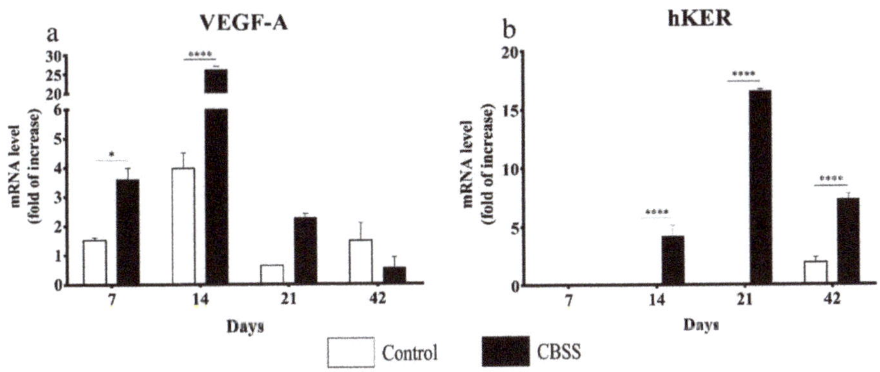

Figure 10. Gene expression analysis for VEGF and hKER genes (**a**) Relative expression of the vascular endothelial growth factor A (VEGF-A) gene and (**b**) hair-Keratin (hKER) gene at 7, 14, 21, and 42 days after wounding in control and CBSS-treated wounds. Data are shown as mean ± SEM. Unwounded skin was used as the calibrator sample. Statistical differences were measured between the two experimental groups at the same time point. * $p < 0.05$; **** $p < 0.0001$.

The gene expression of hKER, a marker for the expression of hair follicles, was first detected at day 14 in the CBSS-treated wounds only (Figure 10b). At 21 days, the mRNA level of hKER increased in treated wounds while in the control wounds no expression was observed. At 42 days, the hKER gene expression was lower with respect to day 21 in treated wounds while in untreated wounds it was observed for the first time during the experimental study.

4. Discussion

Wound care management of hard-to-heal or chronic ulcers is a critical issue in veterinary medicine since conventional treatments are often associated with a poor prognosis, with a consequent impact on the economic sphere [1,9,10,39,40]. Furthermore, inadequately healing wounds are the second most common cause of euthanasia or death in equine patients [41]. In this context, to overcome this issue, innovative strategies such as tissue engineered skin substitutes have been widely investigated in the recent years [42–44].

Collagen is the most abundant component of the ECM, therefore a collagen-based biomaterial could be considered as the best option for mimicking the microenvironment of the skin [45]. Collagen derived from marine sources has demonstrated to be a valuable alternative for manufacturing collagenous biomaterials because of its bioactive properties, biocompatibility, and efficiency, as they can stimulate the healing process of the wounded skin [25,46,47]. The collagen used in the present work comes from sea urchin (*P. lividus*) food industry by-products and the herein described collagen-based skin-like scaffold (CBSS) is composed of two layers: the upper one, "epidermal like," a thin 2D membrane to protect the skin from bacterial infiltration as well to prevent wound dehydration and protein loss, and the lower one, "dermal like," a 3D porous scaffold that mimics the structure and function of the skin dermis [27]. The cytocompatibility and efficiency of this novel biomaterial has been previously tested in in vitro studies [27,31,32]. The objective of this work was to assess the effects of this innovative bi-layered CBSS on the healing of experimental skin wounds in in vivo sheep model.

Inflammation is a critical step during wound healing. The inflammatory phase is fundamental for clearing the wound area from contaminating pathogens so that the healing process can progress. Nonetheless, an intense and/or prolonged inflammatory response may lead to a delay of healing without further progression to the next phases, hence contributing to wound chronicity [48–50]. In the present work, inflammation was examined at the histopathological level. Histopathological findings showed that wounds treated with the CBSS presented a higher presence of inflammatory infiltrate during the first week after wounding compared to wounds left untreated. This scenario might correspond to an anticipated activation of the inflammatory phase because of the biomaterial implantation. One reason underlying this condition could be the earlier activation of platelets by direct interaction with the collagen of the biomaterial [51]. Indeed, it is important to underline that platelets are activated only by collagen in its native fibrillar conformation [52], as the sea urchin-derived collagen used in the CBSS. This property potentially provides these biomaterials also with optimal hemostatic features. After activation, platelets begin to release granules containing several soluble factors; among them, there are different chemotactic molecules (e.g., chemokines, PDGF etc.,) that might have attracted the observed higher amount of inflammatory cells to the wound site during the first week of wound healing with respect to control wounds [53–55]. Moreover, collagen-based wound dressings have demonstrated to possess chemotactic properties to inflammatory cells, in particular macrophages [56,57]. However, inflammation diminished throughout the experimentation completely disappearing between 21 and 42 days in treated wounds. On the contrary, it was still mildly present in untreated wounds at day 42 in the deeper layer. The CBSS did not hamper the wound-healing process; on the contrary, it might have anticipated, by inducing it, and shortened the inflammatory phase therefore hastening the progression to the next phase of healing.

Wounds that heal by second intention, like in the current work, are characterized by the formation of granulation tissue: a provisional matrix that covers the wound bed and acts as a scaffold for migrating cells (epithelium, dermis, and vessels). Indeed, the relevant cellularity observed in the granulation tissue is due to the high number of cells proliferating in it [58]. During the first two weeks, CBSS-treated wounds exhibited a higher presence of granulation tissue than untreated wounds, showing how the marine collagen-based biomaterial might have stimulated the production of granulation tissue matrix by fibroblasts. Concomitantly, the higher deposition of granulation tissue in treated wounds might be related to the elevated number of cells in active proliferation observed in the same wounds (Ki67 positive cells). The observed beneficial effects on granulation tissue formation and cells proliferation might be due to the structural characteristics of the CBSS lower layer. Indeed, in our previous work we demonstrated that the 3D sponge-like scaffold was biocompatible plus able to support cell infiltration and proliferation in vitro as it presents peculiar structural characteristics to recreate the dermis in case of injury [27]. Hence, its biocompatibility and structure could have efficiently supported migration, adhesion, and proliferation of fibroblasts in vivo leading to the higher deposition of granulation tissue as observed with other collagenous biomaterials [59–63].

In order to bring nourishment and oxygen to the upper part of the wound, a proper vascularization is vital for the maturation of the granulation tissue, hence for the healing process. Several researchers have proposed skin substitutes containing growth factors and ECM components that could promote wound vascularization [64]. Neovascularization of the injured skin may be the effect of either proliferation of endothelial cells in existing blood vessels (angiogenesis) or differentiation of endothelial progenitor cells (vasculogenesis). Nevertheless, it has been demonstrate that a new vascular network is restored at the wound site, mainly through the mechanism of angiogenesis [65]. Indeed, the pro-angiogenetic activity of CBSS may positively influence the survival of the substitute itself. In the wound model, the CBSS treatment always showed an upregulation of VEGF-A gene expression, a growth factor that supports neoangiogenesis, especially during the first two weeks. Thereby, it could have improved the formation of new vessels and promoted healing through the development of the granulation tissue [47,66,67].

Along with the fibroblasts migrating to the injured site, the keratinocytes surrounding the wound begin to migrate as well to cover it and recreate the epithelial layer [68]. A newly formed epithelium was histologically observed since day 14 in CBSS-treated wounds along with the gene expression of hKER, hence the biomaterial actually accelerated the re-epithelialization process [69–72]. Indeed, a part of the proliferating cells observed at day 14 was in the basal layer of the epithelium, proving how those cells were actively dividing to recreate a proper epithelial barrier after injury. On the contrary, the epithelium histologically appeared in untreated wounds at day 21 while the gene expression of hKER was detected only after 6 weeks (day 42).

Besides covering the wound with a new epithelium, the organism exploits another healing mechanism to diminish the open wound area: wound contraction, with wound edges being pulled to the center of the wound by specialized cells [73]. This process is mediated by α-SMA-expressing myofibroblasts, a cell population that differentiate from fibroblasts, and is specialized in secretion of ECM components and granulation tissue contraction [38]. Clinically, the percentage of wound contraction was similar in both groups, but slightly minor in wound treated with the CBSS during the last two weeks. However, the immunohistochemistry staining revealed that the presence of myofibroblasts was always higher in untreated wounds, except at day 14. This observation may be explained by the better parallel organization of myofibroblasts and deposition of collagen fibrils in the treated wounds as compared to untreated wounds. As a matter of fact, for a proper and efficient wound contraction myofibroblasts should be parallelly co-aligned with the collagen fibrils to the wound surface [74], hence increasing the mechanical strength of the tissue and efficiently contracting [75,76].

While contracting, the ECM of the wounded skin begins to remodel itself into a mature dermis with a decreased level of cellularity and proliferation, as well as replacement of the ECM components [77,78]. In the current study, we observed that the skin substitute was able to induce the gene expression of the mature type of collagen, collagen type I, at 14 and 21 days, while in the control wounds it was observed a similar trend but for the collagen type III, the immature type. These differences in levels of expression were reflected in the histological appearance of the skin during the last phases of healing. While treated wounds started to show a compact ECM, a peculiar characteristic of mature dermis, along with the presence of skin appendages (hair follicles and glands), since day 21, the untreated wounds presented a higher ratio of loose ECM (i.e., granulation tissue) until day 42. Collagen-based biomaterials have shown to possess the ability to induce the collagen synthesis [69] and a better organization of collagen fibrils [71]. Apart from starting to express a mature type of collagen earlier than untreated wounds, CBSS application showed a decline in the number of α-SMA positive cells and amount of granulation tissue since day 21, with the complete disappearance of both features by day 42. On the contrary, wounds left untreated still showed an intense positivity for α-SMA protein at day 21 with a moderate presence of granulation tissue. Although the latter was not observed at day 42, α-SMA positive cells were still present in the upper layer of the dermis and organized in a parallel pattern to the wound surface. In addition, histological observations highlighted a high cellularity in the same area with compact collagen fibrils. These two characteristics, plus the upregulation of the gene expression of collagen type III observed at day 21, might be the outcome of an excessive deposition of immature ECM that might have evolved into a pathological fibrotic tissue [79,80]. In particular, it resembles the histological features of a hypertrophic scar, because of the higher thickness of the epidermis with respect to day 0 (epidermal thickness index, ETI, Figure 5) together with parallel myofibroblasts and dense collagen fibrils arrangement [81–84]. On the contrary, the skin substitute led to a physiological maturation of the granulation tissue with a proper development and organization of the mature dermis (collagen fibrils) and skin appendages. Hence, the treatment reduced the dermal fibrosis, and scarring, in wounds healing by second intention [85,86].

Skin substitutes must possess certain structural characteristics (e.g., biocompatible, control wound contraction, certain porosity for cell infiltration, etc.,) that can provide a support to the wounded skin for the regenerative process [87]; indeed, the CBSS possesses these characteristics as described by Ferrario and colleagues [27]. In addition, the degradation of the biomaterial should be simultaneous to the remodelling of the newly formed dermis in order to sustain its maturation and re-assembling [88,89]. During the experimental period, the presence of the CBSS (upper layer) was evident until day 14 at the macroscopic level (Figure 3). On the contrary, in histological figures (Figure 5), the presence of the CBSS (lower layer) was evident only at day 7 (Figure 5b). After this time point, it was not possible to distinguish between the host dermal collagen and the biomaterial one. As a matter of fact, starting from day 14 we observed a partial regeneration of the neo-dermis, which it might be considered synchronized with the progressive degradation of the biomaterial from our observations [90].

The research in the tissue engineering and regenerative medicine (TERM) field is rapidly evolving and developing new tools for tissue regeneration. The herein presented positive application of a skin substitute (CBSS) might be further improved by combining the marine collagen-based biomaterial with bioactive factors (e.g., growth factors) and/or cells (e.g., mesenchymal stem cells, MSCs). An addition of platelet-rich plasma (PRP), rich in growth factors, might support the healing process by releasing different soluble factors that in combination with skin substitute might limit inflammation, stimulate angiogenesis and re-epithelialization, hence favoring the physiological reparative mechanisms of the skin [91,92]. However, future studies will be necessary to validate the observed results. In addition, a larger number of animals and the comparison of the application of the CBSS with gold standards techniques (skin autograft) or the usage of a commercially available skin substitute will be fundamental to further assess the beneficial outcomes of the CBSS treatment for large skin wounds in the veterinary clinical practice.

5. Conclusions

In this study, the application of an innovative skin substitute made up of collagen derived from marine sources, i.e., sea urchin food wastes, was assessed in an in vivo wound healing model. The skin substitute supported and stimulated wound healing throughout the whole process: it controlled inflammation, promoted the deposition and maturation of granulation tissue, enhanced re-epithelialization, and induced the formation of skin appendages. At the same time, it showed resistance to wound contraction and dermal fibrosis. Overall, the collagen-based biomaterial served as a template for skin wound healing, i.e., as a scaffold for cell migration and growth. All these features led to a better quality of the repaired skin.

Supplementary Materials: The following are available online at https://www.mdpi.com/article/10.3390/ani11051219/s1. Table S1: List of primer sequences used for the gene expression analysis (RT-PCR).

Author Contributions: Conceptualization, I.I., M.S., and M.P.; methodology, L.M. and T.M.; formal analysis, L.M.; investigation, L.M.; resources, all authors; data curation, L.M., T.M., A.P., R.S., and C.F.; writing—original draft preparation, L.M., T.M., and M.P.; writing—review and editing, all authors; supervision, M.S. and M.P.; project administration, L.M. and M.P.; funding acquisition, M.S. and M.P. All authors have read and agreed to the published version of the manuscript.

Funding: This research was funded by Italian MIUR, grant number 2017FNZPNN (PRIN 2017, PI: Marco Patruno).

Institutional Review Board Statement: The experimental procedure was approved by the Italian Ministry of Health (n°51/2015-PR), in accordance with the Body for the Protection of Animals (OPBA) of the University of Padova.

Informed Consent Statement: Not applicable.

Data Availability Statement: The data presented in this study are available within the article.

Acknowledgments: The Authors are grateful to the technician Rosa Leone for helping in histology procedures and to Owen O'Shea (CORE, Gregory Town Eleuthera, The Bahamas) for the language revision of the manuscript.

Conflicts of Interest: The authors declare no conflict of interest.

References

1. Theoret, C.L. The pathophysiology of wound repair. *Vet. Clin. N. Am. Equine Pract.* **2005**, *21*, 1–13. [CrossRef]
2. Frykfors von Hekkel, A.K.; Pegram, C.; Halfacree, Z.J. Thoracic dog bite wounds in dogs: A retrospective study of 123 cases (2003–2016). *Vet. Surg.* **2020**, *49*, 694–703. [CrossRef] [PubMed]
3. Rothe, K.; Tsokos, M.; Handrick, W. Animal and Human Bite Wounds. *Dtsch. Arztebl. Int.* **2015**, *112*, 433–443. [CrossRef] [PubMed]
4. Cerqueira, M.T.; Pirraco, R.P.; Marques, A.P. Stem Cells in Skin Wound Healing: Are We There Yet? *Adv. Wound Care* **2016**, *5*, 164–175. [CrossRef] [PubMed]
5. Sparks, H.D.; Sigaeva, T.; Tarraf, S.; Mandla, S.; Pope, H.; Hee, O.; Di Martino, E.S.; Biernaskie, J.; Radisic, M.; Scott, W.M. Biomechanics of Wound Healing in an Equine Limb Model: Effect of Location and Treatment with a Peptide-Modified Collagen-Chitosan Hydrogel. *ACS Biomater. Sci. Eng.* **2021**, *7*, 265–278. [CrossRef]
6. Fowler, D. Distal Limb and Paw Injuries. *Vet. Clin. N. Am. Small Anim. Pract.* **2006**, *36*, 819–845. [CrossRef]
7. Dernell, W.S. Initial Wound Management. *Vet. Clin. N. Am. Small Anim. Pract.* **2006**, *36*, 713–738. [CrossRef]
8. Prpich, C.Y.; Santamaria, A.C.; Simcock, J.O.; Wong, H.K.; Nimmo, J.S.; Kuntz, C.A. Second intention healing after wide local excision of soft tissue sarcomas in the distal aspects of the limbs in dogs: 31 cases (2005–2012). *J. Am. Vet. Med. Assoc.* **2014**, *244*, 187–194. [CrossRef]
9. Theoret, C. Tissue engineering in wound repair: The three "r"s—Repair, replace, regenerate. *Vet. Surg.* **2009**, *38*, 905–913. [CrossRef]
10. Riggs, J.; Jennings, J.L.F.; Friend, E.J.; Halfacree, Z.; Nelissen, P.; Holmes, M.A.; Demetriou, J.L. Outcome of full-thickness skin grafts used to close skin defects involving the distal aspects of the limbs in cats and dogs: 52 cases (2005–2012). *J. Am. Vet. Med. Assoc.* **2015**, *247*, 1042–1047. [CrossRef]
11. Ferreira, M.C.; Paggiaro, A.O.; Isaac, C.; Teixeira Neto, N.; Santos, G.B. dos Substitutos cutâneos: Conceitos atuais e proposta de classificação. *Rev. Bras. Cir. Plástica* **2011**, *26*, 696–702. [CrossRef]
12. Cuono, C.B.; Langdon, R.; Birchall, N.; Barttelbort, S.; McGuire, J. Composite autologous-allogeneic skin replacement: Development and clinical application. *Plast. Reconstr. Surg.* **1987**, *80*, 626–635. [CrossRef]
13. Balasubramani, M.; Kumar, T.R.; Babu, M. Skin substitutes: A review. *Burns* **2001**, *27*, 534–544. [CrossRef]
14. Hansbrough, J.F.; Morgan, J.; Greenleaf, G.; Underwood, J. Development of a temporary living skin replacement composed of human neonatal fibroblasts cultured in Biobrane, a synthetic dressing material. *Surgery* **1994**, *115*, 633–644.
15. Jones, I.; Currie, L.; Martin, R. A guide to biological skin substitutes. *Br. J. Plast. Surg.* **2002**, *55*, 185–193. [CrossRef]
16. Su, T.; Zhang, M.; Zeng, Q.; Pan, W.; Huang, Y.; Qian, Y.; Dong, W.; Qi, X.; Shen, J. Mussel-inspired agarose hydrogel scaffolds for skin tissue engineering. *Bioact. Mater.* **2021**, *6*, 579–588. [CrossRef]
17. Elango, J.; Zhang, J.; Bao, B.; Palaniyandi, K.; Wang, S.; Wenhui, W.; Robinson, J.S. Rheological, biocompatibility and osteogenesis assessment of fish collagen scaffold for bone tissue engineering. *Int. J. Biol. Macromol.* **2016**, *91*, 51–59. [CrossRef]
18. Araujo, T.A.T.; Almeida, M.C.; Avanzi, I.; Parisi, J.; Simon Sales, A.F.; Na, Y.; Renno, A. Collagen membranes for skin wound repair: A systematic review. *J. Biomater. Appl.* **2020**. [CrossRef]

19. Jansen, L.A.; De Caigny, P.; Guay, N.A.; Lineaweaver, W.C.; Shokrollahi, K. The evidence base for the acellular dermal matrix AlloDerm: A systematic review. *Ann. Plast. Surg.* **2013**, *70*, 587–594. [CrossRef]
20. Wester, J.L.; Pittman, A.L.; Lindau, R.H.; Wax, M.K. AlloDerm with split-thickness skin graft for coverage of the forearm free flap donor site. *Otolaryngol. Head Neck Surg. USA* **2014**, *150*, 47–52. [CrossRef]
21. Fourman, M.S.; Phillips, B.T.; Fritz, J.R.; Conkling, N.; McClain, S.A.; Simon, M.; Dagum, A.B. Laser-Assisted Indocyanine Green Dye Angiography Accurately Predicts the Split-Thickness Graft Timing of Integra Artificial Dermis. *Ann. Plast. Surg.* **2014**, *73*, 150–155. [CrossRef]
22. Notodihardjo, S.C.; Morimoto, N.; Munisso, M.C.; Le, T.M.; Mitsui, T.; Kakudo, N.; Kusumoto, K. A comparison of the wound healing process after the application of three dermal substitutes with or without basic fibroblast growth factor impregnation in diabetic mice. *J. Plast. Reconstr. Aesthetic Surg.* **2020**, *73*, 1547–1555. [CrossRef]
23. Cervelli, V.; Brinci, L.; Spallone, D.; Tati, E.; Palla, L.; Lucarini, L.; De Angelis, B. The use of MatriDerm ®and skin grafting in post-traumatic wounds. *Int. Wound J.* **2011**, *8*, 400–405. [CrossRef]
24. Min, J.H.; Yun, I.S.; Lew, D.H.; Roh, T.S.; Lee, W.J. The use of Matriderm and autologous skin graft in the treatment of full thickness skin defects. *Arch. Plast. Surg.* **2014**, *41*, 330–336. [CrossRef]
25. Silva, T.H.; Moreira-Silva, J.; Marques, A.L.P.; Domingues, A.; Bayon, Y.; Reis, R.L. Marine origin collagens and its potential applications. *Mar. Drugs* **2014**, *12*, 5881–5901. [CrossRef]
26. Azizur Rahman, M. Collagen of extracellular matrix from marine invertebrates and its medical applications. *Mar. Drugs* **2019**, *17*, 118. [CrossRef]
27. Ferrario, C.; Rusconi, F.; Pulaj, A.; Macchi, R.; Landini, P.; Paroni, M.; Colombo, G.; Martinello, T.; Melotti, L.; Gomiero, C.; et al. From Food Waste to Innovative Biomaterial: Sea Urchin-Derived Collagen for Applications in Skin Regenerative Medicine. *Mar. Drugs* **2020**, *18*, 414. [CrossRef]
28. Ellis, D.L.; Yannas, I.V. Recent advances in tissue synthesis in vivo by use of collagen-glycosaminoglycan copolymers. *Biomaterials* **1996**, *17*, 291–299. [CrossRef]
29. Chung, K.-H.; Bhadriraju, K.; Spurlin, T.A.; Cook, R.F.; Plant, A.L. Nanomechanical Properties of Thin Films of Type I Collagen Fibrils. *Langmuir* **2010**, *26*, 3629–3636. [CrossRef]
30. Bozec, L.; Van Der Heijden, G.; Horton, M. Collagen fibrils: Nanoscale ropes. *Biophys. J.* **2007**, *92*, 70–75. [CrossRef]
31. Di Benedetto, C.; Barbaglio, A.; Martinello, T.; Alongi, V.; Fassini, D.; Cullorà, E.; Patruno, M.; Bonasoro, F.; Barbosa, M.A.; Carnevali, M.D.C.; et al. Production, characterization and biocompatibility of marine collagen matrices from an alternative and sustainable source: The sea urchin Paracentrotus lividus. *Mar. Drugs* **2014**, *12*, 4912–4933. [CrossRef] [PubMed]
32. Ferrario, C.; Leggio, L.; Leone, R.; Di Benedetto, C.; Guidetti, L.; Coccè, V.; Ascagni, M.; Bonasoro, F.; La Porta, C.A.M.; Candia Carnevali, M.D.; et al. Marine-derived collagen biomaterials from echinoderm connective tissues. *Mar. Environ. Res.* **2017**, *128*, 46–57. [CrossRef] [PubMed]
33. Tricarico, S.; Barbaglio, A.; Burlini, N.; del Giacco, L.; Ghilardi, A.; Sugni, M.; Di Benedetto, C.; Bonasoro, F.; Wilkie, I.C.; Carnevali, M.D.C. New insights into the mutable collagenous tissue of *Paracentrotus lividus*: Preliminary results. *Zoosymposia* **2012**, *7*, 279–285. [CrossRef]
34. Russel, W.; Burch, R. *The Principles of Human Experimental Technique*; Methiun & Co: London, UK, 1959; ISBN 0900767782 9780900767784.
35. Martinello, T.; Gomiero, C.; Perazzi, A.; Iacopetti, I.; Gemignani, F.; DeBenedictis, G.M.; Ferro, S.; Zuin, M.; Martines, E.; Brun, P.; et al. Allogeneic mesenchymal stem cells improve the wound healing process of sheep skin. *BMC Vet. Res.* **2018**, *14*, 1–9. [CrossRef]
36. Melotti, L.; Martinello, T.; Perazzi, A.; Martines, E.; Zuin, M.; Modenese, D.; Cordaro, L.; Ferro, S.; Maccatrozzo, L.; Iacopetti, I.; et al. Could cold plasma act synergistically with allogeneic mesenchymal stem cells to improve wound skin regeneration in a large size animal model? *Res. Vet. Sci.* **2021**, *136*, 97–110. [CrossRef]
37. Rahmani-Neishaboor, E.; Yau, F.M.; Jalili, R.; Kilani, R.T.; Ghahary, A. Improvement of hypertrophic scarring by using topical anti-fibrogenic/anti-inflammatory factors in a rabbit ear model. *Wound Repair Regen.* **2010**, *18*, 401–408. [CrossRef]
38. Hinz, B.; Gabbiani, G. Fibrosis: Recent advances in myofibroblast biology and new therapeutic perspectives. *F1000 Biol. Rep.* **2010**, *2*, 78. [CrossRef]
39. Balsa, I.M.; Culp, W.T.N. Wound Care. *Vet. Clin. N. Am. Small Anim. Pract.* **2015**, *45*, 1049–1065. [CrossRef]
40. Eggleston, R.B. Wound Management: Wounds with Special Challenges. *Vet. Clin. N. Am. Equine Pract.* **2018**, *34*, 511–538. [CrossRef]
41. Theoret, C.L.; Bolwell, C.F.; Riley, C.B. A cross-sectional survey on wounds in horses in New Zealand. *N. Z. Vet. J.* **2016**, *64*, 90–94. [CrossRef]
42. Chocarro-Wrona, C.; López-Ruiz, E.; Perán, M.; Gálvez-Martín, P.; Marchal, J.A. Therapeutic strategies for skin regeneration based on biomedical substitutes. *J. Eur. Acad. Dermatol. Venereol.* **2019**, *33*, 484–496. [CrossRef]
43. Ramió-Lluch, L.; Cerrato, S.; Brazis, P.; Rabanal, R.M.; Fondevila, D.; Puigdemont, A. Proof of concept of a new autologous skin substitute for the treatment of deep wounds in dogs. *Vet. J.* **2017**, *230*, 36–40. [CrossRef]
44. Kaasi, A.; Lima-Neto, J.F.; Matiello-Filho, J.A.; Calejo, M.H.S.; Jardini, A.L.; Kharmandayan, P. Regenerative collagen biomembrane: Interim results of a phase I veterinary clinical trial for skin repair. *F1000 Res* **2018**, *7*. [CrossRef]

45. Parenteau-Bareil, R.; Gauvin, R.; Berthod, F. Collagen-Based Biomaterials for Tissue Engineering Applications. *Materials* **2010**, *3*, 1863–1887. [CrossRef]
46. Fernández-Cervantes, I.; Rodríguez-Fuentes, N.; León-Deniz, L.V.; Alcántara Quintana, L.E.; Cervantes-Uc, J.M.; Herrera Kao, W.A.; Cerón-Espinosa, J.D.; Cauich-Rodríguez, J.V.; Castaño-Meneses, V.M. Cell-free scaffold from jellyfish Cassiopea andromeda (Cnidaria; Scyphozoa) for skin tissue engineering. *Mater. Sci. Eng. C* **2020**, *111*, 110748. [CrossRef]
47. Cruz, M.A.; Araujo, T.A.; Avanzi, I.R.; Parisi, J.R.; de Andrade, A.L.M.; Rennó, A.C.M. Collagen from Marine Sources and Skin Wound Healing in Animal Experimental Studies: A Systematic Review. *Mar. Biotechnol.* **2021**, *23*, 1–11. [CrossRef]
48. Falanga, V. Wound healing and its impairment in the diabetic foot. *Lancet* **2005**, *366*, 1736–1743. [CrossRef]
49. Menke, N.B.; Ward, K.R.; Witten, T.M.; Bonchev, D.G.; Diegelmann, R.F. Impaired wound healing. *Clin. Dermatol.* **2007**, *25*, 19–25. [CrossRef]
50. Rosique, R.G.; Rosique, M.J.; Farina Junior, J.A. Curbing inflammation in skin wound healing: A review. *Int. J. Inflam.* **2015**, *2015*. [CrossRef]
51. Farndale, R.W. Collagen-induced platelet activation. *Blood Cells Mol. Dis.* **2006**, *36*, 162–165. [CrossRef]
52. Kehrel, B. Platelet receptors for collagens. *Platelets* **1995**, *6*, 11–16. [CrossRef]
53. Harrison, S.; Vavken, P.; Kevy, S.; Jacobson, M.; Zurakowski, D.; Murray, M.M. Platelet Activation by Collagen Provides Sustained Release of Anabolic Cytokines. *Am. J. Sports Med.* **2011**, *39*, 729–734. [CrossRef]
54. Ollivier, V.; Syvannarath, V.; Gros, A.; Butt, A.; Loyau, S. Collagen Can Selectively Trigger a Platelet Secretory Phenotype via Glycoprotein VI. *PLoS ONE* **2014**, *9*, e104712. [CrossRef]
55. Sonmez, O.; Sonmez, M. Role of platelets in immune system and inflammation. *Porto Biomed. J.* **2017**, *2*, 311–314. [CrossRef]
56. Seaman, S. Dressing selection in chronic wound management. *J. Am. Podiatr. Med. Assoc.* **2002**, *92*, 24–33. [CrossRef]
57. Das, A.; Abas, M.; Biswas, N.; Banerjee, P.; Ghosh, N.; Rawat, A.; Khanna, S.; Roy, S.; Sen, C.K. A Modified Collagen Dressing Induces Transition of Inflammatory to Reparative Phenotype of Wound Macrophages. *Sci. Rep.* **2019**, *9*, 1–10. [CrossRef]
58. Häkkinen, L.; Larjava, H.; Koivisto, L. Granulation tissue formation and remodeling. *Endod. Top.* **2011**, *24*, 94–129. [CrossRef]
59. Matsumoto, Y.; Ikeda, K.; Yamaya, Y.; Yamashita, K.; Saito, T.; Hoshino, Y.; Koga, T.; Enari, H.; Suto, S.; Yotsuyanagi, T. The usefulness of the collagen and elastin sponge derived from salmon as an Artificial dermis and scaffold for tissue engineering. *Biomed. Res.* **2011**, *32*, 29–36. [CrossRef]
60. Chattopadhyay, S.; Raines, R.T. Review collagen-based biomaterials for wound healing. *Biopolymers* **2014**, *101*, 821–833. [CrossRef]
61. Kim, K.O.; Lee, Y.; Hwang, J.W.; Kim, H.; Kim, S.M.; Chang, S.W.; Lee, H.S.; Choi, Y.S. Wound healing properties of a 3-D scaffold comprising soluble silkworm gland hydrolysate and human collagen. *Coll. Surf. B Biointerfaces* **2014**, *116*, 318–326. [CrossRef]
62. Meyer, M. Processing of collagen based biomaterials and the resulting materials properties. *Biomed. Eng. Online* **2019**, *18*, 1–74. [CrossRef] [PubMed]
63. Veeruraj, A.; Liu, L.; Zheng, J.; Wu, J.; Arumugam, M. Evaluation of astaxanthin incorporated collagen film developed from the outer skin waste of squid Doryteuthis singhalensis for wound healing and tissue regenerative applications. *Mater. Sci. Eng. C* **2019**, *95*, 29–42. [CrossRef] [PubMed]
64. Eweida, A.M.; Marei, M.K. Naturally Occurring Extracellular Matrix Scaffolds for Dermal Regeneration: Do They Really Need Cells? *Biomed. Res. Int.* **2015**, *2015*, 839694. [CrossRef] [PubMed]
65. Bluff, J.E.; Ferguson, M.W.J.; O'Kane, S.; Ireland, G. Bone marrow-derived endothelial progenitor cells do not contribute significantly to new vessels during incisional wound healing. *Exp. Hematol.* **2007**, *35*, 500–506. [CrossRef] [PubMed]
66. Tan, Q.; Chen, B.; Yan, X.; Lin, Y.; Xiao, Z.; Hou, X.; Dai, J. Promotion of diabetic wound healing by collagen scaffold with collagen-binding vascular endothelial growth factor in a diabetic rat model. *J. Tissue Eng. Regen. Med.* **2014**, *8*, 195–201. [CrossRef] [PubMed]
67. Elbialy, Z.I.; Atiba, A.; Abdelnaby, A.; Al-Hawary, I.I.; Elsheshtawy, A.; El-Serehy, H.A.; Abdel-Daim, M.M.; Fadl, S.E.; Assar, D.H. Collagen extract obtained from Nile tilapia (*Oreochromis niloticus* L.) skin accelerates wound healing in rat model via up regulating VEGF, bFGF, and α-SMA genes expression. *BMC Vet. Res.* **2020**, *16*, 352. [CrossRef] [PubMed]
68. Ben Amar, M.; Wu, M. Re-epithelialization: Advancing epithelium frontier during wound healing. *J. R. Soc. Interface* **2014**, *11*, 20131038. [CrossRef]
69. Dang, Q.F.; Liu, H.; Yan, J.Q.; Liu, C.S.; Liu, Y.; Li, J.; Li, J.J. Characterization of collagen from haddock skin and wound healing properties of its hydrolysates. *Biomed. Mater.* **2015**, *10*, 015022. [CrossRef]
70. Shekhter, A.B.; Rudenko, T.G.; Istranov, L.P.; Guller, A.E.; Borodulin, R.R.; Vanin, A.F. Dinitrosyl iron complexes with glutathione incorporated into a collagen matrix as a base for the design of drugs accelerating skin wound healing. *Eur. J. Pharm. Sci.* **2015**, *78*, 8–18. [CrossRef]
71. Ouyang, Q.Q.; Hu, Z.; Lin, Z.P.; Quan, W.Y.; Deng, Y.F.; Li, S.D.; Li, P.W.; Chen, Y. Chitosan hydrogel in combination with marine peptides from tilapia for burns healing. *Int. J. Biol. Macromol.* **2018**, *112*, 1191–1198. [CrossRef]
72. Lesperance, M.M.; Francis, T.L.; Norton, B. Postsurgical Soft Tissue Healing. In *Postsurgical Orthopedic Sports Rehabilitation*; Elsevier Inc.: Amsterdam, The Netherlands, 2006; pp. 3–18. ISBN 9780323027021.
73. Li, S.; Van Den Diepstraten, C.; D'Souza, S.J.; Chan, B.M.C.; Pickering, J.G. Vascular smooth muscle cells orchestrate the assembly of type I collagen via α2β1 integrin, RhoA, and fibronectin polymerization. *Am. J. Pathol.* **2003**, *163*, 1045–1056. [CrossRef]
74. Schultz, G.S.; Davidson, J.M.; Kirsner, R.S.; Bornstein, P.; Herman, I.M. Dynamic reciprocity in the wound microenvironment. *Wound Repair Regen.* **2011**, *19*, 134–148. [CrossRef]

75. Toriseva, M.; Laato, M.; Carpén, O.; Ruohonen, S.T.; Savontaus, E.; Inada, M.; Krane, S.M.; Kähäri, V.-M. MMP-13 Regulates Growth of Wound Granulation Tissue and Modulates Gene Expression Signatures Involved in Inflammation, Proteolysis, and Cell Viability. *PLoS ONE* **2012**, *7*, e42596. [CrossRef]
76. Rai, N.K.; Tripathi, K.; Sharma, D.; Shukla, V.K. Apoptosis: A basic physiologic process in wound healing. *Int. J. Low. Extrem. Wounds* **2005**, *4*, 138–144. [CrossRef]
77. Landén, N.X.; Li, D.; Ståhle, M. Transition from inflammation to proliferation: A critical step during wound healing. *Cell. Mol. Life Sci.* **2016**, *73*, 3861–3885. [CrossRef]
78. Aravinthan, A.; Park, J.K.; Hossain, M.A.; Sharmila, J.; Kim, H.J.; Kang, C.W.; Kim, N.S.; Kim, J.H. Collagen-based sponge hastens wound healing via decrease of inflammatory cytokines. *3 Biotech* **2018**, *8*, 487. [CrossRef]
79. Xue, M.; Jackson, C.J. Extracellular Matrix Reorganization During Wound Healing and Its Impact on Abnormal Scarring. *Adv. Wound Care* **2015**, *4*, 119–136. [CrossRef]
80. Limandjaja, G.C.; Belien, J.M.; Scheper, R.J.; Niessen, F.B.; Gibbs, S. Hypertrophic and keloid scars fail to progress from the $CD34^-/\alpha$-smooth muscle actin (α-SMA)$^+$ immature scar phenotype and show gradient differences in α-SMA and p16 expression. *Br. J. Dermatol.* **2020**, *182*, 974–986. [CrossRef]
81. Slemp, A.E.; Kirschner, R.E. Keloids and scars: A review of keloids and scars, their pathogenesis, risk factors, and management. *Curr. Opin. Pediatr.* **2006**, *18*, 396–402. [CrossRef]
82. Ng, C.P.; Hinz, B.; Swartz, M.A. Interstitial fluid flow induces myofibroblast differentiation and collagen alignment in vitro. *J. Cell Sci.* **2005**, *118*, 4731–4739. [CrossRef]
83. Gauglitz, G.G.; Korting, H.C.; Pavicic, T.; Ruzicka, T.; Jeschke, M.G. Hypertrophic scarring and keloids: Pathomechanisms and current and emerging treatment strategies. *Mol. Med.* **2011**, *17*, 113–125. [CrossRef]
84. Wells, A.; Nuschke, A.; Yates, C.C. Skin tissue repair: Matrix microenvironmental influences. *Matrix Biol.* **2016**, *49*, 25–36. [CrossRef]
85. Sharpe, J.R.; Martin, Y. Strategies Demonstrating Efficacy in Reducing Wound Contraction In Vivo. *Adv. Wound Care* **2013**, *2*, 167–175. [CrossRef]
86. Greaves, N.S.; Iqbal, S.A.; Hodgkinson, T.; Morris, J.; Benatar, B.; Alonso-Rasgado, T.; Baguneid, M.; Bayat, A. Skin substitute-assisted repair shows reduced dermal fibrosis in acute human wounds validated simultaneously by histology and optical coherence tomography. *Wound Repair Regen.* **2015**, *23*, 483–494. [CrossRef]
87. Urciuolo, F.; Casale, C.; Imparato, G.; Netti, P.A. Bioengineered Skin Substitutes: The Role of Extracellular Matrix and Vascularization in the Healing of Deep Wounds. *J. Clin. Med.* **2019**, *8*, 2083. [CrossRef] [PubMed]
88. Ma, P.X. Scaffolds for tissue fabrication. *Mater. Today* **2004**, *7*, 30–40. [CrossRef]
89. MacNeil, S. Progress and opportunities for tissue-engineered skin. *Nature* **2007**, *445*, 874–880. [CrossRef]
90. De Angelis, B.; Orlandi, F.; Fernandes Lopes Morais D'Autilio, M.; Scioli, M.G.; Orlandi, A.; Cervelli, V.; Gentile, P. Long-term follow-up comparison of two different bi-layer dermal substitutes in tissue regeneration: Clinical outcomes and histological findings. *Int. Wound J.* **2018**, *15*, 695–706. [CrossRef]
91. Lacci, K.M.; Dardik, A. Platelet-rich plasma: Support for its use in wound healing. *Yale J. Biol. Med.* **2010**, *83*, 1–9.
92. Iacopetti, I.; Patruno, M.; Melotti, L.; Martinello, T.; Bedin, S.; Badon, T.; Righetto, E.M.; Perazzi, A. Autologous Platelet-Rich Plasma Enhances the Healing of Large Cutaneous Wounds in Dogs. *Front. Vet. Sci.* **2020**, *7*, 575449. [CrossRef]

animals

MDPI

Article

Plasmatic Profile of Pregnancy-Associated Glycoprotein (PAG) during Gestation and Postpartum in Sarda and Lacaune Sheep Determined with Two Radioimmunoassay Systems

Martina De Carolis [1,†], Olimpia Barbato [1,*,†], Gabriele Acuti [1], Massimo Trabalza-Marinucci [1], Noelita Melo de Sousa [2], Claudio Canali [1] and Livia Moscati [3]

1 Department of Veterinary Medicine, University of Perugia, 06126 Perugia, Italy; martina.decarolis.87@gmail.com (M.D.C.); gabriele.acuti@unipg.it (G.A.); massimo.trabalzamarinucci@unipg.it (M.T.-M.); claudio.canali@unipg.it (C.C.)

2 Laboratory of Animal Endocrinology and Reproduction, Faculty of Veterinary Medicine, University of Liege, 4000 Liege, Belgium; noelitamelo@gmail.com

3 Istituto Zooprofilattico Sperimentale dell'Umbria e delle Marche, via G. Salvemini 1, 06126 Perugia, Italy; l.moscati@izsum.it

* Correspondence: olimpia.barbato@unipg.it; Tel.: +39-075-585-7640; Fax: +39-075-585-7638

† These authors have contributed equally to this work.

Received: 29 July 2020; Accepted: 20 August 2020; Published: 25 August 2020

Simple Summary: Nowadays the need to optimize and maximize the productivity of dairy sheep leads farmers to plan lambing in specific periods in order to avoid economic loss due to nonproductive animals. The goal is to diagnose pregnancy at early gestation in order to minimize the costs of unproductive animals and to properly formulate rations for the energy requirements of gestating or lactating animals at the same time. Moreover, early pregnancy diagnosis, as well as the possibility of distinguishing between single and multiple gestations, enables farmers to plan the management of lambing well in advance thus minimizing manpower requirements. This study showed, for the first time, the plasmatic profile of PAG (pregnancy-associated glycoproteins) in Sarda and Lacaune ewes during gestation and the postpartum period using two different radioimmune assay (RIA) systems, thus enhancing our knowledge regarding PAG concentrations in domestic ruminants. Moreover, it showed that for both breeds these RIA systems are capable of distinguishing pregnant from nonpregnant ewes starting from day 18 of gestation. Furthermore, the rapid disappearance of PAG concentration following lambing did not require the use of a cut off limit in postpartum animals as a means for detecting a new pregnancy.

Abstract: This study was carried out to determine ovine pregnancy-associated glycoprotein (oPAG) levels in the plasma of Sarda and Lacaune ewes throughout gestation and in the first month postpartum, using two heterologous radioimmunoassays (RIA-706 and RIA-srPool) and to study the correlations between PAG levels and fetal gender and number. On Day 18 of pregnancy, PAG concentrations were detected in 90.1% and 80.8% of Sarda pregnant ewes with RIA-706 and RIA-srPool, respectively; and in 90% and in 75% of Lacaune pregnant ewes with RIA-706 and RIA-srPool, respectively. From Day 30, PAG concentrations were detected in all pregnant ewes by using both RIA methods. In the postpartum period, the PAG concentrations in Sarda ewes decreased rapidly reaching minimal levels (<1 ng/mL) on day 28 using both RIA-706 and RIA-srPool. In Lacaune ewes, PAG-706 levels were higher than PAG-srPool from parturition until the last day of observation (Day 28 postpartum). It was also observed that mean concentrations were higher in multiple than in single pregnancies in Sarda and Lacaune ewes. Moreover, due to the rapid disappearance of PAG concentration following lambing, a cut-off limit in postpartum animals was not required as a means for detecting a new pregnancy.

Keywords: RIA; PAG; gestation; sheep; postpartum; breed; multiple pregnancy; single pregnancy

1. Introduction

Nowadays the need to optimize and maximize the productivity of dairy sheep leads farmers to plan lambing in specific periods in order to avoid economic loss due to nonproductive animals. The goal is to diagnose pregnancy at early gestation in order to minimize the fodder wasted on unproductive animals and to properly formulate rations for the energy requirements of gestating or lactating animals at the same time. Moreover, early pregnancy diagnosis, as well as the possibility of distinguishing between single and multiple gestations, enables farmers to plan the management of lambing well in advance thus minimizing manpower requirements.

In this regard, over the past 30 years, techniques such as radioimmunoassay (RIA) for pregnancy detection have been developed using pregnancy-associated glycoproteins (PAGs) which are glycoproteins belonging to the aspartic proteinase family and are synthesized in the ruminant trophectoderm [1]. Glycoproteins associated with pregnancy in sheep are represented by PAGs (OPAG) [2–4] and SBU-3 [5]. In ovine placenta, 11 cDNA coding for distinct PAGs (ovPAG-1 to ovPAG-11) were identified at different gestational periods, thus confirming the multiplicity and temporal expression of PAG molecules in ruminant placenta [1,6–10].

These molecules are present in maternal plasma of sheep in concentrations detectable with RIA methods as early as 18–20 days after conception, [3,11–16] and by the enzyme immune assay (EIA) method [17–19], as well as in other species of ruminants [1,20–22]. Throughout pregnancy, PAG concentrations in these species differ according to the breed, fetal number, sex and birth weight [3,23–29], as in other ruminants [30].

To our knowledge, no studies have yet been carried out on PAG plasma concentration profiles during gestation and the postpartum period in Sarda and Lacaune ewes, two amongst the most important dairy sheep present in Europe.

Sarda is the main sheep breed raised in Italy and it has been selected over the years for milk production [31,32]. The breeding system is typically characterized by one lambing per year and the mating season starts in early summer (or early autumn for ewe lambs). However, especially for yearlings, total fertility rate is not higher than 75%. When lambing occurs in early spring, since the breeding system is based on pasture, milk production is positively affected by high forage availability. However, strategies to favor out-of-season lambing are encouraged to ensure cheese availability for the market over the year [33,34].

The Lacaune breed originates from the Roquefort area, Southern France. It is the main French dairy sheep breed and has been very efficiently selected during the last 40 years [35]. The Lacaune is now a high milk yield sheep that, contrary to what happens for the Sarda breed, is also appreciated for lamb growth rate and the characteristics of its meat. Moreover, out-of-season lambing is common, and the breeding system is not necessarily based on pasture, being characterized by a large use of conserved forages and concentrates.

Both breeds would benefit from an early pregnancy diagnosis to reduce economic losses and increase productivity [36].

The aim of this study was to investigate the concentrations of plasma PAG using two heterologous radioimmunoassays (RIA-706 and RIA-srPool) in Lacaune and Sarda sheep throughout gestation and after parturition, and to study the correlations between PAG levels and fetal gender and number. Correlations between concentrations measured with the aforementioned RIA systems, their ability to detect PAG molecules during pregnancy and their half-life were also investigated.

2. Materials and Methods

2.1. Animals and Samples

Thirty-five Sarda ewes weighing 42 ± 1.4 kg (mean ± SEM) at mating, and thirty-five Lacaune ewes weighing 46 ± 1.2 kg (mean ± SEM) at mating from a single flock were enrolled in this study. All of the ewes were in their first lactation during the period September–January of 2008. The ewes were housed and managed at the Azienda Zootecnica Didattica of the University of Perugia. The experimental site is approximately 15 km southwest of Perugia, Italy (latitude: 41°34′ N and longitude: 14°39′ E) at an elevation of approximately 650 m above sea level. The area has a Mediterranean climate with an annual rainfall of approximately 650 mm, distributed mainly during late autumn and winter, and mean maximum and minimum temperatures ranging between 15.7 and 8.4 °C over the last 40 years.

Blood samples were withdrawn from the jugular vein and placed into EDTA-coated tubes (Sarstedt®, Numbrecht, Germany). The samples were collected on Days 0 (day of mating), 18, 30, 45, 60, 90, 120 after mating and during the postpartum period, starting within 12 h of birth (*p*) and on Days 7, 14, 21, 28. Plasma was obtained by centrifugation (2500× *g* for 15 min) immediately after collection and was stored at −20 °C until assay.

2.2. Experimental Design

The two groups of ewes were housed in two separate 15 × 5 m straw-bedded pens. The animals were subjected to the same management practices and had feed and water readily available in feeders and troughs in order to ensure an adequate intake of food and water. The basal diet was composed of alfalfa hay (2.0–2.5 kg/head/day, depending on the breed) which was supplemented with pelleted concentrate (crude protein: 180 g/kg; neutral detergent fiber: 149 g/kg as fed) during lactation and the last two months of pregnancy (800 g and 600 g/head/day, respectively) (Mignini & Petrini Spa., Petrignano di Assisi, Perugia, Italy) to meet nutrient requirements according to Cannas (2004) [37].

The ewes were synchronized with intravaginal sponges containing 40 mg fluorogestone acetate (Cronogest sponge, Intervet, Milan, Italy) for 12 days. On sponge withdrawal, the ewes were injected (i.m.) with 350 UI PMSG (Pregnant mare serum gonadotropin) (Folligon, Intervet, Milan, Italy) and 36 h later two rams of proven fertility were introduced (September 2008) to each group for one day and the females were then separated after mating.

Conception was assumed to have occurred 149 days before parturition, which is the average gestation period for sheep. The number and sex of lambs were recorded.

The lambs were kept with the mothers until the end of the trial.

The animals in this trial were supervised in compliance with Italian laws (DL 27 January 1992, n°116) and regulations regarding experimental animals. The experimental design was performed according to good veterinary practices under farm conditions.

2.3. Progesterone Assay

The radioimmunoassay analyses (progesterone, PAG) were performed at the University of Perugia (Department of Veterinary Medicine). Samples from Sarda and Lacaune ewes were assayed for progesterone using an extraction step described elsewhere [38,39]. Progesterone was extracted from plasma with diethyl ether and the efficiency of the extraction procedure was monitored by addition of a tracer amount of (3H) progesterone. The efficiency of the ether extraction ranged from 85 to 95%. Extraction was carried out with 0.2 mL plasma and each sample was assayed in duplicate. Volumes of 0.8 mL of distilled water and 3 mL of diethyl ether were then added to each sample and centrifuged at 1000× *g* for 10 min. Following freezing, the supernatant was discarded and 1 mL of borate buffer with 10% ethanol was added to all samples. Standard curve dilution was prepared using plain tubes for total counts and nonspecific binding. A 0.1 mL volume of increasing concentrations of calibrators (P4), 0.1, 0.25, 0.5, 2, 5, 10, 20 and 40 ng/mL was added. Reference samples (0.5 ng/mL and 10 ng/mL) were also added as quality controls. ^{3}H-labeled progesterone (0.1 mL) and 0.1 mL of progesterone antibody were added to the experimental plasma samples extracted. Incubation was carried out at 4 °C for at least 4 h. The radioactivity antibody bound P4 from free P4 was separated by centrifugation following dextran-charcoal adsorption. The tubes were transferred to a beta-counter (Tri-carb 2100 TR, Packard) to be counted. The minimum detection limit (MDL) was 0.09 ng/mL. Intra and interassay coefficients of variation (CV) were 8 and 12%, respectively.

2.4. Pregnancy-Associated Glycoprotein Assays

Two different radioimmunoassay systems (RIA-706 and RIA srPool) obtained from the methods previously described in detail by Perenyi et al. [40,41] and Barbato et al. [12] were used to measure pregnancy-associated glycoprotein concentrations. All the assays were performed in Tris buffer containing 1% BSA (Fraction V, ICN Biochemicals Inc., Aurora, OH, USA). Measurements were performed in polystyrene tubes and all of the incubations were performed at room temperature (20 to 22 °C). Bovine PAG 67 kDa preparation ($boPAG_{67kDa}$, accession number Q29432) was used as standard and tracer for all assays [30,42].

The ewe samples (0.1 mL, in duplicate) were assayed in a preincubated system. In short, 0.1 mL of each sample, or appropriate standard dilution, were aliquoted into duplicate assay tubes and diluted with 0.1 mL and 0.2 mL of Tris-BSA buffer, respectively. In order to minimize nonspecific interference of plasma proteins, 0.1 mL of bovine PAG-free plasma was added to all standard tubes. Subsequently the antisera (1:80,000 for RIA-706 and 1:50,000 for RIA-srPool) were added and the tubes were incubated overnight at room temperature before adding radio-labelled PAG [40]. Iodination ($Na\text{-}I^{125}$, Amersham Pharmacia Biotech, Uppsala, Sweden) was carried out according to the Chloramine T method [43].

Samples with higher PAG concentrations than the estimated standard dose for which the percentage B/B0 was 20% (ED20) were reassayed in nonpreincubated systems in which the standard curves ranged from 0.8 to 100 ng/mL. In these systems, the tracer was added simultaneously with one of the aforementioned first antibodies (AS#706: 1/80,000 and AS#Pool: 1/50,000). The following day, the double antibody precipitation system was added and a further 30 min incubation was carried out before separation of bound and free PAG.

The minimum detection limit (MDL), calculated as the mean concentration minus twice standard deviation (mean – 2 SD) of 20 duplicates of the zero (B_0) standard [44] were, respectively, 0.4 ng/mL and 0.3 ng/mL for RIA-706 and RIA sr-Pool. Intra-assay CVs of RIA-srPool were 5.8% and 9.5%, respectively. Interassay CVs of RIA-706 and srPool were 5.0% and 3.0%, respectively.

2.5. Pregnancy Diagnosis

A cut-off value of 1.0 ng/mL was used for concentrations of progesterone or PAG (RIA-706, RIA-srPool) in order to distinguish between pregnant and nonpregnant females [12].

2.6. Statistical Analysis

Pregnancy-associated glycoprotein and progesterone concentrations were expressed as least square means ± standard error of the mean (±SEM).

The data obtained were analyzed using the GLM procedure of SAS (2013) [45]. A mixed model with repeated measures considering sheep breed (Sarda or Lacaune), RIA method (POV or 706), sampling time (day of mating and 18, 30, 45, 60, 90 and 120 days after mating; during the postpartum period starting within 12 h from lambing and then on days 7, 14, 21 and 28 of lactation), type of delivery (single or multiple) and sex of fetuses as fixed factors, and all possible interactions between main factors was used. The ewe was considered as a random factor.

Due to the small number of twin pregnancies, the effect of fetus gender were estimated for single pregnancies only.

Two interactions (sheep breed × RIA method and sheep breed × RIA method × sampling time) were removed from the model because the results of ANOVA were not statistically significant.

Differences between the least square means were evaluated by Tukey test ($p < 0.05$). Tendencies were discussed when $p > 0.05$ but ≤ 0.10.

The chi-square test was used to assess the agreement between the two RIA methods in early pregnancy (at 18 and 30 days after mating).

Finally, the elimination rate constant was calculated from the slope of the line during the postpartum period by linear regression analysis of the semilogarithmic plot of PAG concentrations versus time, while the half-life was obtained as (ln 2 /elimination rate constant).

3. Results

From a total of 35 synchronized Sarda and 35 Lacaune ewes, 26 and 20 animals became pregnant respectively as shown by the RIA analysis and the lambing rate.

Nine Sarda and 15 Lacaune sheep were diagnosed to be nonpregnant and were considered as negative controls. They gave PAG concentrations below the cut-off value.

Among the Sarda ewes, 20 had single pregnancies while six carried twins. The average length of gestation was 149.5 days (149.1 and 148.1 for ewes carrying one and two fetuses, respectively). There were 14 male fetuses.

In Lacaune ewes, 14 had single pregnancies while six carried twins. The average length of gestation was 146.0. There were 15 male fetuses.

3.1. Progesterone Concentrations

Progesterone concentrations were detectable in 26/26 (100%) and 20/20 (100%) of pregnant Sarda and Lacaune ewes, respectively. The mean progesterone level of nonpregnant Sarda and Lacaune ewes on Day 18 were 0.1 ± 0.1 ng/mL for both breeds. In pregnant females, progesterone levels ranged from 5.26 ± 0.7 ng/mL on Day 18 to 6.05 ± 0.5 ng/mL on Day 60 for Sarda ewes, and from 10.9 ± 1 ng/mL on Day 18 to 12.0 ± 0.7 ng/mL on Day 60 for Lacaune ewes.

3.2. Profiles of RIA-706 and RIA-srPool during Pregnancy and the Postpartum Period in Sarda and Lacaune Ewes

Similar PAG profiles were observed for both breeds. PAG plasma concentrations detected with RIA-706 were lower than the PAG levels detected with RIA-srPool (overall means: 21.82 ± 0.83 vs. 30.49 ng/mL ± 0.83, respectively). However, a significant ($p < 0.001$) interaction between the RIA method and sampling time was detected and differences between the two methods were only recorded on days 45, 60, 75 and 90 during pregnancy, while no differences were observed during the postpartum period (Table 1).

Table 1. Pregnancy-associated glycoprotein (PAG) values obtained before and after delivery according to radioimmune assay (RIA) methods.

Days	RIA-706	RIA-srPool	*p*-Value
Mating	0.148	0.059	0.23
18	3.735	4.775	0.69
30	16.907	24.034	0.59
45	31.298	50.189	<0.001
60	48.709	78.444	<0.001
75	44.836	69.821	<0.001
90	29.183	46.975	<0.001
120	33.852	34.171	0.33
Delivery	62.753	71.873	0.78
7	5.573	8.035	0.55
14	3.576	4.275	0.08
21	2.057	1.778	0.98
28	0.998	0.842	0.09

Regarding the effect of breed, higher plasma PAG values were observed for Lacaune ewes than for Sarda ewes (overall means: 27.41 ± 0.90 vs. 24.90 ± 0.86 ng/mL, respectively). When plotting the values of the interaction between sheep breed and sampling time together ($p < 0.001$), significant differences were only observed on days 75 and 90 (Figure 1) for both RIA systems.

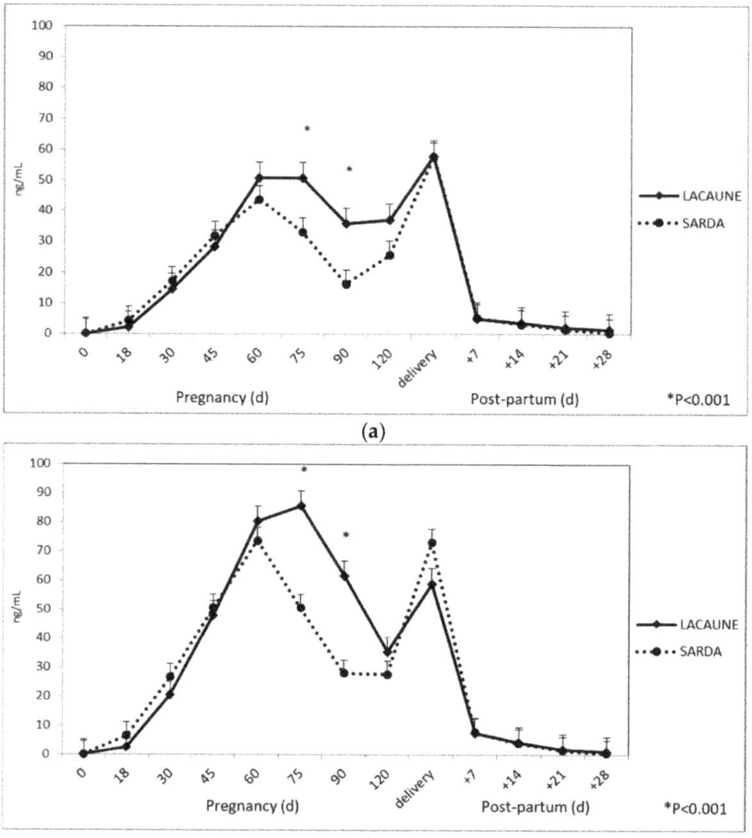

(a)

(b)

Figure 1. Plasma PAG profile in Lacaune and Sarda ewes during pregnancy and postpartum when using (**a**) RIA-706 and (**b**) RIA-srPool.

No differences were observed during the postpartum period. For both RIA systems, at 28 days postpartum recorded values fell below 1 ng/mL (cut-off ≥ 1 ng/mL).

On day 18 of pregnancy the PAG concentrations in Sarda ewes were ≥ 1 ng/mL (cut-off value) in 25/26 (96.1%) and in 22/26 (80.8%) with RIA-706 and RIA-srPool, respectively. As described in Table 2, it was possible to detect PAG in all pregnant animals as early as Day 30.

Table 2. PAG detection (cut-off ≥ 1 ng/mL) in Sarda and Lacaune ewes according to RIA method (RIA-706 vs. RIA-srPool) at day 18 and day 30 of pregnancy.

Day of Pregnancy		Sarda		Lacaune	
		RIA-706	**RIA-srPool**	**RIA-706**	**RIA-srPool**
Day 18	Nonpregnant	1	4	2	5
	Pregnant	25	22	18	15
Day 30	Nonpregnant	0	0	0	0
	Pregnant	26	26	20	20

Mean PAG-706 and PAG-srPool progressively increased from the day of conception (0.08 ± 0.02 and 0.06 ± 0.02 ng/mL, respectively) until day 60 (43.77 ± 5.64 and 73.62 ± 5.64, respectively) then decreased until stabilizing on day 90 (16.25 ± 3.64 and 27.46 ± 3.64, respectively), after which they increased and peaked on the day of delivery (56.77 ± 8.50 and 73.07 ± 8.50 ng/mL, respectively).

In this breed, both PAG-706 and PAG-srPool progressively decreased from parturition reaching values below 1 ng/mL at 28 days after lambing (0.39 ± 0.13 and 0.44 ± 0.13 ng/mL, respectively).

We used scatter plots of ln PAG-706 and ln PAG-srPool concentrations versus postpartum days to calculate the kinetic parameters. The elimination rate constants were 0.12 day^{-1} and 0.13 day^{-1} while the half-lives were 5.8 days and 5.3 days for PAG-706 and PAG-srPool, respectively.

On day 18 of pregnancy, the concentrations of PAG in Lacaune ewes were ≥ 1 ng/mL (cut-off value) in 18/20 (90%) and in 15/20 (75%) measured with RIA-706 and RIA-srPool, respectively. As described in Table 2, it was possible to detect PAG as early as day 30 in all pregnant animals.

Mean PAG-706 and PAG-srPool progressively increased from the day of conception (0.11 ± 0.03 and 0.06 ± 0.03 ng/mL, respectively) to day 75 (50.83 ± 6.65 and 85.51 ± 6.65 ng/mL, respectively), then decreased until day 120 (37.14 ± 3.34 and 35.33 ± 3.34 ng/mL, respectively), after which they increased and peaked on the day of delivery (57.86 ± 9.69 and 58.95 ± 9.69, respectively).

Both PAG-706 and PAG-srPool progressively decreased following parturition, but during the postpartum period at 28 days after lambing, PAG concentrations once again exceeded 1 ng/mL (1.45 ± 0.15 and ± 1.04 ± 0.15 ng/mL, respectively).

We used scatter plots of ln PAG-706 and ln PAG-srPool concentrations versus postpartum days to calculate the kinetic parameters. The elimination rate constants were 0.10 day^{-1} and 0.11 day^{-1} while the half-lives were 6.9 days and 6.3 days for PAG-706 and PAG-srPool, respectively.

3.3. *Effects of Single Versus Multiple Pregnancies and Gender on PAG Concentrations*

At delivery, in Sarda ewes multiple pregnancies were characterized by higher levels of PAG than single pregnancies (Figures 2 and 3) disregarding RIA system (significant interaction breed × type of delivery × sampling time).

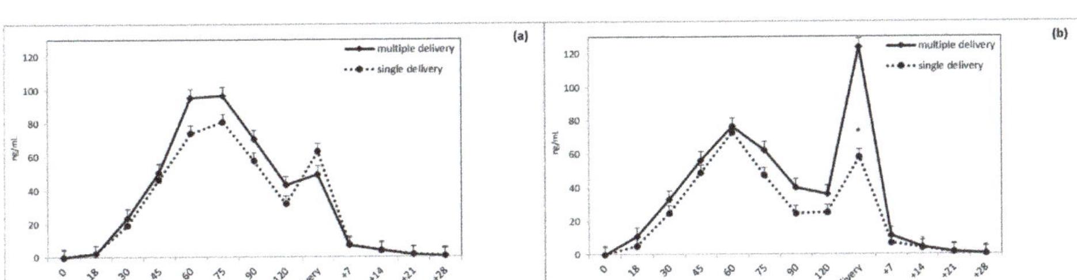

Figure 2. Plasma PAG profile in multiple and single pregnancy at delivery in Sarda ewes when using (**a**) RIA-706 and (**b**) RIA-srPool.

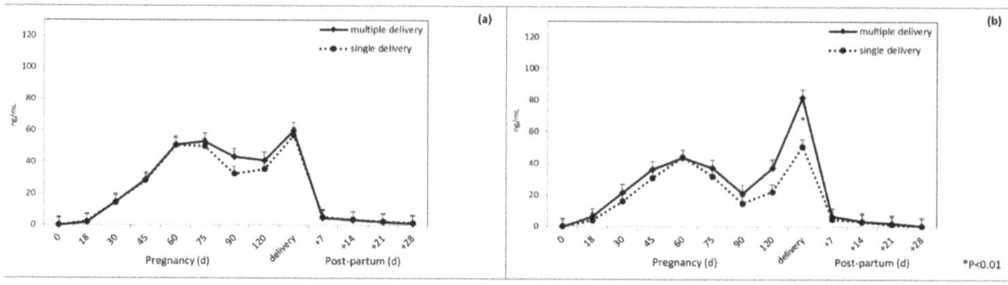

Figure 3. Plasma PAG profile in multiple and single pregnancy at delivery in Lacaune ewes when using (**a**) RIA-706 and (**b**) RIA-srPool.

The PAG levels in plasma tended ($p = 0.064$) to be affected by the gender of the fetus (24.37 ± 0.78 vs. 22.26 ± 0.84 for males and females, respectively). There were no significant interactions between gender of the fetus, sheep breed and RIA method.

4. Discussion

To our knowledge, this is the first study to be carried out on PAG plasma concentration profiles during gestation and the postpartum period in Sarda and Lacaune ewes, using two different radioimmunoassays, during the whole period of gestation and postpartum.

The PAG profile was similar in both breeds and in both RIA systems used. During the gestation period, PAG concentrations increased up to 60 days and then decreased until 120 days. Thereafter there was a significant rise, which peaked at parturition. This is a similar trend to that described by Gajewski et al. [46] for Berrichon ewes, by Ledezma-Torres et al. [14] for various sheep breeds and by Ranilla et al. [11,24] for Churra sheep, yet this trend was not observed for Merinos ewes whose PAG levels initially increased and then dropped to baseline concentrations around mid-pregnancy. Patterns of plasma ovine PAG concentrations differed from those reported for bovine [30], goats [47,48] and buffalo [20,49]. These differences could be explained by the ability of the antisera to distinguish between various epitopes [20].

In the first period of gestation, our data showed that both RIA (RIA-706 and RIA-srPool) are capable of diagnosing pregnancy as early as 18 days of gestation. Similarly, Ayad et al. (2007) [50] reported that it is easier to distinguish between pregnant and nonpregnant cows if a mixture of different antisera is used.

In Lacaune ewes, PAG concentrations measured by using RIA-706 were higher than those measured by RIA-srPool from parturition until the last day of observation (28 days postpartum). In Sarda ewes concentrations decreased rapidly reaching minimal levels (<1 ng/mL) at day 28 when using both RIA-706 and RIA-srPool. A similar decrease in PAG concentrations during the first month postpartum was reported by Ranilla et al. [11,24] for ewes, by Sousa et al. [41] and Gonzalez et al. [51] for goats, and for wild ruminants by Ropstad et al. [52] and Osborn et al. [53]. In these species, PAG concentration drops below 1 ng/mL at day 30 postpartum. In cows, PAG concentrations slowly decreased following parturition and were detected as late as 100 days postpartum [30]. The rapid decrease in PAG concentration during the postpartum period is essential when using PAG as an appropriate marker of pregnancy immediately after postpartum. Unlike in cows, the rapid PAG disappearance in ewes does not require the use of a cut-off limit in postpartum animals as a means for detecting a new pregnancy.

The half-life proved to be longer in Sarda and Lacaune sheep (6.3 to 6.9) than in other ovine breeds as reported by Haugejorden et al., 2006 [54] (4.5 days), or in buffalo females (5.8 days) [21,49]. However, it was shorter than in bovine species (from 7.0 to 8.8 days) [12,54–56] in goats (7.5 days) and Zebu (9.2–10.1 days) [57]. These differences could be due to the presence of N-linked carbohydrate and sialic acid chains on their structure [58].

This study demonstrated that mean PAG concentrations were higher in multiple pregnancies than in single pregnancies, at delivery in Sarda ewes. The higher concentrations in multiple delivery with respect to single delivery are possibly due to the higher member of attachment points, and thus secretory activity, of twin placentas [24]. Regarding the effect of the litter size on PAG concentration, our results showed for the Sarda sheep an evident peak of concentration at delivery, compared to Lacaune sheep, whether the RIA-706 or RIA-srPool was used. This behavior differs from that found in other breeds of sheep [3]. The structure of the placenta could be at the origin of this difference. In sheep before birth, a decrease in the number of binucleated cells is observed [59], and probably in the Sardinian sheep it could undergo a further increase in the prebirth period similar to what happens in cattle [42] and in Churra sheep [11]. However, a different profile during pregnancy was also found in the goat [60] and in the cow [61], depending on the breed.

From our data it was impossible to predict litter size. These results are in agreement with those of various authors [3,12–14,16,24,28] who were unable to predict litter size from PAG concentrations.

Our data show that PAG plasma levels tended to be affected by the gender of the fetus. This difference may be due to the weight of the placenta that differs according to gender [62,63]. Our results are in agreement with those reported by Guilbault et al. [64] and Zoli et al. [30] for bovine species, and by Ranilla et al. [11] for Churra ewes but not for Merinos ewes, and are in contrast to those reported by Vandale et al. [13] for Suffolk and Texel breeds, for which no significant differences were observed between ewes carrying fetuses of different gender. Moreover, our results suggest that the breed and the gender of the fetus could influence ovine PAG production.

5. Conclusions

In conclusion, this study showed, for the first time, the plasmatic profile of PAG in Sarda e Lacaune ewes during gestation and the postpartum period using two different RIA-systems, thus enhancing our knowledge regarding PAG concentrations in domestic ruminants. Moreover, it showed that for both breeds, these RIA systems are capable of distinguishing pregnant from nonpregnant ewes starting from day 18 of gestation. Furthermore, the rapid disappearance of PAG concentration following lambing did not require the use of a cut-off limit in postpartum animals as a means for detecting a new pregnancy.

Author Contributions: Conceptualization and writing, M.D.C.; conceptualization and writing, O.B.; sample collection. G.A.; formal analysis and data curation, M.T.-M.; supervision, N.M.d.S.; laboratory analysis, C.C.; funding acquisition and supervision, L.M.; M.D.C. and O.B.: equal contributors. All authors have read and agreed to the published version of the manuscript.

Funding: This research was funded by the Italian Ministry of Health (IZS 10/08 RC).

Acknowledgments: All authors have read and agreed to the published version of the manuscript. The feed used in the trial was purchased from Mignini & Petrini Spa (Petrignano di Assisi, Italy).

Conflicts of Interest: The authors declare no conflict of interest.

References

1. Wallace, R.M.; Pohler, K.J.; Smith, M.F.; Green, J.A. Placental PAGs: Gene origins, expression patterns, and use as markers of pregnancy. *Reproduction* **2015**, *149*, R115–R126. [CrossRef] [PubMed]
2. Xie, S.; Low, B.G.; Nagel, R.J.; Kramer, K.K.; Anthony, R.V.; Zoli, A.P.; Beckers, J.F.; Roberts, R.M. Identification of the major pregnancy-specific antigens of cattle and sheep as inactive members of the aspartic proteinase family. *Proc. Natl. Acad. Sci. USA* **1991**, *88*, 10247–10251. [CrossRef]
3. Willard, J.M.; White, D.R.; Wesson, C.A.R.; Stellflug, J.; Sasser, R.G. Detection of fetal twins in sheep using a radioimmunoassay for pregnancy-specific protein B. *J. Anim. Sci.* **1995**, *73*, 960–966. [CrossRef] [PubMed]
4. El Amiri, B.; Remy, B.; Sousa, N.M.; Joris, B.; Otthiers, N.G.; Perény, Z.; Banga-Mboko, H.; Beckers, J.F. Isolation and partial characterization of three pregnancy-associated glycoproteins from the ewe placenta. *Mol. Reprod. Dev.* **2003**, *64*, 199–206. [CrossRef] [PubMed]
5. Gogolin-Ewens, K.J.; Lee, C.S.; Mercer, W.R.; Moseby, A.M.; Brandon, M.R. Characterization of a sheep trophoblast-derived antigen first appearing at implantation. *Placenta* **1986**, *7*, 243–255. [CrossRef]

6. Xie, S.; Green, J.A.; Bao, B.; Beckers, J.F.; Valdez, K.E.; Hakami, L.; Roberts, R.M. Multiple pregnancy-associated glycoproteins are secreted by day 100 ovine placental tissue. *Biol. Reprod.* **1997**, *57*, 1384–1393. [CrossRef]
7. Xie, S.; Green, J.; Bixby, J.B.; Szafranska, B.; De Martini, J.C.; Hecht, S.; Roberts, R.M. The diversity and evolutionary relationships of the pregnancy-associated glycoproteins, an aspartic proteinase subfamily consisting of many trophoblast-expressed genes. *Proc. Natl. Acad. Sci. USA* **1997**, *94*, 12809–12816. [CrossRef]
8. Green, J.A.; Xie, S.; Quan, X.; Bao, B.; Gan, X.; Mathialagan, N.; Roberts, R.M. Pregnancy-Associated glycoproteins exhibit spatially and temporally distinct expression patterns during pregnancy. *Biol. Reprod.* **2000**, *62*, 1624–1632. [CrossRef]
9. Garbayo, J.M.; Green, J.A.; Manikkam, M.; Beckers, J.F.; Kiesling, D.O.; Ealy, A.D.; Roberts, R.M. Caprine pregnancy-associated glycoproteins (PAG): Their cloning, expression, and evolutionary relationship to other PAG. *Mol. Reprod. Dev.* **2000**, *57*, 311–322. [CrossRef]
10. Touzard, E.; Reinaud, P.; Dubois, O.; Guyader-Joly, C.; Humblot, P.; Ponsart, C.; Charpigny, J. Specific expression patterns and cell distribution of ancient and modern PAG in bovine placenta during pregnancy. *Reproduction* **2013**, *146*, 347–362. [CrossRef]
11. Ranilla, M.J.; Sulon, J.; Carro, M.D.; Mantecon, A.R.; Beckers, J.F. Plasmatic profiles of pregnancy–associated glycoprotein and progesterone levels during gestation in Churra and Merino sheep. *Theriogenology* **1994**, *42*, 537–545. [CrossRef]
12. Barbato, O.; Sousa, N.M.; Debenedetti, A.; Canali, C.; Todini, L.; Beckers, J.F. Validation of a new pregnancy-associated glycoprotein radioimmunoassay method for the detection of early pregnancy in ewes. *Theriogenology* **2009**, *72*, 993–1000. [CrossRef] [PubMed]
13. Vandaele, L.; Verberckmoes, S.; El Amiri, B.; Sulon, J.; Duchateau, L.; Van Soom, A.; Beckers, J.F.; de Kruift, A. Use of homolougous radioimmunoassay (RIA) to evaluate the effect of maternal and foetal parameters on pregnancy-associated glycoprotein (PAG) concentrations in sheep. *Theriogenology* **2005**, *63*, 1914–1924. [CrossRef] [PubMed]
14. Ledezma-Torres, R.A.; Beckers, J.F.; Holtz, W. Assessment of plasma profile of pregnancy-associated glycoprotein (PAG) in sheep with a heterologous (anti-caPAG22 + 59) RIA and its potential for diagnosing pregnancy. *Theriogenology* **2006**, *66*, 906–912. [CrossRef]
15. El Amiri, B.; Sulon, J.; Karen, A.; Alvarez, A.V.; Cogniè, Y.; Sousa, N.M.; Szenci, O.; Beckers, J.F. Measurement of ovine pregnancy-associated glycoprotein (PAG) during early pregnancy in Lacaune sheep. *Reprod. Domest. Anim.* **2007**, *42*, 257–262. [CrossRef] [PubMed]
16. El Amiri, B.; Sousa, N.M.; Alvarez Oxiley, A.; Hadarbach, D.; Beckers, J.F. Pregnancy-Associated glycoprotein (PAG) concentration in plasma and milk samples for early pregnancy diagnosis in Lacaune dairy sheep. *Res. Vet. Sci.* **2015**, *99*, 30–36. [CrossRef]
17. Rovani, M.T.; Skrebsky Cezar, A.; Lazzari Rigo, M.; Garziera Gasperin, B.; da Nobrega, J.E.; Dias Torres, F.; Bayard Dias Goncalves, P.; Ferreira, R. Evaluation of a bovine pregnancy glycoprotein enzyme-linked immunosorbent assay kit for serological diagnosis of pregnancy in sheep. *Cienc. Rural* **2016**, *46*, 362–367. [CrossRef]
18. De Miranda e Silva Chaves, C.; da Costa, R.L.D.; Duarte, K.M.R.; Manchado, M.C.; de Paz, C.C.P.; Beltrame, R.T. Visual ELISA for detection of pregnancy-associated glycoproteins (PAGs) in ewes serum. *Reprod. Domest. Anim.* **2017**, *97*, 78–82. [CrossRef]
19. Steckeler, P.; Weber, F.; Zerbe, H.; Rieger, A.; Voigt, K. Evaluation of a bovine visual pregnancy test for the detection of pregnancy-associated glycoproteins in sheep. *Reprod. Domest. Anim.* **2019**, *54*, 280–288. [CrossRef]
20. Barbato, O.; Melo de Sousa, N.; Barile, V.L.; Canali, C.; Beckers, J.F. Purification of pregnancy associated glycoproteins from late pregnancy Bubalus bubalis placentas and development of a radioimmunoassay for pregnancy diagnosis in water buffalo females. *BMC Vet. Res.* **2013**, *9*, 89. [CrossRef]
21. Barbato, O.; Menchetti, L.; Sousa, N.M.; Malfatti, A.; Brecchia, G.; Canali, C.; Beckers, J.F.; Barile, V.L. Pregnancy-Associated glycoproteins (PAGs) concentrations in water buffaloes (Bubalus bubalis) during gestation and the postpartum period. *Theriogenology* **2017**, *97*, 73–77. [CrossRef] [PubMed]
22. Beriot, M.; Tchimbou, A.F.; Barbato, O.; Beckers, J.F.; de Sousa, N.M. Identification of pregnancy-associated glycoproteins and alpha protein in fallow deer (Dama dama) placenta. *Acta Vet. Scandinava* **2014**, *56*, 4. [CrossRef] [PubMed]

23. Mialon, M.M.; Camous, S.; Renand, G.; Martal, J.; Ménissier, F. Peripheral gestation of a 60_kDa pregnancy serum protein during gestation and after calving and in relationship to embryonic mortality in cattle. *Reprod. Nutr. Dev.* **1993**, *33*, 269–282. [CrossRef]
24. Ranilla, M.J.; Sulon, J.; Mantecon, A.R.; Beckers, J.F.; Carro, M.F. Plasma pregnancy-associated glycoprotein and progesterone concentrations in pregnant Assaf ewes carrying single and twin lambs. *Small Rumin. Res.* **1997**, *24*, 125–131. [CrossRef]
25. Karen, A.; El Amiri, B.; Beckers, J.F.; Sulon, J.; Taverne, M.A.M.; Szenzi, O. Comparison of accuracy of transabdominal ultrasonography, progesterone and pregnancy-associated glycoproteins tests for discrimination between single or multiple pregnancy in sheep. *Theriogenology* **2006**, *66*, 314–322. [CrossRef]
26. Mercadante, P.M.; Waters, K.M.; Mercadante, V.G.R.; Lamb, G.C.; Elzo, M.A.; Johnson, M.E.; Rae, D.O.; Yelich, J.V.; Ealy, A.D. Subspecies differences in early fetal development and plasma pregnancy-associated glycoprotein concentrations in cattle. *J. Anim. Sci.* **2013**, *91*, 3693–3701. [CrossRef]
27. Lobago, F.; Bekana, M.; Gustafsson, H.; Beckers, J.F.; Yohannes, G.; Aster, Y.; Kindahl, H. Serum Profiles of Pregnancy-Associated Glycoprotein, Oestrone Sulphate and Progesterone During Gestation and Some Factors Influencing the Profiles in Ethiopian Borana and Crossbred Cattle. *Reprod. Domest. Anim.* **2009**, *44*, 685–692. [CrossRef]
28. Patel, O.; Sulon, J.; Beckers, J.F.; Takahashi, T.; Hirako, M.; Sasaki, N.; Domeki, I. Plasma bovine pregnancy-associated glycoprotein concentrations throughout gestation in relationship to fetal number in the cow. *Eur. J. Endocrinol.* **1997**, *137*, 423–428. [CrossRef]
29. Wallace, J.M.; Aitken, R.P.; Cheyne, M.A.; Humblot, P. Pregnancy-Specific protein B and progesterone concentrations in relation to nutritional regimen, placental mass and pregnancy outcome in growing adolescent ewes carrying singleton fetuses. *J. Reprod. Fertil.* **1997**, *109*, 53–58. [CrossRef]
30. Zoli, A.P.; Guilbault, L.A.; Delahaut, P.; Ortiz, W.B.; Beckers, J.F. Radioimmunoassay of a bovine pregnancy-associated glycoprotein in serum: Its application for pregnancy diagnosis. *Biol. Reprod.* **1992**, *46*, 83–92. [CrossRef]
31. Macciotta, N.P.P.; Mele, M.; Cappio-Borlino, A.; Secchiari, P. Issues and perspectives in dairy sheep breeding. *Ital. J. Anim. Sci.* **2005**, *4*, 5–23. [CrossRef]
32. Carta, A.; Casu, S.; Salaris, S. Invited review: Current state of genetic improvement in dairy sheep. *J. Dairy Sci.* **2009**, *92*, 5814–5833. [CrossRef] [PubMed]
33. Sitzia, M.; Bonanno, A.; Todaro, M.; Cannas, A.; Atzori, A.S.; Francesconi, A.H.D.; Trabalza-Marinucci, M. Feeding and management techniques to favour summer sheep milk and cheese production in the Mediterranean environment. *Small Rumin. Res.* **2015**, *126*, 43–58. [CrossRef]
34. Todaro, M.; Dattena, M.; Acciaioli, A.; Bonanno, A.; Bruni, G.; Caroprese, M.; Mele, M.; Sevi, A.; Trabalza-Marinucci, M. Aseasonal sheep and goat milk production in the Mediterranean area: Physiological and technical insights. *Small Rumin. Res.* **2015**, *126*, 59–66. [CrossRef]
35. Barillet, F.; Marie, C.; Jacquin, M.; Lagriffoul, G.; Astruc, J.M. The French Lacaune dairy sheep breed: Use in France and abroad in the last 40 years. *Livest. Prod. Sci.* **2001**, *71*, 17–29. [CrossRef]
36. Karen, A.; Beckers, J.F.; Sulon, J.; El amiri, B.; Szabados, S.; Ismail, S.; Reiczigel, J.; Szenci, O. Evalaution of false transrectal ultrasonographic pregnancy diagnoses in sheep by measuring tha plasma level of pregnancy-associated glycoproteins. *Reprod. Nutr. Dev.* **2003**, *43*, 577–586. [CrossRef] [PubMed]
37. Cannas, A. Feeding of lactating ewes. In *Dairy Sheep Nutrition*; Pulina, G., Ed.; CABI: Bologna, Italy, 2004; pp. 79–108.
38. Todini, L.; Malfatti, A.; Barbato, O.; Costarelli, S.; Debenedetti, A. Progesterone plus PMSG priming in seasonally anovulatory lactating Sarda ewes exposed to the ram effect. *J. Reprod. Dev.* **2007**, *53*, 437–441. [CrossRef]
39. Todini, L.; Marinucci, M.T.; Malfatti, A.; Barbato, O.; Cavallucci, C.; Debenedetti, A. Pre and post-feed ind plasma gastrin-17 and insulin concentrations and feed-intake in female goats during different physiological stages. *Small Rumin. Res.* **2007**, *71*, 38–47. [CrossRef]
40. Perenyi, Z.; Szenci, O.; Sulon, J.; Drion, P.V.; Beckers, J.F. Comparison of the ability of three radioimmunoassays to detect pregnancy-associated glycoproteins in bovine plasma. *Reprod. Domest. Anim.* **2002**, *37*, 100–104. [CrossRef]

41. Perenyi, Z.; Szenci, O.; Drion, P.V.; Banga-Mboko, H.; Sousa, N.M.; El Amiri, B.; Beckers, J.F. Aspartic proteinase members secreted by the ruminant placenta: Specificity of three radioimmunoassay systems for the measurement of pregnancy-associated glycoproteins. *Reprod. Domest. Anim.* **2002**, *37*, 324–329. [CrossRef]
42. Zoli, A.P.; Beckers, J.F.; Wouters-Ballman, P.; Closset, J.; Falmagne, P.; Ectors, F. Purification and characterization of a bovine pregnancy-associated glycoproteins. *Biol. Reprod.* **1991**, *45*, 1–10. [CrossRef]
43. Greenwood, F.C.; Hunter, W.M.; Glover, J.S. The preparation of 131-I labeled human growth hormone of high specific radioactivity. *Biochemistry* **1963**, *89*, 114–123.
44. Skelley, D.S.; Brown, L.P.; Besch, P.K. Radioimmunoassay. *Clin. Chem.* **1973**, *2*, 146–186. [CrossRef]
45. Statistical Analysis Software [SAS]. *SAS/STAT User Guide, Version 9.4*; SAS Institute Inc.: Cary, NC, USA, 2013.
46. Gajewski, Z.; Beckers, J.F.; Sousa, N.M.; Thun, R.; Sulon, J.; Foundez, R. Determination of pregnancy associated glycoprotein concentration in sheep. A retrospective study. *Adv. Cell Biol.* **1999**, *2*, 89–96.
47. Sousa, N.M.; Garbayo, J.M.; Figueiredo, J.R.; Sulon, J.; Goncalvesa, P.B.D.; Beckers, J.F. Pregnancy-Associated glycoprotein and progesterone profiles during pregnancy and postpartum in native goats from the north-east of Brazil. *Small Rumin. Res.* **1999**, *32*, 137–147. [CrossRef]
48. Gonzalez, F.; Sulon, J.; Garbayo, J.M.; Batista, M.; Cabrera, F.; Calero, P.O.; Gracia, A.; Beckers, J.F. Secretory profiles of pregnancy associated glycoproteins at different stages. *Reprod. Domest. Anim.* **2000**, *35*, 79–82. [CrossRef]
49. Barbato, O.; Menchetti, L.; Sousa, N.M.; Brecchia, G.; Malfatti, A.; Canali, C.; Beckers, J.F.; Barile, V.L. Correlation of two radioimmunoassay systems for measuring pregnancy-associated glycoproteins plasma concentration during early pregnancy and post-partum period in water buffalo. *Reprod. Domest. Anim.* **2018**, *53*, 1483–1490. [CrossRef]
50. Ayad, A.; Sousa, N.M.; Sulon, J.; Iguer-Ouada, M.; Beckers, J.F. Comparison of five radioimmunoassay systems for PAG measurement: Ability to detect early pregnancy in cows. *Reprod. Domest. Anim.* **2007**, *4*, 433–440. [CrossRef] [PubMed]
51. Gonzalez, F.; Sulon, J.; Garbayo, J.M.; Batista, M.; Cabrera, F.; Calero, P.O.; Gracia, A.; Beckers, J.F. Early pregnancy diagnosis in goats of pregnancy associated glycoproteins in plasma samples. *Theriogenology.* **1999**, *52*, 717–725. [CrossRef]
52. Ropstad, E.; Veiberg, V.; Sakkinen, H.; Dahl, A.; Kindahl, H.; Holand, O.; Beckers, J.F.; Eloranta, E. Endocrinology of pregnancy and early pregnancy detection by reproductive hormones in reindeer (Rangifer tarandus tarandus). *Theriogenology* **2005**, *63*, 1775–1788. [CrossRef]
53. Osborn, A.; Beckers, J.F.; Sulon, J.; Gassett, J.W.; Muller, L.I.; Murphy, B.P.; Miller, K.V.; Marchinton, L.R. Use of glycoprotein assays for pregnancy diagnois in White-Tailed Deer. *J. Wildl. Manag.* **1996**, *60*, 388–393. [CrossRef]
54. Haugejorden, G.; Wage, S.; Dahl, E.; Karlberg, K.; Beckers, J.F.; Ropstad, E. Pregnancy associated glycoproteins (PAG) in postpartum cows, ewes, goats and their offspring. *Theriogenology* **2006**, *66*, 1976–1984. [CrossRef] [PubMed]
55. Sasser, R.G.; Ruder, C.A.; Kristen, A.I.; Butler, J.E.; Hamilton, W.C. Detection of pregnancy by novel pregnancy-specific protein in serum of cows and a profile of serum concentration during gestation. *Biol. Reprod.* **1986**, *35*, 936–942. [CrossRef] [PubMed]
56. Kirakofe, G.H.; Wright, J.M.; Schalles, R.R.; Ruder, C.A.; Parish, S.; Sasser, R.G. Pregnancy-Specific protein B in serum of postpartum beef cows. *J. Anim. Sci.* **1997**, *71*, 2199–2205. [CrossRef]
57. Sousa, N.M.; Zongo, M.; Pitala, W.; Boly, H.; Sawadogo, L.; Sanon, M.; Figuereido, J.R.; Concalves, P.B.D.; El Amiri, B.; Prenényi, Z.; et al. Pregnancy-Associated glycoprotein concentrations during pregnancy and post partum period in Azawak Zebu Cattle. *Theriogenology* **2003**, *59*, 1131–1142. [CrossRef]
58. Klisch, K.; Boos, A.; Friedrich, M.; Herzog, K.; Feldmann, M.; Sousa, N.M.; Beckers, J.F.; Leiser, R.; Schuler, G. The glycosylation of pregnancy-associated glycoproteinsand prolactin-related protein-I in bovine binucleate trophoblastgiant cells changes before parturition. *Reproduction* **2006**, *132*, 791–798. [CrossRef]
59. Wooding, F.B.P. Current topic: The synepitheliochorial placenta of ruminants: Binucleate cell fusions and hormone production. *Placenta* **1992**, *12*, 101–103. [CrossRef]
60. Batalha, E.S.; Sulon, J.; Figueiredo, E.S.; Beckers, J.F.; Espechit, C.J.B.; Martins, R.; Silva, L.D.M. Plasma profile of pregnancy associated glycoprotein (PAG) in alpine goats using two radioimmunoassay (RIA) systems. *Small Rumin. Res.* **2001**, *42*, 111–118. [CrossRef]

61. Shahin, M.; Friedrich, M.; Gauly, M.; Holtz, W. Pregnancy-Associated glycoprotein (PAG) profile of Holstein–Friesian cows as compared to dual-purpose and beef cows. *Reprod. Domest. Anim.* **2014**, *49*, 618–620. [CrossRef]
62. Greenwood, P.L.; Slepetis, R.M.; Bell, A.W. Influences on fetal and placental weights during mid to late gestation in prolific ewes well nourished throughout pregnancy. *Reprod. Fertil. Dev.* **2000**, *12*, 149–156. [CrossRef]
63. Kaulfusb, K.H.; Schramm, D.; Berttram, M. Effects of genotype, dams age, litter size, birth weight and ram on morphological parameters of the placenta sheep. *DTW* **2000**, *107*, 269–275.
64. Guilbault, L.A.; Roy, G.L.; Beckers, J.F.; Dufour, J.S. Influence of breed of fetus periparturient endocrine responses and subsequent milk production of Ayrshire dams. *J. Dairy Sci.* **1990**, *73*, 2766–2773. [CrossRef]

animals

MDPI

Article

Effects of Feed Supplementation on Nesfatin-1, Insulin, Glucagon, Leptin, T3, Cortisol, and BCS in Milking Ewes Grazing on Semi-Natural Pastures

Olimpia Barbato [1,†], Elena De Felice [2,*,†], Luca Todini [2], Laura Menchetti [3], Alessandro Malfatti [2,*] and Paola Scocco [2]

1 Department of Veterinary Medicine, University of Perugia, via San Costanzo 4, 06126 Perugia, Italy; olimpia.barbato@unipg.it
2 School of Biosciences and Veterinary Medicine, University of Camerino, via Gentile III da Varano, 62032 Camerino, Italy; luca.todini@unicam.it (L.T.); paola.scocco@unicam.it (P.S.)
3 Department of Agricultural and Food Science, University of Bologna, viale G. Fanin 44, 40137 Bologna, Italy; laura.menchetti7@gmail.com
* Correspondence: elena.defelice@unicam.it (E.D.F.); alessandro.malfatti@unicam.it (A.M.)
† Olimpia Barbato and Elena De Felice shared first co-autorship.

Simple Summary: In Central Italy Apennine sheep represent the most bred species, which is still bred today in a semi-extensive manner, exploiting the natural pastures. However, the increase in summer aridity induces a decrease in the pastoral value of the grasslands, resulting in poor animal performance and production. Only research-derived innovation can support farmers' economy in order to maintain the pastoral activities. The objective of this study was to evaluate the effects of cereals supplementation on body condition score and metabolic hormones profile in milking ewes grazing on semi-natural pastures on mid mountain rangelands. Our findings showed that feed supplementation preserves grazing ewes from the usual lowering of the body state associated to the lactation period and sustains the metabolic status of animals. Taken together, our results could represent a helpful instrument in the farm management practices.

Abstract: This study aimed to investigate the effects of feed supplementation on body condition score (BCS) and different metabolic hormones concentration in lactating sheep reared in Italian Central Apennine pastures during the grazing summer period. In this study, 24 Comisana x Appenninica pluriparous ewes from June until August were divided into two homogenous groups: the control group (UNS) was free to graze, while the other group (SUP), in addition to grazing, was supplemented with 600 g/day/head of cereals. At the start of the supplementation and at an interval of 9–10 days until the end of experimentation, BCS evaluation and blood withdrawal to assay nesfatin-1, insulin, glucagon, leptin, triiodothyronine and cortisol levels were performed. Univariable analysis showed no remarkable differences between the groups, while multivariable analysis suggested that the UNS group was characterized by a lower BCS and greater nesfatin-1 than the SUP group. These findings can be considered in relation to milk production, which shows a clear better persistence in the SUP group. Our results indicate that nutritional supplementation has protected ewes from the usual lowering of the body state linked to lactation and provides a good maintenance of milk production, determining also a better overall body and metabolic state of the animals, which is important at the beginning of the sexual season.

Keywords: feed supplementation; *Ovis aries*; metabolic hormones; BCS; drought stress; Italian Central Apennine

Citation: Barbato, O.; De Felice, E.; Todini, L.; Menchetti, L.; Malfatti, A.; Scocco, P. Effects of Feed Supplementation on Nesfatin-1, Insulin, Glucagon, Leptin, T3, Cortisol, and BCS in Milking Ewes Grazing on Semi-Natural Pastures. *Animals* **2021**, *11*, 682. https://doi.org/10.3390/ani11030682

Academic Editor: Mariangela Caroprese

Received: 3 February 2021
Accepted: 28 February 2021
Published: 4 March 2021

1. Introduction

It is well known how the feeding strategies and the state of nutrition are fundamental to improve productions in the livestock species. In the sub-Mediterranean areas, drought stress, due to the increasing summer aridity, is progressively decreasing the pastoral value of natural and semi-natural pastures used as a trophic resource for ovine semi-extensive breeding, thus negatively reflecting on animal status [1–3]. This fact could be taken into account in the management of flocks reared in Apennine pastures during the spring–early autumn period, in which global warming is causing an advance of both the pasture flowering peak and the pasture dryness [4,5].

A better nutrition status can improve the productive and reproductive efficiency through numerous factors, including the circulation of hormones as well as nutrient-sensitive metabolites [6]. It has been stamped that endocrine and neuroendocrine events play a fundamental role in regulating food intake and energy homeostasis. As a consequence of the absorption of nutrients and/or changed metabolism, the blood concentration of metabolic hormones can be changed. Furthermore, the process of digestion can induce changes in different hormones due to mechanical or chemical stimulation of receptors in the digestive tract.

Unlike other metabolic hormones, nesfatin (NES1) was identified only relatively recently, and there is still poor knowledge about its role in metabolism in ruminant animals. It was recognized in 2006 as a potent anorexigenic peptide involved in the regulation of homeostatic feeding [7]. Subsequently, its widespread central and peripheral distribution gave rise to additional effects. In fact, NES1 exerts pleiotropic actions at the levels of digestive systems [8], energy and glucose homeostasis [9]. Moreover, it is involved in stress response [10], sleep [11] and reproduction [12,13] in both humans and different animal species.

Pancreatic insulin (INS) is a main endocrine signal in the control of nutrient partitioning and the metabolism of carbohydrates, proteins and lipids. It has a role in homeorhetic processes, which allow the animal to adapt the nutrient partitioning to changes of physiological states and nutrient requirements [14]. In lactating dairy goats, plasma INS levels were positively correlated with energy intake and negatively with dry matter intake [15]. In lactating dairy cows, insulin positively regulates plasma leptin [16].

Glucagon, the other major pancreatic hormone, plays a key role in glucose homeostasis due to its stimulating effect on hepatic glucose output in response to low blood glucose levels [17]. Glucagon has also been suggested to stimulate lipolysis in adipose tissue [18] and to provide a satiety signal [19].

Leptin (LEP), the hormone mainly produced by the adipose tissue, regulates food intake and energy expenditure [20]. Leptin expression and secretion are correlated with body condition, physiological status (puberty, pregnancy, lactation), age, and level of nutrition [21,22]. Furthermore, the presence of leptin receptors at the mammary level [23] suggests a potential role in stimulating mammary development and functions.

The appropriate thyroid gland function is considered crucial to sustain the productive performance in domestic animals (growth, milk, or hair fiber production). Variations of thyroid hormones bioactivity allow animals to adapt their metabolic balance to different environmental conditions, changes in nutrient requirements and availability, and to homeorhetic changes during different physiological stages [24].

The main functions of adrenal cortisol (COR) are to induce protein and fat catabolism, supporting gluconeogenesis. COR makes available body reserves inducing hyperglycemia and providing energy during the stress response. In the current context of climate changes, heat stress can involve animals also in Mediterranean areas, during the summer hotter months, when dairy small ruminants are lactating and nutrient supplies are not always optimal. The effect of heat and nutritional stress on COR secretion is well documented in sheep [25].

The body condition score (BCS) is a more sensitive indicator of the nutritional status of the animals than body weight, which carries also the contribution of the gastrointestinal contents and can lead to an incorrect estimation (over- or under-) of the animal status and welfare [26].

Therefore, our study was undertaken to investigate the effects of feed supplementation during the second phase of lactation on body condition and different metabolic hormones blood concentration in lactating sheep reared in the Italian Central Apennine pastures during the summer late spring–early autumn grazing period.

In addition, to our knowledge, until now, there are no studies relating to nutritional status and nesfatin plasma concentration in sheep. Findings resulting from our research could have some future benefits, such as improving the evaluation of the opportunity to design the most appropriate diets or if a supplementation strategy can be considered an appropriate investment by the breeders.

2. Materials and Methods

2.1. Location, Animals, and Diets

The trial was carried out on 24 Comisana × Appenninica pluriparous ewes. Lactating ewes were free to graze from the beginning of June to the end of August on a semi-natural pasture located in Central Appennine (Monte Cavallo, altitude 1280 m a.s.l.; longitude 12°58′53″ Est; latitude 42°59′55″ Nord, Marche region, Italy). The animals at the moment of the offspring separation (near 40 d post-partum) were divided in two homogeneous groups as far as body weight (BW) and body condition score (BCS), milk yield, parity, and days of lactation (see Supplementary material). One group (SUP, n = 12, mean BW 49.4 ± 6.1, mean BCS 2.31 ± 0.3) was supplemented with 600 g/day/animal of corn and barley (1:1), while the other group (control unsupplemented group, UNS, n = 12, mean BW 50.5 ± 5.1, mean BCS 2.27 ± 0.3) was fed only with the pasture (Figure S1). The feed supplementation was administered during the morning milking when each animal was in its milking post. The feed supplementation was chosen taking into account the management habits of the farmer and in order to avoid too expensive actions.

The food intake of grazing animals was evaluated by the pasture phytomass removal estimation [27].

Milk production was monitored at three different times: at the beginning (7 July 2016, near 75 days after parturitions, morning milking mean yield 297 ± 63 mL by UNS and 306 ± 64 mL by SUP ewes), in the middle (21 July 2016, 134 ± 47 mL by UNS and 240 ± 52 mL by SUP ewes), and in the end (9 August 2016, 125 ± 51 mL by UNS and 192 ± 52 mL by SUP ewes) of the period of nutritional supplementation as previously described [28].

2.2. BCS, Blood Collection, and Assays

Hormone levels and BCS were evaluated before the supplementation on 7 July (T0), and they were then measured every 9–10 days until 22 August (T1–T5) (Figure S1). The last date corresponded to the pre-mating period, when males were introduced in the flock and all females were dried. BCS was evaluated on the basis of a specific sheep method previously described [2–29]. Briefly, BCS is a comprehensive assessment of animal's body status based on muscle/fat relative proportions, and it is considered a useful management tool in determining the welfare of domestic animals. The steps in ovine BCS evaluation are palpation of body structures (processes of the thoracic and lumbar vertebrae), the state of the dorso-medial area, and the general status of animals. Each step, which was done by 3–5 trained valuators, receives a score (0–5) [2], and the mean of the scores for the four parameters constitutes the BCS value. Blood samples were withdrawn by jugular venipuncture in evacuated tubes containing K3-EDTA as anticoagulant (Sarsted, Numbrecht, Germany). Tubes were immediately centrifuged (2500× *g* for 15 min), and the plasma aliquots were stored at −20 °C until assayed. The samplings were carried out in all the dates at 7:00–8:00 a.m.

All hormone concentrations were determined as the average of duplicate determinants. To minimize the systematic error owing to inter-assay variability, all samples of each animal were analyzed for each hormone within the same assay session, in which an equal number of animals belonging to the two groups was present.

Enzyme immunoassays (EIA) were performed using the automated processor Brio 2 reader (Seac, Firenze, Italy). NES1 was assayed on T0, T2, T4 and T5, due to the lack of plasma amount on the other dates. NES1 level was measured using EIA kits (E-E-H2373, Biotechnology Inc., Human NES1, Wuhan, China). The intra-and inter-assay coefficients of variation (CV) were 9.4% and 11.6%, respectively. Sensitivity (DL) indicated by the manufacturer is 9.38 pg/mL. Plasma INS was determined using the Sheep Insulin ELISA kit (EIA-4739, DRG, Marburg, Germany) and the intra- and inter-assay CVs were 4.3% and 9.7%, respectively. Sensitivity (DL) indicated by the manufacturer is 0.49 μLU/mL. Radioimmunoassay was used to measure plasma glucagon (Glucagon RIA Kit, RB310, Euro Diagnostica AB, Lundavagen, Malmo, Sweden). The intra- and inter-assay coefficients of variation for control samples were 7.6% and 5.4%, respectively. Sensitivity (DL) indicated by the manufacturer is 3 pmol/L. Plasma LEP concentrations were measured by the multi-species Leptin RIA kit (XL-85K, EMD Millipore Co., Billerica, MA, USA) and the intra- and inter-assay coefficients of variation were 3.4 and 8.7%, respectively. Sensitivity (DL) indicated by the manufacturer is 0.801 ng/mL Human Equivalent (HE). Total concentrations of 3-3′-5-triiodothyronine (TH) in plasma were assayed using a radioimmunoassay kit (Immunotech, Prague, Czech Republic, IM 3287). Intra- and inter-assay coefficients of variation (CVs) were 6.1% and 7.8%, respectively. Sensitivity (Detection Limit DL) indicated by the manufacturer is 0.26 nmol/L. Plasma cortisol concentrations were determined using commercial RIA kits (Immunotech, Prague, Czech Republic, IM 1841) and the intra- and inter-assay CVs were 7.3% and 9.1%, respectively. Sensitivity (DL) indicated by the manufacturer is 5 nM/L.

2.3. Statistical Analysis

Diagnostic graphics and Shapiro–Wilk were used for testing assumptions and outliers. Since non-normality of the data was detected, insulin was log transformed, while NES1 was categorized into two levels using the median [30], as its distribution did not improve after transformation.

First, the effect of time and supplementation were analyzed for each hormone (except for NES1) and BCS (treated as a continuous variable) by using univariable approaches and Linear Mixed Models (LMMs) procedures. Animals and days were included as random and repeated factors, respectively. The LMMs evaluated the main effects of Group (2 levels: UNS and SUP), Time (6 levels: T0–T5), and the interaction between Group and Time. Sidak adjustment was used for carrying out multiple comparisons. Results were expressed as estimated marginal means ± standard error (SE) while raw data were presented in figures as means ± SE.

After categorization, NES1 was analysed by a Generalized Linear Model (GLM) using binomial as the probability distribution and logit as the link function. The effects of group (2 levels: UNS and SUP), sampling time (6 levels: T0–T5), and their interaction were evaluated. Data were presented as percentages, medians (Mdn), and interquartile ranges (IQR).

Finally, a multivariable approach was used by Discriminant Analysis (DA) to find the combinations of variables (BCS and hormones) that distinguish sheep that received supplementation (SUP) from the control group (UNS). The relative importance of each variable on this discrimination was expressed by the Wilks' lambda (the smaller the Wilks' lambda, the more important the variable to the Df) and by the discriminant loadings (correlations between each independent variable and the discriminant scores associated with the Df) [31,32]. Mahalanobis distance was used to identify the presence of multivariate outliers. The performance of the DA was evaluated by leave-one-out cross-validation, calculating the probability for each sample to be accurately classified in the correct group. The centroids (mean discriminant scores of the groups) were used to establish the cutting point for classifying samples during the cross-validation.

Statistical analyses were performed with SPSS Statistics version 25 (IBM, SPSS Inc., Chicago, IL, USA). Statistical significance occurred when $p < 0.05$.

3. Results

3.1. Univariable Approach

Only the time effect was found for BCS, which was significantly reduced at T4 compared to T0 (from 1.7 ± 0.1 at T0 to 1.5 ± 0.1 at T4; $p < 0.001$) (Figure 1) and then returned to basal values T5 (1.8 ± 0.1).

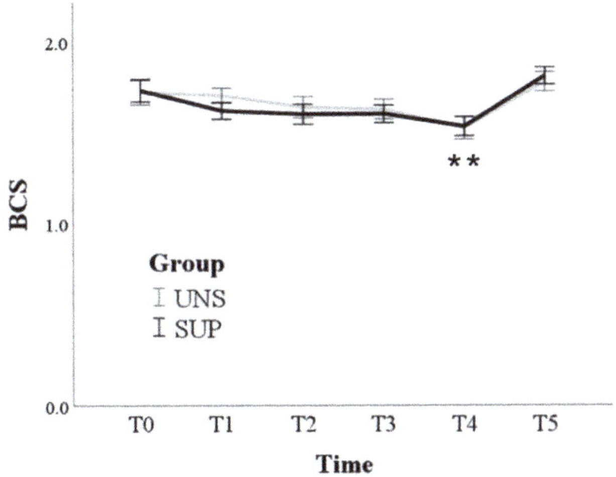

Figure 1. Means and standard errors of body condition score (BCS) in control (UNS) and supplemented (SUP) lactating sheep monitored before the administration of supplementation (T0), and then every 10 days until dry off (T1–T5). ** $p < 0.01$ T4 vs. T0 in SUP and UNS.

Due to the high variability of the NES1 concentrations (mean ± SE = 121.96 ± 29.09 pg/mL; Mdn (IQR) = 15.86 pg/mL (10.00–65.68 pg/mL); range = 10.00–1000.00 pg/mL), we decided to categorize these data according to their median value. Thus, two categories were created for NES1: "low level" if NES1 < 15.86 pg/mL and "high level" if NES1 ≥ 15.86 pg/mL. No change over time was found for this binary variable ($p = 0.949$), while a significant effect of group was found ($p = 0.033$). Indeed, a higher percentage of NES1 samples from the UNS group were included in the high-level category compared to the SUP group (60.2 ± 7.7% and 36.1 ± 7.7% of samples included in the high-level for UNS and SUP groups, respectively) (Figure 2). Mean values, standard error, and median with an interquartile range for NES1 concentrations in UNS and SUP groups are shown in Table S1.

The insulin log values increased during T2–T4 times compared to T0 ($p < 0.001$) (Figure 3) in both groups ($p = 0.795$).

Glucagon levels were not affected by either time ($p = 0.331$) or group ($p = 0.229$) (Figure 4).

Overall, leptin levels increased from a marginal mean of 5.1 ± 0.4 ng/mL at T0 to 5.8 ± 0.4 ng/mL at T5 ($p = 0.016$). A significant group x time effect was also found ($p = 0.028$). Indeed, pairwise comparisons revealed that there were no significant changes over time in the UNS group ($p = 0.252$), while the SUP group showed a progressive increase from T2 to the last time point ($p = 0.002$) (Figure 5).

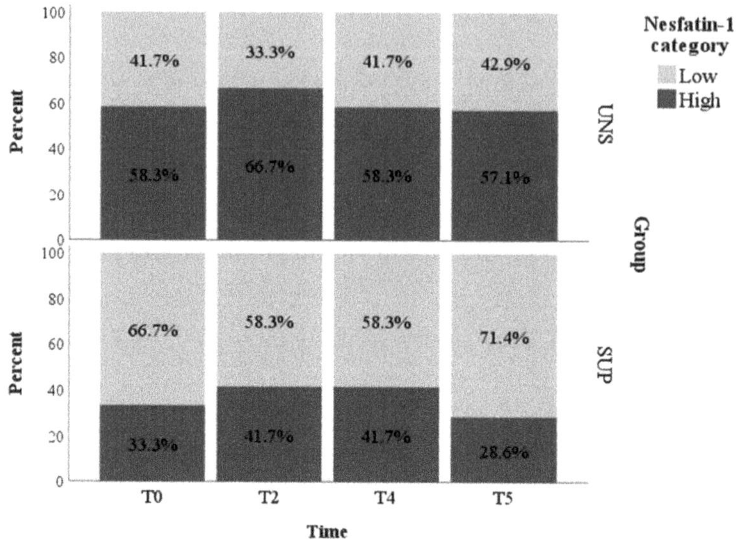

Figure 2. Relative frequencies of Nesfatin-1 categorized into "low" and "high levels" according to group (UNS = control ewes, SUP = supplemented ewes) and time (T0 = before the administration of supplementation; T2 = 20th day of supplementation; T4 = 40th day of supplementation; T5 = 50th day of supplementation).

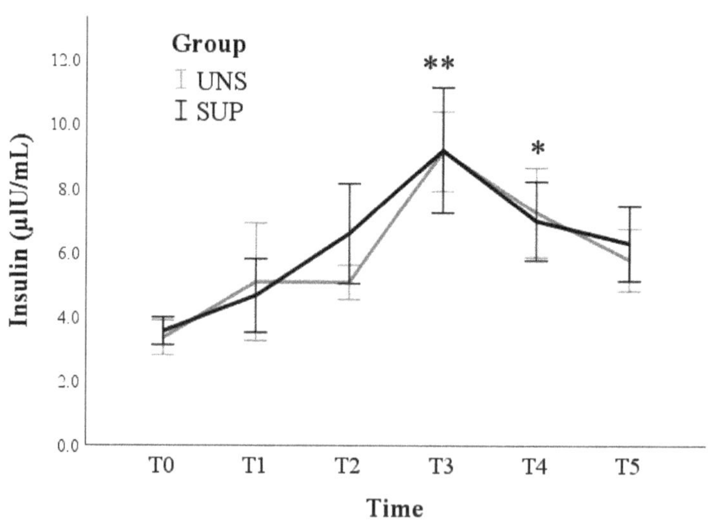

Figure 3. Means and standard errors of insulin plasma concentrations in control (UNS) and supplemented (SUP) lactating sheep monitored before the administration of supplementation (T0), and then every 10 days until dry off (T1–T5). * $p < 0.05$, ** $p < 0.01$: each time vs. T0 in SUP and UNS.

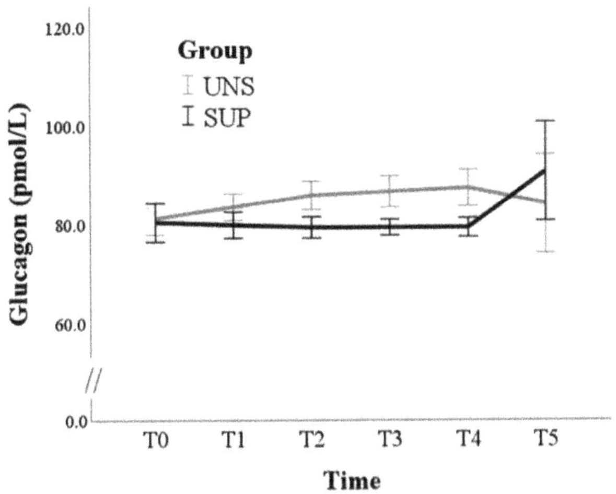

Figure 4. Means and standard errors of glucagon plasma concentrations in control (UNS) and supplemented (SUP) lactating sheep monitored before the administration of supplementation (T0), and then every 10 days until dry off (T1–T5).

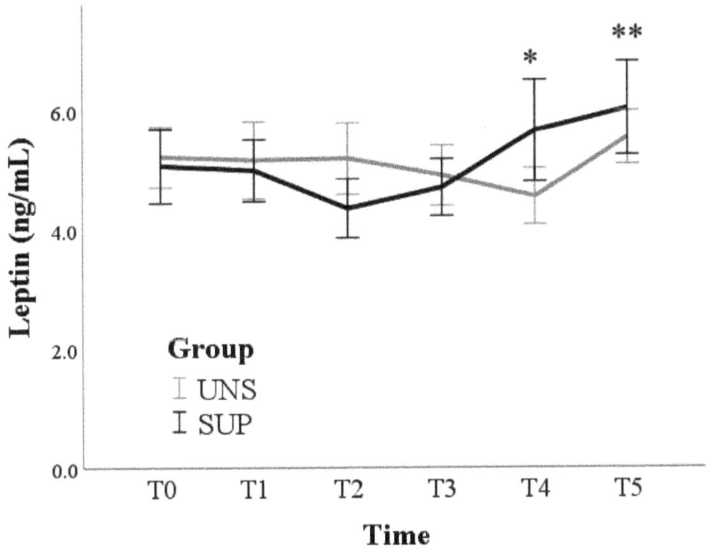

Figure 5. Means and standard errors of leptin plasma concentrations in control (UNS) and supplemented (SUP) lactating sheep monitored before the administration of supplementation (T0), and then every 10 days until dry off (T1–T5). * $p < 0.05$, ** $p < 0.01$: each time vs. T0 in SUP.

TH concentrations reduced after 10 days (from 3.2 ± 0.1 nmol/L at T0 to 2.8 ± 0.1 at T1; $p < 0.001$) but subsequently stabilized without differences between groups ($p = 0.491$) (Figure 6).

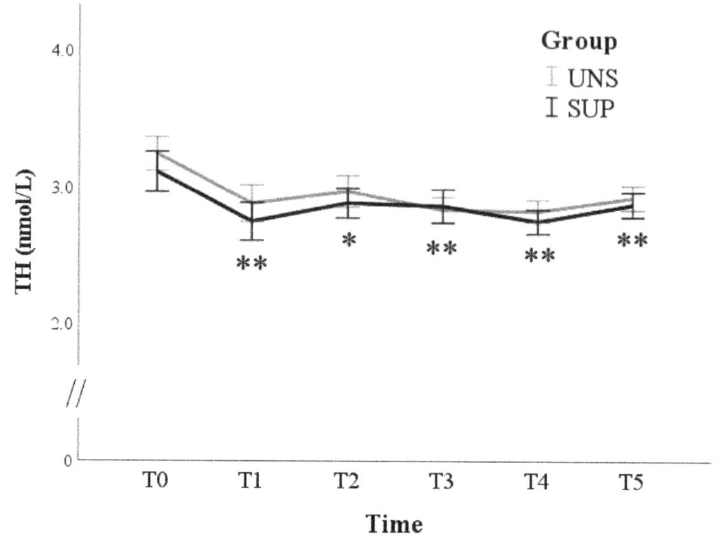

Figure 6. Means and standard errors of 3-3′-5-triiodothyronine (TH) plasma concentrations in control (UNS) and supplemented (SUP) lactating sheep monitored before the administration of supplementation (T0), and then every 10 days until dry off (T1–T5). * $p < 0.05$, ** $p < 0.01$: each time vs. T0 in SUP and UNS.

Nutrition treatment affected the trend of cortisol: changes over time were not significant in the UNS group ($p = 0.111$), while the SUP group showed higher values than UNS at T0 ($p = 0.033$), followed by a significant reduction compared to T0 from T2 until T4 ($p < 0.01$) (Figure 7).

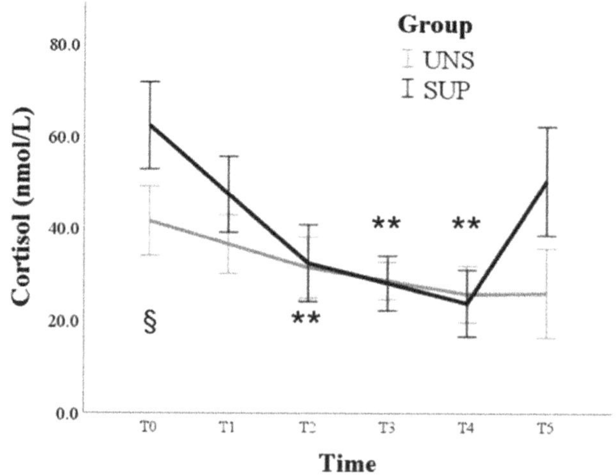

Figure 7. Means and standard errors of cortisol plasma concentrations in control (UNS) and supplemented (SUP) lactating sheep monitored before the administration of supplementation (T0) and then every 10 days until dry off (T1–T5). § $p < 0.05$ UNS vs. SUP; ** $p < 0.01$ vs. T0 in SUP.

3.2. Multivariable Approach

Seven variables were included in the DA describing the body condition and hormonal profile of the ewes. One item was eliminated because it was identified as a multivariate outlier by the Mahalanobis distance. The variables that most discriminated the two groups were BCS and NES1 (p-value < 0.05) (Table 1). The discriminant loadings (negative for BCS and positive for NES1) (Table 1) and centroids (0.666 and −0.610 for UNS and SUP, respectively) showed that lower BCS and higher NES1 characterized the UNS group compared to the SUP one. Overall, the DA showed a moderate discriminating ability: the model Wilks' Lambda was significant at $p < 0.05$ level ($p = 0.045$) and the extracted discriminant function correctly classified 63.0% of the samples (cross-validation procedure).

Table 1. Parameters indicating the relative importance of variables in classifying the ewes receiving nutritional supplementation: discriminant loadings, Wilks' Lambda, and significance of the F test (p-value). The variables with a significant F-test are in bold.

Variable	Discriminant Loadings	Wilks' Lambda	p Value
BCS	−0.566	0.880	**0.018**
Nesfatin-1 *	0.515	0.899	**0.031**
Cortisol	−0.249	0.974	0.288
TH	0.126	0.993	0.588
Insulin #	0.078	0.997	0.737
Leptin	0.042	0.999	0.857
Glucagon	0.037	0.999	0.875

* included as continuous variable; # included after log-transformation.

4. Discussion

The two ewes groups observed in the present study show very similar changes of the mean BCS value while, considering the univariable model analysis, no significant differences in the BCS changes and hormonal profiles between the two groups were noted.

These findings can be considered in relation to the milk production of the groups, which shows a clear better persistence in the SUP group, as reported in another paper by the same research project [28]. Milk yield was considered as relatively equal on day T0, while it was 78% vs. 45% at day 14 and 63% vs. 42% at day 33, in SUP and UNS, respectively. This observation can drive the hypothesis that the supplementation has protected the ewes from the usual lowering of the body state linked to the lactation, because it could be expected, on the contrary, that a higher and persistent milk production induces a more intense body reserve consumption. Indeed, the different hormones change with very scarce differences between the two groups and without any significant relationships with the body state (except NES1).

The multivariable analysis offers a more overall view of the changes in the considered parameters, weighing up the different effects and influences. In our analysis, the relationship between BCS and NES1 indicates the high probability that a better BCS level, as found in the SUP group, is related to a lower NES1 hematic concentration. This finding confirms that the nutritional supplementation had an effect on the body status, balancing the higher milk production in respect to the UNS group, and that these ewes, probably as a consequence of the nutritional supplementation, maintained a hunger motivation higher than the UNS: the lower blood Nesfatin-1 concentrations drive to suppose a higher hunger sense.

Regarding the changes of the other hormone concentrations, which can be of some interest in sheep grazing in our conditions, we can observe that concentrate supplementation significantly sustained milk production [28], which is often without affecting systemic concentrations of the hormones investigated. Therefore, in such conditions, the larger availability of nutrients in SUP ewes seems addressed toward milk synthesis by adaptations mostly driven at the peripheral (local) level [27].

Insulin participates in sustaining a general anabolic state during pregnancy, while the decrease in INS secretion and/or tissue responsiveness at the beginning of lactation allows the catabolic processes to dominate, and most of the nutrients are shifted toward the mammary gland for milk synthesis. The variations of INS concentrations observed over time in our study well agree with previous findings: an increase in plasma INS concentration has been widely described as lactation progresses, with decreasing milk yield and increasing energy balance [33,34]. The lowest concentrations of INS were found in ewes in early lactation (20 days); thereafter, plasma insulin significantly increased with the advancing stages of lactation (40 and 60 days) [35].

In our trial, glucagon changes were very scarce, and no significant difference was detected between groups. We can speculate that the slight increases in SUP opposite to the slight decrease in UNS can be consistent with a tendency to increase the available energy by its effects on hepatic glucose output and, possibly, lipolysis by adipose tissue.

In the SUP group, LEP concentrations progressively increase after two weeks of supplementation toward the end of lactation (T4) and dry off (T5). Similar to our results of the SUP group, in the ewe, leptinemia is reported to be very low during the first 3 weeks of lactation [36,37]; then, it gradually increases until about 3 months after lambing [38]. The lack of LEP increase with advancing lactation in UNS ewes likely mirrors the lack of improvements of the energy balance in these animals.

The plasma TH concentrations in both groups of ewes significantly lowered at day 85 of lactation, accordingly with previous reports. Thyroid hormone concentrations tended to decrease from 36 h to 21 days post-partum and thereafter constantly rose until day 51 post-partum [39], and slightly higher concentrations of TH were observed in the blood of ewes at the 40th day of lactation than at the 20th day [40]. After the start of our trial, animals were also exposed to rising summer ambient temperatures, which exert a well-known depressive effect on thyroid gland activity and TH action [41] and are inversely correlated with plasma thyroid hormone levels [25–43]. In the present study, high ambient temperatures may have prevented the physiological recovery of thyroid hormone concentrations, which are often observed toward the end of lactation in small ruminants, in parallel with the reduction of milk production and the increase of energy balance. On the other hand, in early lactation, low TH, leptin and insulin levels can indicate an energy deficiency of ewes, even despite the concentrate supplementation. Blood metabolic hormones and BCS are very useful tools for the assessment of ewes' nutritional status in very demanding physiological states, such as late pregnancy and lactation [40–44].

COR changes over time were not significant in the UNS group, while the SUP one showed higher values than the UNS group at T0, which was followed by a significant reduction until T4. The higher COR values found in the SUP group at the first sampling are likely due to the very large physiological variations among individuals when hormone concentrations increase facing different situations. For example, the blood-sampling procedure itself in unaccustomed animals could have played a role. Relatively low COR levels during lactation are reported in goats [45]. Regarding the COR peak observed in SUP ewes at the last sampling, different papers report similar effects by heat stress in small ruminants [46,47]. The serum COR levels were the highest during summer [43–49] and positively correlated with ambient temperature, showing reverse trends and negative correlations with TH [42–50]. Moreover, during thermal stress, feed-restricted ewes [51] and rams [25] showed less increase in cortisol blood concentrations: therefore, if the nutritional deficit is added to thermal stress, the animals can adjust their response, avoiding the potential adverse effects of extreme cortisol action, as it could be the case of the UNS group of the present study.

The trial planning foresaw performing further controls, especially aimed to know how the ewes of both groups cope with the successive reproductive season, in the hypothesis that the SUP group could face this demanding phase of the productive cycle performing better than the UNS one. Unfortunately, just at the start of the sexual season, when the males were to be joined in the female flock, in Central Italy a powerful earthquake took place (two strong events at the end of August and at the end of October). The farm where our ewes were kept suffered a lot of damage, as well as the houses of the farmers, so it was not possible to carry on the intended further controls, and the overall situation after the earthquake could have direct and indirect effects influencing the results. Despite this accident, fortunately, we had the possibility to have some indication from research performed on other ewes of the same flock, which showed that the feed supplementation had a positive effect on resistin production in the sheep uterine glands [52] and in the apelinergic system expression in both mammary glands and female reproductive apparatus [27–54].

5. Conclusions

Our results, notwithstanding the lack of the possibility to perform all the forecasted controls, indicate that the nutritional supplementation of lactating ewes reared in semi-extensive conditions on mid mountain rangelands can determine an overall body and metabolic state better than expected in milking animals approaching reproductive activity. Our findings were especially revealed by the combined observations of (a) the ability by the SUP ewes to maintain a BCS value similar to the UNS one, despite higher milk production, and (b) the lower blood levels of NES1 that suggest a better penchant to restore their body reserve in view to face the reproductive engagement. On the contrary, the UNS ewes, due to a higher level of the anorexigenic peptide NES1, could have more difficulties to recover a good nutritional status in view of the sexual activity.

The fact that the circulating levels of some of the studied hormones did not show differences between unsupplemented and supplemented ewes do not indicate necessarily that the supplementation could not had favorable localized effects at the peripheral level. Indeed, recent studies carried out on the same animal groups suggested that the feed supplementation had a positive effect on some adipokine expression in different organ tissues.

Supplementary Materials: The following are available online at https://www.mdpi.com/2076-2615/11/3/682/s1, Figure S1. Experimental trial scheme; Figure S2. Means and standard errors of milk production at morning milking (mL) in control (UNS) and supplemented (SUP) lactating sheep. Milk yield was recorded as manually collected milk by each animal; Table S1. Mean and standard error (SE), and median (Mdn) with interquartile range (IQR) of Nesfatin-1 concentrations in control (UNS) and supplemented ewe (SUP) ewe before the administration of supplementation (T0), and after 20 (T2), 40 (T4), and 50 (T5) days of supplementation.

Author Contributions: Conceptualization, P.S. and A.M.; Methodology and validation, O.B., L.T. and A.M.; Formal analysis, L.M.; Investigation, E.D.F. and O.B.; Data curation, O.B., E.D.F. and L.M.; Writing—original draft preparation, O.B., L.T. and A.M.; Writing—review and editing, E.D.F. and O.B.; Supervision, P.S. All authors have read and agreed to the published version of the manuscript.

Funding: This research was funded by the UNIVERSITY OF CAMERINO by the grant FAR 2014-15 assigned to Prof. Andrea Catorci.

Institutional Review Board Statement: The study was conducted according to the guidelines of the Declaration of Helsinki, and approved by the Ministry of Health (No. of approval 95/2018-PR).

Informed Consent Statement: Not applicable.

Data Availability Statement: Datasets used in the analyses are stored at the authors' home institution and will be provided upon request.

Acknowledgments: Authors wish to thank C. Canali for his technical assistance and the staff of Azienda Di Pietrantonio for collaboration in animal management.

Conflicts of Interest: The authors declare no conflict of interest.

References

1. Scocco, P.; Piermarteri, K.; Malfatti, A.; Tardella, F.M.; Catorci, A. Effects of summer rainfall variations on sheep body state and farming sustainability in sub-Mediterranean pastoral systems. *Span. J. Agric. Res.* **2016**, *14*, 1–4. [CrossRef]

2. Scocco, P.; Piermarteri, K.; Malfatti, A.; Tardella, F.M.; Catorci, A. Increase of drought stress negatively affects the sustainability of extensive sheep farming in sub-Mediterranean climate. *J. Arid. Environ.* **2016**, *128*, 50–58. [CrossRef]
3. Scocco, P.; Mercati, F.; Tardella, F.M.; Catorci, A. Increase of forage dryness indices differentiated anatomical response in the sheep rumen compartment. *Microsc. Res. Tech.* **2016**, *79*, 738–743. [CrossRef] [PubMed]
4. Giorgi, F.; Lionello, P. Climate change projections for the Mediterranean region. *Glob. Planet. Chang.* **2008**, *63*, 90–104. [CrossRef]
5. Arzeni, A.; Storti, D. Le strategie per lo sviluppo rurale nelle Aree interne colpite dal sisma. *Agriregionieuropa* **2017**, *51*, 1–8.
6. Robinson, J.; Ashworth, C.; Ashfort, J.; Rooke, L.; McEvoy, T. Nutrition and fertility in ruminant livestock. *Anim. Feed Sci. Technol.* **2006**, *126*, 259–276. [CrossRef]
7. Dore, R.; Levata, L.; Lehnert, H.; Schulz, C. Nesfatin-1: Functions and physiology of a novel regulatory peptide. *J. Endocrinol.* **2017**, *232*, R45–R65. [CrossRef] [PubMed]
8. Goebel, M.; Stengel, A.; Wang, A.; Tachè, Y. Central nesfatin-1 reduces the nocturnal food intake in mice be reducing meal size and increasing inter-meal intervals. *Peptides* **2011**, *32*, 36–43. [CrossRef]
9. Li, Z.; Gao, L.; Tang, H.; Yin, Y.; Zao, J.; Mulholland, M.; Zangh, W. Peripheral effects of nesfatin-1 on glucose homeostasis. *PLoS ONE* **2013**, *8*, e71513. [CrossRef]
10. Yoshida, N.; Maejima, Y.; Sedbazar, U.; Ando, A.; Kurita, H.; Damdinorj, B.; Takano, E.; Gantulga, D.; Iwasaki, Y.; Kurashina, T. Stressor-responsive central nesfatin-1 activates corticotropin-realising hormone, noradrenaline and serotonine neurons and evokes hipotalamic-pituitary-adrenal axis. *Aging* **2010**, *2*, 775–784. [CrossRef] [PubMed]
11. Vas, S.; Adori, C.; Konczol, K.; Katai, Z.; Pap, D.; Papp, R.S.; Bagdy, G.; Palkovitis, M.; Thot, Z.E. Nesfatin-1/NUCB2 as a potential new element of sleep regulation in rats. *PLoS ONE* **2013**, *6*, e59809. [CrossRef]
12. Gao, X.; Zhang, K.; Song, M.; Li, X.; Luo, L.; Tian, Y.; Zhang, Y.; Li, Y.; Zhangh, X.; Ling, Y. Role of nesfatin-1 in the reproductive axis of male rat. *Sci. Rep.* **2016**, *6*, 32877. [CrossRef] [PubMed]
13. Prajapati, G.N.; Kumar, A.; Laxmi, A.N.; Singh, N.K.; Mathur, A.K. Parallelism of nesfatin-1 and leptin with metabolic parameters and progesterone at puberty in Murrah buffalo heifers. *J. Anim. Res.* **2018**, *7*, 1–6.
14. Laarveld, B.; Christensen, D.A.; Brockman, R.P. The effect of insulin on net metabolism of glucose and amino acids by the bovine mammary gland. *Endocrinology* **1981**, *108*, 2217–2221. [CrossRef] [PubMed]
15. Todini, L.; Trabalza-Marinucci, M.; Malfatti, A.; Barbato, O.; Stradaioli, G.; Cavallucci, C.; Debenedetti, A. Pre- and post-feeding plasma Gastrin-17 and Insulin concentrations and feed intake in female goats during different physiological stages. *Small Rumin. Res.* **2007**, *71*, 38–47. [CrossRef]
16. Block, S.S.; Rhoads, R.P.; Bauman, D.E.; Ehrhardt, R.A.; McGuire, M.A.; Crooker, B.A.; Griinari, J.M.; Mackle, T.R.; Weber, W.J.; Amburgh, M.E.; et al. Demonstration of a role for insulin in the regulation of leptin in lactating dairy cows. *J. Dairy Sci.* **2003**, *86*, 3508–3515. [CrossRef]
17. Jiang, G.; Zhang, B.B. Glucagon and regulation of glucose metabolism. *Am. J. Physiol. Metab.* **2003**, *284*, E671–E678. [CrossRef]
18. Vaughan, M.; Steinberg, G.D. Effect of hormones on lipolysis and esterification of free fatty acid during incubation of adipose tissue in vitro. *J. Lipid Res.* **1963**, *4*, 193–199. [CrossRef]
19. Woods, S.C.; Lutz, T.A.; Geary, N.; Langhans, W. Pancreatic signals controlling food intake; Isulin, Glucagon and Amilyn. *Philos. Trans. R. Soc. B Biol. Sci.* **2006**, *311*, 1219–1235. [CrossRef]
20. Houseknecht, K.L.; Baile, C.A.; Matteri, R.L.; Spurlock, M.E. The biology of leptin: A review. *J. Anim. Sci.* **1998**, *76*, 1405–1420. [CrossRef]
21. Krasnow, S.M.; Steiner, R.A. Physiological Mechanisms Integrating Metabolism and Reproduction. In *Knobil and Neill's Physiology of Reproduction*; Neill, J.D., Ed.; Academic Press: San Diego, CA, USA, 1998; pp. 2553–2626.
22. Blache, D.; Tellam, R.L.; Chagas, L.M.; Blackberry, M.A.; Vercoe, P.E.; Martin, G.B. Level of nutrition affects leptin concentrations in plasma and cerebrospinal fluid in sheep. *J. Endocrinol.* **2000**, *165*, 625–637. [CrossRef] [PubMed]
23. Laud, K.; Gourdou, I.; Belair, L.; Keisler, D.H.; Djiane, J. Detection and regulation of leptin receptor mRNA in ovine mammary epithelial cells during pregnancy and lactation. *FEBS Lett.* **1999**, *463*, 194–198. [CrossRef]
24. Todini, L. Thyroid hormones in small ruminants: Effects of endogenous, environmental and nutritional factors. *Animal* **2007**, *1*, 997–1008. [CrossRef]
25. Maurya, V.P.; Sejian, V.; Kumar, D.; Naqvi, S.M.K. Impact of heat stress, nutritional stress and their combinations on the adaptive capability of Malpura sheep under hot semi-arid tropical environment. *J. Anim. Behav. Biometeorol.* **2019**, *7*, 31–38.
26. Caldeira, R.M.; Belo, A.T.; Santos, C.C.; Vazques, M.I.; Portugal, A.V. The effect of long-term feed restriction and over-nutrition on body condition score, blood metabolites and hormonal profiles in ewes. *Small Rumin. Res.* **2007**, *68*, 242–255. [CrossRef]
27. Mercati, F.; Maranesi, M.; Dall'Aglio, C.; Petrucci, L.; Pasquariello, R.; Tardella, F.M.; de Felice, E.; Scocco, P. Apelin System in Mammary Gland of Sheep Reared in Semi-Natural Pastures of the Central Apennines. *Animals* **2018**, *8*, 223. [CrossRef] [PubMed]
28. Scocco, P.; Rivaroli, S.; Mercati, F.; Tardella, F.M.; Malfatti, A.; de Felice, E.; Catorci, A. Anatomy for economy: Starting from the rumen keratinization degree to enhance the farm income. *Econ. Agro Aliment.* **2018**, *20*, 261–272. [CrossRef]
29. Scocco, P.; Mercati, F.; Brusaferro, A.; Ceccarelli, P.; Belardinelli, C.; Malfatti, A. Keratinization degree of rumen epithelium and body condition score in sheep grazing on Brachipodium rupestre. *Vet. Ital.* **2013**, *49*, 211–217.
30. Altman, D.G.; Bland, J.J. Uncertainty and sampling error. *BMJ* **2014**, *349*, g7064. [CrossRef]
31. Garson, G.D. *Discriminant Function Analysis*; Blue Book; Statistical Associates Publishers: Asheboro, NC, USA, 2012.

32. Agradi, S.; Curone, G.; Negroni, D.; Vigo, D.; Brecchia, G.; Bronzo, V.; Chiesa, L.; Peric, T.; Danes, D.; Menchetti, L. Determination of fatty acids profile in original brown cows dairy products and relationship with alpine pasture farming system. *Animals* **2020**, *10*, 1231. [CrossRef]
33. Herbein, J.H.; Aiello, R.J.; Eckler, L.I.; Pearson, R.E.; Akers, R.M. Glucagon, Insulin, Growth Hormone, and Glucose Concentrations in Blood Plasma of Lactating Dairy Cows. *J. Dairy Sci.* **1985**, *68*, 320–325. [CrossRef]
34. Brockman, R.P.; Laarveld, B. Hormonal regulation of metabolism in ruminants: A review. *Livest. Prod. Sci.* **1986**, *14*, 313–334. [CrossRef]
35. Antunovic, Z.; Novoselec, J.; Sauerwein, H.; Speranda, M.; Vegara, M.; Pavic, V. Blood metabolic profile and some of hormones concentration in ewes during different physiological status. *Bulg. J. Agric. Sci.* **2011**, *17*, 687–695.
36. Ehrhardt, R.A.; Slepetis, R.M.; Bell, A.W.; Boisclair, Y.R. Maternal leptin is elevated during pregnancy in sheep. *Domest. Anim. Endocrinol.* **2001**, *21*, 85–96. [CrossRef]
37. Sorensen, A.; Adam, C.L.; Findlay, P.A.; Marie, M.; Thomas, L.; Travers, M.T.; Vernon, R.G. Leptin secretion and hypothalamic neuropeptide and receptor gene expression in sheep. *Am. J. Physiol. Integr. Comp. Physiol.* **2002**, *282*, R1227–R1235. [CrossRef] [PubMed]
38. González-García, E.; Tesniere, A.; Camous, S.; Bocquier, F.; Barillet, F.; Hassoun, P. The effects of parity, litter size, physiological state, and milking frequency on the metabolic profile of Lacaune dairy ewes. *Domest. Anim. Endocrinol.* **2015**, *50*, 32–44. [CrossRef] [PubMed]
39. Bekeova, E.; Elecko, J.; Krajnicakova, M.; Hendrichovsky, V.; Maracek, I. Dynamics of changes in concentrations of cholesterol and thyroid and ovarian hormones in blood-serum during postparturient period of ewes. *Vet. Med. UVTIZ CSFR* **1991**, *36*, 673–684.
40. Antunović, Z.; Novoselec, J.; Speranda, M.; Vegara, M.; Pavić, V.; Mioč, B.; Djidara, M. Changes in biochemical and hematological parameters and metabolic hormones in Tsigai ewes blood in the first third of lactation. *Arch. Anim. Breed.* **2011**, *4*, 535–545. [CrossRef]
41. Todini, L.; Malfatti, A.; Valbonesi, A.; Trabalza-Marinucci, M.; Debenedetti, A. Plasma total T3 and T4 concentrations in goats at different physiological stages, as affected by the energy intake. *Small Rumin. Res.* **2007**, *68*, 285–290. [CrossRef]
42. Da Silva, W.E.; Leite, J.H.G.M.; da Silva, W.S.T.; Paiva, R.D.M.; de Sousa, J.E.R.; Facanha, D.A.E. Seasonal variations in thermoregulatory patterns enable Morada Nova sheep to adapt to Brazilian semi-arid. *Semin. Agrar. Sci.* **2019**, *40*, 1577–1594. [CrossRef]
43. Rathwa, S.D.; Vasava, A.A.; Pathan, M.M.; Madhira, S.P.; Patel, Y.G.; Pande, A.M. Effect of season on physiological, biochemical, hormonal, and oxidative stress parameters of indigenous sheep. *Vet. World* **2017**, *10*, 650–654. [CrossRef]
44. Menchetti, L.; Curone, G.; Andoni, E.; Barbato, O.; Troisi, A.; Fioretti, B.; Polisca, A.; Codini, M.; Canali, C.; Vigo, D.; et al. Impact of Goji Berries (*Lycium barbarum*) Supplementation on the Energy Homeostasis of Rabbit Does: Uni- and Multivariate Approach. *Animals* **2020**, *10*, 2000. [CrossRef]
45. Gupta, V.K.; Sharma, S.D.; Vihan, V.S.; Kumar, A. Serum enzymes and thyroid hormone in sub-clinical ketosis in goats and sheep reared under organized farming system. *Ind. J. Anim. Sci.* **2008**, *78*, 1199–1201.
46. Ribeiro, M.N.; Ribeiro, N.L.; Bozzi, R.; Costa, R.G. Physiological and biochemical blood variables of goats subjected to heat stress—A review. *J. Appl. Anim. Res.* **2008**, *46*, 1036–1041. [CrossRef]
47. Maurya, V.P.; Sejian, V.; Kumar, D.; Naqvi, S.M.K. Biological ability of Malpura rams to counter heat stress challenges and its consequences on production performance in a semi-arid tropical environment. *Biol. Rhythm. Res.* **2018**, *49*, 479–493. [CrossRef]
48. Inbaraj, S.; Kundu, A.; De, A.K.; Sunder, J.; Sejian, V. Seasonal changes in blood biochemical and endocrine responses of different indigenous goat breeds of tropical island agro-ecological environment. *Biol. Rhythm. Res.* **2018**, *49*, 412–421. [CrossRef]
49. Yilmaz, M.; Altin, T.; Kiral, F.; Taskin, T.; Asici, G.S.E.; Kizilkaya, K. Effects of the season on physiological and endocrine traits and on HSP70 in Saanen goats under Mediterranean climate conditions. *Med. Weter.* **2018**, *74*, 39–47. [CrossRef]
50. Sejian, V.; Indu, S.; Naqvi, S.M.K. Impact of short term exposure to different environmental temperature on the blood biochemical and endocrine responses of Malpura ewes under semi-arid tropical environment. *Ind. J. Anim. Sci.* **2013**, *83*, 1155–1159.
51. Sejian, V.; Maurya, V.P.; Naqvi, S.M.K. Adaptive capability as indicated by endocrine and biochemical responses of Malpura ewes subjected to combined stresses (thermal and nutritional) in a semi-arid tropical environment. *Int. J. Biometeorol.* **2010**, *54*, 653–661. [CrossRef]
52. Dall'Aglio, C.; Scocco, P.; Maranesi, M.; Petrucci, L.; Acuti, G.; de Felice, E.; Mercati, F. Immunohistochemical identification of resistin in the uterus of ewes subjected to different diets: Preliminary results. *Eur. J. Histochem.* **2019**, *63*, 127–130. [CrossRef] [PubMed]
53. Mercati, F.; Scocco, P.; Maranesi, M.; Acuti, G.; Petrucci, L.; Cocci, P.; Renzi, A.; de Felice, E.; Dall'Aglio, C. Apelin system detection in the reproductive apparatus of ewes grazing on semi-natural pasture. *Theriogenology* **2019**, *139*, 156–166. [CrossRef]
54. Scocco, P.; Mercati, F.; Catorci, A.; Maranesi, M.; Acuti, G.; de Felice, E.; Petrucci, L.; Dall'Aglio, C. Histochemistry as supporting tool in grassland ecosystem management: Apelin system detection in ewe reproductive apparatus. *Eur. J. Histochem.* **2019**, *63*, 31.

animals

MDPI

Article

The Expression of ERK1/2 in Female Yak (*Bos grunniens*) Reproductive Organs

Jiangfeng Fan [1,*], Xiaohong Han [1], Honghong He [1], Yuzhu Luo [2], Sijiu Yu [1], Yan Cui [3], Gengquan Xu [3], Libin Wang [3] and Yangyang Pan [3]

1 College of Veterinary Medicine, Gansu Agricultural University, Lanzhou 730070, China; hanxiaohong11@126.com (X.H.); honghong3h@126.com (H.H.); yusj@gsau.edu.cn (S.Y.)

2 College of Animal Science and Technology, Gansu Agricultural University, Lanzhou 730070, China; luoyz@gsau.edu.cn

3 Technology and Research Center of Gansu Province for Embryonic Engineering of Bovine and Sheep & Goat, Lanzhou 730070, China; cuiyan369@sina.com (Y.C.); xugq0307@aliyun.com (G.X.); Wanglb@gsau.edu.cn (L.W.); panyy@gsau.edu.cn (Y.P.)

* Correspondence: fanjf@gsau.edu.cn; Tel.: +86-13609360354; Fax: +86-931-7631229

Received: 3 February 2020; Accepted: 16 February 2020; Published: 20 February 2020

Simple Summary: Extracellular signal-regulated kinases1/2 (ERK1/2) plays a significant role in regulating the reproductive processes of mammals. The goal of our research is to investigate the expression and distribution of ERK1/2 in the main reproductive organs of the yak during different stages. Using immunohistochemistry, western blot, and relative quantitative real-time polymerase chain reaction techniques, we found that the expression of ERK1 and ERK2 proteins and their mRNA in the yak's ovary, oviduct, and uterus varies with the stage of the reproductive cycle. The variation character of ERK1 and ERK 2 expression in the yak's main reproductive organs during different stages implies that they play an important role in regulating the reproductive functions under different physiological statuses.

Abstract: The main reproductive organs undergo different histological appearances and physiological processes under different reproductive statuses. The variation of these organs depends on a delicate regulation of cell proliferation, differentiation, and apoptosis. Extracellular signal-regulated kinases1/2 (ERK1/2) are members of the mitogen-activated protein kinase (MAPK) super family. They have important roles in regulating various biological processes of different cells, tissues, and organ types. Activated ERK1/2 generally promotes cell survival, but under certain conditions, ERK1/2 also have the function of inducing apoptosis. It is widely believed that ERK1/2 play a significant role in regulating the reproductive processes of mammals. The goal of our research is to investigate the expression and distribution of ERK1/2 in the yak's main reproductive organs during different stages. In the present study, samples of the ovary, oviduct, and uterus of 15 adult female yak were collected and used in the experiment. The ERK1/2 proteins, localization, and quantitative expression of their mRNA were investigated using immunohistochemistry (IHC), western blot (WB) and relative quantitative real-time polymerase chain reaction (RT-PCR). The results indicated that ERK1/2 proteins and their mRNA were highly expressed in the ovary of the luteal phase and gestation period, in the oviduct of the luteal phase, and in the uterus of the luteal phase and gestation period. Immunohistochemical analysis revealed a strong distribution of ERK1/2 proteins in follicular granulosa cells, granular luteal cells, villous epithelial cells of the oviduct, endometrial glandular epithelium, and luminal epithelium. These results demonstrated that the expression of ERK1 and ERK2 proteins and their mRNA in the yak's ovary, oviduct, and uterus varies with the stage of the reproductive cycle. The variation character of ERK1 and ERK 2 expression in the yak's main reproductive organs during different stages implies that they play an important role in regulating the reproductive function under different physiological statuses.

Keywords: yak; ERK1/2; ovary; oviduct; uterus

1. Introduction

Extracellular signal-regulated kinases (ERKs) are an important subfamily of mitogen-activated protein kinases (MAPKs), which regulate various cellular activities and physiological processes. Originally, the ERK gene was isolated from the expression library of human gastric cancer in 1993 [1,2], and it was recognized as a genomic DNA, encoding a sequence of the receptor protein-tyrosine kinase. The ERK gene is evolutionarily conserved, and is found in all eukaryotes, from yeast to humans [3]. Activated ERK mediates extracellular signals (transferring from cell membrane receptors to cytoplasm and nuclear effectors) and regulates some specific gene expression by phosphorylating transcription factors; thus, participates in the regulation of cell growth, development, differentiation, and proliferation. By now, it is well known that the classical ERK cascade consists of Rafs (MAP3K), mitogen-activated protein kinase 1/2 (MEK1/2), extracellular signal-regulated kinases1/2 (ERK1/2), and several MAPK-activating protein kinases (MAPKAPKs) [4]. ERK1/2, also known as mitogen-activated protein kinases 3 and 1 (MAPK3/1), is one of the major MAPK cascades. Thus, MEK1/2, MAP3Ks, and other upstream signals can persistently, or transiently, activate ERK1/2. Activated ERK1/2 generally promote cell survival, but under certain conditions, ERK1/2 also have the function of inducing apoptosis [5].Many different stimuli, including growth factors [6], radiation [7,8], osmotic stress [9,10], Fas ligand [11], nitric oxide [12], and hydrogen peroxide [13] activate the ERK1/2 pathway. Remarkably, some researchers have indicated that activated ERK1/2 can inhibit apoptosis induced by hypoxia condition [14–17]. Thus, it is widely believed that ERK1/2 play an important role in mammalian organ development, incorporating with cellular proliferation, differentiation, migration, fate determination, growth, and apoptosis [18]. There are a number of materials reporting the expression of ERK1 and ERK2 in almost all kinds of tissues and cells, such as the brain, lung, gastrointestinal tract, testis, and some kind of cancer cells [19]. In an overall view, the function and mechanism of ERK1/2 vary under different circumstances and physiological statuses [20]. Similarly, some experiments have shown that the expression of ERK1 and ERK2 are widely presented in different parts of the reproductive organs in mammals and poultry [21–32]. However, the whole profile and specific role of ERK1/2 expression in reproductive organs, along with the different reproductive cycle stages, have not yet been detected and analyzed clearly in mammals.

The yak (*Bos grunniens*) is a kind of seasonal breeding-mammal that is mostly allocated around the Qinghai-Tibet Plateau [33]. Because of their predominant adaptability to high altitude, cold conditions, nutrition deficiency, and hypoxia environments, yaks have always been believed to be the most important means of production and livelihood of local herdsmen [34]. However, affected by formidable natural conditions, the yak presents a very low reproductive efficiency. The majority of them can give birth only once every two years or twice every three years [35,36]. Therefore, it is an important aspect—for improving the reproductive efficiency—to investigate the regulation of breeding activities [37–39]. Similar to other mammals, the female reproductive organs of adult yaks experience cyclic variation during different stages of their reproductive cycles. Ovary follicle development, corpus luteum generation, luteolysis, uterine distention, and placentation take place at specific periods [22,40,41]. Along with the reproductive cycle, numerous cells, such as ovarian granular cells, luteal cells, endometrial epithelium cells, and endometrial stromal cells experience proliferation or apoptosis [42]. As previously noted, ERK1/2 are involved in cell growth, differentiation, and apoptosis. All of these processes occur in the ovary and uterus during normal reproductive cycles. Therefore, it should be an attractive prospect to explore the roles and regulatory mechanisms of ERK1/2 in yak reproduction. We hypothesized that the abundance of ERKs would fluctuate in the reproductive tract during the different reproductive situations.

2. Materials and Methods

2.1. Samples of Yak's Reproducitve Organs

All procedures involving animals were approved by the Animal Care and Use Committee of Gansu Agricultural University. Samples of ovaries, oviducts, and uterus were incised from yaks (5 yaks per group) within 10 min after being slaughtered, in Xining abattoir of Qinghai Province, China. The yaks used for sample collection were estimated to be between 5 and 8 years old, were inspected to have no clinical disease, and no obvious pathological changes. For mRNA and protein analysis of ERK1/2 expression using WB and RT-PCR, tissue samples were immersed into liquid nitrogen for storage immediately after washing with 0.1% diethylpyrocarbonate (DEPC) reated water. Another tissue sample of the same yak was cut into small pieces and fixed with 4% paraformaldehyde phosphate buffer (pH 7.3) in 4 °C, at least 2 weeks before subsequent use.

According to the status of the yak's reproductive organ, we divided these tissue samples into 3 groups. The follicular phase group: there was only one ≥10.0 mm follicle and no macroscopically corpus luteum in both ovaries; the two uterine horns were symmetric and not dilated. The luteal phase group: there was only one ≥10.0 mm functional corpus luteum and no >8.0 mm follicle. Smaller luteum existed in both ovaries, and the two uterine horns were symmetric, not dilated. The gestation period group: functional corpus luteum was present in one of the two ovaries; one side of the uterine horns was obviously dilated and contained fetal.

2.2. *erk1* and *erk2* Gene Expression Analysis

Total RNA of yak tissue samples of every group were extracted using TRIzol reagent (Invitrogen, Carlsbad, CA, USA). Experion RNA StdSens Analysis Kit (BioRad, Munich, Germany) was used for total RNA quality and quantity assessment on Experion Automated Electrophoresis Station (BioRad, Munich, Germany). The value of RNA quality indicator (RQI) adopted was between 5 and 10. To avoid genomic DNA contamination, enzymatic digestion was performed with RNase-free DNase I (Omega, Norcross, GA, USA). Subsequently, with MOligo-dT18 primers, total RNA was reverse transcribed into cDNA using RevertAid first Strand cDNA Synthesis Kit (Promega, Mannheim, Germany). For relative quantitative analysis of gene expression, the *erk1* and *erk2* primers were designed according to bovine sequences (NM001110018.1, NM175793.2), and *β-actin* primers were designed based on the yak sequences (NM001034034.2). *erk1* (F: 5′-ATCCCTTGGCTGTCG-3′, R: 5′–AGGCGTTTCCATTCGT-3′). *erk2* (F: 5′-ATCCCTTGGCTGTCG-3′, R: 5′-AGGCGTTTCCATTCGT-3′). *β-actin* (F: 5′-AGGCTGTGCTGTCCCTGTATG-3′, R: 5′–GCTCGGCTGTGGTGGTAAA-3′). The predicted product length of these primers was 107, 111, and 187 bp, respectively. Real-time fluorescent quantitative PCR system (Light Cycler 480, Roche, Germany) was used to perform Real Time PCR analysis, as previously described [41]. Briefly, 200 ng of total cDNA was amplified in a 20 μL reaction mixture containing 10 μL SYBR Premix Dimer Eraser (Promega, Mannheim, Germany) and 100 nM of forward and reverse primer. The optimized 0conditions of RT-PCR were listed as below: pre-denaturation at 95 °C for 4 min, denaturing 40 cycles at 95 °C for 30 s, annealing at 58 °C for 30 s, and extension fluorescence acquisition at 72 °C for 25 s. The specificity analysis was performed through melting curve from 65 °C to 95 °C in 0.5 °C steps, each lasting 5 s, and the product was conduct electrophoresis in 2.5% agarose gel to confirm correct size.

2.3. The Quantitative Analysis of ERK1/2 Proteins Expression

For quantitative analysis of ERK1 and ERK2 proteins expression, frozen tissues were taken out of liquid nitrogen. After, they were washed with cold phosphate buffer saline (PBS) for three times. The total proteins of samples were extracted using Beyotime extraction buffer, and the protein concentration was detected via Bradford assay kit (Bio-Rad, Hercules, CA, USA). Then, the protein suspension was conducted, electrophoresis (PAGE), filled on 10% SDS-PAGE gel. Subsequently, separated proteins were electrophoretically transferred onto enhanced chemiluminescence (ECL) polyvinylidene fluoride (PVDF) membranes (Amersham, Piscataway, NJ, USA), using a mini transfer instrument for electrophoresis (Bio-Rad, Hercules, CA, USA) at 300 mA for 90 min. After, it was blocked with Tris-buffered saline (contained 5% non-fat dry milk and 0.1% Tween-20) the protein was incubated with a primary antibody against ERK1/2 (p44/42 MAPKRabbit mAb, 4695 s, CST, Danvers, MA, USA) and β-actin (ß-Actin Rabbit mAb, 4970 s, CST, Danvers, MA, USA) in Tris-buffered saline at 37 °C for 2 h. After, it was washed in Tris-buffered saline, the protein reacted with a secondary antibody (goat anti-rabbit immunoglobulin G conjugated with horseradish peroxidase, sc-2030, Santa Cruz Biotechnology, Santa Cruz, CA, USA) under 37 °C for another 2 h. After it was washed with Tris-buffered saline, the membrane reacted with ECL Substrate kit (ab65623, Abcam, Cambridge, UK). For negative controls (were conducted using normal IgG reagent) replaced the primary antibodies. The protein band's intensity on the membranes was measured using the densitometric analysis system (Bio-Rad, Hercules, CA, USA). Relative intensity of ERK1/2 proteins were normalized with β-actin bands.

2.4. Localizational Analysis of ERK1/2 Proteins Expression

The localization of ERK1/2 proteins was analyzed on paraffin-embedded 4 μm tissue sections using the immunohistochemical technique. At first, paraffin-embedded tissue sections were dewaxed and rehydrated using gradient acetone and alcohol solution, and eliminated endogenous peroxidase activity by incubating with 3% H_2O_2 for 5 min. Then, the tissue sections were immersed in citrate buffer (pH 6.0) and boiled for 10 min to enhance the antigen activity. Subsequently, for reducing non-specific binding of the primary antibody, sections were incubated with normal goat blocking serum for 15 min after being thoroughly washed with PBS. After that, tissue sections were incubated with primary antibody (1:450 diluted p44/42 MAPK Rabbit mAb, 4695 s, Cell Signaling Technology, Inc., Danvers, MA, USA) at 37 °C for 120 min, followed by incubation with secondary antibody (biotinylated goat anti-rabbit IgG, Invitrogen Zymed Laboratories, Carlsbad, CA, USA) for 15 min at 37 °C. Then, sections were washed another three times in PBS and reacted with avidin-biotin peroxidase. The immunoperoxidase color reaction was completed by adding 3,3′-diaminobenzidine agent (DAB, Invitrogen Zymed Laboratories, Carlsbad, CA, USA) as substrate. After a moderate brown reaction product presented, tissue sections were timely washed with PBS to stop color reaction. In the end, after being counterstained with haematoxylin, tissue sections were mounted with resin, for observation under microscope and storage. We also performed the negative controls experiment by replacing the primary antibody with normal non-immune IgG agent.

2.5. Data Analysis

The relative quantity of *erk1* and *erk2* mRNA and protein expression are presented as fold increase, where expression of target mRNA or proteins was divided by the expression of ß-Actin. The difference of expression quantity between reproductive stages was analyzed by one-way ANOVA procedure using SPSS 19.0 (SPSS Inc., Chicago, USA). It was considered to be statistically significant at a probability of $p < 0.05$; meanwhile, extremely significant was defined as $p < 0.01$.

3. Results

3.1. mRNA Expression of erk1 *and* erk2 *in Female Yak Reproductive Organs*

A variation of *erk1* and *erk2* gene expression was detected in female yak reproduction organs during different stages of the reproductive cycle. In the ovary, the *erk1* gene expression was significantly higher in the luteal phase ($p < 0.05$) and gestation period ($p < 0.01$) than the follicular phase. Similar to *erk1*, the mRNA level of *erk2* was also higher in the luteal phase compared with the follicular phase and gestation period ($p < 0.05$), but no significant difference existed between gestation period and follicular phase ($p > 0.05$). In the oviduct, the level of the *erk1* gene expression was highest in the luteal phase, which was significantly different with that in the follicular phase and gestation period ($p < 0.05$).

However, the *erk2* gene expression was extremely higher in the follicular phase than in the luteal phase ($p < 0.05$) and gestation period ($p < 0.01$). In the uterus, the *erk1* gene expression was extremely higher in the luteal phase and gestation period than in the follicular phase ($p < 0.05$), although no significant difference was found between the luteal phase and gestation period. Similarly, the level of *erk2* gene expression was also higher in the luteal phase and gestation period than that in the follicular phase ($p < 0.05$) (Figure 1).

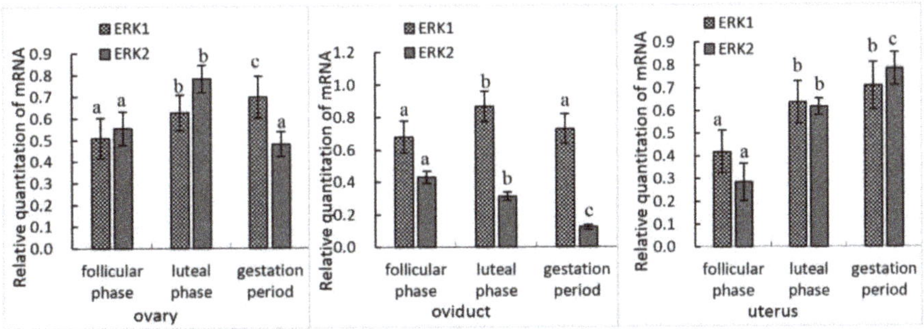

Figure 1. Expression of extracellular signal-regulated kinase (*ERK)1* and *ERK2* mRNA in female reproductive organs at different stages. In a certain reproductive organ, the difference of the same protein between a and b or b and c is significant ($0.05 > p > 0.01$), and the difference of the same protein between a and c is extremely significant ($p < 0.01$).

3.2. ERK1 and ERK2 Protein Expression in the Female Yak's Reproductive Organs

Both ERK1 and ERK2 proteins were detected using the western blotting technique. The characteristics of ERK1 and ERK2 protein expression in the female yak's reproductive organs at different reproductive statuses were similar as their mRNA expression. In ovaries, the level of ERK1 protein expression was nearly twofold of β-actin during the gestation period, while it only approximated one-half of β-actin in the follicular phase. The difference between these two stages was extremely significant ($p < 0.01$). The expression of ERK2 was higher in the luteal phase than that in the follicular phase and gestation period ($p < 0.05$). A little higher ERK2 expression was also found in the gestation period compared with the follicular phase, although there was no significant difference between these two groups ($p > 0.05$). In the oviduct, the level of ERK1 protein expression was highest in the luteal phase, which was significantly different with that in the follicular phase and gestation period ($p < 0.01$). However, ERK2 protein expression was extremely higher, in both the follicular phase and luteal phase, than that in the gestation period ($p < 0.01$). In the uterus, both ERK1 and ERK2 protein expressions were extremely higher in the luteal phase and gestation period, compared with the follicular phase ($p < 0.01$), although the difference between the luteal phase and gestation period was not significant (Figure 2).

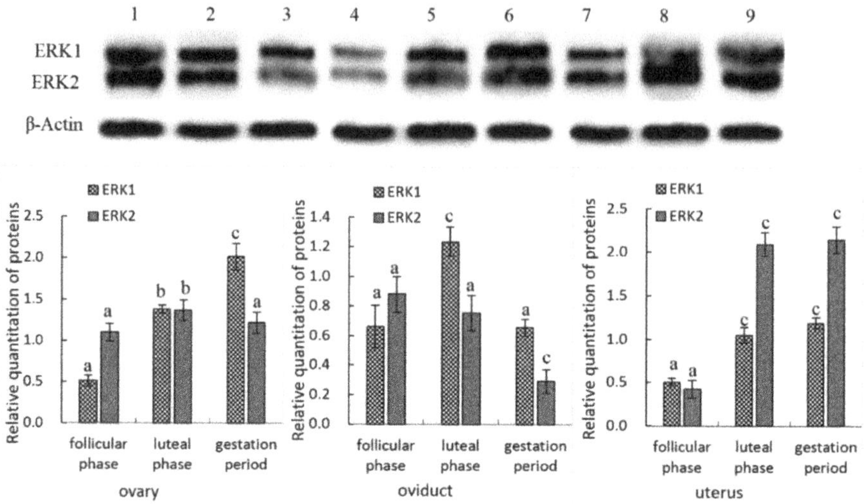

Figure 2. The western blot bands and expression analysis of ERK1, ERK2, and β-Actin proteins; 1, 2, 3: ovary; 4, 5, 6: oviduct; 7, 8, 9: uterus; 1, 4, 7: follicular phase; 2, 5, 8: luteal phase; 3, 6, 9: gestation period. Note: In certain reproductive organs, the difference of the same protein between a and b or b and c is significant ($0.05 > p > 0.01$), and the difference of the same protein between a and c is extremely significant ($p < 0.01$).

3.3. Immunolocalization of ERK1/2 Proteins in the Female Yak's Reproductive Organs

Immunohistochemical analysis revealed a light to dense positive reaction accumulation of signals for ERK1/2 proteins in the female yak's reproductive organs under different reproductive statuses. In the ovary, ERK1/2 proteins were mainly expressed in surface epithelium, follicular granulosa cells, ovarian stroma, vascular endothelium, and corpus luteum. In particular, in the corpus luteum of the gestation period, most granular luteal cells were stained into strong brown, while membranous luteal cells were not stained. However, the overall signal intensity of ERK1/2 proteins in the ovary of the luteal phase was less pronounced compared with that of the follicular phase and gestation period (Figure 3A–C). In the oviduct, the positive accumulations of signals for ERK1/2 proteins localized moderately in villous epithelial cells, and some dispersed villous stromal cells. The intensity of ERK1/2 proteins expression was slightly higher in the oviduct of the luteal phase and the follicular phase compared with that of the gestation period (Figure 3E–G). In the uterus, the main compartments of ERK1/2 protein expressions were endometrial luminal epithelium, glandular epithelium, endometrial stroma, and vascular endothelium. In addition, ERK1/2 protein signals also moderately appeared in the myometrium of the luteal phase and gestation period. The signal intensity of ERK1/2 proteins in the uterus of luteal phase and gestation period was obviously higher than that of the follicular phase (Figure 3I–K).

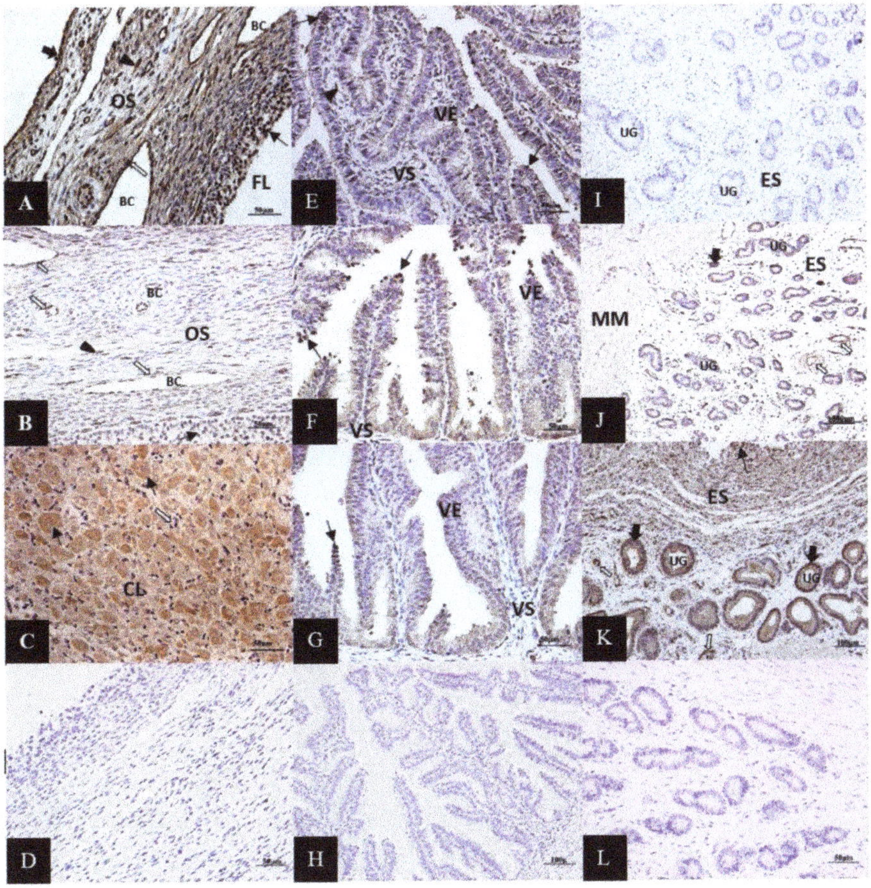

Figure 3. Immunolocalization of ERK1/2 proteins (brown stained regions) in the female yak's ovary (**A**,**B**,**C**), oviduct (**E**–**G**), and uterus (**I**,**J**,**K**) during the follicular phase (**A**,**E**,**I**), luteal phase (**B**,**F**,**J**), and gestation period (**C**,**G**,**K**). (**A**) ERK1/2 appeared in follicular granulosa cells (thin arrow), ovarian stroma cells (arrowhead), vascular endothelium (thick arrow), and surface epithelium (white thick arrow) of the ovary during the follicular phase. (**B**) ERK1/2 proteins moderately expressed in vascular endothelium (thick arrow), vascular endothelium (thick arrow), and follicular granulosa cells (thin arrow) of the ovary during the luteal phase. (**C**) Most of granular luteal cells (thin arrow) were stained into strong brown, while membranous luteal cells (white thick arrow) in corpus luteum of gestation period. (**D**) Negative control, ovary tissue sections were incubated with an equivalent non-immune IgG agent instead of a rabbit polyclonal antibody to ERK1/2. (**E**,**F**,**G**). The positive reaction of ERK1/2 proteins moderately localized in villous epithelial cells (thin arrow) and some dispersed villous stromal cells (arrowhead) of the oviduct. The intensity of ERK1/2 proteins expression was slight higher in the luteal phase (**F**) and follicular phase (**E**) compared with that of the gestation period (**F**). (**H**) Negative control of oviduct sections. (**I**,**J**,**K**) The main compartments of the ERK1/2 protein expression were endometrial epithelium (thin arrow), glandular epithelium (thick arrow), endometrial stroma cells (arrowhead), and vascular endothelium (white thick arrow) of the uterus. The signal intensity of ERK1/2 proteins in the uterus of the luteal phase (**J**) and gestation period (**K**) was obviously higher than that of the follicular phase (**I**). (**L**) Negative control of the uterus section. FL: follicle, BC: blood capillary, OS: ovarian stroma, CL: corpus luteum, VE: villous epithelium, VS: villous stroma, UG: uterine gland, ES: endometrial stroma, MM: myometrium.

4. Discussion

The present study described the expression of ERK1 and ERK2 proteins and mRNA in female reproductive organs of the adult yak during different the stage of the reproductive cycle based on RT-PCR, western blot, and immunohistochemistry analysis. The results showed that both the intensity and distribution of ERK1 and ERK2 expression in the ovary, oviduct, and uterus varied with the stage of the reproductive cycle. In general, the expression of ERK1 and ERK2 proteins and mRNA was most pronounced in the ovary of the luteal phase and gestation period, the oviduct of the luteal phase and the uterus of the gestation period. The histological appearance and physiological process of the main reproductive organs also varies with the different reproductive stages [43].

In the ovary, the results of the present study indicated that both the ERK1 and ERK2 proteins and mRNA were highly expressed during the luteal phase and gestation period, and the immunohistochemical analysis revealed an intense distribution of the ERK1/2 protein in follicular granulosa cells and corpus luteum cells. It is well know that the ovary is a highly organized composite of oocytes, granulosa cells, stromal cells, and sometimes corpus luteum cells whose interaction induce development of follicles, ovulation, formation of corpus luteum, and luteolysis during different stages of the reproductive cycle; many factors involved in the regulation mechanism of the ovary's physiological process. The classical point of view believes that the development of the ovary is mainly regulated by the hypothalamic-pituitary-ovarian axis. Therefore, various reproductive hormones play a very important role as external signals. Recently, studies have focused on the role and pathway of intra-ovarian signaling cascades in regulating ovary development [44]. As we discuss in the present study, ERK1/2 is one of the important intra-ovarian regulators of ovary development in mammals. The ERK1/2 pathway is intensively studied in follicle development of mice and other mammal species [45–48]. It has been reported that ERK1/2, and some other protein kinases, impact the cumulus expansion and oocyte maturation of porcine cumulus oocyte complexes (COCs) by inducing the expression of both the epidermal growth factor (EGF)-like factor and its protease [49]. Using cultured primary rat granulosa, Wayne et al. proved that follicle-stimulating hormone (FSH) and luteinizing hormone (LH) control granulosa cell function and differentiation by activating Ras protein, and some downstream kinases, especially MEK1 and ERK1/2 [50]. On the contrary, the experiment conducted on gene knockout mice and pharmacologically inhibited materials also demonstrated a pivotal role of the ERK1/2 pathway in ovulatory processes [51–53]. Most recently, by injection of pharmacological inhibitor (PD0325901) of ERK1/2 into the pre-ovulatory dominant follicle, Yasmin et al. found that four of five cows failed to ovulate, and 285 differentially expressed genes were identified from granulosa cells of drug treated follicles. Based on the analysis of the differentially expressed genes, they further affirmed a significant role of ERK1/2 in mediating LH induced gene expression in ovulating follicles, and the physiological process of ovulation dependent on proper ERK1/2 signaling in bovine [25]. Ryan et al. also found that the level of ERK signal proteins was different between dominant and subordinate follicles early in the stage of dominant follicle selection in cows [26]. However, there is no report about the expression of ERK1/2 in the ovary of the yak. The mass emergence of the ERK1/2 protein in follicular granulosa cells, demonstrated by the present study, probably means that the follicular development and further ovulation is a physiological process dependent on the regulation of ERK1/2.

Our study also demonstrates an intense expression of ERK1/2 in corpus luteum of the luteal phase and gestation period. It is well known that corpus luteum is a transient endocrine tissue that is derived from a pre-ovulatory follicle (Graffian follicle). It plays a significant role in controlling the reproductive cycle of mammals. The degeneration of corpus luteum, in case of un-pregnancy, is a key event for the initiation of a new reproductive cycle, re-ovulation, and obtaining the next chance of conception. On the contrary, the prolongation of the luteal function is also essential for the development of the embryo and maintenance of pregnancy relationship [54]. Generally, prostaglandin $F_{2\alpha}$ ($PGF_{2\alpha}$), as an extra-ovarian physiological luteolysin derived from uterus, is the key factor inducing the degeneration of the corpus luteum. It is well recognized that apoptosis is the main mechanism of $PGF_{2\alpha}$ induced regression of corpus luteum in bovine and other mammals [54]. As an important regulatory factor of apoptosis and proliferation of many types of cells, ERK1/2 is also involved in the mechanism of regression and functional maintenance of corpus luteum. For example, Maekawa et al. have found that human chorionic gonadotrophin (hCG) increased the StAR gene (coding histone modification enzymes) expression in granulosa cells through the ERK1/2 mediated signal pathway in the physiological process of follicular luteinization [55]. This is one of a few reports about the role of ERK1/2 in luteal formation. Strikingly, numerous research indicates that ERK1/2 is related to the regression of corpus luteum. Based on the experiment performed on Sprague-Dawley rats, Choi et al. found that PGF2α inducted luteal cell autophagy was accompanied with the activation of ERK1/2 during corpus luteum regression, and it is not regulated by the mammalian target of rapamycin (mTOR) signal [27]. Qi L et al. also proved that prostaglandin F (PGF) treatment increased ERK1/2 and signal transducers and activators of transcription 3 (STAT3) phosphorylation, and this is a probable molecular mechanism of luteal regression in pseudopregnant rats [56]. A similar conclusion was obtained by Chen, who demonstrated that ERK1/2 signaling cascade can be activated by other molecular mechanisms in bovine luteal cells [57].

In addition, many different stimuli can induce the change of intracellular signal transduction. To analyze the change of intracellular mitogen-activated protein kinase (MAPK) signaling cascade induced by stress-related signaling events. Rueda et al. detected the phosphorylation level of main MAPK members in cultured luteal cells. They found that both jun-n-terminal kinase (JNK) and p38MAPK were highly phosphorylated after UV treatment, but the phosphorylation of ERK1 and ERK2 was low. In addition, all of these changes were related to a high apoptosis rate of in vitro cultured luteal cells. Based on these experimental results, they believed that stress signals induced regression of corpus luteum, perhaps mediated by the activation of MAPK cascade [58]. As we have mentioned above, the yak is a kind of livestock living in a very harsh environment. In particular, the gestation period of the female yak is mainly spent in winter and spring, when low temperature, hypoxia, and nutrition deficiency threaten them at all times. The high expression of ERK1/2 in corpus luteum perhaps is a mechanism to maintain the luteal function and pregnancy; thus, yaks developed a prominent adaptability to the rigorous natural environment.

In the oviduct, our present study has also found a more intense expression of both ERK1 protein and mRNA in the luteal phase compared with the follicular phase and gestation period of the female yak reproductive cycle. Further immunolocalization analysis indicated that ERK1/2 proteins mainly localized in villous epithelial cells. As we all know, the mammalian oviduct is a convoluted tube bridging the ovary and the uterus in the reproductive process of mammals. It also provides a suitable place for fertilization and development of preimplantation embryos. The function of the ovary and uterus in reproductive activities is always an important issue attracting great attention, and has widely been investigated for a long times. However, it seems that we neglect the importance of the oviduct in reproduction, and relative information about the oviduct is insufficiently. Recently, some researchers begin to focus on the important role of the oviduct, both in natural fertilization and in pre-implantation embryo development. The correct biophysical function, such as a moderate and rhythmic smooth muscle contraction, ensures the transportation of the gametes and zygote in the oviduct. Meanwhile, a suitable biochemical component of the oviduct content provides an adapt circumstance for the development of the preimplantation embryo. On the contrary, the incorrect function of the oviduct may be the result in infertility, or deficiency of the embryonal development [59,60]. The oviduct content, also named as oviduct fluid, consists of proteins secreted by the secretory cells of the oviduct epithelium and other plasma-derived constituents [61,62]. Therefore, a normal function of this epithelium and its secretions is a basic requirement for successful fertilization and establishment of pregnancy. Recently, numerous genes and proteins were identified involving in the regulation of oviduct functions, using transcriptomic and proteomic techniques [63]. For instance, Cerny et al. performed a transcriptome analysis of bovine oviduct epithelial cells, and they found that a large number of genes differentially expressed during different estrus stages [64]. In line with these observations, significant differences of the profile of the gene and protein expression in the oviduct, between different estrus cycle stages, have also been reported in ewes [65], bovine [66], pigs [67,68], and humans [69,70]. For example, in spontaneous estrous ewes, a total of 280 proteins were identified in the oviduct, and 64 proteins upregulated during estrus, while 17 proteins were upregulated in the luteal phase [65]. However, there is no report about the expression of ERK1 and ERK2 in the oviduct of the mammalian species.

The result obtained from the oviduct of different stages of the yak's reproductive cycle in the present study suggests that, as an intracellular signal transduction pathway, ERK1/2 plays a significant role in regulating the secretory activity of the oviduct epithelium, and provides an optimal oviduct microenvironment for oocyte fertilization and embryo development before implantation.

In the uterus, we also observed a more strong expression of ERK1 and ERK2 during the luteal phase and gestation period. The positive reaction of the ERK1/2 proteins were located strongly in the endometrial epithelium cells, stroma cells, and moderately in the uterine smooth muscle cells. In the follicular phase, under the stimulation of high concentrations of FSH and LH, follicular recruitment and abundant E_2 secretion take place in the ovary, but the endometrium thickness and P_4 level are minimal. In the same times, numerous of endometrial epithelium cells undergo degeneration and necrosis [71,72]. After ovulation (luteal phase), a new corpus luteum begins to develop and produces a large amount of progesterone. Then, the differentiation and proliferation of the endometrial epithelium cells take place under the stimulation of progesterone, to make preparation for embryo implantation [71,72].

The essential role of P_4 and E_2 in regulating endometrial differentiation, growth, and receptivity to implantation during primate menstrual cycle has been widely recognized [73–76]. Numbers of reports also demonstrated that MAPK signaling pathways are activated in the uterus during the process of embryos implantation in the rat [75,77] and human [78]. Sayem's experiment indicated that in female rats, thyroxin treatment increased the expression of ERK1/2 proteins in uterine stroma, which could help the uterine to adopt the implantation of embryos [79]. However, most of the materials are obtained from the estrus cycle or early gestation period (before embryo implantation). A few researchers also pay attention to the role of ERK in the uterus during the whole gestation period. For example, Welshet al. proved that estrogenic actions were mediated by the estrogen receptor α via activating ERK signals in human myometrium during pregnancy [80].

According to the analysis above, we consider that the expression of ERK1 and ERK2 in yak's endometrial glandular and luminal epithelial cells and uterine smooth muscle cells during the luteal phase and gestation period may play an important role in glandular secretion, and maintain an appropriate muscle tension of the uterus. This is essential for embryo implantation and subsequent development.

5. Conclusions

The present study firstly demonstrates the wide expression of ERK1 and ERK2 proteins and mRNA in female reproductive organs of the adult yak. The intensity of ERK1 and ERK2 proteins and their mRNA expression in the yak's ovary, oviduct, and uterus varies with the stage of the reproductive cycle. The variation character of ERK1 and ERK 2 expression in the yak's main reproductive organs during different stages implies that they play an important role in the regulation of reproductive functions under different physiological situations.

Author Contributions: Conceptualization, J.F. and S.Y.; methodology, Y.P. and H.H.; software, X.H.; validation, J.F., Y.L. and S.Y.; formal analysis, Y.C.; investigation, X.H. and H.H.; resources, G.X. and L.W.; data curation, X.H.; writing—original draft preparation, X.H.; writing—review and editing, J.F.; visualization, Y.C.; supervision, S.Y.; project administration, J.F.; funding acquisition, J.F. All authors have read and agreed to the published version of the manuscript.

Funding: This study was supported by the National Natural Science Foundation of China (31660732), the Project of Preeminent Youth Fund of Gansu Province (1506RJDA002), and the Innovation Foundation of Gansu Agriculture University (GAU-QDFC-2018-10). Science Research Fund of Animal Husbandry Post-doctor of Gansu Agricultural University, Agricultural Science and Technology Innovational Fund of Gansu Province (GNCX-2014-30), and Basic Scientific Research Service Fee for Colleges and Universities of Gansu Province.

Conflicts of Interest: The authors declare no conflicts of interest.

References

1. Kiyokawa, E.; Takai, S.; Tanaka, M.; Iwase, T.; Suzuki, M.; Xiang, Y.; Naito, Y.; Yamada, K.; Sugimura, H.; Kino, I. Overexpression of ERK, an EPH family receptor protein tyrosine kinase, in various human tumors. *Cancer Res.* **1994**, *54*, 3645–3650.
2. Iwase, T.; Tanaka, M.; Suzuki, M.; Naito, Y.; Sugimura, H.; Kino, I. Identification of Protein-Tyrosine Kinase Genes Preferentially Expressed in Embryo Stomach and Gastric Cancer. *Biochem. Biophys. Res. Commun.* **1993**, *194*, 700–705. [CrossRef]
3. Widmann, C.; Gibson, S.; Jarpe, M.B.; Johnson, G.L. Mitogen-activated protein kinase: Conservation of a three-kinase module from yeast to human. *Physiol. Rev.* **1999**, *79*, 143–180. [CrossRef] [PubMed]
4. Johnson, G.L.; Lapadat, R. Mitogen-activated protein kinase pathways mediated by ERK, JNK, and p38 protein kinases. *Science* **2002**, *298*, 1911–1912. [CrossRef] [PubMed]
5. Lu, Z.; Xu, S. ERK1/2 MAP kinases in cell survival and apoptosis . *IUBMB Life* **2008**, *58*, 621–631.
6. Erhardt, P.; Schremser, E.J.; Cooper, G.M. B-Raf inhibits programmed cell death downstream of cytochrome c release from mitochondria by activating the MEK/Erk pathway. *Mol. Cell. Biol.* **1999**, *19*, 5308–5315. [CrossRef] [PubMed]
7. Park, J.S.; Carter, S.; Reardon, D.B.; Schmidt-Ullrich, R.; Dent, P.; Fisher, P.B. Roles for basal and stimulated p21(Cip-1/WAF1/MDA6) expression and mitogen-activated protein kinase signaling in radiation-induced cell cycle checkpoint control in carcinoma cells. *Mol. Biol. Cell* **1999**, *10*, 4231–4246. [CrossRef] [PubMed]
8. Kitagawa, D.; Tanemura, S.; Ohata, S.; Shimizu, N.; Seo, J.; Nishitai, G.; Watanabe, T.; Nakagawa, K.; Kishimoto, H.; Wada, T.; et al. Activation of extracellular signal-regulated kinase by ultraviolet is mediated through Src-dependent epidermal growth factor receptor phosphorylation. Its implication in an anti-apoptotic function. *J. Biol. Chem.* **2002**, *277*, 366–371. [CrossRef]
9. Lu, Z.; Xu, S.; Joazeiro, C.; Cobb, M.H.; Hunter, T. The PHD domain of MEKK1 acts as an E3 ubiquitin ligase and mediates ubiquitination and degradation of ERK1/2. *Mol. Cell* **2002**, *9*, 945–956. [CrossRef]
10. Nagata, Y.; Todokoro, K. Requirement of activation of JNK and p38 for environmental stress-induced erythroid differentiation and apoptosis and of inhibition of ERK for apoptosis. *Blood* **1999**, *94*, 853–863. [CrossRef]
11. Tran, S.E.; Holmstrom, T.H.; Ahonen, M.; Kahari, V.M.; Eriksson, J.E. MAPK/ERK overrides the apoptotic signaling from Fas, TNF, and TRAIL receptors. *J. Biol. Chem.* **2001**, *276*, 16484–16490. [CrossRef] [PubMed]
12. Kim, S.J.; Ju, J.W.; Oh, C.D.; Yoon, Y.M.; Song, W.K.; Kim, J.H.; Yoo, Y.J.; Bang, O.S.; Kang, S.S.; Chun, J.S. ERK-1/2 and p38 kinase oppositely regulate nitric oxide-induced apoptosis of chondrocytes in association with p53, caspase-3, and differentiation status. *J. Biol. Chem.* **2002**, *277*, 1332–1339. [CrossRef] [PubMed]
13. Wang, X.; Martindale, J.L.; Liu, Y.; Holbrook, N.J. The cellular response to oxidative stress: Influences of mitogen-activated protein kinase signalling pathways on cell survival. *Biochem. J.* **1998**, *333 Pt 2*, 291–300. [CrossRef]
14. Buckley, S.; Driscoll, B.; Barsky, L.; Weinberg, K.; Anderson, K.; Warburton, D. ERK activation protects against DNA damage and apoptosis in hyperoxic rat AEC2. *Am. J. Physiol.* **1999**, *277*, L159–L166. [CrossRef] [PubMed]
15. Karagiota, A.; Kourti, M.; Simos, G.; Mylonis, I. HIF-1α-derived cell-penetrating peptides inhibit ERK-dependent activation of HIF-1 and trigger apoptosis of cancer cells under hypoxia. *Cell. Mol. Life Sci.* **2018**, *76*, 809–825. [CrossRef]
16. Mottet, D.; Michel, G.; Renard, P.; Ninane, N.; Raes, M.; Michiels, C. Role of ERK and calcium in the hypoxia-induced activation of HIF-1. *J. Cell Physiol.* **2003**, *194*, 30–44. [CrossRef]
17. Minet, E.; Arnould, T.; Michel, G.; Roland, I.; Mottet, D.; Raes, M.; Remacle, J.; Michiels, C. ERK activation upon hypoxia: Involvement in HIF-1 activation. *FEBS Lett.* **2000**, *468*, 53–58. [CrossRef]
18. Almog, T.; Naor, Z. Mitogen activated protein kinases (MAPKs) as regulators of spermatogenesis and spermatozoa functions. *Mol. Cell. Endocrinol.* **2008**, *282*, 39–44. [CrossRef]
19. Tong, J.S.; Zhang, Q.H.; Huang, X.; Fu, X.Q.; Qi, S.T.; Wang, Y.P.; Hou, Y.; Sheng, J.; Sun, Q.Y. Icaritin causes sustained ERK1/2 activation and induces apoptosis in human endometrial cancer cells. *PLoS ONE* **2011**, *6*, e16781. [CrossRef]
20. Samuels, I.S.; Karlo, J.C.; Faruzzi, A.N.; Pickering, K.; Herrup, K.; Sweatt, J.D.; Saitta, S.C.; Landreth, G.E. Deletion of ERK2 Mitogen-Activated Protein Kinase Identifies Its Key Roles in Cortical Neurogenesis and Cognitive Function. *J. Neurosci.* **2008**, *28*, 6983–6995. [CrossRef]

21. Wang, Y.; Kristensen, G.B.; Helland, A.; Nesland, J.M.; Borresen-Dale, A.L.; Holm, R. Protein expression and prognostic value of genes in the erb-b signaling pathway in advanced ovarian carcinomas. *Am. J. Clin. Pathol.* **2005**, *124*, 392–401. [CrossRef]
22. Wong, C.H.; Cheng, C.Y. Mitogen-activated protein kinases, adherens junction dynamics, and spermatogenesis: A review of recent data. *Dev. Biol.* **2005**, *286*, 1–15. [CrossRef] [PubMed]
23. Li, Q.; Bao, F.; Zhi, D.; Liu, M.; Yan, Q.; Zheng, X.; Ren, L.; Cong, S.; Li, Y.; Cao, G. Lipopolysaccharide induces SBD-1 expression via the P38 MAPK signaling pathway in ovine oviduct epithelial cells. *Lipids Health Dis.* **2016**, *15*, 127–136. [CrossRef] [PubMed]
24. Pan, H.; Cui, H.; Liu, S.; Yu, Q.; Wu, H.; Liang, L.; Guan, Y.; Xin, G.; Long, Z.; Fan, H.Y. Lgr4 Gene Regulates Corpus Luteum Maturation Through Modulation of the WNT-Mediated EGFR-ERK Signaling Pathway. *Endocrinology* **2014**, *155*, 3624–3637. [CrossRef] [PubMed]
25. Schuermann, Y.; Rovani, M.T.; Gasperin, B.; Ferreira, R.; Ferst, J.; Madogwe, E.; Gonçalves, P.B.; Bordignon, V.; Duggavathi, R. ERK1/2-dependent gene expression in the bovine ovulating follicle. *Sci. Rep.* **2018**, *8*, 16170. [CrossRef] [PubMed]
26. Ryan, K.E.; Casey, S.M.; Canty, M.J.; Crowe, M.A.; Martin, F.; Evans, A.C.O. Akt and Erk signal transduction pathways are early markers of differentiation in dominant and subordinate ovarian follicles in cattle. *Reproduction* **2007**, *133*, 617–626. [CrossRef]
27. Choi, J.; Jo, M.; Lee, E.; Choi, D. ERK1/2 is involved in luteal cell autophagy regulation during corpus luteum regression via an mTOR-independent pathway. *Mol. Hum. Reprod.* **2014**, *20*, 972–980. [CrossRef]
28. Lee, J.H.; Banu, S.K.; McCracken, J.A.; Arosh, J.A. Early pregnancy modulates survival and apoptosis pathways in the corpus luteum in sheep. *Reproduction* **2016**, *151*, 187–202. [CrossRef]
29. Wang, J.; Huang, X.; Zhang, K.; Mao, X.; Ding, X.; Zeng, Q.; Bai, S.; Xuan, Y.; Peng, H. Vanadate oxidative and apoptotic effects are mediated by the MAPK-Nrf2 pathway in layer oviduct magnum epithelial cells. *Metallomics* **2017**, *9*, 1562–1575. [CrossRef]
30. Crepieux, P.; Marion, S.; Martinat, N.; Fafeur, V.; Vern, Y.L.; Kerboeuf, D.; Guillou, F.; Reiter, E. The ERK-dependent signalling is stage-specifically modulated by FSH, during primary Sertoli cell maturation. *Oncogene* **2001**, *20*, 4696–4709. [CrossRef]
31. Lu, Q.; Sun, Q.Y.; Breitbart, H.; Chen, D.Y. Expression and phosphorylation of mitogen-activated protein kinases during spermatogenesis and epididymal sperm maturation in mice. *Arch. Androl.* **1999**, *43*, 55–66. [CrossRef] [PubMed]
32. Sette, C.; Barchi, M.; Bianchini, A.; Conti, M.; Rossi, P.; Geremia, R. Activation of the mitogen-activated protein kinase ERK1 during meiotic progression of mouse pachytene spermatocytes. *J. Biol. Chem.* **1999**, *274*, 33571–33579. [CrossRef] [PubMed]
33. Ji, Q.M. Advances in research of yak resources in China. *J. Nat. Resour.* **2001**, *16*, 564–569.
34. Qiu, Q.; Zhang, G.; Ma, T.; Qian, W.; Wang, J.; Ye, Z.; Cao, C.; Hu, Q.; Kim, J.; Larkin, D.M.; et al. The yak genome and adaptation to life at high altitude. *Nat. Genet.* **2012**, *44*, 946–949. [CrossRef]
35. Mann, G.E. Reproduction in the yak. *Br. Vet. J.* **1993**, *149*, 513–514. [CrossRef]
36. Zi, X.D. Reproduction in female yaks (*Bos grunniens*) and opportunities for improvement. *Theriogenology* **2003**, *59*, 1303–1312. [CrossRef]
37. Lan, D.; Xiong, X.; Huang, C.; Mipam, T.D.; Li, J. Toward Understanding the Genetic Basis of Yak Ovary Reproduction: A Characterization and Comparative Analyses of Estrus Ovary Transcriptiome in Yak and Cattle. *PLoS ONE* **2016**, *11*, e0152675. [CrossRef]
38. Yu, S.J.; Huang, Y.M.; Chen, B.X. Reproductive patterns of the yak. III. Levels of progesterone and oestradiol-17β during pregnancy and the periparturient period. *Br. Vet. J.* **1993**, *149*, 595–602. [CrossRef]
39. Yu, S.J.; Huang, Y.M.; Chen, B.X. Reproductive patterns of the yak. II. Progesterone and oestradiol-17β levels in plasma and milk just before the breeding season; also during normal and short oestrous cycles. *Br. Vet. J.* **1993**, *149*, 585–593. [CrossRef]
40. JiangFeng, F.; Jiu, Y.; Wen, Z.; Ben, L. The expression of Fas/FasL and apoptosis in yak placentomes. *Anim. Reprod. Sci.* **2011**, *128*, 107–116. [CrossRef]
41. Fan, J.; Yu, S.; Cui, Y.; Xu, G.; Wang, L.; Pan, Y.; He, H. Bcl-2/Bax protein and mRNA expression in yak (*Bos grunniens*) placentomes. *Theriogenology* **2017**, *104*, 23–29. [CrossRef] [PubMed]
42. Radi, Z.; Khan, N. Comparative Expression and Distribution of c-fos, Estrogen Receptorα (ERα), and p38α in the Uterus of Rats, Monkeys, and Humans. *Toxicol. Pathol.* **2006**, *34*, 327–335. [CrossRef] [PubMed]

43. Radi, Z.; Marusak, R.; Morris, D. Species Comparison of the Role of p38 MAP Kinase in the Female Reproductive System. *J. Toxicol. Pathol.* **2009**, *22*, 109–124. [CrossRef] [PubMed]
44. Richards, J.S.; Pangas, S.A. The ovary: Basic biology and clinical implications. *J. Clin. Investig.* **2010**, *120*, 963–972. [CrossRef] [PubMed]
45. Su, Y.Q.; Wigglesworth, K.; Pendola, F.L.; O'Brien, M.J.; Eppig, J.J. Mitogen-activated protein kinase activity in cumulus cells is essential for gonadotropin-induced oocyte meiotic resumption and cumulus expansion in the mouse. *Endocrinology* **2002**, *143*, 2221–2232. [CrossRef] [PubMed]
46. Ryan, K.E.; Glister, C.; Lonergan, P.; Martin, F.; Knight, P.G.; Evans, A.C. Functional significance of the signal transduction pathways Akt and Erk in ovarian follicles: In vitro and in vivo studies in cattle and sheep. *J. Ovarian Res.* **2008**, *1*, 2. [CrossRef]
47. Henriquez, S.; Kohen, P.; Munoz, A.; Godoy, A.; Orge, F.; Strauss, J.F., 3rd; Devoto, L. In-vitro study of gonadotrophin signaling pathways in human granulosa cells in relation to progesterone receptor expression. *Reprod. Biomed. Online* **2017**, *35*, 363–371. [CrossRef]
48. Zhang, X.Y.; Chang, H.M.; Taylor, E.L.; Liu, R.Z.; Leung, P.C.K. BMP6 Downregulates GDNF Expression Through SMAD1/5 and ERK1/2 Signaling Pathways in Human Granulosa-Lutein Cells. *Endocrinology* **2018**, *159*, 2926–2938. [CrossRef]
49. Yamashita, Y.; Hishinuma, M.; Shimada, M. Activation of PKA, p38 MAPK and ERK1/2 by gonadotropins in cumulus cells is critical for induction of EGF-like factor and TACE/ADAM17 gene expression during in vitro maturation of porcine COCs. *J. Ovarian Res.* **2009**, *2*, 20. [CrossRef]
50. Wayne, C.M.; Fan, H.Y.; Cheng, X.; Richards, J.S. Follicle-stimulating hormone induces multiple signaling cascades: Evidence that activation of Rous sarcoma oncogene, RAS, and the epidermal growth factor receptor are critical for granulosa cell differentiation. *Mol. Endocrinol.* **2007**, *21*, 1940–1957. [CrossRef]
51. Siddappa, D.; Beaulieu, E.; Gevry, N.; Roux, P.P.; Bordignon, V.; Duggavathi, R. Effect of the transient pharmacological inhibition of Mapk3/1 pathway on ovulation in mice. *PLoS ONE* **2015**, *10*, e0119387. [CrossRef] [PubMed]
52. Su, Y.Q.; Nyegaard, M.; Overgaard, M.T.; Qiao, J.; Giudice, L.C. Participation of mitogen-activated protein kinase in luteinizing hormone-induced differential regulation of steroidogenesis and steroidogenic gene expression in mural and cumulus granulosa cells of mouse preovulatory follicles. *Biol. Reprod.* **2006**, *75*, 859–867. [CrossRef] [PubMed]
53. Fan, H.Y.; Liu, Z.; Shimada, M.; Sterneck, E.; Johnson, P.F.; Hedrick, S.M.; Richards, J.S. MAPK3/1 (ERK1/2) in ovarian granulosa cells are essential for female fertility. *Science* **2009**, *324*, 938–941. [CrossRef] [PubMed]
54. Cooke, I.D. The corpus luteum. *Hum. Reprod.* **1988**, *3*, 153–156. [CrossRef]
55. Maekawa, R.; Lee, L.; Okada, M.; Asada, H.; Shinagawa, M.; Tamura, I.; Sato, S.; Tamura, H.; Sugino, N. Changes in gene expression of histone modification enzymes in rat granulosa cells undergoing luteinization during ovulation. *J. Ovarian Res.* **2016**, *9*, 15. [CrossRef] [PubMed]
56. Qi, L.; Jiang, J.; Jin, P.; Kuang, M.; Wei, Q.; Shi, F.; Mao, D. Expression patterns of claudin 5 and its related signals during luteal regression in pseudopregnant rats: The enhanced effect of additional PGF treatment. *Acta Histochem.* **2018**, *120*, 221–227. [CrossRef]
57. Chen, D.B.; Davis, J.S. Epidermal growth factor induces c-fos and c-jun mRNA via Raf-1/MEK1/ERK-dependent and -independent pathways in bovine luteal cells. *Mol. Cell. Endocrinol.* **2003**, *200*, 141–154. [CrossRef]
58. Rueda, B.R.; Hendry, I.R.; Ndjountche, L.; Suter, J.; Davis, J.S. Stress-induced mitogen-activated protein kinase signaling in the corpus luteum. *Mol. Cell. Endocrinol.* **2000**, *164*, 59–67. [CrossRef]
59. Wang, H.; Guo, Y.; Wang, D.; Kingsley, P.J.; Marnett, L.J.; Das, S.K.; DuBois, R.N.; Dey, S.K. Aberrant cannabinoid signaling impairs oviductal transport of embryos. *Nat. Med.* **2004**, *10*, 1074–1080. [CrossRef]
60. Lopez-Cardona, A.P.; Perez-Cerezales, S.; Fernandez-Gonzalez, R.; Laguna-Barraza, R.; Pericuesta, E.; Agirregoitia, N.; Gutierrez-Adan, A.; Agirregoitia, E. CB1 cannabinoid receptor drives oocyte maturation and embryo development via PI3K/Akt and MAPK pathways. *FASEB J.* **2017**, *31*, 3372–3382. [CrossRef]
61. Leese, H.J.; Hugentobler, S.A.; Gray, S.M.; Morris, D.G.; Sturmey, R.G.; Whitear, S.-L.; Sreenan, J.M. Female reproductive tract fluids: Composition, mechanism of formation and potential role in the developmental origins of health and disease. *Reprod. Fertil. Dev.* **2007**, *20*, 1–8. [CrossRef] [PubMed]
62. Buhi, W.C.; Alvarez, I.M.; Kouba, A.J. Secreted proteins of the oviduct. *Cells Tissues Organs* **2000**, *166*, 165–179. [CrossRef] [PubMed]

63. Bauersachs, S.; Rehfeld, S.; Ulbrich, S.E.; Mallok, S.; Prelle, K.; Wenigerkind, H.; Einspanier, R.; Blum, H.; Wolf, E. Monitoring gene expression changes in bovine oviduct epithelial cells during the oestrous cycle. *J. Mol. Endocrinol.* **2004**, *32*, 449–466. [CrossRef] [PubMed]
64. Cerny, K.L.; Garrett, E.; Walton, A.J.; Anderson, L.H.; Bridges, P.J. A transcriptomal analysis of bovine oviductal epithelial cells collected during the follicular phase versus the luteal phase of the estrous cycle. *Reprod. Biol. Endocrinol.* **2015**, *13*, 84. [CrossRef] [PubMed]
65. Soleilhavoup, C.; Riou, C.; Tsikis, G.; Labas, V.; Harichaux, G.; Kohnke, P.; Reynaud, K.; de Graaf, S.P.; Gerard, N.; Druart, X. Proteomes of the Female Genital Tract During the Oestrous Cycle. *Mol. Cell. Proteom.* **2016**, *15*, 93–108. [CrossRef] [PubMed]
66. Lamy, J.; Labas, V.; Harichaux, G.; Tsikis, G.; Mermillod, P.; Saint-Dizier, M. Regulation of the bovine oviductal fluid proteome. *Reproduction* **2016**, *152*, 629–644. [CrossRef]
67. Acuña, O.; Avilés, M.; López-Úbeda, R.; Guillen-Martinez, A.; Soriano-Úbeda, C.; Torrecillas, A.; Coy, P.; Izquierdo Rico, M.J. Differential gene expression in porcine oviduct during the oestrous cycle. *Reprod. Fertil. Dev.* **2017**, *29*, 2387–2399. [CrossRef]
68. Seytanoglu, A.; Georgiou, A.S.; Sostaric, E.; Watson, P.F.; Holt, W.V.; Fazeli, A. Oviductal cell proteome alterations during the reproductive cycle in pigs. *J. Proteome Res.* **2008**, *7*, 2825–2833. [CrossRef]
69. Tone, A.A.; Begley, H.; Sharma, M.; Murphy, J.; Rosen, B.; Brown, T.J.; Shaw, P.A. Gene expression profiles of luteal phase fallopian tube epithelium from BRCA mutation carriers resemble high-grade serous carcinoma. *Clin. Cancer Res.* **2008**, *14*, 4067–4078. [CrossRef]
70. George, S.H.; Greenaway, J.; Milea, A.; Clary, V.; Shaw, S.; Sharma, M.; Virtanen, C.; Shaw, P.A. Identification of abrogated pathways in fallopian tube epithelium from BRCA1 mutation carriers. *J. Pathol.* **2011**, *225*, 106–117. [CrossRef]
71. Haschek, W.M.; Rousseaux, C.G.; Wallig, M.A. *Handbook of Toxicologic Pathology*; Academic Press: San Diego, CA, USA, 2002; pp. 47–894.
72. Marusak, R.A.; Radi, Z.A.; Obert, L. Expression of Ki-67 in the uterus during various stages of the estrous cycle in rats. *Exp. Toxicol. Pathol.* **2007**, *59*, 151–155. [CrossRef] [PubMed]
73. Chen, B.; Pan, H.; Zhu, L.; Deng, Y.; Pollard, J.W. Progesterone inhibits the estrogen-induced phosphoinositide 3-kinase–>AKT–>GSK-3beta–>cyclin D1–>pRB pathway to block uterine epithelial cell proliferation. *Mol. Endocrinol.* **2005**, *19*, 1978–1990. [CrossRef] [PubMed]
74. Lessey, B. Two pathways of progesterone action in the human endometrium: Implications for implantation and contraception. *Steroids* **2003**, *68*, 809–815. [CrossRef] [PubMed]
75. Kokatam, S.; Blesson, C.; Fatima, I.; Kitchlu, S.; Jain, S.; Mehrotra, P.; Dwivedi, A. Expression of $\alpha V\beta 3$ integrin in rat endometrial epithelial cells and its functional role during implantation. *Gen. Comp. Endocrinol.* **2008**, *160*, 124–133.
76. Okulicz, W.C.; Ace, C.I.; Longcope, C.; Tast, J. Analysis of differential gene regulation in adequate versus inadequate secretory-phase endometrial complementary deoxyribonucleic acid populations from the rhesus monkey. *Endocrinology* **1996**, *137*, 4844–4850. [CrossRef]
77. Thienel, T.; Chwalisz, K.; Winterhager, E. Expression of MAPkinases (Erk1/2) during decidualization in the rat: Regulation by progesterone and nitric oxide. *Mol. Hum. Reprod.* **2002**, *8*, 465–474. [CrossRef]
78. Casals, G.; Ordi, J.; Creus, M.; Fabregues, F.; Casamitjana, R.; Quinto, L.; Campo, E.; Balasch, J. Osteopontin and alphavbeta3 integrin expression in the endometrium of infertile and fertile women. *Reprod. Biomed. Online* **2008**, *16*, 808–816. [CrossRef]
79. Sayem, A.S.M.; Giribabu, N.; Muniandy, S.; Salleh, N. Effects of thyroxine on expression of proteins related to thyroid hormone functions (TR-alpha, TR-beta, RXR and ERK1/2) in uterus during peri-implantation period. *Biomed. Pharm.* **2017**, *96*, 1016–1021. [CrossRef]
80. Welsh, T.; Johnson, M.; Yi, L.; Tan, H.; Rahman, R.; Merlino, A.; Zakar, T.; Mesiano, S. Estrogen receptor (ER) expression and function in the pregnant human myometrium: Estradiol via ERalpha activates ERK1/2 signaling in term myometrium. *J. Endocrinol.* **2012**, *212*, 227–238. [CrossRef] [PubMed]

animals

MDPI

Article

Influence of Different Feed Physical Forms on Mandibular Gland in Growing Pigs

Cecilia Dall'Aglio [1], Francesca Mercati [1,*], Elena De Felice [2], Federico Maria Tardella [2], Josef Kamphues [3], Maria Grazia Cappai [4,†] and Paola Scocco [2,†]

1. Department of Veterinary Medicine, University of Perugia, Via San Costanzo 4, 06126 Perugia, Italy; cecilia.dallaglio@unipg.it
2. School of Biosciences and Veterinary Medicine, University of Camerino, Via Pontoni 5, 62032 Camerino, Italy; elena.defelice@unicam.it (E.D.F.); dtfederico.tardella@unicam.it (F.M.T.); paola.scocco@unicam.it (P.S.)
3. Institute of Animal Nutrition, University of Veterinary Medicine Hannover, Foundation, BischofsholerDamm 15, D-30173 Hannover, Germany; josef.kamphues@tiho-hannover.de
4. Department of Veterinary Medicine, University of Sassari, Via Vienna 2, 07100 Sassari, Italy; mgcappai@uniss.it

* Correspondence: francesca.mercati@unipg.it; Tel.: +0039-075-5857-636

† These authors share senior co-authorship.

Received: 9 May 2020; Accepted: 21 May 2020; Published: 24 May 2020

Simple Summary: The study was carried out on growing pigs fed with different dietary treatments based on different grinding intensities and compactions of the same diet. Chewing acts are associated with salivary production and different extents of saliva fluidity also depend on the specific glycoconjugate content. Therefore, in order to have information about the modifications induced by different feed physical forms in the pig mandibular gland, the glycohistochemical profile and the presence of aquaporin 5, a channel protein modulating the saliva fluidity, were investigated. In addition, to have wider information about the apelinergic system function, presence and localization of both apelin and its receptor were studied. Findings suggest that the different mechanical stimuli in the mouth linked to different feed physical forms likely allow one to diverse physiological behavior of the pig mandibular gland. The intense chewing activity linked to the highest feed compaction and hardness promotes an increase in pig mandibular gland secretion. In addition, saliva becomes more fluid and richer in acid glycoconjugates in order to better lubricate the bolus and protect the mouth mucosae. The apelinergic system is likely involved in the above modifications enhancing both the fluidity and the quantity of serous saliva by the pig mandibular gland.

Abstract: A study was performed on the mandibular gland obtained from growing pigs enrolled in a wide research project aiming to test the effects of different feed physical forms on animal health, production and welfare. We used 48 pigs fed for four weeks with different dietary treatments based on different grinding intensities and compactions of the same diet, namely coarsely ground meal (CM), finely ground pelleted (FP) and coarsely ground pelleted (CP) diets. Samples were analyzed by conventional histochemistry to identify the glycohistochemical profile and by immunohistochemistry to localize aquaporin 5, apelin and apelin receptor. Statistical elaborations were performed using the Stats R-package, version 3.5.3. Pig mandibular gland adenomere increased both the quantity and acidity of produced glycoconjugates from CM to FP and CP diets. This probably calls forth higher watery saliva, thus promoting a better feed softening facilitating the amalgamation of the bolus. Mandibular gland increased aquaporin 5 positivity in the CP diet, supporting the hypothesis of an augmented demand for water. Based on apelin/receptor localization, it was hypothesized that in pig mandibular gland the apelinergic system likely performs an endocrine control on the demilunes activity and a paracrine control on ducts, facilitating the production of a more fluid saliva.

Keywords: pig; mandibular gland; feed physical form; apelin; aquaporin 5; complex carbohydrates

1. Introduction

A number of scientific investigations have established that the feed physical form can have an impact on the morphological characteristics and functional activity of the gastrointestinal tracts as well as on associated organs, especially salivary glands of domestic animals [1,2]. In addition, different levels of moisture content in the diet appear involved and may be reasonably expected if liquid or dry diets are provided. In particular, in laboratory animals, a diet with increasing dry matter content is capable of stimulating a progressively more intense chewing activity, with a consistent increase in the absolute fresh weight of mandibular glands (MG). On the contrary, wet to liquid diets were observed to be associated with a reduced organ weight [3–6]. Such morphological changes are also often associated with qualitative changes in the saliva composition that can be related to different functionalities of the gland, as a result of more or less intense chewing activity [7,8].

More recently, a series of studies involving food-producing animals, especially pigs, were carried out to explore the dietary modulation of salivary gland morpho-functional traits. The findings reported in the literature showed that diets with different moisture contents, thus with different physical characteristics requiring different chewing activity to allow the softening and swallowing of the bolus, are able to determine morphological modifications of the MG in these animals, whereas parotid glands appeared not to be involved [4–8]. Such effects are also associated with variations in the expression of some molecules likely involved in the functional control of the gland itself, such as leptin production and the expression of its receptor, as well as the expression of endocannabinoid receptors [1,2].

Apelin, likewise leptin, is an adipokine produced predominantly, but not exclusively, by the adipose tissue and binding to the specific G-protein-coupled receptors for endogenous ligands. The apelinergic system is a complex system including the peptide apelin (APLN) and its related receptor (APLNR) [9]. APLN and APLNR are extensively expressed in many tissues either in human and different animal species, both laboratory and farm ones [10–15]. The literature regarding APLN is scanty and the presence and distribution of the apelinergic system in the salivary glands of the different animal species is not available to date, even though the presence of APLN in human saliva has been recently demonstrated [16]. So far, our work meant to provide additional insights regarding the APLN system in the MG of the pig in view of the effect of different feed physical forms of one same diet. It was, therefore, considered to explore the presence and distribution of the apelinergic system in the MG of growing pigs and its variation as a consequence of more or less intense chewing of the feed, stimulating a consistent effect of the functional activity of the MG. It is well known that chewing acts are associated with salivary production and that different extents of saliva fluidity also depend on the specific glycoconjugate content [17,18]. Therefore, in order to have additional information about the modifications induced in MG functions of the pig in view of different feed physical forms, MG's glycohistochemical profile was investigated. In fact, complex carbohydrates composing the saliva are able to call forth different amounts of water into the saliva, depending on their chemical functional groups. To support this, we also investigated the presence of aquaporin 5 (AQP5), a water protein channel modulating the water amount in saliva production [19–21]. AQP5 pertains to Mammalian aquaporins (AQPs), a class of integral membrane proteins facilitating the rapid and passive moving of water [22]. AQPs have a sequence of about 270 amino acids that form two hemichannels derived by six helical domains spanning the lipid bilayer and resulting in an hourglass-shaped channel for water [23–25]. Some of the AQPs are mainly water selective, but there are some AQPs, named aquaglyceroporins, that also transport glycerol, urea and neutral solutes [26,27]. Mandibular gland represents a good organ model being a mixed exocrine gland, whose ademomeres are constituted by a mucous acinar preterminal portion and a serous demilunar portion.

This study is aimed to investigate the influence of different feed physical forms in growing pig's mandibular gland analyzing the glycohistochemical profile, the presence of AQP5 and localizing the apelinergic system components.

2. Materials and Methods

2.1. Dietary Treatments and Sampling Collection

The project was approved by the Ethics Committee on Animal Welfare of the Hannover District Government in accordance with the German legislation on animal welfare.

The experiment was conducted using 48 castrated male growing pigs (German Landrace x Large White on Duroc sires) which received a control diet, namely coarsely ground meal diet (CM, dMEAN, 0.88 mm). Raw ingredients were differently processed to obtain pelleted feeds, in particular, finely ground pelleted diet (FP, dMEAN, 0.46 mm); coarsely ground pelleted diet (CP, dMEAN, 0.84 mm) [1]. Feed components and chemical composition are reported in Supplementary Table S1.

The initial total number of animals was divided into three groups of 16 animals and each group received one of the three diets, i.e., CM, FP and CP, from a six-week age for four weeks. [1]. The numerousness of animals for each group was calculated and considered optimal for a significance level of 0.05, a test power of 0.8 and an effect size of 1. Animals were fed ad libitum and had free access to water. At the end of the trial the animals were euthanized according to the European Union regulation on the protection of animals at the time of slaughter (Council Regulation EC No. 1099/2009).

The MG specimens were immediately removed and fixed in buffered formaldehyde (2.5% *v/v*) for 24 h at room temperature and subsequently processed for embedding in paraffin, following routine tissue preparation procedures [28,29].

2.2. Immuno- and Glycohistochemical Treatments

The immunohistochemical reactions were visualized on 5 μm sections, collected on poly-L-lysine-coated glass slides. Briefly, sections, dewaxed and brought to water, were microwaved for 15 min in 10 mM citric acid (pH 6.0) for antigen retrieval. All subsequent steps were carried out in a moist chamber at room temperature (RT), to prevent the evaporation of reagents; while, to prevent non-specific binding of primary antibody and before using the primary antibody, the sections properly cooled were preincubated with the normal serum for 30 min. Subsequently, once the excess of reagent has been removed from the sections, they were incubated, overnight at RT, each with one of the primary antibodies: rabbit polyclonal antibody anti-APLN (1:100, NBP2-31176, Novus Biologicals, Littleton, CO, USA), mouse monoclonal antibody anti-APLNR (1:100, sc-517300, Santa Cruz Biotechnology, Santa Cruz, CA, USA), and rabbit polyclonal antibody anti-AQP5 (1:100, AQP-005, Alomone Labs, Jerusalem 9104201, Israel). The specificity of each primary antibody used, verified by blasting the full protein sequences with corresponding swine ones, is shown in Table S2.

The next day, after washing in PBS, the sections were incubated at RT for 30 min with the corresponding secondary antibody and subsequently processed for 30 min using the avidin–biotin complex (ABC KIT, PK-6100, Vector Laboratories, Burlingame, CA, USA) and the DAB (SK-4100, Vector Laboratories, Burlingame, CA, USA) as the chromogen. The corresponding secondary biotinylated antibodies were: horse anti-rabbit (1:200, BP-1100, Vector Laboratories, Burlingame, CA, USA) and horse anti-mouse IgG antibodies (1:200, BP-2000, Vector Laboratories, Burlingame, CA, USA). At the end of the immunoreaction, the sections were rinsed in PBS, counterstained or not with hematoxylin, dehydrated and mounted in Eukitt.

As positive controls sheep abomasum was used for both APLN and APLNR [14], while sheep MG was used for AQP5 [30]. Sections in which the primary antibodies were omitted or substituted with preimmune gamma globulin were used as controls of non-specific staining.

Carbohydrate characterization was performed on 5 μm sections by staining with periodic acid–Schiff (PAS, evidencing vicinal diols), Alcian blue (AB) pH 2.5 (evidencing acid groups), AB/PAS, AB pH 1 (evidencing sulfated groups), AB pH 0.5 (evidencing highly sulfated groups) [31].

All tissue analyses were carried out on coded slides using a light microscope (Nikon Eclipse E800, Nikon Corporation, Tokyo, Japan) connected to a digital camera (Dxm 1200 Nikon digital camera). Images were processed using an image analysis system (Lucia, Laboratory Imaging Ltd., Prague, Czech Republic). The settings for image capture were standardized by subtracting the background signals obtained from the marched tissue sections which had not reacted with the primary antibodies and which were used as immunohistochemical controls.

For each animal, three independent observers, unaware of the treatments carried out, evaluated five microscopic fields of each experimental group and the intensity of the staining was graded in arbitrary units as follows: negative (−), weak (±), moderate (+), strong (++) and very strong (+++).

Variations in the intensity of immunopositivity for APLN, APLNR and AQP5 were observed among different groups of animals, probably reflecting the expression of the corresponding antigens. Even if the immunohistochemical technique shows a prevalently qualitative nature [32], a semiquantitative evaluation of immunopositivity was performed using the same scale applied to conventional glycohistochemistry.

2.3. Statistical Analyses

To test the null hypothesis of no difference among different experimental groups for each immune- and glycohistochemical treatment, we performed one-way ANOVAs if the variables satisfied conditions for parametric tests (normality was tested using the Shapiro–Wilk test); homogeneity of variance was ANOVA/Kruskal–Wallis test. We ran pairwise comparisons using the independent samples *t*-test or Wilcoxon–Mann–Whitney test to identify which groups were significantly different from each other. To test the null hypothesis that the location shift between groups is equal to 0, we performed two Wilcoxon signed-rank tests between the serial treatments AB pH 2.5 vs. AB pH 1 and AB pH 1 vs. AB pH 0.5. In both the analyses, a Holm correction for multiple comparisons was used to avoid type I error.

Statistical elaborations were performed using the R version 3.5.3 (R Core Team 2019, Vienna, Austria), the stats R-package, version 3.5.3 (shapiro.test, aov, kruskal.test, t.test, wilcox.test, Vienna, Austria), and the *car* R-package, version 3.0-2 (leveneTest function, Vienna, Austria) [33].

3. Results

3.1. Immunohistochemistry

The immunohistochemistry showed APLN binding sites at the duct cell level in the pig mandibular gland for the three diet groups. In particular, the moderate APLN reactivity observed in CM diet (Figure 1 CM) was slightly decreased in both FP and CP diets showing a similar reactivity (Figure 1 FP, CP).

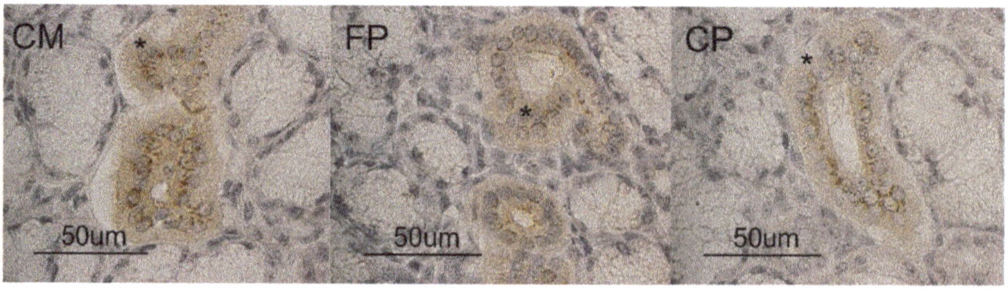

Figure 1. Pig mandibular gland. Apelin (APLN) binding sites at duct (*) level in coarsely ground meal (CM), finely ground pelleted (FP) and coarsely ground pelleted (CP) groups.

With regards to APLNR reactivity in the pig mandibular gland, in the CM diet, only a weak positivity was observed in ductal cells (Figure 2 CM). In FP and CP diets, a moderate reactivity to APLNR appeared in the mandibular gland demilunes and ducts. In addition, in both FP and CP samples, at duct level, it was possible to observe a few cells strongly APLNR reactive (Figure 2 FP, CP). In addition, the morphological observation of the different diet samples seemed to suggest an increase in demilune size.

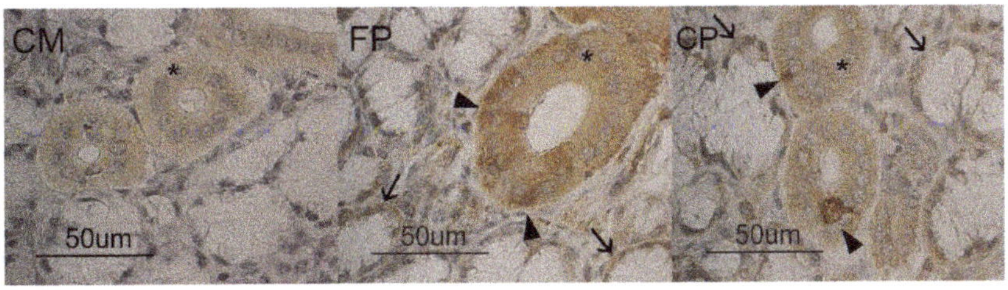

Figure 2. Pig mandibular gland. APLNR binding sites at duct (*) and demilune (↑) level in CM, FP and CP groups. In FP and CP samples, some cells (arrowhead) have a higher reactivity than others.

As for the other immunohistochemical treatments, pig mandibular acini did not react to AQP5 antibody. On the contrary, demilunes showed a weak positivity in CM and FP samples, which became strong in the CP diet samples. Additionally, in the CP diet, a slight AQP5 positivity was seen in the ducts, above all at cell coat level (Figure 3).

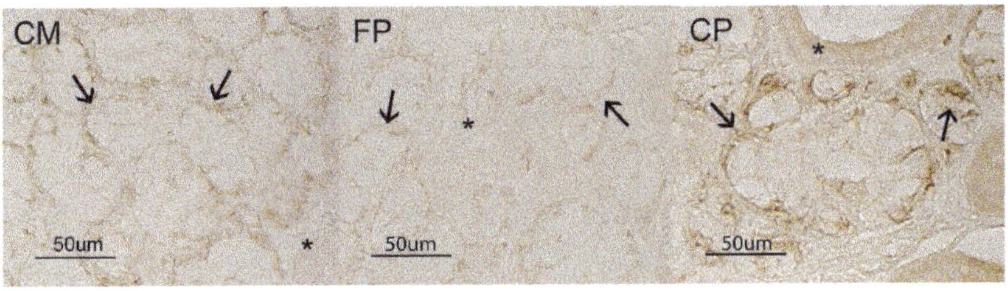

Figure 3. Pig mandibular gland. AQP5 binding sites at duct (*) and demilune (↑) level in CM, FP and CP groups.

Sample reactivities to immunohistochemical treatments are summarized in Table 1.

Table 1. Sample reaction intensity expressed in arbitrary units toward immunohistochemical targets.

Antigen	Secretory Structures	Diet Groups		
		CM	FP	CP
APLN	Acini	−	−	−
	Demilunes	−	−	−
	Ducts	+	±/+	±/+
APLNR	Acini	−	−	−
	Demilunes	−	+	+
	Ducts [a]	±	+	+
	Ducts [b]	−	++	++
APQ5	Acini	−	−	−
	Demilunes	±	±	+
	Ducts	−	−	±

CM: coarse meal diet; FP: fine pellet diet; CP: coarse pellet diet. [a] Epithelial lining; [b] epithelial more reactive cells. APLN = Apelin; APLNR = Apelin Receptor; AQP5 = Aquaporin 5. Intensity of the staining: − = negative, ± = weak, + = moderate, ++ = strong.

Sheep abomasum and MG, used as positive controls, showed a binding pattern for APLN, APLNR and AQP5 (Figure S1), while staining was completely absent in the control sections where the primary antibodies were omitted and in sections incubated with normal rabbit IgG (Figure S2).

3.2. Glycohistochemistry

Glycohistochemical treatments evidenced a higher production of complex carbohydrates by the pig mandibular acini in FP and CP diets with respect to CM. In addition, comparing the reactivity of AB pH 2.5, AB pH1 and AB pH 0.5 among the three diets showed an increased production of acid glycoconjugates, particularly in the CP diet where acinar cells showed a weak positivity also to AB pH 0.5 (Figure 4).

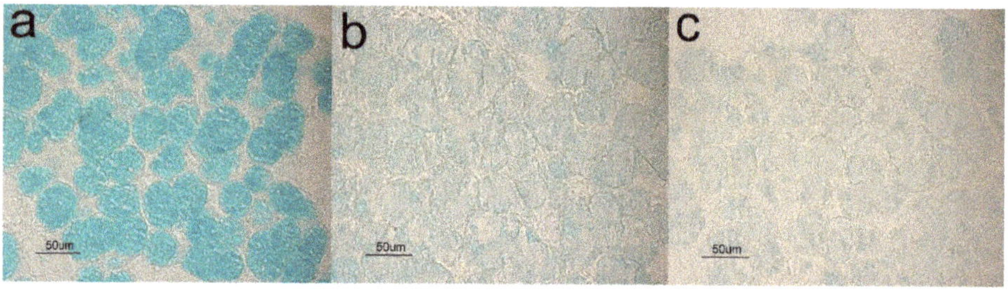

Figure 4. Pig mandibular gland. CP group shows mucous acini strongly reactive to Alcian Blue (AB) pH 2.5 (**a**), moderately positive to AB pH 1 (**b**) and weakly stained with AB pH 0.5 (**c**).

Sample reactivities to glycohistochemical treatments are summarized in Table 2.

Table 2. Sample reaction intensity expressed in arbitrary units toward glycohistochemical treatments.

Glycohistochemical Treatments	Secretory Structures	Diet Groups		
		CM	FP	CP
AB pH2.5	Acini	+/++	++	++
	Demilunes	−/±	±/+	±
	Ducts	−	−	−
AB pH1	Acini	±/+	±/+	±/+
	Demilunes	−	−	−
	Ducts	−	−	−
AB pH0.5	Acini	−	−	−/±
	Demilunes	−	−	−
	Ducts	−	−	−
PAS	Acini	++	++	+++
	Demilunes	±	±	±
	Ducts	−	−	−
AB/PAS	Acini	B+/R++	B++/R++	B+++/R++
	Demilunes	B+/R±	B±/R±	B+/R±
	Ducts	−	−	−

CM: coarse meal diet; FP: fine pellet diet; CP: coarse pellet diet. B = blue; R = red. AB = Alcian Blue; PAS = Periodic Acid Schiff.

3.3. Statistical Analysis

The significance of differences among the diet groups for each immuno- and glycohistochemical treatment are shown in Table 3.

Table 3. Statistical significance of differences ($p \leq 0.01$) for each histochemical treatment among different experimental treatments, as performed by one-way ANOVA and Kruskal–Wallis tests and respective pairwise comparisons, as performed by independent samples *t*-test and Wilcoxon–Mann–Whitney tests. *P*-values were adjusted for multiple testing using the Holm correction.

Histochemical Treatments		ANOVA/Kruskal–Wallis Test	*t*-test/Wilcoxon–Mann–Whitney Test		
			P CM vs. FP	*P* FP vs. CP	*P* CM vs. CP
APLN	Acini	–	–	–	–
	Demilunes	–	–	–	–
	Ducts	4.26×10^{-11}	1.76×10^{-7}	1.00	1.76×10^{-7}
APLNR	Acini	–	–	–	–
	Demilunes	1.92×10^{-6}	5.50×10^{-6}	1.00	5.50×10^{-6}
	Ducts [a]	$<10^{-16}$	8.36×10^{-13}	1.00	1.61×10^{-12}
	Ducts [b]	1.60×10^{-6}	5.50×10^{-6}	1.00	5.50×10^{-6}
APQ5	Acini	–	–	–	–
	Demilunes	3.16×10^{-6}	1.00	1.92×10^{-5}	1.92×10^{-5}
	Ducts	6.57×10^{-9}	–	5.50×10^{-6}	5.50×10^{-6}
AB pH2.5	Acini	3.16×10^{-6}	1.92×10^{-5}	1.00	1.92×10^{-5}
	Demilunes	2.33×10^{-8}	1.80×10^{-5}	1.90×10^{-5}	1.92×10^{-5}
	Ducts	–	–	–	–
AB pH1	Acini	1.00	–	–	–
	Demilunes	–	–	–	–
	Ducts	–	–	–	–
AB pH0.5	Acini	6.12×10^{-9}	–	5.28×10^{-6}	5.28×10^{-6}
	Demilunes	–	–	–	–
	Ducts	–	–	–	–
PAS	Acini	$<10^{-16}$	1.00	$<10^{-16}$	$<10^{-16}$
	Demilunes	1.00	–	–	–
	Ducts	–	–	–	–
AB/AS	Acini	$<10^{-16}$	1.08×10^{-13}	4.08×10^{-14}	$<10^{-16}$
	Demilunes	1.03×10^{-12}	1.24×10^{-9}	1.05×10^{-10}	1.00
	Ducts	–	–	–	–

CM: coarse meal diet; FP: fine pellet diet; CP: coarse pellet diet. [a] Epithelial lining; [b] epithelial more reactive cells. APLN = Apelin; APLNR = Apelin Receptor; AQP5 = Aquaporin 5; AB = Alcian Blue; PAS = Periodic Acid Schiff.

The significance of the differences among the serial treatments AB pH 2.5, AB pH 1 and AB pH 0.5 for the three diet groups are shown in Table 4.

Table 4. Statistical significance of differences ($p \leq 0.01$) among different pH AB serial treatments as performed by Wilcoxon signed-rank tests. *P*-values were adjusted for multiple testing using the Holm correction.

Diet Groups	Secretory Structures	Wilcoxon Signed-Rank Test	
		AB pH 2.5 vs. AB pH 1	AB pH 1 vs. AB pH 0.5
CM	Acini	0.0035	0.0035
	Demilunes	0.0035	–
	Ducts	–	–
FP	Acini	0.0035	0.0035
	Demilunes	0.0035	–
	Ducts	–	–
CP	Acini	0.0035	0.0035
	Demilunes	0.0035	–
	Ducts	–	–

CM: coarse meal diet; FP: fine pellet diet; CP: coarse pellet diet. AB = Alcian Blue

Glycohistochemical evidence allows one to hypothesize that secretory structures of pig MG produce the complex carbohydrate types listed, in descending semi-quantitative order, in Table 5.

Table 5. Kinds of glycoconjugates produced by the pig mandibular gland (MG) secretory structures, listed in descending order as also evidenced by a differentiated style.

Secretory Structures	CM	FP	CP
Acini	Hyaluronic acid and/or Chondroitin-like GAGs Chondroitin sulfate A/B/C-like GAGs	**Hyaluronic acid and/or Chondroitin-like GAGs** Chondroitin sulfate A/B/C-like GAGs	**Hyaluronic acid and/or Chondroitin-like GAGs** Chondroitin sulfate A/B/C-like GAGs Heparin and/or heparan-sulfate-like GAGs
Demilunes	Neutral and acid glycoproteins	Neutral and acid glycoproteins	Acid glycoproteins

CM: coarse meal diet; FP: fine pellet diet; CP: coarse pellet diet. GAGs = Glycosaminoglycans.

4. Discussion

The glycohistochemical and immunohistochemical investigations were carried out on the MG obtained from pigs fed with different dietary treatments based on different grinding intensities and compactions of the same diet [1]. Animals were enrolled in a wide research project, running in the Institute of Animal Nutrition of the University of Hannover, aiming to test the effects of different feed of physical forms on the health, production and welfare of pigs. This research extends the knowledge about the diet modifications on pig welfare [31,34] and could be useful for further investigation also related to diet and farm income relationships [35].

Our findings allow considering that the different feed physical form is capable of inducing morphological and functional modification of the MG, on the basis of presence and distribution of molecules differently modulating the saliva composition.

The different compaction of the diet can reasonably produce different perception extents of the physical form of the bolus, with consequent different voluntary chewing and adaptation of salivary production. In response to the feed physical form, demilunes of pig MG increase in size and the acidity of produced glycoproteins change in view of CM vs. FP vs. CP diets. Acinar cells increase their production of hyaluronic acid-/chondroitin-like Glycosaminoglycans (GAGs) in FP and CP groups; in addition, in the MG of pigs fed with the CP diet a low amount of heparin/heparan-sulfate-like GAGs production is stimulated. Additional production of complex carbohydrates observed in MG from pigs of FP and CP groups is able to call forth higher watery saliva, thus promoting a better feed softening facilitating the formation and amalgamation of the bolus. A handful of evidence pointed out that the increased glycoconjugate production, especially when they are highly acid, allows drawing a large amount of water in many organ tissues and in different animal species [36–41]. In addition, complex carbohydrates could envelope pathogenic bacteria, viruses and parasites acting as hapten-like binding sites, preventing their attachment to the mucosae potentially damaged by the harder feed [34,42,43].

Pig MG demilunes are weakly reactive to AQP5 in CM and FP diet but increase their positivity in the CP diet samples. The latter also shows AQP5 binding sites in the ducts. This evidence further supports the hypothesis of an augmented demand for water in the mouth. It was stated that AQP5 in the salivary glands is probably involved in providing for a suitable amount of fluid to be secreted [20,44]; in addition, its expression can be affected by feed and environmental modification, in particular at serous adenomere level [30].

APLN binding sites observed in ducts of MG from the pigs fed with the CM diet decrease their reactivity in MG of samples from FP and CP dietary groups. On the contrary, such structures increased their reactivity to APLNR from CM to FP and CP samples. In addition, in the two last groups, there was a strong positivity for APLNR in some cells. Demilunes also showed binding sites for APLNR in FP and CP diet samples. Comparing the specific APLN and APLNR localization and the differences in the binding sites reactivity among the CM, FP and CP diet samples, it is likely to hypothesize that in pig MG the apelinergic system performs an endocrine control on the demilunes activity facilitating the production of a more fluid saliva. In addition, pig MG ducts seem to be affected by a paracrine control by the apelinergic system, which could allow an inhibition in the water absorption, enhancing the saliva water content.

According to the above considerations, the literature states that adipokines can act through endocrine action as well as paracrine and autocrine ones. The interaction of those mechanisms has already been described in a wide range of physiological and physio-pathological processes in different animal species [45–47]. A paracrine action for apelin was already suggested for sheep uterus [15] and mammary gland [14] where the molecule aimed at regulating the gland secretive action. Up to now the apelin action on salivary glands has not yet been described. However, the role of this adipokine on the gastroenteric apparatus is well known and numerous studies suggest a role for apelin in both exocrine and endocrine functions [48–50]. Apelin performs important functions in the pancreas, an organ that shows, as its exocrine parenchyma regards, similar anatomical structures and physiological functions with salivary glands [51]. Particularly, apelin injected intravenously decreased pancreatic juice volume, protein and trypsin outputs in a dose-dependent manner while intraduodenal administration of apelin increases pancreatic protein and trypsin secretion [52].

Finally, in pig mandibular gland it was stated that different feed physical forms affect the expression of both the leptinergic system and cannabinoid type 1 and 2 receptors [1,2], as now observed for the apelinergic system.

5. Conclusions

In conclusion, obtained data suggest that the diverse feed physical forms, performing differentiated mechanical stimuli in the mouth, likely allow a different physiological behavior of pig MG. The intense chewing activity linked to the highest feed compaction and hardness promotes an increase in pig MG secretion; in addition, saliva becomes more fluid and richer in acid glycoconjugates in order to better lubricate the bolus and protect the mouth mucosae. The apelinergic system is likely involved in the above modifications enhancing both the fluidity and the quantity of serous saliva by the pig mandibular gland.

Supplementary Materials: The following are available online at http://www.mdpi.com/2076-2615/10/5/910/s1, Table S1: Feed components and chemical composition. Table S2. Specificity of the primary antibodies used. Figure S1: Immunohistochemistry positive controls. Figure S2: Immunohistochemistry negative controls.

Author Contributions: Conceptualization, P.S. and C.D.; methodology and validation, P.S., F.M. and C.D.; formal analysis, F.M.T.; investigation, E.D.F., M.G.C. and J.K.; data curation, F.M.; writing—original draft preparation, P.S., C.D. and E.D.F.; writing—review and editing, E.D.F. and C.D.; supervision, P.S. and M.G.C. All authors have read and agreed to the published version of the manuscript.

Funding: This research received no external funding.

Acknowledgments: The authors wish to thank Paola Coliolo and Maria Gabriella Mancini for their excellent technical assistance.

Conflicts of Interest: The authors declare no conflicts of interest.

References

1. Cappai, M.G.; Dall'Aglio, C.; Sander, S.J.; Ratert, C.; Dimauro, C.; Pinna, W.; Kamphues, J. Different physical forms of one diet fed to growing pigs induce morphological changes in mandibular glands and local leptin (Ob) production and receptor (ObR) expression. *J. Anim. Physiol. Anim. Nutr.* **2016**, *100*, 1067–1072. [CrossRef] [PubMed]

2. Pirino, C.; Cappai, M.G.; Maranesi, M.; Tomassoni, D.; Giontella, A.; Pinna, W.; Boiti, C.; Kamphues, J.; Dall'Aglio, C. The presence and distribution of cannabinoid type 1 and 2 receptors in the mandibular gland: The influence of different physical forms of diets on their expression in piglets. *J. Anim. Physiol. Anim. Nutr.* **2018**, *102*, 870–876. [CrossRef] [PubMed]
3. Ito, k. The effect of food consistency and dehydration on reflex parotid and submandibular salivary secretion in conscious rats. *Arch. Oral. Biol.* **2001**, *46*, 353–363. [CrossRef]
4. Johnson, D.A. Effect of a ground versus a pelleted bulk diet on the rat parotid gland. *Arch. Oral Biol.* **1981**, *26*, 1091–1093. [CrossRef]
5. Johnson, D.A.; Cardenas, H. Effects of food mastication on rat parotid gland adrenergic and cholinergic cell surface receptors. *Crit. Rev. Oral Biol. Med.* **1993**, *4*, 591–597. [CrossRef]
6. Kurahashi, M. The effect of dietary consistency and water content on the parotid glands of submandibular and sublingual duct-ligated rats. *Arch. Oral Biol.* **2002**, *47*, 369–374. [CrossRef]
7. Johnson, D.A. Effect of a liquid diet on the protein composition of rat parotid saliva. *J. Nutr.* **1982**, *112*, 175–181. [CrossRef]
8. Johnson, D.A.; Lopez, H.; Navia, J.M. Effects of protein deficiency and diet consistency on the parotid gland and parotid saliva of rats. *J. Dent. Res.* **1995**, *74*, 1444–1452. [CrossRef]
9. Tatemoto, K.; Hosoya, M.; Habata, Y.; Fujii, R.; Kakegawa, T.; Zou, M.X.; Kawamata, Y.; Fukusumi, S.; Hinuma, S.; Kitada, C.; et al. Isolation and characterization of a novel endogenous peptide ligand for the human APJ receptor. *Biochem. Biophys. Res. Commun.* **1998**, *251*, 471–476. [CrossRef]
10. Kawamata, Y.; Habata, Y.; Fukusumi, S.; Hosoya, M.; Fujii, R.; Hinuma, S.; Nishizawa, N.; Kitada, C.; Onda, H.; Nishimura, O.; et al. Molecular properties of apelin: Tissue distribution and receptor binding. *Biochim. Biophys. Acta* **2001**, *1538*, 162–171. [CrossRef]
11. Kleinz, M.J.; Skepper, J.N.; Davenport, A.P. Immunocytochemical localisation of the apelin receptor, APJ, to human cardiomyocytes, vascular smooth muscle and endothelial cells. *Regul. Pept.* **2005**, *126*, 233–240. [CrossRef] [PubMed]
12. Medhurst, A.D.; Jennings, C.A.; Robbins, M.J.; Davis, R.P.; Ellis, C.; Winborn, K.Y.; Hervieu, G.; Riley, G.; Bolaky, J.E.; Herrity, N.C.; et al. Pharmacological and immunohistochemical characterization of the APJ receptor and its endogenous ligand apelin. *J. Neurochem.* **2003**, *84*, 1162–1172. [CrossRef] [PubMed]
13. O' Carroll, A.M.; Selby, T.L.; Palkovits, M.; Lolait, S.J. Distribution of mRNA encoding B78/apj, the rat homologue of the human APJ receptor, and its endogenous ligand apelin in brain and peripheral tissues. *Biochim. Biophys. Acta* **2000**, *1492*, 72–80. [CrossRef]
14. Mercati, F.; Maranesi, M.; Dall'Aglio, C.; Petrucci, L.; Pasquariello, R.; Tardella, F.M.; De Felice, E.; Scocco, P. Apelin system in mammary gland of sheep reared in semi-natural pastures of the central apennines. *Animals* **2018**, *8*, 223. [CrossRef] [PubMed]
15. Mercati, F.; Scocco, P.; Maranesi, M.; Acuti, G.; Petrucci, L.; Cocci, P.; Renzi, A.; De Felice, E.; Dall'Aglio, C. Apelin system detection in the reproductive apparatus of ewes grazing on semi-natural pasture. *Theriogenology* **2019**, *139*, 156–166. [CrossRef]
16. Nizam, M.; Basoglu, O.K.; Tasbakan, M.S.; Lappin, D.F.; Buduneli, N. Is there an association between obstructive sleep apnea syndrome and periodontal inflammation? *Clin. Oral Investig.* **2015**, *20*, 659–668. [CrossRef]
17. Scocco, P.; Menghi, G.; Ceccarelli, P.; Pedini, V. Lectin histochemistry and identification of O-acetilated sialoderivatives in the horse sublingual gland. *Eur. J. Histochem.* **1999**, *43*, 47–54.
18. Lynge Pedersen, A.M.; Belstrøm, D. The role of natural salivary defences in maintaining a healthy oral microbiota. *J. Dent.* **2019**, *80*, S3–S12. [CrossRef]
19. Nagahama, M.; Ma, N.; Semba, R.; Naruse, S. Aquaporin 1 immunoreactive enteric neurons in the rat ileum. *Neurosci. Lett.* **2006**, *395*, 206–210. [CrossRef]
20. Ishikawa, Y.; Cho, G.; Yuan, Z.; Skowronski, M.T.; Pan, Y.; Ishida, H. Water channels and zymogen granules in salivary glands. *J. Pharmacol. Sci.* **2006**, *100*, 495–512. [CrossRef]
21. Krane, C.M.; Melvin, J.E.; Nguyen, H.V.; Richardson, L.; Towne, J.E.; Doetschman, T.; Menon, A.G. Salivary acinar cells from aquaporin 5-deficient mice have decreased membrane water permeability and altered cell volume regulation. *J. Biol. Chem.* **2001**, *276*, 23413–23420. [CrossRef] [PubMed]

22. Kawedia, J.D.; Nieman, M.L.; Boivin, G.P.; Melvin, J.E.; Kikuchi, K.I.; Hand, A.R.; Lorenz, J.N.; Menon, A.G. Interaction between transcellular and paracellular water transport pathways through Aquaporin 5 and the tight junction complex. *Proc. Natl. Acad. Sci. USA* **2007**, *104*, 3621–3626. [CrossRef] [PubMed]
23. Jung, J.S.; Preston, G.M.; Smith, B.L.; Guggino, W.B.; Agre, P. Molecular structure of the water channel through aquaporin CHIP. The hourglass model. *J. Biol. Chem.* **1994**, *269*, 14648–14654. [PubMed]
24. Murata, K.; Mitsuoka, K.; Hirai, T.; Walz, T.; Agre, P.; Heymann, J.B.; Engel, A.; Fujiyoshi, Y. Structural determinants of permeation through aquaporin-1. *Nature* **2000**, *407*, 599–605. [CrossRef]
25. Sui, H.; Han, B.G.; Lee, K.J.; Walian, P.; Jap, B.K. Structural basis of water-specific transport through the AQP1 water channel. *Nature* **2001**, *414*, 872–878. [CrossRef]
26. Agre, P.; King, L.S.; Yasui, M.; Guggino, W.B.; Ottersen, O.P.; Fujiyoshi, Y.; Engel, A.; Nielsen, S. Aquaporin water channels— from atomic structure to clinical medicine. *J. Physiol.* **2002**, *542*, 3–16. [CrossRef]
27. Yasui, M. Molecular mechanisms and drug development in aquaporin water channel diseases: Structure and function of aquaporins. *J.Pharmacol. Sci.* **2004**, *96*, 260–263. [CrossRef]
28. Dall'Aglio, C.; Scocco, P.; Maranesi, M.; Petrucci, L.; Acuti, G.; De Felice, E.; Mercati, F. Immunohistocehmical identification of resistin in the uterus of ewes subjected to different diets: Preliminary results. *Europ. J. Histochem.* **2019**, *63*, 3020.
29. Pappalardo, A.; Porreca, I.; Caputi, L.; De Felice, E.; Schulte-Merker, S.; Zannini, M.; Sordino, P. Thyroid development in zebrafish lacking Taz. *Mech. Dev.* **2015**, *138*, 268–278. [CrossRef]
30. Scocco, P.; Aralla, M.; Catorci, A.; Belardinelli, C.; Arrighi, S. Immunodetection of aquaporin 5 in sheep salivary glands related to pasture vegetative cycle. *Folia Histochem. Cytobiol.* **2011**, *49*, 458–464. [CrossRef]
31. Dall'Aglio, C.; Mercati, F.; Faeti, V.; Acuti, G.; Trabalza-Marinucci, M.; De Felice, E.; Tardella, F.M.; Franciosini, M.P.; Casagrande Proietti, P.; Catorci, D.; et al. Immuno- and glyco-histochemistry as a tool to evaluate the oregano supplemented feed effects in pig gut. *Eur. J. Histochem.* **2020**, *64*, 3110. [CrossRef] [PubMed]
32. Dall'Aglio, C.; Pascucci, L.; Mercati, F.; Boiti, C.; Ceccarelli, P. Localization of the orexin system in the gastrointestinal tract of fallow deer. *Acta Histochem.* **2012**, *114*, 74–78.
33. R Core Team. *R: A Language and Environment for Statistical Computing*; R Foundation for Statistical Computing: Vienna, Austria, 2019; Available online: http://www.R-project.org/ (accessed on 29 February 2020).
34. Mercati, F.; Dall'Aglio, C.; Acuti, G.; Faeti, V.; Tardella, F.M.; Pirino, C.; De Felice, E.; Scocco, P. Oregano feed supplementation affects glycoconjugates production in swine gut. *Animals* **2020**, *10*, 149. [CrossRef]
35. Scocco, P.; Rivaroli, S.; Mercati, F.; Tardella, F.M.; Malfatti, A.; De Felice, E.; Catorci, A. Anatomy for economy: Starting from the rumen keratinization degree to enhance the farm income. *Econ. Agro-Aliment.* **2018**, *20*, 261–272. [CrossRef]
36. Domeneghini, C.; Arrighi, S.; Radaelli, G.; Bosi, G.; Veggetti, A. Histochemical analysis of glycoconjugate secretion in the alimentary canal of *Anguilla anguilla* L. *Acta Histochem.* **2005**, *106*, 477–487. [CrossRef] [PubMed]
37. Scocco, P.; Pedini, V. Histochemical characterisation of complex carbohydrates expressed in the alimentary tract of chickens. *Vet. J.* **2010**, *185*, 228–230. [CrossRef]
38. Schulte, B.A.; Spicer, S.S. Light microscopic detection of sugar residues in glycoconjugates of salivary glands and the pancreas with lectin-horseradish peroxidase conjugates. I. Mouse. *Histochem. J.* **1983**, *15*, 1217–1238. [CrossRef]
39. Menghi, G.; Bondi, A.M.; Accili, D.; Materazzi, G. Visualization of carbohydrate chains in rabbit salivary glands by means of enzymatic degradation and plant lectins. *Acta Histochem.* **1988**, *84*, 163–177. [CrossRef]
40. Scocco, P.; Pedini, V. Localization of influenza virus sialoreceptors in equine respiratory tract. *Histol. Histopathol.* **2008**, *23*, 973–978.
41. Pedini, V.; Dall'Aglio, C.; Parillo, F.; Scocco, P. Glycoconjugate distribution in gastric fundic mucosa of *Umbrina cirrosa* L. revealed by lectin histochemistry. *J. Fish. Biol.* **2005**, *66*, 222–229. [CrossRef]
42. Nieuw-Amerongen, A.V.; Bolscher, J.G.M.; Veerman, E.C.I. Salivary mucins: Protective functions in relation to their diversity. *Glycobiology* **1995**, *5*, 733–740. [CrossRef] [PubMed]
43. Scocco, P.; Pedini, V. Equine mandibular gland: In situ characterization of sialoderivatives. *Equine Vet. J.* **2006**, *38*, 410–415. [CrossRef] [PubMed]
44. Sugiya, H.; Matsuki-Fukushima, M.; Hashimoto, S. Role of aquaporins and regulation of secretory vesicle volume in cell secretion. *J. Cell. Mol. Med.* **2008**, *12*, 1486–1494. [CrossRef] [PubMed]

45. Lago, F.; Dieguez, C.; Gómez-Reino, J.; Gualillo, O. The emerging role of adipokines as mediators of inflammation and immune responses. *Cytokine Growth Factor Rev.* **2007**, *8*, 313–325. [CrossRef] [PubMed]
46. Kälin, R.E.; Kretz, M.P.; Meyer, A.M.; Kispert, A.; Heppner, F.L.; Brändli, A.W. Paracrine and autocrine mechanisms of apelin signalling govern embryonic and tumor angiogenesis. *Dev. Biol.* **2007**, *305*, 599–614. [CrossRef]
47. Masri, B.; Knibiehler, B.; Audigier, Y. Apelin signalling: A promising pathway from cloning to pharmacology. *Cell Signal.* **2005**, *17*, 415–426. [CrossRef]
48. Reaux-Le Goazigo, A.; Iturrioz, X.; Llorens-Cortens, C. *Encyclopedia of Neuroscience*; Elsevier: Amsterdam, NL, USA, 2009; pp. 509–516.
49. Wang, G.; Anini, Y.; Wei, W.; Qi, X.; O'Carroll, A.M.; Mochizuki, T.; Wang, H.Q.; Hellmich, M.R.; Englander, E.W.; Greeley, G.H., Jr. Apelin, a new enteric peptide: Localization in the gastrointestinal tract, ontogeny, and stimulation of gastric cell proliferation and of cholecystokinin secretion. *Endocrinol.* **2004**, *145*, 1342–1348. [CrossRef]
50. Han, S.; Wang, G.; Qi, X.; Lee, H.M.; Englander, E.W.; Greeley, G.H., Jr. A possible role for hypoxia-induced apelin expression in enteric cell proliferation. *Am. J. Physiol. Regul. Integr. Comp. Physiol.* **2008**, *294*, R1832–R1839. [CrossRef]
51. Rakonczay, Z., Jr.; Vág, J.; Földes, A.; Nagy, K.; Nagy, Á.; Hegyi, P.; Varga, G. Chronic inflammation in the pancreas and salivary glands-lessons from similarities and differences in pathophysiology and treatment modalities. *Curr. Pharm. Des.* **2014**, *20*, 1104–1120.
52. Kapica, M.; Puzio, I.; Kato, I.; Kuwahara, A.; Zabielski, R. Role of feed-regulating peptides on pancreatic exocrine secretion. *J. Physiol. Pharmacol.* **2008**, *59*, 145–159.

animals

MDPI

Article

Effect of Acrylamide Supplementation on the CART-, VAChT-, and nNOS-Immunoreactive Nervous Structures in the Porcine Stomach

Katarzyna Palus *, Michał Bulc and Jarosław Całka

Department of Clinical Physiology, Faculty of Veterinary Medicine, University of Warmia and Mazury in Olsztyn, 10-718 Olsztyn, Poland; michal.bulc@uwm.edu.pl (M.B.); calkaj@uwm.edu.pl (J.C.)
* Correspondence: katarzyna.palus@uwm.edu.pl

Received: 19 February 2020; Accepted: 25 March 2020; Published: 26 March 2020

Simple Summary: The progress of civilization has provided people with virtually unlimited access to food products. However, while the pace of life has increased, the consumption of products with high levels of acrylamide (e.g., chips, corn flakes or coffee) has also increased. The gastrointestinal tract is the first-exposure site for noxious substances ingested with food and it is also often the first defence mechanism. Changes in the expression of neuroactive substances in the intramural neurons of the enteric nervous system (ENS) are a common preclinical symptom of the harmful effect of pathological factors on the body. Using the double immunofluorescence staining method, it was established that supplementation with low and high doses of acrylamide resulted in alterations of the porcine stomach neuron phenotype, which was reflected in an increased number of the cocaine- and amphetamine-regulated transcript (CART)-, vesicular acetylcholine transporter (VAChT)-, and neuronal isoform of nitric oxide synthase (nNOS)-immunoreactive neurons. The recorded changes revealed that even low doses of acrylamide influence the nervous structures located in the porcine gastric wall. This may result from the neurotoxicity of acrylamide or from the response of the ENS to acrylamide-induced inflammation and suggests an important role of the ENS in protecting the gastrointestinal tract during acrylamide intoxication.

Abstract: Acrylamide is found in food products manufactured with high-temperature processing, and exposure to acrylamide contained in food products may cause a potential risk to human health. The aim of this investigation was to demonstrate the changes in the population of CART-, nNOS-, and VAChT-immunoreactive enteric neurons in the porcine stomach in response to supplementation of low and high acrylamide doses. The study was carried out with 15 Danish landrace gilts divided into three experimental groups: the control group—animals were administered empty gelatine capsules; the low-dose group—animals were administrated a tolerable daily intake (TDI) dose (0.5 μg/kg of body weight (b.w.)/day) of acrylamide capsules, and the high-dose group—animals were administrated high-dose (ten times higher than TDI: 5 μg/kg b.w./day) acrylamide capsules for 28 days. Using the double immunofluorescence staining method, it was established that supplementation with low and high doses of acrylamide resulted in alterations of the porcine stomach neuron phenotype, which was reflected in an increased number of CART-, VAChT-, and nNOS-immunoreactive neurons. These changes were accompanied by an increased density of CART-, VAChT-, and nNOS-positive fibres. The results suggest that the enteric nervous system plays an important role in protecting the gastrointestinal tract during acrylamide intoxication.

Keywords: acrylamide; stomach; enteric nervous system; CART; VAChT; nNOS; pig

1. Introduction

The oesophageal, gastric, and intestinal wall encompasses a dense network of nerve cells called the enteric nervous system (ENS) [1]. These neurons are organised in the enteric ganglions that are interlinked with a dense network of fibres and are thereby arranged in the plexuses. It is known that in the porcine stomach and in other large animals, as well as in humans, there are two plexuses: the myenteric plexus, situated between the circular and the longitudinal muscular layer, and the submucous plexus located between the tunica muscularis and lamina propria of the mucosa. In the small and large intestines, there are two submucous plexuses: the outer submucous plexus (OSP), located on the inner side of the circular muscular layer, and the inner submucous plexus (ISP), situated near the tunica muscularis in the intestinal mucosa [2–4]. It has been demonstrated that the ENS neurons control numerous physiological functions of the gastrointestinal tract (GIT), such as the secretion of digestive fluids, gastric and intestinal motility, and the absorption of nutrients, which are predominantly regulated independently of the central nervous system (CNS) [5]. Each ENS neuron synthesises and secretes a variety of neuroactive substances which are involved in both controlling the physiological functions and in protecting the neurons against the harmful effects of many irritant factors [2,5]. The physiological function of the cocaine- and amphetamine-regulated transcript (CART) in the GIT has not yet been fully elucidated. However, it has been shown that CART enhances gastric acid secretion, causes an inhibitory effect on gastric emptying, and stimulates peristalsis in the colon [6]. The presence of CART in enteric neurons in the porcine stomach, small intestine, and descending colon was described in previous studies [2,7,8]. In contrast, a neuronal isoform of nitric oxide synthase (nNOS) is an enzyme that catalyses the production of nitric oxide (NO) from L-arginine and is considered the major marker of nitrergic neurons. NO is a gaseous inhibitory neurotransmitter. It has a vasodilatory effect, controls the mesenteric blood flow, inhibits intestinal hormone and digestive fluid secretion, and demonstrates a protective effect on the gastric mucosa [9]. Nitric oxide as a non-adrenergic, non-cholinergic inhibitory neurotransmitter detected in different part of the porcine GIT, such as the stomach, duodenum, jejunum, ileum, and descending colon [10–12]. In turn, vesicular acetylcholine transporter (VAChT) is believed to be the major marker of cholinergic neurons. It stimulates the secretion of digestive enzymes and intestinal hormones and has a stimulating effect on the gastrointestinal motility [13]. Previous reports confirmed that VACHT-containing cholinergic neurons constitute a significant population of ENS neurons in the porcine GIT. Their presence was noted in the stomach and small intestines [14,15].

Furthermore, numerous studies have demonstrated that the ENS neurons exhibit neuronal plasticity. This kind of plasticity is defined as the adaptation to changes in the external and internal environment, encompassing functional changes, such as a change in the number and transmission of synaptic connections, modification of intracellular signalling cascades, regulation of gene expression, and changes in neurotransmitter secretion [5,16]. Previous studies in animal models and in humans have shown that different pathological conditions in the gastrointestinal tract (inflammation, gastric ulcers, treatment with nonsteroidal anti-inflammatory drugs (NSAIDs), diabetes mellitus, mycotoxins in feed or bisphenol-A intoxication) changed the neurochemical profile of the enteric neurons and had a significant effect on modifying the number of CART-, nNOS- or VAChT-immunoreactive neurons [2,7,10,11,17–19].

Acrylamide (ACM) (an α, β-unsaturated reactive monomer) is used in many industrial branches, such as polyacrylamide production, the manufacturing of beauty products, glues, paper, and polyacrylamide gels [20]. ACM is known to have carcinogenic and genotoxic properties, a harmful effect on reproduction, and is said to be neurotoxic [20,21]. Among the many adverse effects of acrylamide, only its neurotoxic effects have been documented in humans [21]. However, only after discovering that ACM is found in food products manufactured with high-temperature processing was a series of studies conducted to investigate food-related exposure as a potential risk to human health [22]. ACM is synthesized in the Maillard reaction between L-asparagine and reducing sugars. A very high level of acrylamide is found in potato chips, French fries, corn flakes, crackers, and coffee, all food products which are very popular among young consumers [23]. Epidemiological studies have demonstrated that the average acrylamide intake in adults varies between 0.3 and 0.8 μg/kg body weight (b.w.) per day [24]. The maximum safe intake levels have not yet been determined for acrylamide found in food products. Despite numerous studies and confirmation that the gastrointestinal tract is the main absorption route of ACM, little is known about its effect on the ENS neurons.

The aim of this study was to demonstrate the changes in the population of CART-, nNOS-, and VAChT-immunoreactive enteric neurons in the porcine stomach in response to supplementation of low and high acrylamide doses. The study was conducted on pigs which (as an omnivorous species) have similar anatomy and physiological processes to humans [25]. For years, pigs have been used in biomedical research, especially in studies investigating the gastrointestinal system [26].

2. Materials and Methods

The study was carried out with 15 Danish landrace gilts with an approximate body weight of 15 kg, originating from a farm. The animals were kept in common pens suitable for their age, were provided water ad libitum, and were fed a commercial pig feed mixture. After a one-week acclimatization period, the gilts were divided into three experimental groups: 1. control group (C group, n = 5): the animals were administered empty gelatine capsules; 2. low-dose group (LD group, n = 5): the animals were administrated a tolerable daily intake (TDI) dose (0.5 μg/kg b.w./day) of acrylamide (>99%; Sigma–Aldrich, Poznań, Poland) capsules; 3. high-dose group (HD group, n = 5): the animals were administrated high-dose (ten times higher than TDI: 5 μg/kg b.w./day) acrylamide capsules. All experimental procedures were approved by the Local Ethical Committee for Experiments on Animals in Olsztyn (Approval No. 11/2017). In each group, the animals were administered the specific capsules with morning feeding for four weeks. After a 28-day treatment, all gilts were euthanized with a lethal anaesthetic injection (sodium pentobarbital, Morbital, Biowet Puławy, Puławy, Poland), and stomach samples from the cardia, corpus, and pylorus region (Figure 1) were immediately collected and fixed by immersion in a 4% buffered solution of paraformaldehyde (pH = 7.4) for 1 h and then placed in a phosphate buffer solution (PBS, pH 7.4) for three days (the buffer was changed daily). Finally, samples were placed into an 18% buffered solution of sucrose (pH 7.4) for two weeks.

The illustration shows the porcine stomach with marked fragments collected for further study.

Frozen sections (14 μm thick) from the collected stomach samples were then processed with the double immunofluorescent staining method (as described previously by Palus et al. [27]). On day 1, the sections were dried at room temperature, rinsed three times in PBS (10 min), and blocked with the blocking mixture (10% horse serum, 0.1% bovine serum albumin in 0.1 M PBS, 1% Triton X-100, 0.05% thimerosal, and 0.01% sodium azide) for 1 h, rinsed again three times in PBS (10 min). The primary antibody solution was then added (Table 1), and the sections were incubated overnight in a humid chamber. On day 2, the sections were rinsed three times in PBS (10 min) and incubated with the secondary antibody solution (Table 1) for one hour, and after three rinsing steps in PBS (10 min), the tissue sections were immersed in a glycerol solution and covered with a coverslip. As a negative control, the following tests were used: pre-absorption for antisera with appropriate antigens and an omission and replacement test. No fluorescence was observed in any of the tests.

After staining, the sections were examined with an Olympus BX51 fluorescent microscope. To determine the number of individual neuron populations, a minimum of 500 PGP 9.5-positive neurons were analysed for each of the investigated neuroactive substances in both types of the plexuses (myenteric and submucous), and the number of PGP 9.5-positive neurons was assumed as 100%. The sections were separated by a distance of at least 200 μm to avoid counting the same neurons. To estimate the density of nerve fibres immunoreactive to CART, nNOS, and VAChT in the circular muscle layer (CML) and the submucosal/mucosal layer (S/ML), an arbitrary scale was used, in which (-) indicated a lack of fibres immunoreactive to CART, nNOS or VACHT, and (+++) indicated a dense network of nerve fibres. The results were statistically processed with Statistica 12 software (StatSoft Inc., Tulsa, OK, USA) and were expressed as the mean ± standard error of the mean (SEM). Statistically significant differences were evaluated with one-way analysis of variance (ANOVA) with Dunnett's test (* $p < 0.05$, ** $p < 0.01$, *** $p < 0.001$).

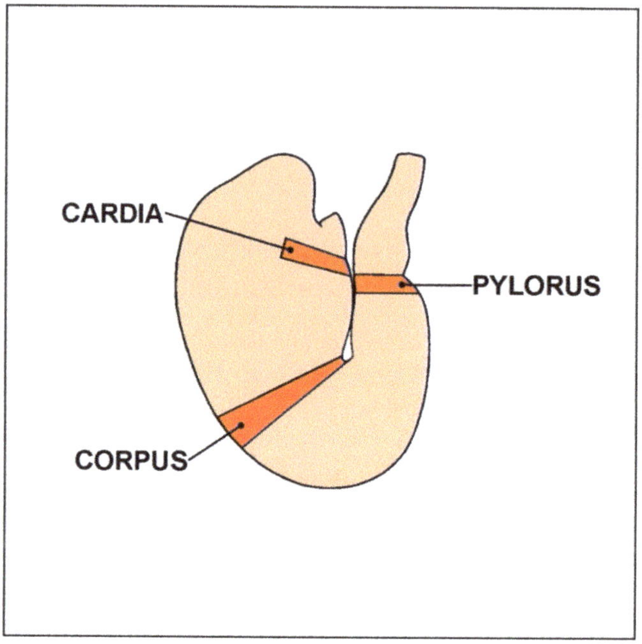

Figure 1. The porcine stomach.

Table 1. List of immunoreagents used in the present study.

Antigen	Host Species	Code	Dilution	Manufacturer/Supplier
		Primary antibodies		
PGP 9.5	Mouse	7863-2004	1:1000	Bio-Rad, Hercules, CA, USA
CART	Rabbit	H-003-61	1:16000	Phoenix Pharmaceuticals, Burlingame, CA, USA
VAChT	Rabbit	H-V007	1:2000	Phoenix Pharmaceuticals, Burlingame, CA, USA
nNOS	Rabbit	AB5380	1:2000	Sigma–Aldrich, Saint Louis, MO, USA
		Secondary antibodies		
Alexa Fluor 488 nm donkey anti-mouse IgG		A21202	1:1000	Thermo Fisher Scientific, Waltham, MA, USA
Alexa Fluor 546 nm goat anti-rabbit IgG		A11010	1:1000	Thermo Fisher Scientific, Waltham, MA, USA

PGP 9.5-protein gene product 9.5, CART- the cocaine- and amphetamine-regulated transcript, nNOS- the neuronal isoform of nitric oxide synthase, the cocaine vesicular acetylcholine transporter (VAChT).

3. Results

3.1. Myenteric Plexus

In the myenteric plexus, the administration of acrylamide increased the CART-, VAChT-, and nNOS-immunoreactive (IR) neuronal populations in all investigated stomach fragments (Figure 2). For CART, the major changes were found in the cardia, i.e., an increase from 22.72 ± 1.45% to 38.75 ± 1.18% and to 50.73 ± 2.61% in the LD and HD groups, respectively (Figure 2A, Figure 3A–C). In the stomach corpus region, the changes were also significant for low-dose and high-dose acrylamide supplementation (an increase from 29.86 ± 1.35% to 34.45 ± 0.72% and 49.80 ± 2.47%, respectively) (Figure 2B, Figure 3D–F). The smallest changes were demonstrated in the pylorus, where the number of CART-immunoreactive (CART-IR) neurons increased only in the HD group (from 15.15 ± 0.80% to 26.16 ± 1.39%) (Figure 2C, Figure 3G,I). In contrast, for VAChT, the greatest changes were found in the pylorus, where the number of VAChT-IR neurons increased from 23.41 ± 1.11% in the control group to 26.32 ± 0.69% in the LD group and 30.08 ± 1.23% in the HD group (Figure 2C, Figure 4G–I). In the corpus, the changes were statistically significant in both experimental groups (an increase from 18.10 ± 1.01% to 22.73 ± 0.86% and 29.42 ± 0.87%) (Figure 2B, Figure 4D–F), whereas in the cardia, a substantial increase in the VAChT-IR neuronal population (from 13.01 ± 0.87% to 19.23 ± 1.23%) was only found in the HD group (Figure 2A, Figure 4A,C). In the nNOS case, the greatest changes were demonstrated in the cardia, where the number of nNOS- immunoreactive (NOS-IR) neurons increased from 35.27 ± 1.85% to 43.69 ± 1.47% in the LD group and to 58.21 ± 1.04% in the HD group (Figure 2A, Figure 5A–C). A slightly smaller alteration was recorded in the pylorus: from 17.16 ± 1.53% in the control group to 22.89 ± 1.06% and to 36.03 ± 1.23% in the experimental groups (Figure 2C, Figure 5G–I). The smallest increase was detected in the corpus, as statistically significant differences were demonstrated only in the group supplemented with high acrylamide doses (an increase from 21.38 ± 1.10% to 33.36 ± 1.06%) (Figure 2D, Figure 5D,F).

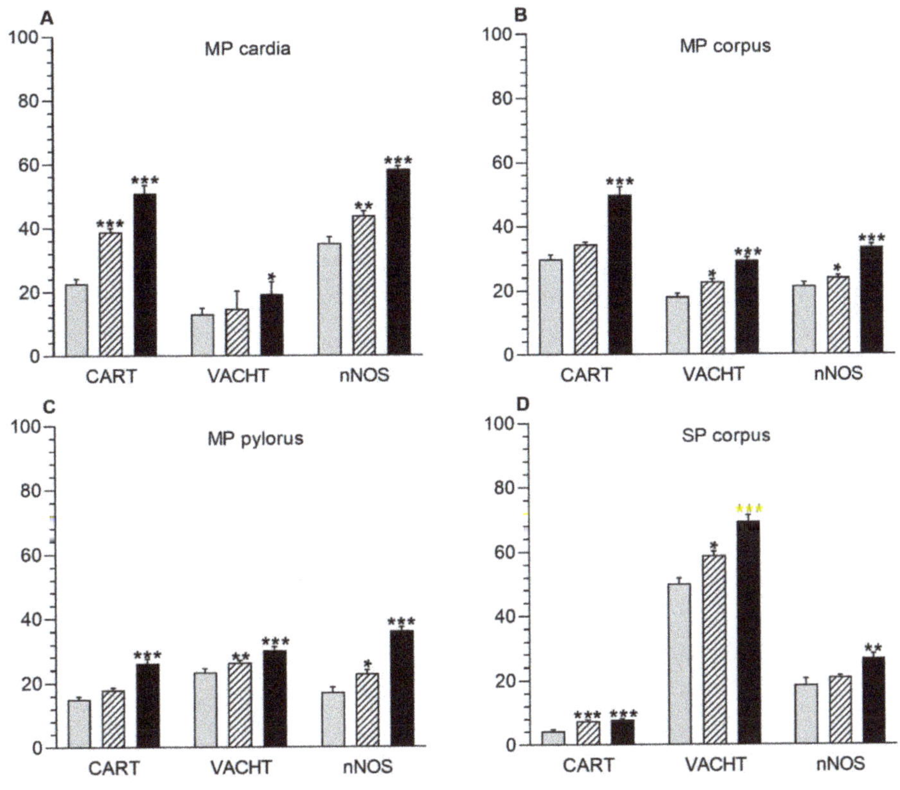

Figure 2. Acrylamide-induced changes in number of stomach enteric neurons immunoreactive to the cocaine- and amphetamine-regulated transcript (CART), the vesicular acetylcholine transporter (VAChT) and the neuronal isoform of nitric oxide synthase (nNOS). Enteric neurons immunoreactive to CART, VAChT, and nNOS in the myenteric plexuses (MP) of cardia (**A**), corpus (**B**), and pylorus (**C**) and in the submucous plexuses (SP) of corpus (**D**) in animals from the control (grey bar), LD (lined bar), and HD (black bar) groups. * $p < 0.05$,** $p < 0.01$,*** $p < 0.001$ indicate differences in the expression of particular substance studied in comparison to the control animals.

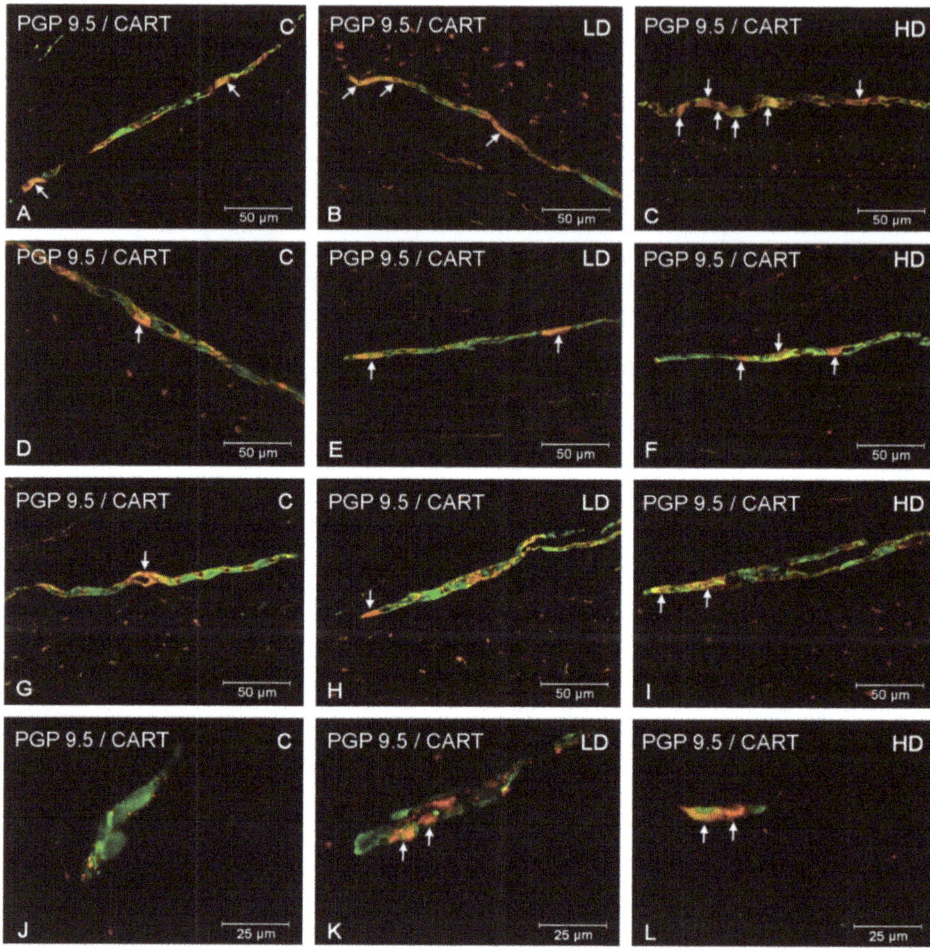

Figure 3. CART-immunoreactive neurons in the porcine stomach. The pictures show the cocaine- and amphetamine-regulated transcript (CART)-immunoreactive intramural neurons in the wall of the porcine stomach in the control group and in animals receiving low and high doses of acrylamide. (**A**)—myenteric neurons immunoreactive to protein gene-product 9.5 (PGP 9.5), used as a pan-neuronal marker, and CART in the cardia of control animals; (**B**)—myenteric neurons immunoreactive to PGP 9.5 and CART in the cardia of animals from LD group; (**C**)—myenteric neurons immunoreactive to PGP 9.5 and CART in the cardia of animals from HD group; (**D**)—myenteric neurons immunoreactive to PGP 9.5 and CART in the corpus of control animals; (**E**)—myenteric neurons immunoreactive to PGP 9.5 and CART in the corpus of animals from LD group; (**F**)—myenteric neurons immunoreactive to PGP 9.5 and CART in the corpus of animals from HD group; (**G**)—myenteric neurons immunoreactive to PGP 9.5 and CART in the pylorus of control animals; (**H**)—myenteric neurons immunoreactive to PGP 9.5 and CART in the pylorus of animals from LD group; (**I**)—myenteric neurons immunoreactive to PGP 9.5 and CART in the pylorus of animals from HD group; (**J**)—submucous neurons immunoreactive to PGP 9.5 and CART in the corpus of control animals; (**K**)—submucous neurons immunoreactive to PGP 9.5 and CART in the corpus of animals from LD group; (**L**)—submucous neurons immunoreactive to PGP 9.5 and CART in the corpus of animals from HD group.

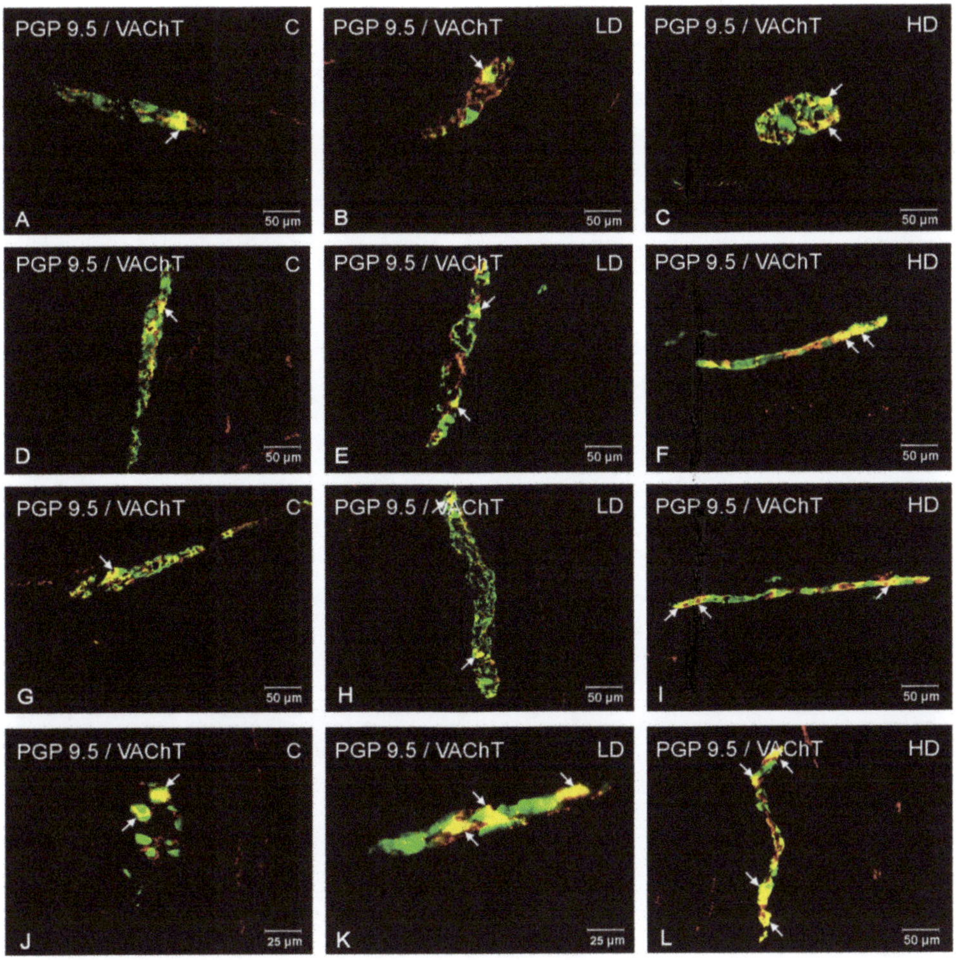

Figure 4. VAChT-immunoreactive neurons in the porcine stomach. The pictures show the vesicular acetylcholine transporter (VAChT)-immunoreactive intramural neurons in the wall of the porcine stomach in the control group and in animals receiving low and high doses of acrylamide. (**A**)—myenteric neurons immunoreactive to protein gene-product 9.5 (PGP 9.5), used as a pan-neuronal marker, and VAChT in the cardia of control animals; (**B**)—myenteric neurons immunoreactive to PGP 9.5 and VAChT in the cardia of animals from LD group; (**C**)—myenteric neurons immunoreactive to PGP 9.5 and VAChT in the cardia of animals from HD group; (**D**)—myenteric neurons immunoreactive to PGP 9.5 and VAChT in the corpus of control animals; (**E**)—myenteric neurons immunoreactive to PGP 9.5 and VAChT in the corpus of animals from LD group; (**F**)—myenteric neurons immunoreactive to PGP 9.5 and VAChT in the corpus of animals from HD group; (**G**)—myenteric neurons immunoreactive to PGP 9.5 and VAChT in the pylorus of control animals; (**H**)—myenteric neurons immunoreactive to PGP 9.5 and VAChT in the pylorus of animals from LD group; (**I**)—myenteric neurons immunoreactive to PGP 9.5 and VAChT in the pylorus of animals from HD group; (**J**)—submucous neurons immunoreactive to PGP 9.5 and VAChT in the corpus of control animals; (**K**)—submucous neurons immunoreactive to PGP 9.5 and VAChT in the corpus of animals from LD group; (**L**)—submucous neurons immunoreactive to PGP 9.5 and VAChT in the corpus of animals from HD group.

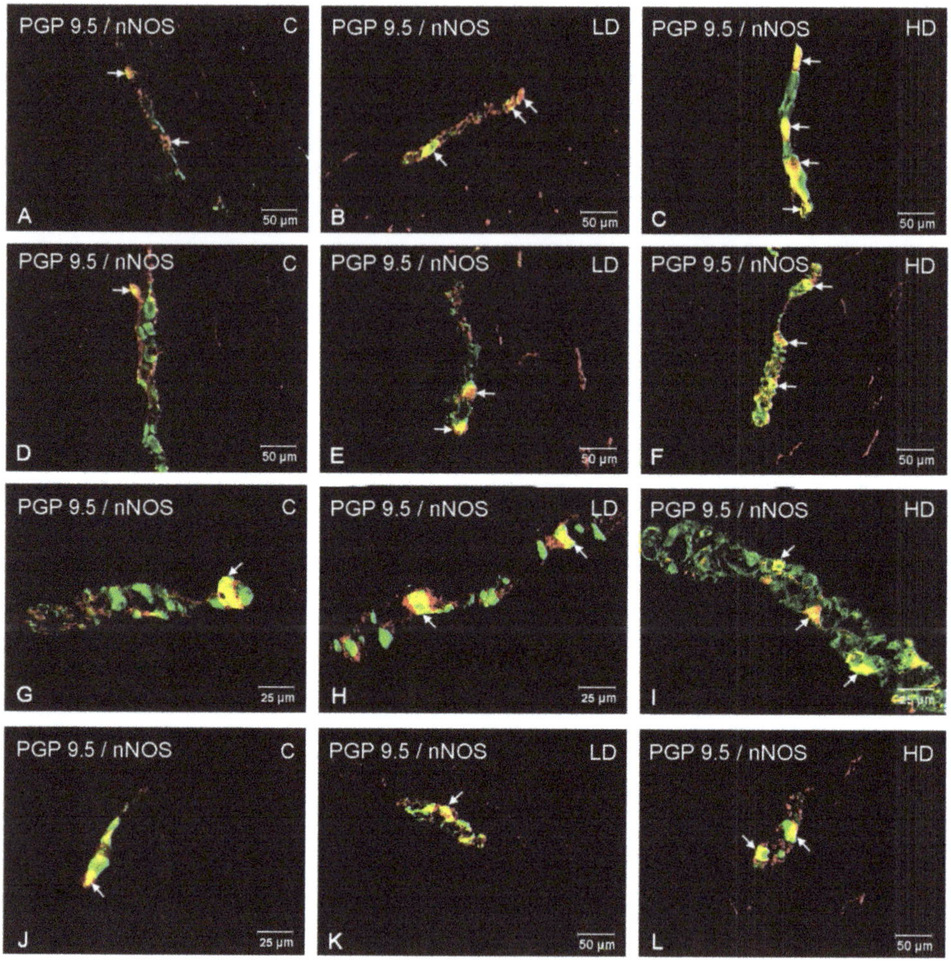

Figure 5. nNOS-immunoreactive neurons in the porcine stomach. The pictures show the neuronal isoform of nitric oxide synthase (nNOS)- immunoreactive intramural neurons in the wall of the porcine stomach in the control group and in animals receiving low and high doses of acrylamide. (**A**)—myenteric neurons immunoreactive to protein gene-product 9.5 (PGP 9.5), used as a pan-neuronal marker, and nNOS in the cardia of control animals; (**B**)—myenteric neurons immunoreactive to PGP 9.5 and nNOS in the cardia of animals from LD group; (**C**)—myenteric neurons immunoreactive to PGP 9.5 and nNOS in the cardia of animals from HD group; (**D**)—myenteric neurons immunoreactive to PGP 9.5 and nNOS in the corpus of control animals; (**E**)—myenteric neurons immunoreactive to PGP 9.5 and nNOS in the corpus of animals from LD group; (**F**)—myenteric neurons immunoreactive to PGP 9.5 and nNOS in the corpus of animals from HD group; (**G**)—myenteric neurons immunoreactive to PGP 9.5 and nNOS in the pylorus of control animals; (**H**)—myenteric neurons immunoreactive to PGP 9.5 and nNOS in the pylorus of animals from LD group; (**I**)—myenteric neurons immunoreactive to PGP 9.5 and nNOS in the pylorus of animals from HD group; (**J**)—submucous neurons immunoreactive to PGP 9.5 and nNOS in the corpus of control animals; (**K**)—submucous neurons immunoreactive to PGP 9.5 and nNOS in the corpus of animals from LD group; (**L**)—submucous neurons immunoreactive to PGP 9.5 and nNOS in the corpus of animals from HD group.

3.2. Submucous Plexus

Acrylamide also induced a substantial change in the CART-, VACHT-, and nNOS-IR neuronal populations in the submucous plexuses in the stomach corpus (Figure 2). For CART, a highly significant increase in the number of CART-IR neurons was found in the LD group (from 4.32 ± 0.53% to 7.52 ± 0.36%) and HD group (to 7.76 ± 0.42%) (Figure 2D, Figure 3J–L). An increase in the VAChT-IR neurons was also recorded in both experimental groups: from 50.05 ± 1.83% to 58.81 ± 1.38% and to 69.39 ± 1.95%, respectively (Figure 2D, Figure 4J–L). While for nNOS, a statistically significant increase in the nNOS-IR neurons was shown only in the HD group (from 18.64 ± 1.96% to 26.87 ± 1.37%) (Figure 2D, Figure 5J,L). In the submucous plexuses of the cardia and pylorus region, no neurons immunoreactive to the investigated neuroactive substances were found.

3.3. Nerve Fibres

CART-, VAChT-, and nNOS-immunoreactive nervous fibres were detected in both the circular muscle layer (CML) and the submucosal/mucosal layer (S/ML) in all investigated gastric regions (Table 2). In the CML, a dense network of CART-IR nerve fibres was demonstrated in the cardia (++), corpus (+++) (Figure 6A) and pylorus (++). Fibres with slightly smaller density were found in the S/ML (+/++) (Figure 7A), although few varicose VACHT (+) and nNOS (+) fibres were detected in the control group or in the CLM (Figure 6D,G) and the S/ML (Figure 7D,G). The VACHT- and nNOS-immunoreactive fibre density was comparable in all investigated gastric regions (cardia, corpus, and pylorus). Acrylamide produced a substantial increase in the CART-, VAChT-, and nNOS-immunoreactive fibre density in both the CML (Figure 6B,C,E,F,H,I) and S/ML (Figure 7B,C,E,F,H,I). The greatest changes were demonstrated in the HD acrylamide group; however, even in the LD group, changes were also noticeable (Table 2).

Table 2. Density of nerve fibres immunoreactive to the cocaine- and amphetamine-regulated transcript (CART), the vesicular acetylcholine transporter (VAChT) and the neuronal isoform of nitric oxide synthase (nNOS) in the stomach wall.

Part of the Stomach	Cardia			Corpus			Pylorus		
	C Group	LD Group	HD Group	C Group	LD Group	HD Group	C Group	LD Group	HD Group
				CART					
CML	++	++	+++	+++	+++	++++	++	++	+++
S/ML	+	+	++	+	++	+++	++	++	+++
				VACHT					
CML	+	+	++	+	+	++	+	++	++
S/ML	+	+	+	+	++	++	+	++	++
				nNOS					
CML	+	+	++	+	++	++	+	++	+++
S/ML	+	+	++	+	+	++	+	++	++

CML-the circular muscle layer, S/ML-the submucous/mucous layer.

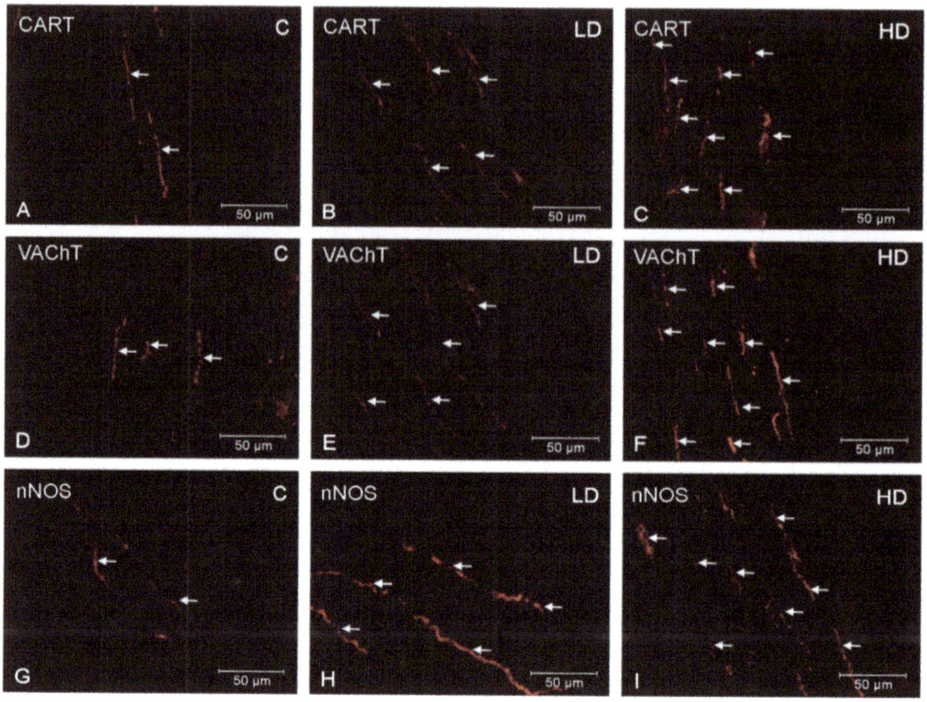

Figure 6. Nerve fibres in the circular muscle layer. The pictures show nerve fibres immunoreactive to the the cocaine- and amphetamine-regulated transcript (CART), the vesicular acetylcholine transporter (VAChT) and the neuronal isoform of nitric oxide synthase (nNOS) in the circular muscle layer (CML) in the porcine stomach in the control group and in animals receiving low and high doses of acrylamide. (**A**)—CART-immunoreactive nerve fibres in the CML in the corpus of control animals; (**B**)—CART-immunoreactive nerve fibres in the CML in the corpus of animals from LD group; (**C**)—CART-immunoreactive nerve fibres in the CML in the corpus of animals from HD group; (**D**)—VAChT-immunoreactive nerve fibres in the CML in the pylorus of control animals, (**E**)—VAChT-immunoreactive nerve fibres in the CML in the pylorus of animals from LD group, (**F**)—VAChT-immunoreactive nerve fibres in the CML in the pylorus of animals from HD group, (**G**)—nNOS-immunoreactive nerve fibres in the CML in the pylorus of control animals, (**H**)—nNOS-immunoreactive nerve fibres in the CML in the pylorus of animals from LD group, (**I**)—nNOS-immunoreactive nerve fibres in the CML in the pylorus of animals from HD group.

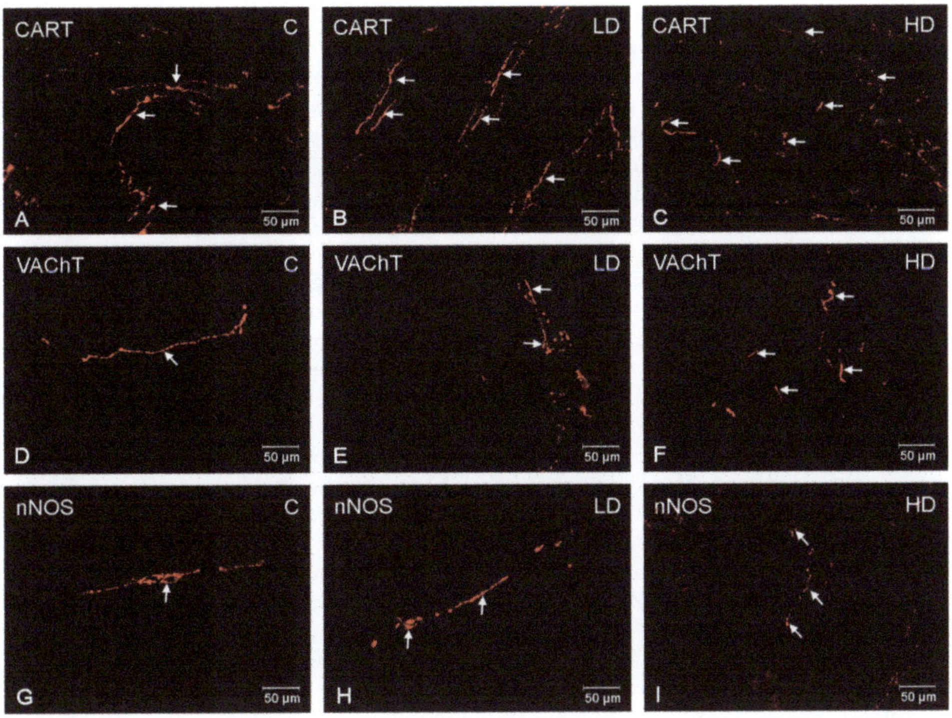

Figure 7. Nerve fibres in the submucous/mucous layer. The pictures show nerve fibres immunoreactive to the the cocaine- and amphetamine-regulated transcript (CART), the vesicular acetylcholine transporter (VAChT) and the neuronal isoform of nitric oxide synthase (nNOS in the submucous/mucous layer (S/ML) in the porcine stomach in the control group and in animals receiving low and high doses of acrylamide. (**A**)—CART-immunoreactive nerve fibres in the S/ML in the corpus of control animals; (**B**)—CART-immunoreactive nerve fibres in the S/ML in the corpus of animals from LD group; (**C**)—CART-immunoreactive nerve fibres in the S/ML in the corpus of animals from HD group; (**D**)—VAChT-immunoreactive nerve fibres in the S/ML in the corpus of control animals, (**E**)—VAChT-immunoreactive nerve fibres in the S/ML in the corpus of animals from LD group, (**F**)—VAChT-immunoreactive nerve fibres in the S/ML in the corpus of animals from HD group, (**G**)—nNOS-immunoreactive nerve fibres in the S/ML in the pylorus of control animals, (**H**)—nNOS-immunoreactive nerve fibres in the S/ML in the pylorus of animals from LD group, (**I**)—nNOS-immunoreactive nerve fibres in the S/ML in the pylorus of animals from HD group.

4. Discussion

The progress of civilization and the development of industry have provided people with virtually unlimited access to food products. However, while the pace of life has increased, the consumption of products with high levels of acrylamide is also on the rise. The gastrointestinal tract is the first-exposure site to noxious substances ingested with food, and it is also often the first defence mechanism. Changes in the expression of neuroactive substances in the intramural neurons of the ENS are a common preclinical symptom of the harmful effect of pathological factors on the body. The investigation showed that the supplementation of both low and high doses of acrylamide had a substantial effect on the population of the enteric CART-, VAChT-, and nNOS-immunoreactive neurons and the fibre density in the porcine stomach. The degree of these changes was variable, depending on the specific stomach region and the type of investigated plexus. Nevertheless, the findings may also indicate that the alimentary exposure to acrylamide causes significant alterations in the neurochemical phenotype of the enteric neurons and is therefore not neutral in the body.

Acrylamide is categorised as a neurotoxic compound, with toxic effects on the central nervous system (CNS) and peripheral nervous system (PNS), both in laboratory animals and in humans [28]. The reported symptoms of acrylamide exposure include limb numbness, muscle weakness or ataxia, and they most probably result from an inhibition of the kinesin-based fast axonal transport and alterations in the neurotransmitter levels [29–31]. Acrylamide binds with protein receptors rich in cysteine that are involved in the presynaptic release of the neurotransmitters, membrane reuptake, and vesicular nerve conduction, which results in disturbed synaptic transmission [32]. It has also been demonstrated that acrylamide causes damage to nerve ends and Purkinje cells, as well as axonal oedema [33]. In acrylamide intoxication, an increase in dopamine expression has also been shown in the rat striatum [34]. Furthermore, acrylamide results in a reduction of enzymatic and non-enzymatic antioxidants and in lipid peroxidation, which leads to nerve cell apoptosis, and the latter phenomenon is involved in the pathogenesis of many neurodegenerative diseases [35,36]. This has also been demonstrated in some studies on acrylamide-induced reactive gliosis, presenting with a potentiated synthesis of free radicals and stimulating amino acids and proinflammatory cytokines in the glial cells, which results in neuronal death [35]. Previous studies have not elucidated how acrylamide impacts the enteric neurons. However, the above-mentioned studies on different parts of the nervous system may indicate that these findings are a result of neurotoxicity.

Changes in neurotransmitter expression in the enteric neurons are a result of adaptation to irritants to which the cell is exposed, and it should accommodate the neurons to survive under variable, often unfavourable conditions. Previous studies investigating the ENS have shown that many of the neuroactive substances synthesized by these neurons have neuroprotective properties [11,16,18]. Undoubtedly, CART is one of them. In the investigation, the reported increase in the CART-IR neuron population is consistent with previous studies that demonstrated an elevated CART expression in the enteric neurons due to mycotoxin intoxication, neuron damage, diabetes or hypertension [2,7,37,38]. The severity of these alterations was determined by the section of the gastrointestinal (GI) tract and the given pathology. The neuroprotective effect of CART has been also evidenced in in vitro mice studies since it found increased survival of neurons in a CART-supplemented myenteric neuron culture [39]. Acetylcholine (ACh) is also an important endogenous neurotransmitter, and its neuroprotective effects have been demonstrated in both the central and peripheral nervous systems [40,41]. Recent studies have found that the population of neurons immunoreactive to VAChT (a marker of the cholinergic neurons) in the enteric neuronal population increased during irinotecan treatment [42]. The elevated number of VAChT-immunoreactive neurons reported in the presented investigation is also consistent with previous studies that demonstrated an elevation of acetylcholinesterase (AChE) levels in the peripheral cholinergic neurons during acrylamide intoxication [43]. Interestingly, VAChT is also believed to be an excellent morphological indicator to investigate the regenerative process of motor neuron endings [44]. In turn, nNOS may present a two-pronged effect, depending on the site of a pathological process in the GI tract and the nature of the noxious stimulus. An elevated number of intramural nNOS-immunoreactive neurons has been reported in many GI dysfunctions, such as hyperacidity of the stomach and naproxen treatment [12,18]. Furthermore, it has been shown that the levels of transcription factors involved in protecting the myenteric neurons against ischaemia–reperfusion injury increased, and the survival of the rat myenteric neurons in an nNOS-supplemented culture was elevated, which provides evidence of the neuroprotective effect of NO [45,46]. There are also reports on the reduced expression of nNOS in enteric neurons, for example, in diabetes or Crohn disease [10,47]. However, they prove the above-mentioned bidirectionality of NO effects in response to different neurotoxic stimuli.

Although the neurotoxicity of acrylamide has been well discussed, it cannot be precluded that the changes in the neurochemical phenotype of the stomach enteric neurons, reported in the present study, may result from a response to inflammation. It has been shown that acrylamide has proinflammatory properties. There is some evidence that acrylamide increases interleukin-6 (Il-6) and tumour necrosis factor α (TNF-α) levels in the blood serum via overproduction of reactive oxygen species (ROS) and therefore leads to neuroinflammation [35]. Previous studies conducted by the authors have also demonstrated that supplementation with acrylamide results in local ileitis with such symptoms as the increased synthesis of proinflammatory cytokines by gut-associated lymphoid tissue (GALT) (Interleukin 1β (IL-1β), IL-6, TNF-α) [48]. The authors' findings correlate with the previous studies in which an elevated number of CART-, VAChT-, and nNOS-IR neurons was detected in the course of different inflammatory GIT conditions. In particular, an increased immunoreactivity to CART in the ENS structures was recorded in ulcerative colitis and experimentally induced colitis in pigs [38,49]. It has been proven that the cholinergic system is involved in the immune response mechanisms by stimulating the nicotine receptors and thereby leading to a reduction in the inflammatory response in the CNS [50]. Additionally, it decreases lipid peroxidation by removing reactive oxygen species (ROS) [51]. A direct inhibitory effect of ACh on the synthesis of proinflammatory cytokines by macrophages has been demonstrated [52]. NO is known for its anti-inflammatory and proinflammatory properties in the gastrointestinal tract. The outcome depends on the type of inflammation and its site. Nevertheless, it has been shown that NO is involved in protecting the gastric mucosa against damage, supporting ulcer healing, and increasing cytokine synthesis in the mucosa [53–55]. Considering these data, it may be speculated that the investigated neuroactive substances are involved in controlling acrylamide-induced inflammation in the porcine stomach.

It is also acknowledged that despite the lack of data on the toxic mechanism of acrylamide in the ENS structures, the demonstrated changes may result from an elevated synthesis of the investigated neuroactive substances on its different stages, such as transcription, translation or modification of the activity of enzymes involved in this synthesis, or these changes may be a consequence of the inhibition of axonal transport or slowed degradation of neurotransmitters. However, the reported increased density of the CART-, VAChT-, and nNOS-positive fibres associated with the elevated number of neurons, which are immunoreactive to the investigated substances, supports the increased synthesis scenario.

5. Conclusions

To conclude, the discussed investigation has demonstrated that supplementation with low and high doses of acrylamide resulted in alterations of the porcine stomach neuron phenotype, which was reflected in an increased number of CART-, VAChT-, and nNOS-immunoreactive neurons. These changes were accompanied by an increased density of CART-, VAChT-, and nNOS-positive fibres. The recorded changes revealed that even low doses of acrylamide influenced the nervous structures located in the porcine gastric wall. This may result from the neurotoxicity of acrylamide or from the response of the ENS to acrylamide-induced inflammation. The authors' studies also suggest that the ENS plays an important role in protecting the gastrointestinal tract during acrylamide intoxication. Considering the role of the pig as an important model in biomedical research, these results may become a topic of further toxicological and clinical studies on reducing the harmful effects of acrylamide found in food products in the body.

Author Contributions: Conceptualization, K.P.; Investigation, K.P., M.B.; Methodology, K.P.; Writing—original draft, K.P.; Writing—review & editing, J.C. All authors have read and agreed to the published version of this manuscript.

Funding: This study was funded by the KNOW (Leading National Research Centre) Scientific Consortium "Healthy Animal–Safe Food", the Ministry of Science and Higher Education, No. 05-1/KNOW2/2015, and the University of Warmia and Mazury in Olsztyn (statutory research) grant No 15.610.003-300. The project was financially co-supported by Minister of Science and Higher Education in the range of the program entitled "Regional Initiative of Excellence" for the years 2019-2022, Project No. 010/RID/2018/19, amount of funding 12.000.000 PLN.

Acknowledgments: The authors wish to express their deep thanks to mgr. Andrzej Pobiedziński, for his skillful technical assistance.

Conflicts of Interest: The authors declare no conflicts of interest.

References

1. Furness, J.B. The enteric nervous system and neurogastroenterology. *Nat. Rev. Gastroenterol. Hepatol.* **2012**, *69*, 286–294. [CrossRef] [PubMed]
2. Makowska, K.; Gonkowski, S.; Zielonka, L.; Dabrowski, M.; Calka, J. T2 toxin-induced changes in cocaine- and amphetamine-regulated transcript (CART)-Like immunoreactivity in the enteric nervous system within selected fragments of the porcine digestive tract. *Neurotox. Res.* **2017**, *31*, 136–147. [CrossRef] [PubMed]
3. Palus, K.; Makowska, K.; Całka, J. Alterations in Galanin-Like Immunoreactivity in the Enteric Nervous System of the Porcine Stomach Following Acrylamide Supplementation. *Int. J. Mol. Sci.* **2019**, *20*, 3345. [CrossRef]
4. Miller, S.M.; Reed, D.; Sarr, M.G.; Farrugia, G.; Szurszewski, J.H. Haem oxygenase in enteric nervous system of human stomach and jejunum and co-localization with nitric oxide synthase. *Neurogastroenterol Motil.* **2001**, *13*, 121–131. [CrossRef] [PubMed]
5. Furness, J.B.; Callaghan, B.P.; Rivera, L.R.; Cho, H.J. The enteric nervous system and gastrointestinal innervation: Integrated local and central control. *Adv. Exp. Med. Biol.* **2014**, *817*, 39–71. [CrossRef]
6. Okumura, T.; Yamada, H.; Motomura, W.; Kohgo, Y. Cocaine amphetamine-regulated transcript (CART) acts in the central nervous system to inhibit gastric acid secretion via brain corticotropin-releasing factor system. *Endocrinology* **2000**, *141*, 2854–2860. [CrossRef]
7. Bulc, M.; Gonkowski, S.; Całka, J. Expression of Cocaine and Amphetamine Regulated Transcript (CART) in the Porcine Intramural Neurons of Stomach in the Course of Experimentally Induced Diabetes Mellitus. *J. Mol. Neurosci.* **2015**, *57*, 376–385. [CrossRef]
8. Palus, K.; Makowska, K.; Całka, J. Acrylamide-induced alterations in the cocaine- and amphetamine-regulated peptide transcript (CART)-like immunoreactivity within the enteric nervous system of the porcine small intestines. *Ann. Anat.* **2018**, *219*, 94–101. [CrossRef]
9. Umathe, S.N.; Kochar, N.I.; Jain, N.S.; Dixit, P.V. Gastrointestinal dysfunction in diabetic rats relates with a decline in tissue L-arginine content and consequent low levels of nitric oxide. *Nitric Oxide* **2009**, *20*, 129–133. [CrossRef]
10. Bulc, M.; Palus, K.; Dąbrowski, M.; Całka, J. Hyperglycaemia-induced downregulation in expression of nNOS intramural neurons of the small intestine in the pig. *Int. J. Mol. Sci.* **2019**, *20*, 1681. [CrossRef]
11. Szymanska, K.; Calka, J.; Gonkowski, S. Nitric oxide as an active substance in the enteric neurons of the porcine digestive tract in physiological conditions and under intoxication with bisphenol A (BPA). *Nitric Oxide* **2018**, *80*, 1–11. [CrossRef] [PubMed]
12. Calka, J. Increased expression of CART, nNOS, VIP, PACAP, SP and GAL in enteric neurons of the porcine stomach prepyloric region following hydrochloric acid infusion. *Folia Histochem. Cytobiol.* **2019**, *57*(4), 179–187. [CrossRef]
13. Arvidsson, U.; Riedl, M.; Elde, R.; Meister, B. Vesicular acetylcholine transporter (VAChT) protein: A novel and unique marker for cholinergic neurons in the central and peripheral nervous systems. *J. Comp. Neurol.* **1997**, *378*, 454–467. [CrossRef]
14. Kaleczyc, J.; Klimczuk, M.; Franke-Radowiecka, A.; Sienkiewicz, W.; Majewski, M.; Łakomy, M. The distribution and chemical coding of intramural neurons supplying the porcine stomach—The study on normal pigs and on animals suffering from swine dysentery. *Anat. Histol. Embryol.* **2007**, *36*, 186–193. [CrossRef] [PubMed]
15. Szymanska, K.; Gonkowski, S. Bisphenol A-Induced changes in the enteric nervous system of the porcine duodenum. *Neurotoxicology* **2018**, *66*, 78–86. [CrossRef]
16. Ekblad, E.; Bauer, A.J. Role of vasoactive intestinal peptide and inflammatory mediators in enteric neuronal plasticity. *Neuro Gastroenterol. Motil.* **2004**, *16*, 123–128. [CrossRef]
17. Kolgazi, M.; Uslu, U.; Yuksel, M.; Velioglu-Ogunc, A.; Ercan, F.; Alican, I. The role of cholinergic anti-inflammatory pathway in acetic acid-induced colonic inflammation in the rat. *Chem. Biol. Interact.* **2013**, *205*, 72–80. [CrossRef]
18. Czajkowska, M.; Rychlik, A.; Całka, J. Long-term treatment with naproxen changes the chemical coding of the porcine intramural duodenum neurons. *Ann Anat.* **2020**, *227*, 151425. [CrossRef]

19. Rychlik, A.; Gonkowski, S.; Nowicki, M.; Calka, J. Inflammatory bowel disease affects density of nitrergic nerve fibers in the mucosal layer of the canine gastrointestinal tract. *Can. J. Vet. Res.* **2017**, *81*, 129–136.
20. Shipp, A.; Lawrence, G.; Gentry, R.; McDonald, T.; Bartow, H.; Bounds, J.; Macdonald, N.; Clewell, H.; Allen, B.; van Landingham, C. Acrylamide: Review of toxicity data and dose-response analyses for cancer and noncancer effects. *Crit. Rev. Toxicol.* **2006**, *36*, 481–608. [CrossRef]
21. Lo Pachin, R.M. The changing view of acrylamide neurotoxicity. *Toxicol. Vitr.* **2010**, *25*, 573–579. [CrossRef]
22. Zödl, B.; Schmid, D.; Wassler, G.; Gundacker, C.; Leibetseder, V.; Thalhammer, T.; Ekmekcioglu, C. Intestinal transport and metabolism of acrylamide. *Toxicology* **2007**, *232*, 99–108. [CrossRef] [PubMed]
23. Van Lancker, F.; Adams, A.; De Kimpe, N. Chemical modifications of peptides and their impact on food properties. *Chem. Rev.* **2011**, *111*, 7876–7903. [CrossRef] [PubMed]
24. WHO. *Health Implications of Acrylamide in Food*; FAO: Rome, Italy; WHO: Geneva, Switzerland, 2002; Available online: http://apps.who.int/iris/handle/10665/42563 (accessed on 15 November 2016).
25. Kararli, T.T. Comparison of the gastrointestinal anatomy, physiology, and biochemistry of humans and commonly used laboratory animals. *Biopharm. Drug Dispos.* **1995**, *16*, 351–38066. [CrossRef] [PubMed]
26. Kobayashi, E.; Hanazono, Y.; Kunita, S. Swine used in the medical university: Overview of 20 years of experience. *Exp. Anim.* **2018**, *67*, 7–13. [CrossRef]
27. Palus, K.; Bulc, M.; Całka, J. Changes in Somatostatin-Like Immunoreactivity in the Sympathetic Neurons Projecting to the Prepyloric Area of the Porcine Stomach Induced by Selected Pathological Conditions. *Biomed. Res. Int.* **2017**, *2017*, 9037476. [CrossRef]
28. Smith, E.A.; Oehme, F.W. Acrylamide and polyacrylamide: A review of production, use, environmental fate and neurotoxicity. *Rev. Environ. Health* **1991**, *9*, 215–228. [CrossRef]
29. Abelli, L.; Ferri, G.L.; Astolfi, M.; Conte, B.; Geppetti, P.; Parlani, M.; Dahl, D.; Polak, J.M.; Maggi, C.A. Acrylamide-induced visceral neuropathy: Evidence for the involvement of capsaicin-sensitive nerves of the rat urinary bladder. *Neuroscience* **1991**, *41*, 311–321. [CrossRef]
30. Spencer, P.S.; Schaumburg, H.H. A review of acrylamide neurotoxicity. Part II. Experimental animal neurotoxicity and pathologic mechanisms. *Can. J. Neurol. Sci.* **1974**, *1*, 152–169. [CrossRef]
31. Radad, K.; Al-Shraim, M.; Al-Emam, A.; Moldzio, R.; Rausch, W.D. Neurotoxic effects of acrylamide on dopaminergic neurons in primary mesencephalic cell culture. *Folia Neuropathol.* **2019**, *57*, 196–204. [CrossRef] [PubMed]
32. LoPachin, R.M.; Gavin, T. Molecular mechanism of acrylamide neurotoxicity: Lessons earned from organic chemistry. *Environ. Health Perspect.* **2012**, *120*, 1650. [CrossRef] [PubMed]
33. Erkekoglu, P.; Baydar, T. Acrylamide neurotoxicity. *Nutr. Neurosci.* **2014**, *17*, 49–57. [CrossRef] [PubMed]
34. Pan, X.; Guo, X.; Xiong, F.; Cheng, G.; Lu, Q.; Yan, H. Acrylamide increases dopamine levels by affecting dopamine transport and metabolism related genes in the striatal dopaminergic system. *Toxicol. Lett.* **2015**, *236*, 60–68. [CrossRef] [PubMed]
35. Elblehi, S.S.; El Euony, O.I.; El-Sayed, Y.S. Apoptosis and astrogliosis perturbations and expression of regulatory inflammatory factors and neurotransmitters in acrylamide-induced neurotoxicity under ω3 fatty acids protection in rats. *Neurotoxicology* **2019**, *76*, 44–57. [CrossRef] [PubMed]
36. Komoike, Y.; Matsuoka, M. Endoplasmic reticulum stress-mediated neuronal apoptosis by acrylamide exposure. *Toxicol. Appl. Pharmacol.* **2016**, *310*, 68–77. [CrossRef]
37. Kasacka, I.; Piotrowska, Z. Evaluation of density and distribution of CART-immunoreactive structures in gastrointestinal tract of hypertensive rats. *Biofactors* **2012**, *38*, 407–415. [CrossRef]
38. Burliński, P.J. Inflammation- and axotomy-induced changes in cocaine- andamphetamine-regulated transcript peptide-like immunoreactive (CART-LI) nervous structures in the porcine descending colon. *Pol. J. Vet. Sci.* **2012**, *15*, 517–524. [CrossRef]
39. Ekblad, E. CART in the enteric nervous system. *Peptides* **2006**, *27*, 2024–2030. [CrossRef]
40. Dajas-Bailador, F.A.; Lima, P.A.; Wonnacott, S. The α7 nicotinic acetylcholine receptor subtype mediates nicotine protection against NMDA excitotoxicity in primary hippocampal cultures through a calcium ion dependent mechanism. *Neuropharmacology* **2000**, *39*, 2799–2807. [CrossRef]
41. Thompson, S.A.; Smith, O.; Linn, D.M.; Linn, C.L. Acetylcholine neuroprotection against glutamate-induced excitotoxicity in adult pig retinal ganglion cells is partially mediated through alpha4 nAChRs. *Exp. Eye. Res.* **2006**, *83*, 1135–1145. [CrossRef]

42. McQuade, R.M.; Stojanovska, V.; Donald, E.L.; Rahman, A.A.; Campelj, D.G.; Abalo, R.; Rybalka, E.; Bornstein, J.C.; Nurgali, K. Irinotecan-Induced Gastrointestinal Dysfunction Is Associated with Enteric Neuropathy, but Increased Numbers of Cholinergic Myenteric Neurons. *Front Physiol.* **2017**, *8*, 391. [CrossRef]
43. Pennisi, M.; Malaguarnera, G.; Puglisi, V.; Vinciguerra, L.; Vacante, M.; Malaguarnera, M. Neurotoxicity of acrylamide in exposed workers. *Int. J. Environ. Res. Public Health.* **2013**, *10*, 3843–3854. [CrossRef] [PubMed]
44. Maeda, M.; Ohba, N.; Nakagomi, S.; Suzuki, Y.; Kiryu-Seo, S.; Namikawa, K.; Kondoh, W.; Tanaka, W.; Kiyama, H. Vesicular acetylcholine transporter can be a morphological marker for the reinnervation to muscle of regenerating motor axons. *Neurosci Res.* **2004**, *48*, 305–314. [CrossRef]
45. Filpa, V.; Carpanese, E.; Marchet, S.; Pirrone, C.; Conti, A.; Rainero, A.; Moro, E.; Chiaravalli, A.M.; Zucchi, I.; Moriondo, A.; et al. Nitric oxide regulates homeoprotein OTX1 and OTX2 expression in the rat myenteric plexus after intestinal ischemia-reperfusion injury. *Am. J. Physiol. Gastrointest. Liver Physiol.* **2017**, *312*, G374–G389. [CrossRef] [PubMed]
46. Sandgren, K.; Lin, Z.; Fex Svenningsen, A.; Ekblad, E. Vasoactive intestinal peptide and nitric oxide promote survival of adult rat myenteric neurons in culture. *J. Neurosci. Res.* **2003**, *72*, 595–602. [CrossRef] [PubMed]
47. Winston, J.H.; Li, Q.; Sarna, S.K. Paradoxical regulation of ChAT and nNOS expression in animal models of Crohn's colitis and ulcerative colitis. *Am. J. Physiol. Gastrointest. Liver Physiol.* **2013**, *305*, G295–G302. [CrossRef]
48. Palus, K.; Obremski, K.; Bulc, M.; Całka, J. The impact of low and high doses of acrylamide on the intramural neurons of the porcine ileum. *Food Chem. Toxicol.* **2019**, *132*, 110673. [CrossRef]
49. Gonkowski, S.; Burlinski, P.; Szwajca, P.; Całka, J. Changes in cocaine- andamphetamine-regulated transcript-like immunoreactive (CART-LI) nerve structures of the porcine descending colon during proliferative enteropathy. *Bull. Vet. Inst. Pulawy* **2012**, *56*, 199–203. [CrossRef]
50. Hao, J.; Simard, A.R.; Turner, G.H.; Wu, J.; Whiteaker, P.; Lukas, R.J.; Shi, F.D. Attenuation of CNS inflammatory responses by nicotine involves α7 and non-α7 nicotinic receptors. *Exp. Neurol.* **2011**, *227*, 110–119. [CrossRef]
51. Lebda, M.A.; El-Neweshy, M.S.; El-Sayed, Y.S. Neurohepatic toxicity of subacute manganese chloride exposure and potential chemoprotective effects of lycopene. *Neurotoxicology* **2012**, *33*, 98–104. [CrossRef]
52. Borovikova, L.V.; Ivanova, S.; Zhang, M.; Yang, H.; Botchkina, G.I.; Watkins, L.R.; Wang, H.; Abumrad, N.; Eaton, J.W.; Tracey, K.J. Vagus nerve stimulation attenuates the systemic inflammatory response to endotoxin. *Nature* **2000**, *405*, 458–462. [CrossRef] [PubMed]
53. Kubes, P.; Wallace, J.L. Nitric oxide as a mediator of gastrointestinal mucosal injury?—Say it ain't so. *Mediat. Inflamm.* **1995**, *4*, 397–405. [CrossRef] [PubMed]
54. Palus, K.; Całka, J. The Influence of Prolonged Acetylsalicylic Acid Supplementation-Induced Gastritis on the Neurochemistry of the Sympathetic Neurons Supplying Prepyloric Region of the Porcine Stomach. *PLoS ONE* **2015**, *10*, e0143661. [CrossRef]
55. Barrachina, M.D.; Panés, J.; Esplugues, J.V. Role of nitric oxide in gastrointestinal inflammatory and ulcerative diseases: Perspective for drugs development. *Curr. Pharm. Des.* **2001**, *7*, 31–48. [CrossRef] [PubMed]

animals

MDPI

Article

Cannabinoid and Cannabinoid-Related Receptors in the Myenteric Plexus of the Porcine Ileum

Andrea Toschi [1], Giorgia Galiazzo [1], Andrea Piva [1,2], Claudio Tagliavia [1], Gemma Mazzuoli-Weber [3], Roberto Chiocchetti [1,*,†] and Ester Grilli [1,4,†]

1 Department of Veterinary Medical Sciences (UNI EN ISO 9001:2008), University of Bologna, Via Tolara di Sopra, 50, 40064 Ozzano dell'Emilia, Italy; andrea.toschi5@unibo.it (A.T.); giorgia.galiazzo2@unibo.it (G.G.); andrea.piva@unibo.it (A.P.); claudio.tagliavia2@unibo.it (C.T.); ester.grilli@unibo.it (E.G.)
2 R&D Division, Vetagro S.p.A., via Porro 2, 42124 Reggio Emilia, Italy
3 Institute for Physiology and Cell Biology, University of Veterinary Medicine Hannover, Foundation, Bischofsholer Damm 15, 30173 Hannover, Germany; Gemma.Mazzuoli-Weber@tiho-hannover.de
4 R&D Division, Vetagro, Inc., 116 W. Jackson Blvd., Suite #320, Chicago, IL 60604, USA
* Correspondence: roberto.chiocchetti@unibo.it; Tel.: +39-051-2097946
† These authors shared the senior authorship of this study.

Simple Summary: The endocannabinoid system (ECS) has opened the door to novel therapeutical approaches targeting cancer, pain, anxiety, stress, and inflammatory diseases. The ECS is ubiquitously expressed in almost all members of *Animalia*, but its precise localization outside the central nervous system is still under investigation. In this study, the localization of the main and related cannabinoid receptors in the myenteric plexus of the porcine ileum was immunohistochemically analyzed. The myenteric plexus neurons were found to be positive for cannabinoid receptor 1 (CB1R) and the cannabinoid-related receptors transient potential vanilloid receptor 1 (TRPV1), transient potential ankyrin receptor 1 (TRPA1), and serotonin receptor 5-HT1a (5-HT1aR). In addition, the ECS receptors were also located on nerve fibers, the *tunica muscularis*, and the endothelium. The wide distribution of cannabinoid and cannabinoid-related receptors in the myenteric plexus provides the anatomical basis for additional investigation, suggesting the possible role of the ECS in treating pathological conditions in livestock.

Abstract: An important piece of evidence has shown that molecules acting on cannabinoid receptors influence gastrointestinal motility and induce beneficial effects on gastrointestinal inflammation and visceral pain. The aim of this investigation was to immunohistochemically localize the distribution of canonical cannabinoid receptor type 1 (CB1R) and type 2 (CB2R) and the cannabinoid-related receptors transient potential vanilloid receptor 1 (TRPV1), transient potential ankyrin receptor 1 (TRPA1), and serotonin receptor 5-HT1a (5-HT1aR) in the myenteric plexus (MP) of pig ileum. CB1R, TRPV1, TRPA1, and 5-HT1aR were expressed, with different intensities in the cytoplasm of MP neurons. For each receptor, the proportions of the immunoreactive neurons were evaluated using the anti-HuC/HuD antibody. These receptors were also localized on nerve fibers (CB1R, TRPA1), smooth muscle cells of *tunica muscularis* (CB1R, 5-HT1aR), and endothelial cells of blood vessels (TRPV1, TRPA1, 5-HT1aR). The nerve varicosities were also found to be immunoreactive for both TRPV1 and 5-HT1aR. No immunoreactivity was documented for CB2R. Cannabinoid and cannabinoid-related receptors herein investigated showed a wide distribution in the enteric neurons and nerve fibers of the pig MP. These results could provide an anatomical basis for additional research, supporting the therapeutic use of cannabinoid receptor agonists in relieving motility disorders in porcine enteropathies.

Keywords: CB1R; CB2R; TRPV1; TRPA1; 5-HT1aR; endocannabinoid system; immunohistochemistry; pig; enteric nervous system

Citation: Toschi, A.; Galiazzo, G.; Piva, A.; Tagliavia, C.; Mazzuoli-Weber, G.; Chiocchetti, R.; Grilli, E. Cannabinoid and Cannabinoid-Related Receptors in the Myenteric Plexus of the Porcine Ileum. *Animals* **2021**, *11*, 263. https://doi.org/10.3390/ani11020263

Academic Editors: Paola Scocco, Elena De Felice and Alessandro Malfatti

Received: 29 December 2020
Accepted: 19 January 2021
Published: 21 January 2021

1. Introduction

The network of sensory neurons, motor neurons, interneurons, and glial cells embedded in the gut walls is called the enteric nervous system (ENS) [1]. The ENS is responsible for controlling various functions of the gastrointestinal tract (GIT) such as motility, absorption, and secretion in physiological and pathological conditions [2]. There is a strict interaction between the ENS and the central nervous system (CNS), with a bidirectional information flow between these two systems; however, the ENS can control digestive functions independently of the CNS [3]. The ENS also cooperates with the immune and endocrine systems by adapting nutrient absorption depending on the condition of the gut, thereby preserving mucosal barrier functionality [3].

The endocannabinoid system (ECS) is constituted of three fundamental components: receptors, signaling molecules, and the enzymes responsible for ligand biosynthesis and degradation. It typically comprises the prototypical cannabinoid receptors types 1 and 2 (CB1R and CB2R), endocannabinoids anandamide (AEA) and 2-arachidonylglycerol (2-AG), and the enzymes involved in their biosynthesis and degradation [4–7]. The ECS is typically localized at the CNS level [8]. CB1R was proven to be the most widely expressed receptor protein from the G protein-coupled receptors (GPCRs) family in the brain, mainly in the basal ganglia, hippocampus, olfactory bulb, and cerebellum [9]. In contrast, CB2R is mainly expressed in immune tissues such as the microglia, leukocytes, and cells of macrophage lineage [10,11]. The broad localization of the CB1R in the CNS represents a limit to its potential as a pharmacological target for CNS pathologies due to the undesired psychotropic side effects related to its activation from agonists and antagonists [12]. On the other hand, CB2R may constitute a promising pharmacological target for inflammatory disorders, thanks to its anti-inflammatory properties [13]. It has recently been clarified that the localization of the ECS is not limited only to the CNS since it was found ubiquitously expressed throughout the body, serving a multiplicity of physiological roles including the regulation of gastrointestinal functions [14,15]. In particular, the ECS is supposed to regulate gastrointestinal secretion and motility via the ENS [16]. Various studies have suggested a possible implication of CB1R and CB2R in inflammatory bowel disease (IBD), exerting a protective effect, thus suggesting the potential of pharmacological agents capable of targeting and modulating these pathways [15,17,18]. Moreover, additional cannabinoid-related receptors and endocannabinoid-like molecules may also be involved. In particular, among the secondary receptors belonging to the ECS are found G-protein coupled receptors (GPRs), transient receptor potential (TRP) channels, serotonin (5-HT) receptors, and nuclear peroxisome proliferator-activated receptors (PPARs) [19]. In particular, TRP channels are sensitive to harmful stimuli, pungent compounds, acid, temperature, and inflammation mediators, qualifying these receptors as being suitable candidates and novel targets for gastrointestinal pain [20].

Concerning the endocannabinoid-like mediators, growing interest is driven by palmitoylethanolamide (PEA) and cannabidiol (CBD) [6,21]. Growing research regarding this topic indicates that activation of the cannabinoid and cannabinoid-related receptors, mediated by endogenous or plant-derived cannabinoids, may influence GIT motility and secretion, with a reduction in inflammation and visceral pain [17,22–27].

To the authors' knowledge, only a few studies have described the presence of the ECS in the porcine GIT, limited to the mucosa [28] or, in the ENS, only to CB1R [29]. For this reason, the aim of this study was to immunohistochemically characterize the distribution of the canonical cannabinoid receptors CB1R and CB2R, and the cannabinoid-related receptors TRP vanilloid 1 (TRPV1), TRP ankyrin 1 (TRPA1), and 5-HT1a serotonin receptor (5-HT1aR) in the myenteric plexus (MP) of the pig ileum.

2. Materials and Methods

2.1. Animals

Intestinal tissues were collected from six pigs at the slaughterhouse. All animals were 7-month-old genetic hybrids (Landrace × Large White). The animals did not have a history of gastrointestinal disorders and did not show gross alterations of the gastrointestinal wall.

Italian legislation (D. Lgs. n. 26/2014), according to Directive 2010/63/EU of the European Parliament and the Council of 22/09/2010 regarding the protection of animals used for scientific purposes, does not require any approval by the appropriate authorities or ethics committees since this research did not influence any therapeutic decisions.

2.2. Tissue Collection

The ileum was harvested within 30 min from the animals' deaths and was longitudinally opened along the mesenteric border. The tissues were then washed in phosphate-buffered saline (PBS), fixed, and processed to obtain longitudinal (2.0 cm × 0.5 cm) and tangential cryosections (2.0 cm × 1.0 cm), which were later processed for immunohistochemistry, as previously described [30].

2.3. Immunofluorescence

After hydration in PBS, the cryosections were processed for immunostaining. To prevent non-specific bindings, the cryosections were incubated in a solution containing 20% normal donkey serum (Colorado Serum Co., Denver, CO, USA), 0.5% Triton X-100 (Sigma Aldrich, Milan, Italy, Europe), and bovine serum albumin (1%) in PBS for 1 h at room temperature (RT). The cryosections were then incubated in a humid chamber overnight at RT, and single or double immunostaining was carried out. In the single immunostaining, the cryosections were incubated with only one of the primary antibodies (Table 1) directed against the cannabinoid and cannabinoid-related receptors. For double immunostaining, the cryosections were incubated with a cocktail of primary antibodies (Table 1). Since double immunostaining was carried out to identify the enteric neurons, the cryosections were co-incubated with one of the anti-cannabinoid receptors or anti-cannabinoid-related antibodies and the anti-HuC/HuD antibody. All the primary antibodies were diluted in 1.8% NaCl in 0.01 M PBS containing 0.1% sodium azide. After washing the cryosections in PBS (3 × 10 min), they were incubated for 1 h at RT in a humid chamber with the secondary antisera (Table 2) diluted in PBS. The cryosections were then washed in PBS (3 × 10 min) and mounted in buffered glycerol at pH 8.6 with 4′,6-diamidino-2-phenylindole (DAPI) (Santa Cruz Biotechnology, Santa Cruz, CA, USA).

Table 1. The primary antibodies used in the study.

Primary Antibodies	Host	Code	Dilution	Source
CB1R	Rabbit	ab23703	1:100	Abcam
CB2R	Rabbit	ab45942	1:200	Abcam
CB2R	Mouse	sc-293188	1:50	Santa Cruz
TRPV1	Rabbit	ACC-030	1:200	Alomone
TRPA1	Rabbit	ab58844	1:100	Abcam
5-HT1aR	Rabbit	ab85615	1:100	Abcam
HuC/HuD	Mouse	A21271	1:200	Life Technologies

Table 2. The secondary antibodies used in the study.

Secondary Antibodies	Host	Code	Dilution	Source
Anti-rabbit 488	Donkey	A-21206	1:1000	Thermo Fisher
Anti-rabbit 594	Donkey	ab150076	1:1000	Abcam
Anti-mouse 594	Donkey	A-21203	1:500	Thermo Fisher
Anti-mouse 488	Donkey	A-21202	1:500	Thermo Fisher

The proportion of neurons that were HuC/HuD immunoreactive and that were also immunoreactive for CB1R, CB2R, TRPV1, TRPA1, and 5-HT1aR was determined by examining fluorescently labelled, double-stained preparations. The neurons were first identified by the presence of a fluorophore labeling one antigen (HuC/HuD), and the microscope filter was subsequently switched to determine whether or not the neuron expressed a second antigen (CB1R, CB2R, TRPV1, TRPA1, and 5-HT1aR), identified with a different-colored fluorophore. In doing so, the proportion of neurons labeled for pairs of antigens was determined.

A minimum of one hundred HuC/HuD immunoreactive MP neurons was counted for each marker expressed by nerve cell bodies. Data were collected from preparations obtained from at least three animals ($n = 5$). The percentages of immunoreactive neurons were expressed as mean $\pm$ standard deviation.

2.4. Specificity of the Primary Antibodies

CB1R, the synthetic peptide MSVSTDTSAEAL, corresponding to carboxy-terminal amino acids 461-472 of human CB1R, was used as an immunogen to obtain the anti-CB1R antiserum. The homology between the full amino acid sequences of the pig (F1S0E6) and the human (P21554) CB1R was 97.9% (https://blast.ncbi.nlm.nih.gov/Blast.cgi); correspondence with the specific sequence of the immunogen was 100%. Therefore, the antibody anti-CB1R should also recognize the same receptor in pig. Since this antibody is human specific, it was applied on a submucosal wholemount preparation of human descending colon as a positive control, having previously obtained donor consent. The wholemount preparation was prepared and analyzed using pre-validated immunohistochemical protocols [31].

CB2R, the synthetic peptide conjugated to keyhole limpet hemocyanin (KLH) derived from within residues 200–300 of rat CB2, was used as an immunogen to obtain the rabbit ant-CB2R antibody (ab45942). The homology between the full amino acid sequences of pig (I3LUS5) and rat CB2R (Q9QZN9) was 76.3%; the correspondence with the specific sequence of the immunogen was 76%. The amino acid sequence 302–360 of CB2 of human origin (P34972) was used as an immunogen to obtain the mouse anti-CB2R antibody (sc-293188). The homology between the full amino acid sequences of pig and human CB2R was 81.9%.

TRPV1, the peptide (C)EDAEVFK DSMVPGEK, corresponding to residues 824–838 of rat TRPV1, was used as an immunogen to obtain the anti-TRPV1 antibody. The homology between the full amino acid sequences of pig (A0A4X1UCR0) and rat (O35433) TRPV1 was 84.52% (https://blast.ncbi.nlm.nih.gov/Blast.cgi), and the correspondence with the specific sequence of the immunogen was 93%. However, this antibody was tested on the porcine nervous system (dorsal root ganglia) using western blot (Wb) analysis [32].

TRPA1, the synthetic peptide CEKQHELIKLIIQKME, corresponding to amino acids 1060-1075 of rat TRPA1, was used as an immunogen to obtain the anti-TRPA1 antibody. The alignment of the immunogen sequence with the target protein in the pig was 93% (https://blast.ncbi.nlm.nih.gov/Blast.cgi). It is plausible that the antibody anti-rat TRPA1 should also recognize the same receptor in the pig.

5-HT1aR, the synthetic peptide, corresponding to amino acids 100–200 (conjugated to keyhole limpet hemocyanin) of rat 5-HT1aR, was used as an immunogen to obtain the anti-5-HT1aR antibody. The alignment of the immunogen with the target protein sequence in the pig was 100% (https://blast.ncbi.nlm.nih.gov/Blast.cgi). Therefore, the antibody anti-rat 5-HT1aR should also recognize the same receptor in pig.

The suppliers of the anti-CB2R, -TRPV1, -TRPA1, and -5-HT1aR antibodies employed in the present study stated that they were rat specific; thus, for comparison purposes, the anti-CB2R and -TRPV1 antibodies were applied on the positive control tissues, in particular on wholemount preparations of rat ileum (authorization no. 112/2018-PR of 12 February 2018). Data related to the anti-TRPA1 and 5-HT1aR antibodies have recently been published [33].

2.5. Specificity of the Secondary Antibodies

The specificity of the secondary antibodies (Table 2) was tested by the absence of signal after the exclusion of the primary antibody on pig intestinal tissues.

2.6. Fluorescence Microscopy

Cryosections and wholemount preparations were examined by the same observer (Dr. R. Chiocchetti) using a Nikon Eclipse Ni microscope (Nikon Instruments Europe BV, Amsterdam, The Netherlands) equipped with the appropriate filter cubes to differentiate the fluorochromes used. The images were recorded using a Nikon DS-Qi1Nc digital camera and NIS Elements software BR 4.20.01 (Nikon Instruments Europe BV, Amsterdam, Netherlands). Enteric neuron counts were carried out at 40× magnification. Slight adjustments to contrast and brightness were made using Corel Photo Paint whereas the figure panels were prepared using Corel Draw (Corel Photo Paint and Corel Draw, Ottawa, ON, Canada).

3. Results

3.1. CB1R Immunoreactivity

Weak-to-moderate granular and diffuse CB1R immunoreactivity (CB1R-IR) was expressed by the cytoplasm of the MP neurons; the brightest CB1R immunoreactive neurons showed large dimensions and a smooth outline (Figure 1a–c). The percentages of HuC/HuD immunoreactive neurons co-expressing CB1R-IR was 57 ± 19% (377/713 cells counted, n = 5). Nerve fibers within the MP ganglia, distributed in the interganglionic strands and scattered within the muscular layers, showed weak CB1R-IR. Weak CB1R-IR was also observed in the smooth muscle cells of the *tunica muscularis* (longitudinal muscle layer, LML > circular muscle layer, CML) (data not shown).

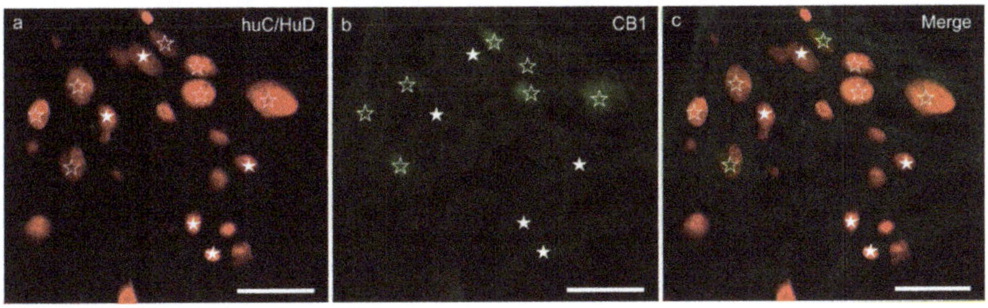

Figure 1. CB1 receptor immunoreactivity in the myenteric plexus of the pig ileum: (**a**) HuC/HuD immunoreactive neurons, (**b**) CB1 receptor immunoreactivity, (**c**) merge image. The open stars indicate HuC/HuD immunoreactive neurons co-expressing weak-to-moderate CB1 receptor immunoreactivity. The white stars indicate HuC/HuD immunoreactive neurons, which were CB1 negative. Scale bar: 50 μm.

Moderate CB1R-IR was expressed by the cytoplasm of the submucosal plexus neurons of the human colon (Figure S1).

3.2. CB2R Immunoreactivity

No immunolabeling was observed in the porcine MP with either of the anti-CB2 receptor antibodies. In the rat ileum, MP neurons expressed weak-to-moderate CB2R-IR (Figure S2).

3.3. TRPV1 Immunoreactivity

Moderate-to-bright granular TRPV1-IR was expressed by the cytoplasm of the majority of the MP neurons (71 ± 14%; 462/602 cells counted, n = 5). The TRPV1 immunolabelling, which was mainly confined to the cell bodies of neurons showing a smooth outline, was more intense in neurons of large dimensions (Figure 2a–c) whereas it was very faint or undetectable in neurons of small dimensions. However, TRPV1 immunoreactive nerve processes arising from large neurons were also visible (Figure 2d–f). Few TRPV1 immunoreactive nerve fibers were seen in the MP ganglia, in the interganglionic nerve strands and within the muscular layers; nevertheless, bright and small TRPV1 immunoreactive varicosities were seen in the neuropil of the ganglia and around some MP neurons. In one subject, TRPV1 was also brightly expressed by the enteric glial cells (Figure 2g–i). Moderate TRPV1-IR was expressed by the endothelial cells of thin blood vessels (capillaries) distributed in the *tunica muscularis* (data not shown). In the rat ileum, MP and enteric glial cells (EGCs) expressed TRPV1-IR (EGCs > neurons) (Figure S3).

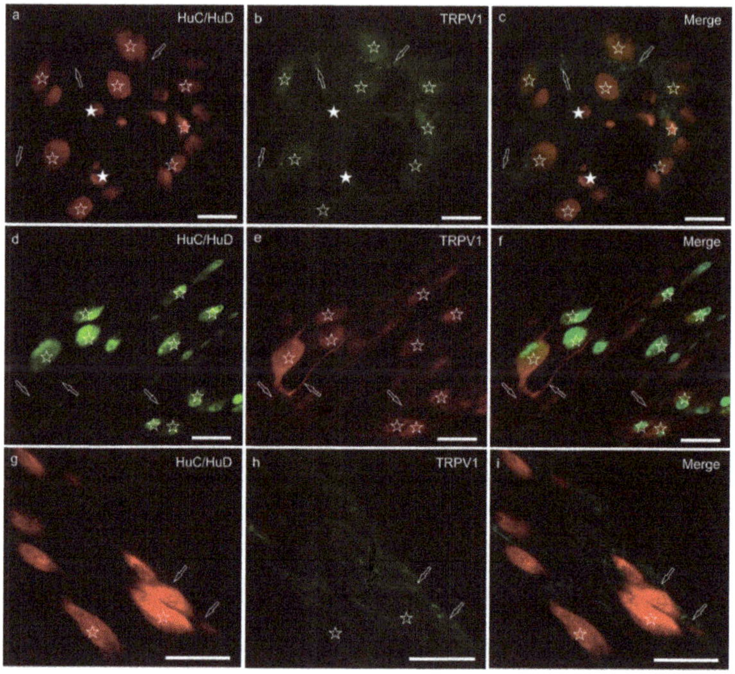

Figure 2. TRPV1 immunoreactivity in the myenteric plexus of the pig ileum: (**a**) HuC/HuD immunoreactive neurons, (**b**) TRPV1 immunoreactivity, (**c**) merge image. The open stars indicate the HuC/HuD immunoreactive neurons co-expressing moderate TRPV1 immunoreactivity. The white stars indicate HuC/HuD immunoreactive neurons, which were TRPV1 negative. The arrows indicate the TRPV1 immunoreactive varicosities encircling the neuronal cell bodies. (**d**) HuC/HuD immunoreactive neurons, (**e**) TRPV1 immunoreactivity, (**f**) merge image. The stars indicate the HuC/HuD immunoreactive neurons co-expressing moderate-to-bright TRPV1 immunoreactivity. The arrows indicate TRPV1 immunoreactive neuronal processes. (**g**) HuC/HuD immunoreactive neurons, (**h**) TRPV1 immunoreactivity, (**i**) merge image. The stars indicate two HuC/HuD immunoreactive myenteric plexus neurons co-expressing weak and diffuse TRPV1 immunoreactivity; the arrows indicate two perineuronal enteric glial cells expressing bright TRPV1 immunoreactivity. Scale bar: 50 μm.

3.4. *TRPA1 Immunoreactivity*

Diffuse and moderate cytoplasmic TRPA1-IR was shown by a large percentage of MP neurons (66 ± 23%; 336/527 cells counted, n = 5) and was brighter in the cytoplasm of large neurons (Figure 3a–c). Nerve fibers within the ganglia and those distributed along the nerve strands and musculature showed moderate TRPA1-IR (data not shown). Bright TRPA1-IR was displayed by the endothelial cells of the blood vessels (Figure 3d–f).

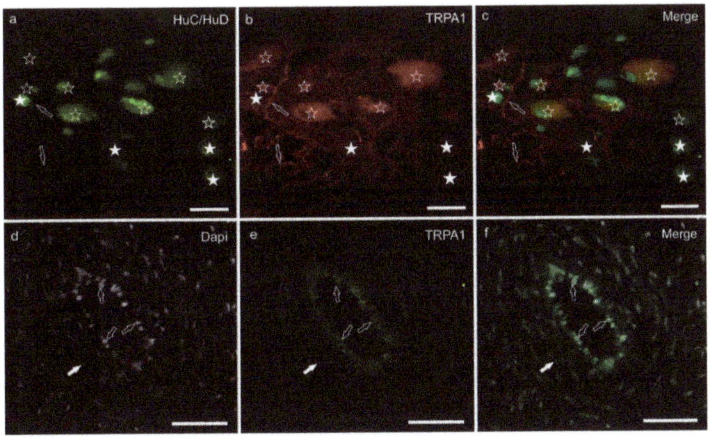

Figure 3. TRPA1 immunoreactivity in the myenteric plexus and blood vessels of the *tunica muscularis* of the pig ileum: (**a**) HuC/HuD immunoreactive neurons, (**b**) TRPA1 immunoreactivity, (**c**) merge image. The open stars indicate HuC/HuD immunoreactive neurons co-expressing moderate-to-bright TRPA1 immunoreactivity. The white stars indicate HuC/HuD immunoreactive neurons, which were TRPA1 negative. The arrows indicate a TRPA1 immunoreactive neuronal process. (**d**) Dapi stained nuclei of endothelial cells and vascular smooth muscle cells, (**e**) TRPA1 immunoreactivity, (**f**) merge image. The open arrows indicate the Dapi stained nuclei of endothelial cells expressing bright TRPA1 immunoreactivity. The white arrow indicates the elongated Dapi stained nucleus of one smooth muscle cell of the arterial *tunica media*. Scale bar: 50 μm.

Transient potential ankyrin receptor 1 immunoreactivity was also expressed by the MP neurons in the ileum of a control rat [33].

3.5. 5-HT1aR Immunoreactivity

Weak and diffuse 5-HT1aR immunolabelling was expressed by approximately half the MP neurons (51 ± 6%; 345/682 cells counted, *n* = 5) (Figure 4a–c). In general, neurons of large dimensions showed brighter immunofluorescence. The 5-HT1aR-IR was expressed by nerve varicosities. Weak 5-HT1aR-IR was displayed by the smooth muscle cells of the blood vessels and the *tunica muscularis* (data not shown).

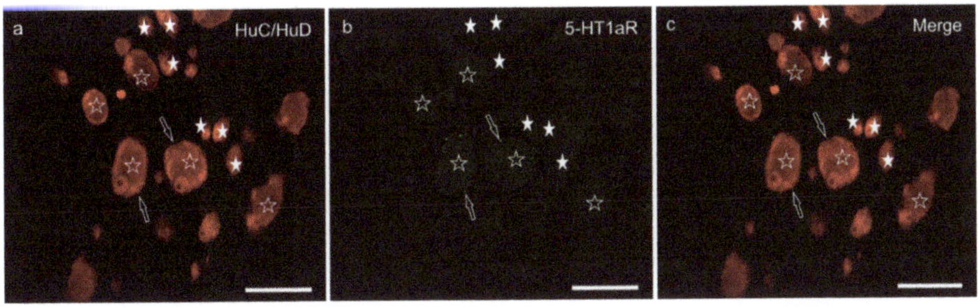

Figure 4. 5-HT1a receptor immunoreactivity in the myenteric plexus of the pig ileum: (**a**) HuC/HuD immunoreactive neurons, (**b**) 5-HT1a receptor immunoreactivity, (**c**) merge image. The open stars indicate HuC/HuD immunoreactive neurons co-expressing weak 5-HT1a receptor immunoreactivity. The white stars indicate the HuC/HuD immunoreactive neurons, which were 5-HT1a receptor negative. The arrows indicate 5-HT1a receptor immunoreactive varicosities encircling the neuronal cell bodies. Scale bar: 50 μm.

The 5-HT1aR was also expressed by MP neurons in the rat ileum [33].

The results of the cellular distribution and intensity of immunolabeling in the pig ileum are summarized in semiquantitative Table 3.

Table 3. Semiquantitative evaluation of the density of CB1R, CB2R, TRPV1, TRPA1, and 5-HT1aR immunoreactivity in different cellular elements (myenteric plexus neurons, nerve fibers, enteric glial cells, *tunica muscularis*, and blood vessels) of the pig ileum.

Receptors	MP Neurons	Nerve Fibers	EGCs	*Tunica Muscularis*	Blood Vessels
CB1R	C +/++	+	−	+	−
CB2R	−	−	−	−	−
TRPV1	++/+++	+/++	−/+++	−	E ++
TRPA1	C ++	++	−	−	E +++
5-HT1aR	C +	−	−	+	SMC +

The immunoreactive cells were graded as: −, negative; +, weakly stained, ++, moderately stained and +++, brightly stained. C: cytoplasmic; E: endothelium; EGCs: enteric glial cells; M: membrane; MP: myenteric plexus; SMCs: vascular smooth muscle cells.

4. Discussion

In the GIT, cannabinoid receptors regulate motility, secretion, emesis, food intake, and inflammation [15,17,24,26]. In this paper, the authors focused their attention on the presence of the ECS in the ileal MP of pigs, with particular emphasis on both the cannabinoid receptors, namely CB1R and CB2R, and the cannabinoid-related receptors TRPV1, TRPA1, and 5-HT1aR by carrying out immunohistochemical analysis.

The observation of the CB1R in the MP neurons and nerve fibers in the porcine ileum is consistent with the findings of Kulkarni-Narla and Brown [29]. The expression of CB1R-IR in the enteric neurons has been observed in many other species such as rodents, ferrets, dogs, cats, and humans [33–37]. Previous studies regarding various species including pigs have indicated that CB1R immunoreactive neurons show a cholinergic phenotype, exerting an inhibitory effect on MP cholinergic neurotransmission [29,38–41].

The evaluation of the proportion of myenteric CB1R immunoreactive neurons of the pig is not comparable with other data available in the literature. However, the percentage of CB1R immunoreactive neurons (57 $\pm$ 19%) observed in the present study was similar to the percentage of cholinergic neurons observed in the guinea-pig (approximately 80%; [42]), but was greater than the proportions of the ChAT-IR neurons counted in other species such as horses (64%; [43]), sheep (62%; [44]) or pigs (58%; [45]). In effect, Brehmer and colleagues [45] observed that there was a subclass of enteric cholinergic neurons, which could be identified only by the use of the antibody directed against the peripheral form of ChAT (i.e., pChAT). Thus, the percentage of cholinergic neurons of the pig ileum should be greater than that identified only with the anti-ChAT antibody. However, the great percentage of CB1R immunoreactive neurons observed in the present study indicated that this receptor might also be expressed by other neuronal subpopulations as well as by the cholinergic neurons. The influence on intestinal motility and contractility mediated by cannabinoids might be confirmed by the presence of CB1R on the smooth muscle cells, suggesting a direct muscular mechanism of cannabinoids [41].

CB2R is mainly expressed in immune tissues and cells of macrophage lineage [11]. The lack of immunoreactivity to the CB2R in enteric neurons is in line with the results obtained in the MP of dogs and cats [33,37] in contrast with rats (present study) and mice [25]. In the pig ileum, the lack of results could also depend on the low homology between the full amino acid sequences of pig and rat CB2R (76%), and pig and human CB2R (81.9%), original immunogens of the antibodies employed in the present study.

Transient potential vanilloid receptor 1 is a non-selective cation channel expressed by peptidergic and non-peptidergic nociceptors in rodents and large mammals [32,46,47]; TRPV1-IR was observed in the MP neurons of the pig (and rat) ileum, according to Poonyachoti et al. [48], who indicated that the majority of TRPV1 immunoreactive neurons were cholinergic. The expression of TRPV1-IR in the intramural neurons is a matter of debate.

In fact, there are numerous investigations reporting TRPV1-IR expressed by extrinsic sensory neurons [49–56]. However, there are publications supporting the expression of TRPV1 immunolabelling in intramural neurons of different mammalian species [29,35,48,53,56–59] as has been shown in the present study. In addition, functional studies have supported the existence of enteric TRPV1 immunoreactive neurons in mice [60] and pigs [61].

The use of different anti-TRPV1 antibodies seems to be the reason for the discrepancy in the expression of TRPV1. In fact, Buckinx and colleagues [62] found that different distribution patterns of TRPV1 in the ENS were due to the antibodies discriminating between different modulated forms of TRPV1, which influence the recognition of the intracellular forms of TRPV1.

The large percentages of CB1R and TRPV1immunoreactive neurons, which we found in the porcine MP ileum, allowed us to speculate that CB1R and TRPV1 may co-exist on the same subclass of cholinergic neurons as substantiated by functional and immunohistochemical studies [35,63]. Double labeling of CB1R with TRPV1 (and the other receptors) was not examined due to incompatibility of the species in which the antisera were raised.

The expression of TRPV1 by enteric neurons could constitute a target for the development of new therapies against nociceptive and inflammatory intestinal stimuli [48]. Moreover, TRPV1 seems to be involved in protection against pathogenic bacteria such as *Salmonella enterica*, releasing the calcitonin gene-related peptide (CGRP), which regulates the number of microfold (M) cells and the levels of segmented filamentous bacteria that fight pathogen colonization [64].

The expression of TRPV1-IR in EGCs might be involved in their differentiation/maturation as suggested by Yamamoto et al. [65], or might be upregulated in different conditions of gut homeostasis/physiology as observed in the porcine tissues. The localization of TRPV1 on endothelial cells of the capillaries in the *tunica muscularis* suggested a modulation of vasocontraction and vasorelaxation in an endothelium-dependent manner, supporting the therapeutic potential of TRPV1 as a target for improving vascular functionality [66].

Transient potential ankyrin receptor 1 has been successfully found in the GIT neuronal [33,60,67] and non-neuronal cells [33,68,69] in which the receptor can detect specific food chemicals such as cinnamaldehyde, allyl isothiocyanate (AITC), allicin, and thymol [70,71]. In the present study, TRPA1 was observed in the enteric neurons as described in the MP neurons of rodents [33,67]. Moreover, functional investigations have indicated that TRPA1 may regulate gastrointestinal motility by means of the 5-HT release from enterochromaffin cells [72]. However, the localization of TRPA1 in the enteric neurons indicated that it could directly modulate intestinal contraction/motility, as suggested by Sandor et al. [73]. In addition, localization on the endothelial cells of the blood vessels suggested a role in controlling vasodilatation and vasoconstriction, as for TRPV1. The ability of botanicals to act on TRPA1 in the GIT seemed to modulate the majority of its functionality. In fact, AITC modulated the gastrointestinal contractions in mice [60,74], guinea-pigs [72], and dogs [75], and inhibited colonic transit [67] via TRPA1 activation. Moreover, AITC was also capable of inducing blood vessel dilatation due to the activation of TRPA1, as reported by Earley et al. [76,77] and Sullivan et al. [78]. It is possible to assume that the use of botanicals capable of modulating TRPA1 and/or TRPV1 could also play a role in reducing or controlling inflammatory stimuli. In fact, Blackshaw et al. [79] suggested that TRPA1-IR expressed by intramural neurons might not contribute to normal ENS functions, exerting its role only during inflammation or injury, or in response to exogenous agonism.

In the porcine ENS, 5-HT has been found in MP neurons of the pig colon [80] and perineuronal varicosities of the ileum [81]. Given the variety and the complexity of the effects that 5-HT exerts in the gut, it is not surprising that there is more than one type of enteric neuronal 5-HT receptor [82]. Many of the effects of CBD are mediated through 5-HT receptor activation in the CNS and peripheral nervous system, which regulate neuronal excitability and neurotransmitter release. Of the 5-HT receptors, CBD acts as an agonist on the 5-HT1aR, as a partial agonist on the 5-HT2aR, and as an antagonist on the 5-HT3R [5,83].

The full agonism of CBD at the 5-HT1aR is responsible for the anxiolytic/antidepressant and analgesic effects of CBD in animals [84–86].

In the present study, 5-HT1aR-IR was observed in the MP neurons and smooth muscle cells of the *tunica muscularis* in line with Youn et al. [87] and Delesalle et al. [88], who observed 5-HT1aR-IR in the MP neurons and *tunica muscularis* of the guinea-pig stomach and muscular smooth muscle cells of the equine jejunum, respectively. Electrophysiological studies have indicated that the 5-HT1aR is primarily involved in the presynaptic inhibition of transmitter release [89]. The location of 5-HT1aR-IR in varicosities around the MP neurons observed in the present study supported the idea that the receptor might be involved in the 5-HT mediated inhibition of cholinergic neurotransmission. The expression of 5-HT1aR-IR in approximately 50% of the MP neurons also suggested that some of the cells capable of producing the 5-HT1aR were cholinergic. The expression of the 5-HT1aR in the vascular smooth muscle cells of the pig ileum observed in the present study may support the direct action of 5-HT on the vascular smooth muscle. In pigs, direct vascular smooth muscle relaxation may be the predominant mechanism involved in the vasodilatation action of serotonin [90].

The localization of cannabinoid and cannabinoid-related receptors in the MP of pigs implied a possible role of phytocannabinoids and botanicals in the control and support of various gastrointestinal activities. For example, CBD has been found to act as an agonist on the 5-HT1aR, exerting a neuroprotective effect by modulating oxidative stress and inflammation [91]. Thymol also seemed to be capable of modulating the expression of the ECS in the porcine GIT [28], representing a therapeutic approach to several gastrointestinal diseases. Additional investigation is required to obtain a better understanding of the localization of the ECS receptors in the GIT of pigs including other tracts, receptors, and phenotypes of the enteric neurons in an attempt to overcome the limitations posed by the absence of specific antibodies for pigs.

5. Conclusions

In conclusion, the data in the present study highlighted the expression of cannabinoid (CB1R and CB2R) and cannabinoid-related receptors (TRPV1, TRPA1, and 5-HT1aR) not only in the MP neurons and enteric glial cells, but also on the smooth muscle cells and the blood vessels of the porcine ileum. These morphological findings could be of particular relevance for future functional, pre-clinical, and clinical studies assessing the effects of cannabinoids in pigs in order to manage the hypermotility associated with gastrointestinal inflammatory diseases and pain. In fact, this could justify the use of phytocannabinoids or natural molecules capable of modulating the ECS in the diet of pigs. By modulating the activation of cannabinoid and cannabinoid-related receptors, it seems possible to regulate gastrointestinal functionality at different levels. Of particular interest, TRPV1 can interfere with pathogen proliferation and, together with TRPA1, could play a role in reducing the inflammation that occurs during weaning.

Supplementary Materials: The following are available online at https://www.mdpi.com/2076-2615/11/2/263/s1, Figure S1: CB1R immunoreactivity in the submucosal plexus neurons of the human colon, Figure S2: CB2R immunoreactivity in the myenteric plexus neurons of the rat ileum, Figure S3: TRPV1 immunoreactivity in the myenteric plexus of the rat ileum.

Author Contributions: Conceptualization, R.C., E.G., and A.P.; Methodology, R.C.; Investigation, A.T., G.G., and C.T.; Writing—original draft preparation, A.T., G.G., and R.C.; Writing—review and editing, R.C., E.G., and G.M.-W.; Supervision, R.C., E.G., and A.P. All authors have read and agreed to the published version of the manuscript.

Funding: This research received no external funding.

Informed Consent Statement: Not applicable.

Data Availability Statement: The data presented in this study are available on request from the corresponding author.

Acknowledgments: This research was supported by a grant from Vetagro SpA.

Conflicts of Interest: Andrea Piva serves as Professor at the University of Bologna and is a member of the Board of Directors of Vetagro SpA (Reggio Emilia, Italy), which funded, conceptualized, and supervised the project. Ester Grilli serves as an advisor of Vetagro SpA, which conceptualized, written, and supervised the project.

References

1. Wood, J.D. Enteric nervous system. In *Encyclopedia of Gastroenterology*; Johnson, L.R., Ed.; Elsevier: New York, NY, USA, 2004; pp. 701–706, ISBN 978-0-12-386860-2.
2. Costa, M.; Brookes, S.J.H.; Hennig, G.W. Anatomy and physiology of the enteric nervous system. *Gut* **2000**, *47*, iv15–iv19. [CrossRef] [PubMed]
3. Furness, J.B.; Callaghan, B.P.; Rivera, L.R.; Cho, H.-J. The enteric nervous system and gastrointestinal innervation: Integrated local and central control. In *Microbial Endocrinology: The Microbiota-Gut-Brain Axis in Health and Disease*; Lyte, M., Cryan, J.F., Eds.; Springer: New York, NY, USA, 2014; pp. 39–71, ISBN 978-1-4939-0897-4.
4. Stella, N. Cannabinoid signaling in glial cells. *Glia* **2004**, *48*, 267–277. [CrossRef] [PubMed]
5. Ligresti, A.; De Petrocellis, L.; Di Marzo, V. From phytocannabinoids to cannabinoid receptors and endocannabinoids: Pleiotropic physiological and pathological roles through complex pharmacology. *Physiol. Rev.* **2016**, *96*, 1593–1659. [CrossRef] [PubMed]
6. Iannotti, F.A.; Di Marzo, V.; Petrosino, S. Endocannabinoids and endocannabinoid-related mediators: Targets, metabolism and role in neurological disorders. *Prog. Lipid Res.* **2016**, *62*, 107–128. [CrossRef]
7. Lu, Y.; Anderson, H.D. Cannabinoid signaling in health and disease. *Can. J. Physiol. Pharmacol.* **2017**, *95*, 311–327. [CrossRef]
8. Maccarrone, M.; Bab, I.; Bíró, T.; Cabral, G.A.; Dey, S.K.; Di Marzo, V.; Konje, J.C.; Kunos, G.; Mechoulam, R.; Pacher, P.; et al. Endocannabinoid signaling at the periphery: 50 years after THC. *Trends Pharmacol. Sci.* **2015**, *36*, 277–296. [CrossRef]
9. Mackie, K. Distribution of cannabinoid receptors in the central and peripheral nervous system. In *Handbook of Experimental Pharmacology*; Cannabinoids; Pertwee, R.G., Ed.; Springer Berlin Heidelberg: Berlin, Heidelberg, 2005; pp. 299–325, ISBN 978-3-540-26573-3.
10. Klegeris, A.; Bissonnette, C.J.; McGeer, P.L. Reduction of human monocytic cell neurotoxicity and cytokine secretion by ligands of the cannabinoid-type CB2 receptor. *Br. J. Pharmacol.* **2003**, *139*, 775–786. [CrossRef]
11. Turcotte, C.; Blanchet, M.-R.; LaViolette, M.; Flamand, N. The CB2 receptor and its role as a regulator of inflammation. *Cell. Mol. Life Sci.* **2016**, *73*, 4449–4470. [CrossRef]
12. Moreira, F.A.; Grieb, M.; Lutz, B. Central side-effects of therapies based on CB1 cannabinoid receptor agonists and antagonists: Focus on anxiety and depression. *Best Pract. Res. Clin. Endocrinol. Metab.* **2009**, *23*, 133–144. [CrossRef]
13. Parlar, A.; Arslan, S.O.; Doğan, M.F.; Elibol, E.; Kara, H. The exogenous administration of CB2 specific agonist, GW405833, inhibits inflammation by reducing cytokine production and oxidative stress. *Exp. Ther. Med.* **2018**, *16*, 4900–4908. [CrossRef]
14. Izzo, A.A.; Sharkey, K.A. Cannabinoids and the gut: New developments and emerging concepts. *Pharmacol. Ther.* **2010**, *126*, 21–38. [CrossRef] [PubMed]
15. DiPatrizio, N.V. Endocannabinoids in the Gut. *Cannabis Cannabinoid Res.* **2016**, *1*, 67–77. [CrossRef] [PubMed]
16. Vera, G.; Fichna, J.; Abalo, R. Chapter 98—Cannabinoids and effects on the gastrointestinal tract: A focus on motility. In *Handbook of Cannabis and Related Pathologies*; Preedy, V.R., Ed.; Academic Press: San Diego, CA, USA, 2017; pp. 947–957, ISBN 978-0-12-800756-3.
17. Izzo, A.A. Cannabinoids and intestinal motility: Welcome to CB2 receptors. *Br. J. Pharmacol.* **2004**, *142*, 1201–1202. [CrossRef] [PubMed]
18. Di Marzo, V. Endocannabinoid overactivity and intestinal inflammation. *Gut* **2006**, *55*, 1373–1376. [CrossRef]
19. Morales, P.; Hurst, D.P.; Reggio, P.H. Molecular Targets of the Phytocannabinoids: A complex picture. *Prog. Chem. Org. Nat. Prod.* **2017**, *103*, 103–131. [CrossRef]
20. Yu, X.; Yu, M.; Liu, Y.; Yu, S. TRP channel functions in the gastrointestinal tract. *Semin. Immunopathol.* **2016**, *38*, 385–396. [CrossRef]
21. Kreitzer, F.R.; Stella, N. The therapeutic potential of novel cannabinoid receptors. *Pharmacol. Ther.* **2009**, *122*, 83–96. [CrossRef]
22. Di Carlo, G.; Izzo, A.A. Cannabinoids for gastrointestinal diseases: Potential therapeutic applications. *Expert Opin. Investig. Drugs* **2003**, *12*, 39–49. [CrossRef]
23. Hornby, P.J.; Prouty, S.M. Involvement of cannabinoid receptors in gut motility and visceral perception. *Br. J. Pharmacol.* **2004**, *141*, 1335–1345. [CrossRef]
24. Duncan, M.; Davison, J.S.; Sharkey, K.A. Review article: Endocannabinoids and their receptors in the enteric nervous system. *Aliment. Pharmacol. Ther.* **2005**, *22*, 667–683. [CrossRef]
25. Duncan, M.; Mouihate, A.; Mackie, K.; Keenan, C.M.; Buckley, N.E.; Davison, J.S.; Patel, K.D.; Pittman, Q.J.; Sharkey, K.A. Cannabinoid CB2 receptors in the enteric nervous system modulate gastrointestinal contractility in lipopolysaccharide-treated rats. *Am. J. Physiol. Gastrointest. Liver Physiol.* **2008**, *295*, G78–G87. [CrossRef] [PubMed]
26. Storr, M.A.; Sharkey, K.A. The endocannabinoid system and gut–brain signalling. *Curr. Opin. Pharmacol.* **2007**, *7*, 575–582. [CrossRef] [PubMed]
27. Wright, K.L.; Duncan, M.; Sharkey, K.A. Cannabinoid CB2 receptors in the gastrointestinal tract: A regulatory system in states of inflammation. *Br. J. Pharmacol.* **2008**, *153*, 263–270. [CrossRef] [PubMed]

28. Toschi, A.; Tugnoli, B.; Rossi, B.; Piva, A.; Grilli, E. Thymol modulates the endocannabinoid system and gut chemosensing of weaning pigs. *BMC Vet. Res.* **2020**, *16*, 289. [CrossRef] [PubMed]
29. Kulkarni-Narla, A.; Brown, D.R. Localization of CB 1 -cannabinoid receptor immunoreactivity in the porcine enteric nervous system. *Cell Tissue Res.* **2000**, *302*, 73–80. [CrossRef]
30. Chiocchetti, R.; Giancola, F.; Mazzoni, M.; Sorteni, C.; Romagnoli, N.; Pietra, M. Excitatory and inhibitory enteric innervation of horse lower esophageal sphincter. *Histochem. Cell Biol.* **2015**, *143*, 625–635. [CrossRef]
31. Giancola, F.; Torresan, F.; Repossi, R.; Bianco, F.; Latorre, R.; Ioannou, A.; Guarino, M.; Volta, U.; Clavenzani, P.; Mazzoni, M.; et al. Downregulation of neuronal vasoactive intestinal polypeptide in Parkinson's disease and chronic constipation. *Neurogastroenterol. Motil.* **2017**, *29*, e12995. [CrossRef]
32. Russo, D.; Clavenzani, P.; Sorteni, C.; Minelli, L.B.; Botti, M.; Gazza, F.; Panu, R.; Ragionieri, L.; Chiocchetti, R. Neurochemical features of boar lumbosacral dorsal root ganglion neurons and characterization of sensory neurons innervating the urinary bladder trigone. *J. Comp. Neurol.* **2013**, *521*, 342–366. [CrossRef]
33. Stanzani, A.; Galiazzo, G.; Giancola, F.; Tagliavia, C.; De Silva, M.; Pietra, M.; Fracassi, F.; Chiocchetti, R. Localization of cannabinoid and cannabinoid related receptors in the cat gastrointestinal tract. *Histochem. Cell Biol.* **2020**, *153*, 339–356. [CrossRef]
34. Van Sickle, M.D.; Oland, L.D.; Ho, W.; Hillard, C.J.; Mackie, K.; Davison, J.S.; Sharkey, K.A. Cannabinoids inhibit emesis through CB1 receptors in the brainstem of the ferret. *Gastroenterology* **2001**, *121*, 767–774. [CrossRef]
35. Coutts, A.A.; Irving, A.J.; Mackie, K.; Pertwee, R.G.; Anavi-Goffer, S. Localisation of cannabinoid CB1 receptor immunoreactivity in the guinea pig and rat myenteric plexus. *J. Comp. Neurol.* **2002**, *448*, 410–422. [CrossRef] [PubMed]
36. Wright, K.L.; Rooney, N.; Feeney, M.; Tate, J.; Robertson, D.; Welham, M.; Ward, S. Differential expression of cannabinoid receptors in the human colon: Cannabinoids promote epithelial wound healing. *Gastroenterology* **2005**, *129*, 437–453. [CrossRef] [PubMed]
37. Galiazzo, G.; Giancola, F.; Stanzani, A.; Fracassi, F.; Bernardini, C.; Forni, M.; Pietra, M.; Chiocchetti, R. Localization of cannabinoid receptors CB1, CB2, GPR55, and PPARα in the canine gastrointestinal tract. *Histochem. Cell Biol.* **2018**, *150*, 187–205. [CrossRef] [PubMed]
38. Coutts, A.A.; Pertwee, R.G. Inhibition by cannabinoid receptor agonists of acetylcholine release from the guinea-pig myenteric plexus. *Br. J. Pharmacol.* **1997**, *121*, 1557–1566. [CrossRef] [PubMed]
39. Izzo, A.A.; Mascolo, N.; Tonini, M.; Capasso, F. Modulation of peristalsis by cannabinoid CB(1) L1 ligands in the isolated guinea-pig ileum. *Br. J. Pharmacol.* **2000**, *129*, 984–990. [CrossRef]
40. Storr, M.A.; Bashashati, M.; Hirota, C.; Vemuri, V.K.; Keenan, C.M.; Duncan, M.; Lutz, B.; Mackie, K.; Makriyannis, A.; Macnaughton, W.K.; et al. Differential effects of CB1 N1 neutral antagonists and inverse agonists on gastrointestinal motility in mice. *Neurogastroenterol. Motil.* **2010**, *22*, 787–796.e233. [CrossRef]
41. Donnerer, J.; Liebmann, I. Effect of CB1 ligands on neurogenic and myogenic contractile responses in the guinea-pig ileum. *Pharmacology* **2018**, *101*, 330–336. [CrossRef]
42. Furness, J.B. *The Enteric Nervous System*; Blackwell Publishing: Malden, MA, USA, 2006; ISBN 978-1-4051-3376-0.
43. Chiocchetti, R.; Bombardi, C.; Mongardi-Fantaguzzi, C.; Venturelli, E.; Russo, M.; Spadari, A.; Montoneri, C.; Romagnoli, N.; Grandis, A. Intrinsic innervation of the horse ileum. *Res. Vet. Sci.* **2009**, *87*, 177–185. [CrossRef]
44. Mazzuoli-Weber, G.; Mazzoni, M.; Albanese, V.; Clavenzani, P.; Lalatta-Costerbosa, G.; Lucchi, M.L.; Furness, J.B.; Chiocchetti, R. Morphology and neurochemistry of descending and ascending myenteric plexus neurons of sheep ileum. *Anat. Rec. (Hoboken)* **2007**, *290*, 1480–1491. [CrossRef]
45. Brehmer, A.; Schrödl, F.; Neuhuber, W.; Tooyama, I.; Kimura, H. Co-expression pattern of neuronal nitric oxide synthase and two variants of choline acetyltransferase in myenteric neurons of porcine ileum. *J. Chem. Neuroanat.* **2004**, *27*, 33–41. [CrossRef]
46. Zwick, M.; Davis, B.M.; Woodbury, C.J.; Burkett, J.N.; Koerber, H.R.; Simpson, J.F.; Albers, K.M. Glial cell line-derived neurotrophic factor is a survival factor for isolectin B4-positive, but not vanilloid receptor 1-positive, neurons in the mouse. *J. Neurosci.* **2002**, *22*, 4057–4065. [CrossRef] [PubMed]
47. Tominaga, M.; Caterina, M.J.; Malmberg, A.B.; Rosen, T.A.; Gilbert, H.; Skinner, K.; Raumann, B.E.; Basbaum, A.I.; Julius, D. The cloned capsaicin receptor integrates multiple pain-producing stimuli. *Neuron* **1998**, *21*, 531–543. [CrossRef]
48. Poonyachoti, S.; Kulkarni-Narla, A.; Brown, D.R. Chemical coding of neurons expressing δ- and κ-opioid receptor and type I vanilloid receptor immunoreactivities in the porcine ileum. *Cell Tissue Res.* **2002**, *307*, 23–33. [CrossRef] [PubMed]
49. Guo, A.; Vulchanova, L.; Wang, J.; Li, X.; Elde, R. Immunocytochemical localization of the vanilloid receptor 1 (VR1): Relationship to neuropeptides, the P2X3purinoceptor and IB4 binding sites. *Eur. J. Neurosci.* **1999**, *11*, 946–958. [CrossRef] [PubMed]
50. Yiangou, Y.; Facer, P.; Dyer, N.; Chan, C.; Knowles, C.; Williams, N.; Anand, P. Vanilloid receptor 1 immunoreactivity in inflamed human bowel. *Lancet* **2001**, *357*, 1338–1339. [CrossRef]
51. Schicho, R.; Florian, W.; Liebmann, I.; Holzer, P.; Lippe, I.T. Increased expression of TRPV1 receptor in dorsal root ganglia by acid insult of the rat gastric mucosa. *Eur. J. Neurosci.* **2004**, *19*, 1811–1818. [CrossRef] [PubMed]
52. Kadowaki, M.; Kuramoto, H.; Takaki, M. Combined determination with functional and morphological studies of origin of nerve fibers expressing transient receptor potential vanilloid 1 in the myenteric plexus of the rat jejunum. *Auton. Neurosci.* **2004**, *116*, 11–18. [CrossRef]
53. Faussone-Pellegrini, M.S.; Taddei, A.; Bizzoco, E.; Lazzeri, M.; Vannucchi, M.G.; Bechi, P. Distribution of the vanilloid (capsaicin) receptor type 1 in the human stomach. *Histochem. Cell Biol.* **2005**, *124*, 61–68. [CrossRef]

54. Matsumoto, K.; Hosoya, T.; Tashima, K.; Namiki, T.; Murayama, T.; Horie, S. Distribution of transient receptor potential vanilloid 1 channel-expressing nerve fibers in mouse rectal and colonic enteric nervous system: Relationship to peptidergic and nitrergic neurons. *Neuroscience* **2011**, *172*, 518–534. [CrossRef]
55. Sharrad, D.; Hibberd, T.J.; Kyloh, M.; Brookes, S.J.H.; Spencer, N.J. Quantitative immunohistochemical co-localization of TRPV1 and CGRP in varicose axons of the murine oesophagus, stomach and colorectum. *Neurosci. Lett.* **2015**, *599*, 164–171. [CrossRef]
56. Chan, C.; Facer, P.; Davis, J.; Smith, G.; Egerton, J.; Bountra, C.; Williams, N.; Anand, P. Sensory fibres expressing capsaicin receptor TRPV1 in patients with rectal hypersensitivity and faecal urgency. *Lancet* **2003**, *361*, 385–391. [CrossRef]
57. Anavi-Goffer, S.; McKay, N.G.; Ashford, M.L.J.; Coutts, A.A. Vanilloid receptor type 1-immunoreactivity is expressed by intrinsic afferent neurones in the guinea-pig myenteric plexus. *Neurosci. Lett.* **2002**, *319*, 53–57. [CrossRef]
58. Filippova, L.V.; Fedorova, A.V.; Nozdrachev, A.D. Mechanism of activation of enteric nociceptive neurons via interaction of TLR4 and TRPV1 receptors. *Dokl. Biol. Sci.* **2018**, *479*, 44–46. [CrossRef]
59. Akiba, Y.; Nakamura, M.; Ishii, H. Immunolocalization of vanilloid receptor-1 (VR-1) in CGRP-postive neurons and inter-stitial cells of cajal in the myenteric ptexus of the rat gastrointestinal tract. *Gastroenterology* **2001**, *120*. [CrossRef]
60. Penuelas, A.; Tashima, K.; Tsuchiya, S.; Matsumoto, K.; Nakamura, T.; Horie, S.; Yano, S. Contractile effect of TRPA1 receptor agonists in the isolated mouse intestine. *Eur. J. Pharmacol.* **2007**, *576*, 143–150. [CrossRef]
61. Schmidt, P.T.; Holst, J.J. Use of antagonists to define tachykininergic control of intestinal motility in pigs. *Peptides* **1997**, *18*, 373–379. [CrossRef]
62. Buckinx, R.; Van Nassauw, L.; Avula, L.R.; Alpaerts, K.; Adriaensen, D.; Timmermans, J.-P. Transient receptor potential vanilloid Type 1 channel (TRPV1) immunolocalization in the murine enteric nervous system is affected by the targeted C-terminal epitope of the applied antibody. *J. Histochem. Cytochem.* **2013**, *61*, 421–432. [CrossRef]
63. Bashashati, M.; Fichna, J.; Piscitelli, F.; Capasso, R.; Izzo, A.A.; Sibaev, A.; Timmermans, J.-P.; Cenac, N.; Vergnolle, N.; Di Marzo, V.; et al. Targeting fatty acid amide hydrolase and transient receptor potential vanilloid-1 simultaneously to modulate colonic motility and visceral sensation in the mouse: A pharmacological intervention with N-arachidonoyl-serotonin (AA-5-HT). *Neurogastroenterol. Motil.* **2017**, *29*, e13148. [CrossRef] [PubMed]
64. Lai, N.Y.; Musser, M.A.; Pinho-Ribeiro, F.A.; Baral, P.; Jacobson, A.; Ma, P.; Potts, D.E.; Chen, Z.; Paik, D.; Soualhi, S.; et al. Gut-Innervating Nociceptor Neurons Regulate Peyer's Patch Microfold Cells and SFB levels to mediate salmonella host defense. *Cell* **2020**, *180*, 33–49.e22. [CrossRef] [PubMed]
65. Yamamoto, M.; Nishiyama, M.; Iizuka, S.; Suzuki, S.; Suzuki, N.; Aiso, S.; Nakahara, J. Transient receptor potential vanilloid 1-immunoreactive signals in murine enteric glial cells. *World J. Gastroenterol.* **2016**, *22*, 9752–9764. [CrossRef]
66. Ives, S.J.; Park, S.Y.; Kwon, O.S.; Gifford, J.R.; Andtbacka, R.H.I.; Hyngstrom, J.R.; Richardson, R.S. TRPV1channels in human skeletal muscle feed arteries: Implications for vascular function. *Exp. Physiol.* **2017**, *102*, 1245–1258. [CrossRef] [PubMed]
67. Poole, D.P.; Pelayo, J.C.; Cattaruzza, F.; Kuo, Y.; Gai, G.; Chiu, J.V.; Bron, R.; Furness, J.B.; Grady, E.F.; Bunnett, N.W. Transient receptor potential ankyrin 1 is expressed by inhibitory motoneurons of the mouse intestine. *Gastroenterology* **2011**, *141*, 565–575.e4. [CrossRef] [PubMed]
68. Kaji, I.; Yasuoka, Y.; Karaki, S.-I.; Kuwahara, A. Activation of TRPA1 by luminal stimuli induces EP4-mediated anion secretion in human and rat colon. *Am. J. Physiol. Gastrointest. Liver Physiol.* **2012**, *302*, G690–G701. [CrossRef]
69. Cho, E.; Callaghan, B.; Bron, R.; Bravo, D.M.; Furness, J.B. Identification of enteroendocrine cells that express TRPA1 channels in the mouse intestine. *Cell Tissue Res.* **2014**, *356*, 77–82. [CrossRef] [PubMed]
70. Lee, S.P.; Buber, M.T.; Yang, Q.; Cerne, R.; Cortés, R.Y.; Sprous, D.G.; Bryant, R.W. Thymol and related alkyl phenols activate the hTRPA1 channel. *Br. J. Pharmacol.* **2008**, *153*, 1739–1749. [CrossRef]
71. Kurganov, E.; Zhou, Y.; Saito, S.; Tominaga, M. Heat and AITC activate green anole TRPA1 in a membrane-delimited manner. *Pflügers Arch.* **2014**, *466*, 1873–1884. [CrossRef]
72. Nozawa, K.; Kawabata-Shoda, E.; Doihara, H.; Kojima, R.; Okada, H.; Mochizuki, S.; Sano, Y.; Inamura, K.; Matsushime, H.; Koizumi, T.; et al. TRPA1 regulates gastrointestinal motility through serotonin release from enterochromaffin cells. *Proc. Natl. Acad. Sci. USA* **2009**, *106*, 3408–3413. [CrossRef]
73. Sandor, Z.I.; Bencsik, T.; Dekany, A.; Bartho, L. Effects of cinnamaldehyde on smooth muscle preparations. *Pharmacology* **2019**, *104*, 207–211. [CrossRef]
74. Capasso, R.; Aviello, G.; Romano, B.; Borrelli, F.; De Petrocellis, L.; Di Marzo, V.; Izzo, A.A. Modulation of mouse gastrointestinal motility by allyl isothiocyanate, a constituent of cruciferous vegetables (Brassicaceae): Evidence for TRPA1-independent effects. *Br. J. Pharmacol.* **2012**, *165*, 1966–1977. [CrossRef]
75. Doihara, H.; Nozawa, K.; Kawabata-Shoda, E.; Kojima, R.; Yokoyama, T.; Ito, H. Molecular cloning and characterization of dog TRPA1 and AITC stimulate the gastrointestinal motility through TRPA1 in conscious dogs. *Eur. J. Pharmacol.* **2009**, *617*, 124–129. [CrossRef]
76. Earley, S.; Gonzales, A.L.; Crnich, R. Endothelium-dependent cerebral artery dilation mediated by TRPA1 and Ca2+-activated K+channels. *Circ. Res.* **2009**, *104*, 987–994. [CrossRef] [PubMed]
77. Earley, S. TRPA1 channels in the vasculature. *Br. J. Pharmacol.* **2012**, *167*, 13–22. [CrossRef] [PubMed]
78. Sullivan, M.N.; Gonzales, A.L.; Pires, P.W.; Bruhl, A.; Leo, M.D.; Li, W.; Oulidi, A.; Boop, F.A.; Feng, Y.; Jaggar, J.H.; et al. Localized TRPA1 channel Ca2+signals stimulated by reactive oxygen species promote cerebral artery dilation. *Sci. Signal.* **2015**, *8*, ra2. [CrossRef] [PubMed]

79. Blackshaw, L.A.; Brierley, S.M.; Hughes, P.A.; Harrington, A.M. The hot mustard receptor's role in gut motor function. *Gastroenterology* **2011**, *141*, 423–427. [CrossRef] [PubMed]
80. Barbiers, M.; Timmermans, J.-P.; Adriansen, D.; Groodt-Lasseel, M.H.A.; Scheuermann, D.W.; Adriaensen, D. Projections of neurochemically specified neurons in the porcine colon. *Histochem. Cell Biol.* **1995**, *103*, 115–126. [CrossRef] [PubMed]
81. Cornelissen, W.; De Laet, A.; Kroese, A.B.; Van Bogaert, P.-P.; Scheuermann, D.W.; Timmermans, J.-P. Excitatory synaptic inputs on myenteric Dogiel type II neurones of the pig ileum. *J. Comp. Neurol.* **2001**, *432*, 137–154. [CrossRef] [PubMed]
82. Gershon, M.D. Review article: Serotonin receptors and transporters—roles in normal and abnormal gastrointestinal motility. *Aliment. Pharmacol. Ther.* **2004**, *20*, 3–14. [CrossRef]
83. Russo, E.B.; Burnett, A.; Hall, B.; Parker, K.K. Agonistic properties of cannabidiol at 5-HT1a receptors. *Neurochem. Res.* **2005**, *30*, 1037–1043. [CrossRef]
84. Alexander, S.P.; Christopoulos, A.; Davenport, A.P.; Kelly, E.; Mathie, A.; Peters, J.A.; Veale, E.L.; Armstrong, J.F.; Faccenda, E.; Harding, S.D.; et al. The concise guide to pharmacology 2019/20: G protein-coupled receptors. *Br. J. Pharmacol.* **2019**, *176*, S21–S141. [CrossRef]
85. Jesus, C.H.A.; Redivo, D.D.B.; Gasparin, A.T.; Sotomaior, B.B.; De Carvalho, M.C.; Genaro, K.; Zuardi, A.W.; Hallak, J.E.C.; Crippa, J.A.; Zanoveli, J.M.; et al. Cannabidiol attenuates mechanical allodynia in streptozotocin-induced diabetic rats via serotonergic system activation through 5-HT1A receptors. *Brain Res.* **2019**, *1715*, 156–164. [CrossRef]
86. De Almeida, D.L.; Devi, L.A. Diversity of molecular targets and signaling pathways for CBD. *Pharmacol. Res. Perspect.* **2020**, *8*, e00682. [CrossRef] [PubMed]
87. Youn, Y.H.; Choi, E.J.; Lee, Y.H.; Oshima, T.; Miwa, H.; Park, H. The effects of 5-hydroxytryptamine1a receptor agonist, buspirone on the gastric fundus accommodation in an animal model using guinea pigs. *Neurogastroenterol. Motil.* **2015**, *27*, 532–541. [CrossRef] [PubMed]
88. Delesalle, C.; Acker, N.; Claes, P.; Deprez, P.; Smet, I.; Dewulf, J.; Lefebvre, R.A. Contractile effects of 5-hydroxytryptamine (5-HT) in the equine jejunum circular muscle: Functional and immunohistochemical identification of a 5-HT1A-like receptor. *Equine Vet. J.* **2008**, *40*, 313–320. [CrossRef] [PubMed]
89. Johnson, S.M.; Katayama, Y.; North, R.A. Multiple actions of 5-hydroxytryptamine on myenteric neurones of the guinea-pig ileum. *J. Physiol. (Lond.)* **1980**, *304*, 459–470. [CrossRef] [PubMed]
90. Mylecharane, E. Mechanisms involved in serotonin-induced vasodilatation. *Blood Vessel.* **1990**, *27*, 116–126. [CrossRef]
91. Pazos, M.R.; Mohammed, N.; Lafuente, H.; Santos, M.; Martínez-Pinilla, E.; Moreno, E.; Valdizán, E.M.; Romero, J.; Pazos, A.; Franco, R.; et al. Mechanisms of cannabidiol neuroprotection in hypoxic–ischemic newborn pigs: Role of 5HT1A and CB2 receptors. *Neuropharmacology* **2013**, *71*, 282–291. [CrossRef]

animals

MDPI

Article

Impaired Mitochondrial Function Results from Oxidative Stress in the Full-Term Placenta of Sows with Excessive Back-Fat

Liang Tian [1,*], Jiahe Huang [2], Aiyou Wen [3] and Peishi Yan [1,*]

1 College of Animal Science and Technology, Nanjing Agricultural University, Nanjing 210095, China
2 College of Animal Science and Technology, Northwest A & F University, Yangling 712100, China; HuangJiaHeNLA@163.com
3 College of Animal Science, Anhui Science and Technology University, Fengyang 233100, China; aywen2008@126.com
* Correspondence: tianliang2013@njau.edu.cn (L.T.); yanps@njau.edu.cn (P.Y.)

Received: 27 December 2019; Accepted: 20 February 2020; Published: 23 February 2020

Simple Summary: Placental dysfunction associated with maternal obesity has been demonstrated to be a possible detrimental determinant for a reproductive disorder in human and animals such as pigs. Moreover, there exists a substantial amount of evidence supporting that mitochondrial dysfunction associated with obesity contributes to dysfunction of highly metabolic tissues, including adipose, skeletal muscle and placenta. Despite previous reports have demonstrated that back-fat thickness of sows is associated with placental dysfunction, the influences of excessive back-fat on mitochondrial structure and function in porcine placenta still remain elusive. In this study, animal (Landrace) and cell in vitro model (pig placental trophoblasts) were employed to evaluate mitochondrial alterations in the placentas of sows with different back-fat depth. We revealed that excessive back-fat of sows is associated with placental mitochondrial abnormalities corresponding to decreased ATP production and impaired mitochondrial respiration in the placenta. Together, our findings develop the understanding about the impact of excessive back-fat induced oxidative stress on mitochondrial alterations in the pig placenta, which may contribute to generate some strategies in future to improve sow reproduction.

Abstract: The aim of this study was to determine the effect of excessive back-fat (BF) of sows on placental oxidative stress, ATP generation, mitochondrial alterations in content and structure, and mitochondrial function in isolated trophoblasts. Placental tissue was collected by vaginal delivery from BFI (15–20 mm, n = 10) and BFII (21–27 mm, n = 10) sows formed according to BF at mating. Our results demonstrated that excessive back-fat contributed to augmented oxidative stress in term placenta, as evidenced by excessive production of ROS, elevated protein carbonylation, and reduced SOD, GSH-PX, and CAT activities ($p < 0.05$). Indicative of mitochondrial dysfunction, reduced mitochondrial respiration in cultured trophoblasts was linked to decreased ATP generation, lower mitochondrial Complex I activity and reduced expression of electron transport chain subunits in placenta of BFII sows ($p < 0.05$). Meanwhile, we observed negative alterations in mitochondrial biogenesis and structure in the placenta from BFII group ($p < 0.05$). Finally, our in vitro studies showed lipid-induced ROS production resulted in mitochondrial alterations in trophoblasts, and these effects were blocked by antioxidant treatment. Together, these data reveal that excessive back-fat aggravates mitochondrial injury induced by increased oxidative stress in pig term placenta, which may have detrimental consequences on placental function and therefore impaired fetal growth and development.

Keywords: NF-κB; mitochondria; oxidative stress; back-fat; placenta; pig

1. Introduction

Pregnancy complicated by maternal obesity is associated with an abnormal intrauterine milieu characterized by increased lipid accumulation, augmented oxidative stress, and significant inflammation within the placenta [1,2], resulting in placental dysfunction and therefore the poor pregnancy outcomes in human beings and animals such as pig [3,4]. Excessive lipid accumulation has been shown to induce cellular dysfunction through the over production of reactive oxygen species (ROS), mitochondrial dysfunction, and the activation of inflammation in highly metabolic tissues like adipose and skeletal muscle [5,6]. As an extremely metabolically active fetal tissue, the placenta is also susceptible to obesity-associated lipotoxicity [1,7]. Recently, lipotoxicity has been demonstrated to inhibit placental development and induce placental dysfunction evidenced by dysregulation of lipid metabolism and transport in the human or pig full-term placenta [8–10].

As a significant marker of placental lipotoxicity, oxidative stress can be increased in pregnancy complicated by maternal obesity resulting in placental oxidative injure [11]. In addition, mitochondria, a key energy source for placental function, are also the major source of ROS under physiologic conditions. However, mitochondrial function itself can be compromised by severe or prolonged oxidative stress [12]. Excessive ROS can damage lipids, proteins, and nucleic acids within the mitochondria, leading to changes in mitochondrial structure and function [13]. Several studies have shown that increased metabolic activity of placental mitochondria and the reduced antioxidant scavenging power may be responsible for increased ROS generation and lower mitochondrial respiration in pregnancies complicated by maternal obesity or diabetes, resulting in detrimental consequences on placental function and therefore fetal developmet [14,15]. Thus, mitochondrial dysfunction and over generation of superoxide could be part of a vicious cycle and can be considered as a potential mechanism of placental dysfunction involved in obese pregnancy. Although several studies reveal that excessive back-fat of sows is associated with placental dysfunction [4,16], it is not currently known whether increased back-fat is coupled with changes in mitochondrial structure and function in the pig placenta.

In this study, we addressed the hypothesis that increased back-fat affect mitochondrial structure and function in porcine placenta. Thus, the purpose of this study was to investigate the effect of excessive back-fat of sows on placental ROS production, mitochondrial biogenesis, mitochondrial respiration, and mitochondria alterations in density and structure. In addition, we assessed the role of lipid-induced oxidative stress in the development of mitochondrial abnormalities in pig placental trophoblast cells in vitro and found that lipid-induced ROS production resulted in mitochondrial alterations in trophoblasts. Consistent with our hypothesis, we demonstrated that excessive back-fat resulted in increased ROS production, reduced mitochondrial biogenesis, impaired mitochondrial respiration, and mitochondrial abnormalities in content and structure in pig full-term placenta, suggesting excessive back-fat of sows may promote placental mitochondrial dysfunction that is associated with impaired placental function and fetal development.

2. Materials and Methods

2.1. Animal Model and Dietary Management

All animal procedures involved in this study were specifically approved by the Laboratory Animal Care and Use Committee of Nan Jing Agricultural University (SYXK2015-0072, 6 September 2015). A total of 20 lean breed sows (Landrace) were investigated in the present study. All the sows were multiparous with same parity (Parity = 2) and inseminated with pooled semen doses containing 3–4 billion sperm cells to produce purebred litters. After the weaning period with a corn soybean diet (3200 Kal/kg of metabolizable energy, 18% of crude protein and 0.9% lysine), the females were dichotomized based on BF (back-fat thickness) at mating into BFI (15–20 mm, n = 10) and BFII (21–27 mm, n = 10) as described previously [17]. All the sows were housed together in passively ventilated collective pens with concrete slatted floors at the Research Farm of Nan Jing Agricultural University. During pregnancy, a standard grain-based diet (89.9% of dry matter, 14.5% of crude protein,

3300 Kcal/kg of metabolizable energy) was offered per day, divided over two meals with ad libitum access to water. The amount of food offered to each sows was adjusted to 1.9, 2.2, 2.4, and 2.9 kg of feed per day at 0–30, 31–60, 61–90, and 90–110 days of gestation (day 0 = day of insemination), respectively, based on the data from the National Research Council (NRC) (2012).

2.2. Data Collection and Sampling

Sows in this study were weighted and measured for back-fat thickness at mating (2 to 3 days before insemination) and at farrowing, respectively, by the research team with the help of farm personnel. Back-fat depth (BF) was determined by A-mode ultrasonography (Renco, Rockledge, FL, USA) as previously described [18]. Briefly, the probe was placed in a point at the right side of the animal located at 4 cm from the midline and transversal to the head of the last rib as determined by palpation. At day 105 of pregnancy, maternal blood (overnight fasting) was collected into 5-mL sterile heparinized vacuum tubes (Greiner, Frickenhausen, German). Immediately after recovery, blood samples were centrifuged at 3500× *g* for 15 min, and the plasma was separated and stored at −80 °C for further analysis. After removal of fetal amnion, placental villous tissues, obtained from vaginal delivery, were rinsed thoroughly in cold Phosphate-buffered saline (PBS) solution before further processing. Placental samples were then cut into approximately 5-cm^2 pieces and flash-frozen in liquid nitrogen and stored at −80 °C until further processing. For each pregnancy, piglets born alive and stillborn were counted and individually weighed within 12 h after birth, and the placentas were also weighed for measuring placental efficiency as a ratio of litter weight to placental weight as previously described [10].

2.3. Cell Culture and Drug Treatment

Porcine placental trophoblast cells (cytotrophoblasts) were isolated from fresh porcine full-term placentas, as previously described [19]. Briefly, placental tissue, obtained from vaginal delivery, was separated from the visible blood vessels as well as connective tissues, thoroughly washed with cold PBS containing 100 U/mL penicillin (Invitrogen, Carlsbad, CA, USA) and 100 μg/mL streptomycin (Invitrogen, Carlsbad, CA, USA), and then excised and cut into 1–3 mm^3 pieces. The tissue fragments were digested three times at 37 °C for 30 min with continuous shaking, followed by filtration through a 70 μm cell strainer. The digestion medium was composed of Ham's F12/Dulbecco's Modified Eagle Medium (DMEM/F12) (HyClone, Logan, UT, USA), 0.125% (*W*/*V*) trypsin (Gibco, Grand Island, NY, USA), 20 U/mL DNase I (Roche, Basel, Switzerland), and 0.1% BSA (Amresco, Solon, OH, USA). The cell suspension of all three digestions was centrifuged at 1000× *g* for 10 min to collect the cell pellet. Pellets were then resuspended in DMEM/F12 and deposited on top of a discontinuous 35% and 45% (*V*/*V*) Percoll (Pharmacia, London, UK) gradient solution, and centrifuged at 2000× *g* for 20 min. Villous cytotrophoblasts were collected from the appropriate layers, and cultured in DMEM/F12 supplemented with 10% FBS (HyClone, Logan, UT, USA), 1% (*V*/*V*) Insulin–Transferrin–Selenium (ITS; Sigma, Saint Louis, MO, USA), 10 ng/mL of epidermal growth factor (EGF; Invitrogen, Carlsbad, CA, USA), 100 U/mL penicillin, and 100 μg/mL streptomycin at 37 °C under 5% CO_2 as previously reported [19,20]. The purity of preparations of cytotrophoblasts isolated from full-term placentas was evaluated by flow cytometry as previously described [20], using FITC fluorescein-labeled antibody against cytokeratin-7 (Santa Cruz Tech, Dallas, CA, USA) as a specific marker of trophoblast cells.

In order to induce oxidative stress in cytotrophoblasts in vitro, cells were treated with 400 μM Fatty Acid (FA) Supplement containing 2 mol of linoleic acid and 2 mol of oleic acid per mole of albumin (L9655; Sigma-Aldrich, St. Louis, MO, USA) in triplicate as previously described [19,21]. Treatment media without fatty acids was added with bovine serum albumin (FA free) to maintain the same osmolarity. In cell experiments in vitro, cells were treated with 400 μM FA or 2 mM Vitamin E (VE) (oxidative stress inhibitor; MCE, Shanghai, China) for the amount of time specified in the individual figures.

2.4. Plasma Assay

The concentrations of triglyceride (TG) and nonesterified fatty acids (NEFA) in plasma samples were determined by using Beckman AU5811 analyzer (Beckman-Coulter, Inc.250S.Kraemer Boulevard Brea, CA, USA) as previously reported [8]. A commercially available porcine specific ELISA kit (RD SYSTEMS, Minneapolis, MN, USA) and an Amplex Red hydrogen peroxide assay kit (Invitrogen, Carlsbad, CA, USA) were utilized to measure plasma leptin and hydrogen peroxide (H_2O_2) content, respectively, according to the manufacturer's instructions as previously described [5,17]. All blood samples were determined in duplicate in single assay. The minimum detection limit was 3.2 pg/dL, 0.1 pmol/L, 0.2 ng/mL, and 0.2 pmol/L for TG, NEFA, Leptin, and H_2O_2, respectively. For Leptin and H_2O_2 analysis, within-assay coefficient of variation (CV) was acceptable when less than 5%.

2.5. Oxidative Stress Assessment

Reactive oxygen species production was detected using a cell-permeable non fluorescent probe 2′, 7′- dichlorofluorescin diacetate (DCFH-DA) (KeyGen BioTECH, Nanjing, China). For visualization and quantification of ROS generation in placental villous tissue, flash-frozen villous tissue sections (7 μm) from 7 placentas in each group were incubated with 10 μM DCFH-DA for 20 min at 37 °C, and staining was performed as previously described [22]. Quantification of the fluorescence intensity from dichlorofluorescein (DCF) was performed using the Image J software (NIH Image). For the intracellular level of ROS test, villous cytotrophoblasts were treated with fatty acid (400 μM) or without fatty acid for 48 h. Cells were then harvested, washed with cold PBS, and subjected to DCFH-DA staining in binding buffer at 37 °C for 30 min. Stained cells were analyzed using flow cytometry or a spectrophotometer by measuring DCF fluorescence- activated cell sorting or the fluorescence intensity of DCF at an excitation wavelength of 488 nm and emission wavelength of 530 nm according to the manufacturer's protocol as previously reported [23].

For the key oxidative enzyme activity detection, placental tissue (100 mg) was homogenized in ice-cold cell and tissue lysis buffer (Beyotime Biotechnology, Nanjing, China) using T18 digital ULTRA-TURRAX Package disperser (IKA, Shanghai, China). The homogenate was centrifuged at 10,000× *g* for 10 min at 4 °C. Then total antioxidant capacity (TAC), superoxide dismutase (SOD) activity, glutathione peroxidase (GSH-PX) activity and catalase (CAT) activity measurements were performed using the commercially available kits from Jiangcheng Bioengineering Institute (Nanjing, China) as previously described [24]. For reduced glutathione (GSH)/oxidative glutathione (GSSG) ratio measurement, the GSH and GSSG Assay Kit (KeyGen BioTECH, Nanjing, China) was used as previously reported [23].

2.6. Mitochondrial Analysis

For mitochondrial biogenesis assessment, mitochondrial DNA (mtDNA) copy number was detected by real-time PCR, using the $2^{-\Delta\Delta Ct}$ method as previously described [25]. Total genomic DNA was isolated from placenta of BFI and BFII sows (n = 10 from each group) or cytotrophoblasts treated with fatty acid (400 μM) or not for 48 h in vitro (n = 3 from each group), using TIANamp Genomic DNA Kit (TIANGEN BIOTECH, Beijing, China). A pair of primers (Table S1) for ND1 and Cyclophilin-A were used to amplify a mitochondrial and nuclear DNA fragment, respectively. Citrate synthase (CS) activity was determined in villous tissue or cytotrophoblasts using a commercial kit from Sigma (CS0720; Shanghai, China) according to the manufacturer's protocol. Placental tissues (200 mg) or cells were homogenized or solubilized in ice-cold extraction buffer, and the homogenate or cell lysate was centrifuged at 16,000× *g* for 20 min at 4 °C. Then CS activity was measured in the supernatant by a spectrophotometer method as previously reported [14].

Fluorescent probe JC-1 (KeyGen BioTECH, Nanjing, China) was used to estimate mitochondrial membrane potential (ΔΨm) as previously described [17,23]. Cytotrophoblasts were harvested, washed with cold PBS, and incubated at 37 °C for 15 min with 5 μg/mL JC-1. After incubation, cells were analyzed by fluorescence- activated cell sorting using flow cytometry. The stained cells with high mitochondrial membrane potential showed red fluorescence and were present in the upper right (UR) quadrant of the FACS histogram.

For mitochondrial complexes activity and ATP production rate measurement, mitochondria was isolated from placental villous tissue or cell homogenate using Mitochondria Isolation Kit (KeyGen BioTECH, Nanjing, China) as previously described [17]. Briefly, placental samples (200 mg) or cytotrophoblasts were homogenized in ice-cold isolation buffer. The homogenate was centrifuged at 1200× *g* for 5 min at 4 °C. The pellet was discarded, supernatant collected and centrifuged at 7000× *g* for 10 min at 4 °C. The resulting supernatant was discarded, and the pellet re-suspended in the ice-cold suspension buffer and centrifuged at 9500× *g* for 5 min at 4 °C. The mitochondria was collected in the sediments. The activities of mitochondrial complexes were determined in isolated mitochondria using the MitoCheck Complex I and II/III Activity Assay Kits (Cayman Chemical Company, Ann Arbor, MI, USA) as previously reported [23]. The mitochondrial ATP production rate was measure using ENLITEN ATP Assay System (Promega; Madison, WI, USA) with 1420 Multilabel Counter (PerkinElmer; Fremont, CA, USA) according to the manufacturer's instruction.

Mitochondrial respiration of the cultured cytotrophoblasts was measured using a Seahorse XF24 analyzer (Seahorse Biosciences, Shanghai, China) as previously described [26]. Briefly, cells, 2×10^6 total, were suspended in 2 mL DMEM/F12 culture medium and transferred into the chamber, which was maintained at 37 °C. After equilibration and stirring, basal respiration was measured as the average oxygen consumption from four baseline oxygen consumption rates (OCR) readings, and mitochondrial respiratory parameters were calculated from OCR readings following sequential injection of oligomycin (1 μM) and p-trifluoromethoxy carbonyl cyanide phenylhydrazone (FCCP, 2 μM), respectively. OCR was normalized to total cellular protein. Protein concentration was determined using Pierce BCA Protein Assay Kit (Thermo Scientific, Waltham, MA, USA).

2.7. Transmission Electron Microscopy

To characterize alterations in the mitochondrial structure and density, placental tissue (small pieces) or cytotrophoblast cultures was fixed in 30 mg/L glutaraldehyde overnight at 4 °C, postfixed in 1% Osmium tetroxide for 1 h at 4 °C, dehydrated, and embedded in Epon-812. The tissue or cell pellet was then cut using an RMC/MTX ultramicrotome (Boeckeler, Tucson, AZ, USA), and ultrathin sections (50 nm) were mounted on copper grids, contrasted with 8% uranyl acetate and lead citrate, and observed with a JEM-1400 Plus transmission electron microscopy (Jeol LTD, Tokyo, Japan). Image analysis was performed using the Image-ProPlus6.0 software from Media Cybernetics (Rockville, MD, USA).

2.8. Real-Time Quantitative PCR Analysis

Total RNA was extracted from placental tissue (n = 10 from each group) or cultured cells (n = 3 from each group) with the High Pure RNA tissue kit (Omega Bio-Tek, Norcross, GA, USA) and 500 ng of total RNA was reverse transcribed using PrimeScript RT Master Mix Kit (TaKaRa, Tokyo, Japan). Primers (Table S1) were synthesized by Invitrogen (Shanghai, China). Quantitative PCR was conducted on the Step One Plus Real-Time PCR System (ABI, Waltham, MA, USA) with the following program: 95 °C for 30 s, 95 °C for 5 s, 60 °C for 30 s, 95 °C for 15 s, 60 °C for 1 min, and 95 °C for 15 s, with 40 cycles of steps 2 and 3. Amplification was performed in 25 μL reaction system containing specific primers (Table S1) and SYBR Premix Ex Taq II (TaKaRa, Tokyo, Japan). Relative gene expression was calculated using the comparative Ct method with the formula $2^{-\Delta\Delta Ct}$ [27]. The two reference genes GAPDH and HPRT1 were used. The geometric mean of relative gene expression was calculated and used for further analysis as previously reported [28].

2.9. Protein Carbonylation and Western Blotting Analysis

Protein carbonyl derivatives (a marker of oxidative stress) were detected using OxiSelect Protein Carbonyl Immunoblot Kit (Cell Biolabs, San Diego, CA, USA) as previously reported [5]. Briefly, after the derivatization of the protein sample on PVDF with dinitrophenylhydrazine (DNPH) following the manufacturer's instructions, the blot was developed using a standard immunoblot protocol. Total protein from cultured trophoblast cells was extracted using cell lysis buffer (Beyotime Biotechnology, Nanjing, China) by procedures as previously described [29]. Mitochondrial fractions for Western blotting were obtained from placental villous tissue using Mitochondrial Protein Extraction Kit (KeyGen BioTECH, Nanjing, China) and suspended in lysis buffer supplemented with a protease and phosphatase inhibitor cocktail (Sigma, Saint Louis, MO, USA). Protein samples (50 μg) were separated by SDS-PAGE and transferred to PVDF membrane (Bio-Rad Laboratories, Hercules, CA, USA). After blocking in 5% fat-free milk for 1 h at room temperature, the membranes were incubated with rabbit anti-Mark4 (4834, dilution 1:1000, Cell Signaling Technology, Danvers, MA, USA), AMPK (5831, dilution 1:1000, Cell Signaling Technology), Phospho-AMPK (2535, dilution 1:1000, Cell Signaling Technology), Phospho-NF-κB (3033, dilution 1:1000, Cell Signaling Technology), GAPDH (2118, dilution 1:1000, Cell Signaling Technology), NF-κB (ab90532, dilution 1:1000, Abcam Biotechnology, Cambridge, MA, USA), ATP5α (ab245580, dilution 1:2000, Abcam Biotechnology) and Phospho-Mark4 (SAB4504258, dilution 1:500, Sigma, Saint Louis, MO, USA) antibody, or Mouse anti- NDUFB8 (ab110242, dilution 1:1000, Abcam Biotechnology), SDHB (ab14714, dilution 1:2000, Abcam Biotechnology), and UQCRC2 (ab14745, dilution 1:2000, Abcam Biotechnology) antibody overnight at 4 °C, followed by incubation with Goat anti- mouse or rabbit IgG horseradish peroxidase (HRP)-conjugated secondary antibodies (HAF007 and HAF008, dilution 1:2000, RD SYSTEMS, Minneapolis, USA) for 1 h at room temperature. Proteins were visualized using the LumiGLO Reagent and Peroxide system (Cell Signaling Technology, Danvers, USA), and then the blots were quantified using Bio-Rad ChemiDoc imaging system (Bio-Rad Laboratories, Hercules, USA). Band density was normalized according to the GAPDH content.

2.10. Statistical Analysis

Statistical analyses were conducted using SPSS Statistics 20.0 software (IBM SPSS, Armonk, NY, USA). Data obtained from sows and placentas were analyzed as a completely randomized design. Each sow and her litter were considered as an experimental unit. Statistical differences between groups were evaluated using Independent-Samples T Test. For all cell culture experiments, data were obtained from at least three independent experiments. Statistical differences among individual means were analyzed using One-way ANOVA for comparisons among groups, followed by Duncan test. Results were expressed as means ± SEM. A p-value < 0.05 was considered statistically significant, and very significant was indicated when $p < 0.01$.

3. Results

3.1. Characteristics of the Studied Sows

BFII sows had markedly greater body weight (BW) and BF than BFI sows ($p < 0.05$), at mating (BW: 173.18 ± 0.52 vs. 167.70 ± 0.30 kg; BF: 24.11 ± 0.33 vs. 16.51 ± 0.21 mm) and at farrowing (BW: 221.22 ± 0.43 vs. 214.38 ± 0.28 kg; BF: 25.09 ± 0.31 vs. 18.63 ± 0.18 mm), respectively (Table 1). The litter size, litter live size, litter weight and placental efficiency were lower in BFII sows than those in BFI group ($p < 0.05$), whereas both groups were similar in terms of birth weight and placental weight. Consistent with increased adiposity, there were 47% and 76% increases ($p < 0.05$) in TG and NEFA levels in maternal plasma from BFII sows, respectively, compared to the BFI group (Table 1). Excessive back-fat was also associated with higher leptin levels ($p < 0.05$) in maternal plasma. Concurrently, we observed increased H_2O_2 levels in BFII sows compared with BFI group ($p < 0.05$).

Table 1. Characteristics of pregnancies for sows studied [†].

	BFI (15–20 mm, n = 10)	BFII (21–27 mm, n = 10)
Parity	2	2
BF at mating, mm	16.51 ± 0.21 [a]	24.11 ± 0.33 [b]
BW at mating, kg	167.70 ± 0.30 [a]	173.18 ± 0.52 [b]
BF at farrowing, mm	18.63 ± 0.18 [a]	25.09 ± 0.31 [b]
BW at farrowing, kg	214.38 ± 0.28 [a]	221.22 ± 0.43 [b]
TG *, mg/dL	21.15 ± 2.14 [a]	31.18 ± 1.36 [b]
NEFA *, mmol/L	0.21 ± 0.04 [a]	0.37 ± 0.03 [b]
Leptin *, ng/mL	13.13 ± 1.20 [a]	20.02 ± 2.31 [b]
H_2O_2 *, μM/L	66.20 ± 3.71 [a]	89.32 ± 3.01 [b]
Litter size #	14.83 ± 0.31 [a]	12.88 ± 0.53 [b]
Litter live size	12.75 ± 0.35 [a]	11.23 ± 0.86 [b]
Placental weight, kg	3.22 ± 0.19	3.38 ± 0.25
Birth weight ‡, kg	1.42 ± 0.05	1.38 ± 0.23
Litter weight, kg	18.04 ± 0.23 [a]	15.42 ± 0.75 [b]
Placental efficiency §	5.59 ± 0.16 [a]	4.61 ± 0.22 [b]

BF, Back-fat thickness; BW, Body weight; TG, triglyceride; NEFA, non-esterified fatty acid; †: Results are expressed as means ± SEM. Means within the same row with different superscripts mean significant difference ($p < 0.05$); *: A value measured in maternal plasma; #: The number of live-born and stillborn piglets in litter; ‡: Birth weight of born alive and stillborn piglets; §: A ratio of litter weight to placental weight.

3.2. Excessive Back-Fat is Associated with Increased Placental Oxidative Stress

Considering BFII sows had higher plasma H_2O_2 levels (an indicator of systemic oxidative stress), we investigated the impact of excessive back-fat on oxidative stress in the placenta of sows. Oxidant status alterations in BFII placenta were marked by higher production of ROS assessed by DCF staining, decreased placental TAC and elevated protein carbonylation (a marker of protein oxidation), compared to the BFI group ($p < 0.05$, Figure 1A–D). The mRNA level of uncoupling protein 2 (UCP2, a marker of increased mitochondrial ROS production) was induced in the placenta of BFII sows, but the mRNA expression of genes associated with antioxidant system, including glutathione peroxidase (GPx) and catalase (CAT), was decreased in BFII group compared with BFI group ($p < 0.05$, Figure 1E), without a significant change in superoxide dismutase 2 (SOD2). Consistent with increased oxidative stress, the antioxidant enzyme activities of superoxide dismutase, glutathione peroxidase and catalase were decreased in BFII group ($p < 0.05$, Figure 1F). In addition, in the placenta of BFII sows, a decrease ($p < 0.05$) was noted in glutathione (GSH)/oxidative glutathione (GSSG) ratio compared to values obtained with placentas from BFI sows (Figure 1G). Thus, we concluded excessive back-fat was associated with increased oxidative stress in the full-term pig placenta.

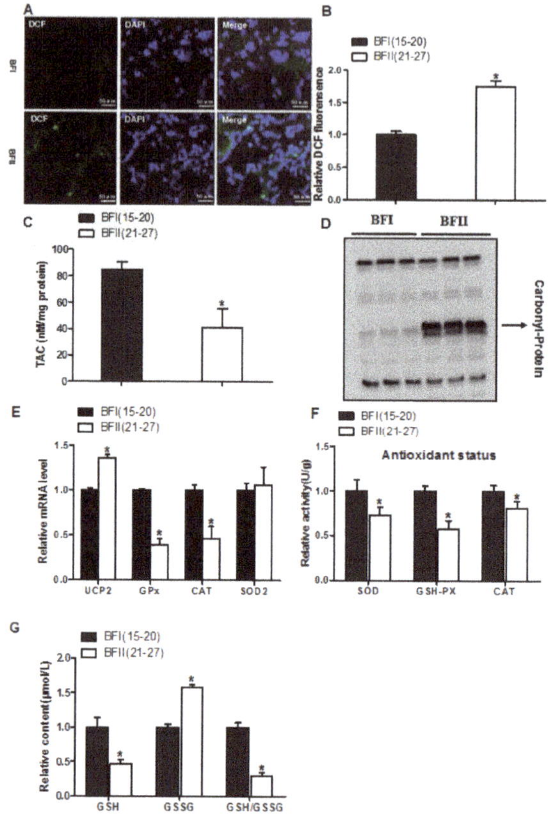

Figure 1. Increased oxidative stress in placentas with increasing back-fat. Representative visualization (**a**) and quantification (n = 7 in each group) (**b**) of dichlorofluorescein (DCF) in cryosections of placenta. Scale bar: 50 μm; (**c**) Total antioxidant capacity (TAC) was measured in term placentas from BFI and BFII sows (n = 10/group). TAC was normalized to total protein level and expressed as nM TAC per mg protein; (**d**) Representative immunoblots showing total protein carbonylation in placentas of BFI and BFII sows; (**e**) mRNA levels of oxidant stress-related genes determined by real-time RT-PCR in placentas from BFI and BFII sows (n = 10/group); (**f**) Placental antioxidant status was estimated by SOD, GSH-PX, and CAT (n = 10/group); (**g**) Relative content of GSH/GSSG in placentas from BFI and BFII sows (n = 10/group). Results were expressed as fold change versus the BFI sow set to 1 unit. Values are expressed as mean ± SEM. * $p < 0.05$ compared with the BFI group. DCF: dichlorofluorescein; SOD: superoxide dismutase; GSH-PX: glutathione peroxidase; CAT: catalase; GSH: reduced glutathione; GSSG: oxidative glutathione; BF: back-fat thickness.

3.3. *Effect of Excessive Back-Fat on Placenta Mitochondrial Oxidative Respiration*

Because ROS is mainly generated from mitochondria, we subsequently tested whether an increase in production of ROS is associated with mitochondrial dysfunction in placenta of BFII sows. We initially evaluated mitochondrial function by detecting the levels of ATP in placental tissue. As illustrated in Figure 2A, the ATP content was lower (68% decrease, $p < 0.05$) in the BFII placentas compared with BFI group. To determine whether changes in placental ATP levels are related to altered mitochondrial respiration, we next investigated the effects of excessive back-fat on mitochondrial function in vitro using cultured cytotrophoblasts (Figure 2B). Cytotrophoblasts from BFII sows cultured in fatty acid demonstrated a significant decrease in basal mitochondrial oxygen consumption, ATP-coupled respiration, maximal respiration and spare capacity compared with cells from BFI group ($p < 0.05$, Figure 2B), suggesting that mitochondrial function was compromised in the placenta of sows with excessive back-fat. Consistent with reduced mitochondrial oxidative respiration, the activity of mitochondrial Complex I was also decreased in BFII group ($p < 0.05$; Figure 2C), whereas there was no difference in coupled mitochondrial Complex II/III activity (Figure 2D). To explore the mechanism that may account for the reduction in placental mitochondrial respiration with increasing back-fat, we utilized a immunoblot approach to measure changes in protein expression of the mitochondrial electron transport complexes. As shown in Figure 2E,F, the expression levels of subunits of complexes I (NDUFB8), II (SDHB), III (UQCRC2), and V (ATP5α) were significantly reduced ($p < 0.05$) in placentas of BFII sows compared with BFI sows.

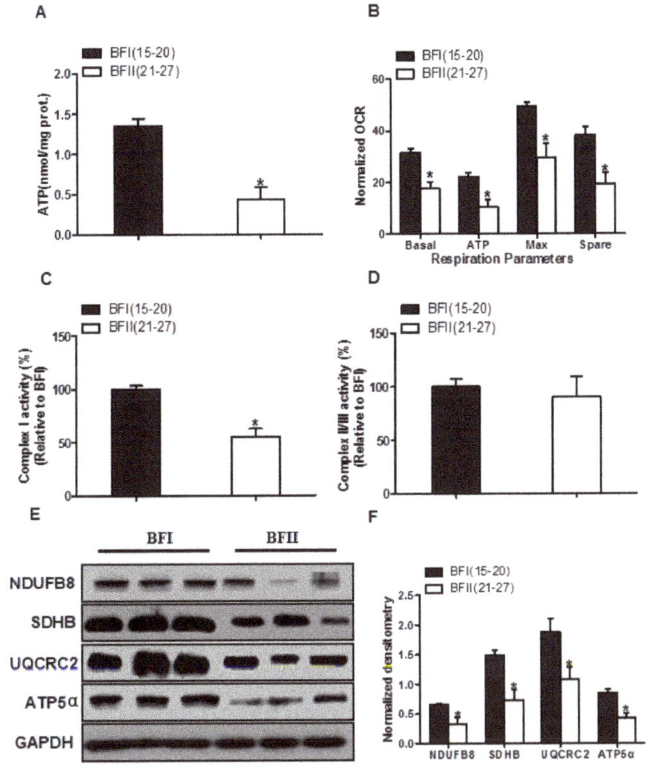

Figure 2. Impairment of placental mitochondrial oxidative respiration in sow with increased back-fat. (**a**) Mitochondrial ATP production in placental tissue from BFI and BFII sows. ATP levels were normalized to total protein level of villous tissue extract; (**b**) Mitochondrial respiratory parameters were measured in cytotrophoblast cultures isolated from placentas of BFI and BFII sows. Before assessment of mitochondrial respiration, cytotrophoblasts were incubated with 400 μM NEFA for 24 h. Oxygen consumption rate (OCR) measurements were normalized to total protein content (pmol O_2/μg protein); (**c** and **d**) Placental mitochondrial complexes I (**c**) and combined II/III (**d**) activity in BFI and BFII sows; (**e**) Representative immunoblot analysis of mitochondrial complexes I (NDUFB8), II (SDHB), III (UQCRC2) and V (ATP5α) in placental mitochondrial fractions from BFI and BFII sows; (**f**) Densitometric analysis of corresponding proteins in E by normalization to GAPDH as an internal control. Results are expressed as mean ± SEM. * $p < 0.05$ compared with the BFI group. n = 10 in each group. BF: back-fat thickness; NEFA: non-esterified fatty acid; Basal: basal respiration; ATP: ATP-coupled respiration; Max: maximal respiration; Spare: spare capacity.

3.4. Effect of Excessive Back-Fat on Placental Mitochondrial Content

We next investigated the effects of excessive back-fat on placenta mitochondrial density. As shown in Figure 3A, the ratio of mtDNA to nuclear DNA in the placenta was significantly more decreased in the BFII sows than in the BFI sows ($p < 0.05$). In agreement, mitochondrial content estimated by citrate synthase (CS) activity showed a significant reduction in placentas of BFII sows ($p < 0.05$) compared with BFI group (Figure 3B). Using transmission electron microscopy, we found that the amount of mitochondria in placental villi was lower in the BFII sows (32% decrease, $p < 0.05$) than in the BFI group (Figure 3C,D). Consistent with reduced mitochondrial DNA (mtDNA) copy number and CS activity, the mRNA levels of mitochondria-encoded genes, including ATP6, ND1, COX3, and CYTB, were lower in placenta of BFII sows ($p < 0.05$, Figure 3E), whereas the mRNA content of several nuclear-encoded mitochondrial genes, including ACADM and CS, was increased in the placenta of BFII sows ($p < 0.05$) compared with BFI group (Figure 3F), without a significant change in CTP1b and COX6c. To clarify the mechanisms implicated in the reduction of mitochondrial density in the placenta of BFII sows, we measured the mRNA levels of genes involved mitochondrial biogenesis. As illustrated in Figure 3G, the mRNA expression of PGC1α, PGC1β, nuclear respiratory factor 1 (NRF1), and mitochondrial transcription factor A (TFAM) was lower in the placenta of BFII sows than that of BFI sows ($p < 0.05$).

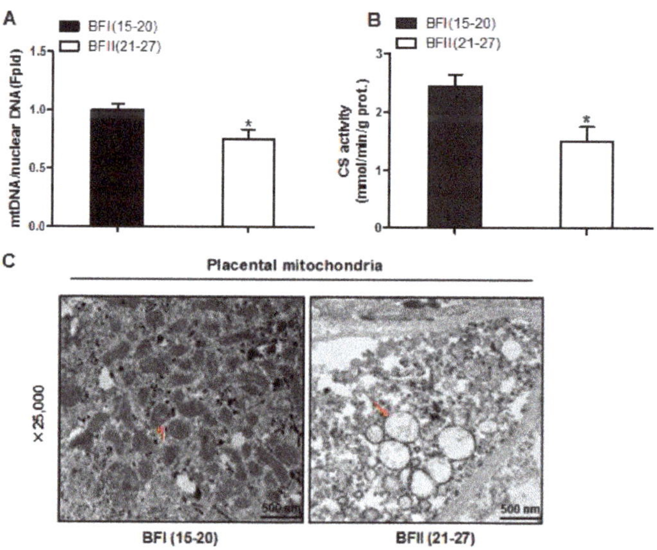

Figure 3. *Cont.*

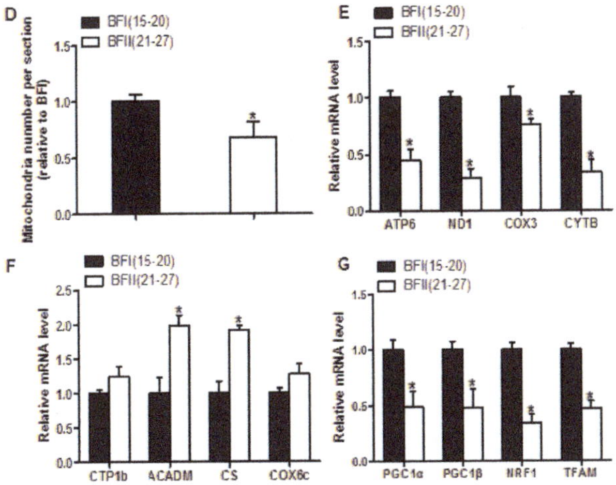

Figure 3. Decreased mitochondrial density in the placenta of sow with excessive back-fat. Mitochondrial biogenesis was estimated by bothmtDNA copy number (**a**) and CS activity (**b**) in the placenta from BFI and BFII sows (n = 10/group); (**c**) Mitochondrial density assessed by electron microscopy in the placentas of BFI and BFII sows. Original magnification, 25,000×; (**d**) Quantification of mitochondrial number per image area in whole placental villous tissue from 7-BFI and 7-BFII sows (analysis of 10 random images per placenta). (**e–g**) The relative mRNA level of various mitochondria-encoded genes (**e**), nuclear-encoded mitochondrial genes (**f**), and mitochondrial biogenesis genes (**g**) was determined by quantitative RT-PCR in placentas from BFI and BFII sows (n = 10/group). Results were normalized by the mean value for the BFI sow set to 1 unit. Values are expressed as mean ± SEM. * $p < 0.05$ compared with the BFI group. BF: back-fat thickness; mtDNA: mitochondrial DNA; CS: citrate synthase; Red arrow: mitochondria.

3.5. Alteration of Mitochondrial Ultrastructure in the Placenta of Sows with Excessive Back-Fat

In addition to the observed reduction in mitochondrial content, the transmission electron microscopy analysis revealed marked alterations in mitochondrial morphology in the placenta of BFII sows. As shown in Figure 4A,C, the area of mitochondria was decreased (43% reduction, $p < 0.05$) in placenta from BFII sows compared to BFI group. Higher magnification (100,000×) demonstrated swelling of mitochondria associated with an increased number of disarrayed cristae and a decreased electron density of the matrix in the placenta of BFII sows (Figure 4B). Concerning mitochondrial morphology regulation, we investigated the mRNA content of genes associated with mitochondrial fission and fusion, including Dynamin 1 (Drp1), mitofusin 1 (Mfn1), mitofusin 2 (Mfn2), and optic atrophy type 1 (OPA1), which play a key role in this process [30]. As illustrated in Figure 4D, excessive back-fat induced an increase ($p < 0.05$) in Drp1 (regulating mitochondrial fission) mRNA level in placenta, whereas reduced mRNA expression of Mfn1 (regulating mitochondrial fusion) was observed in the placenta from BFII sows ($p < 0.05$). Both Mfn2 and OPA1 (regulating mitochondrial fusion) expression were not affected by excessive back-fat.

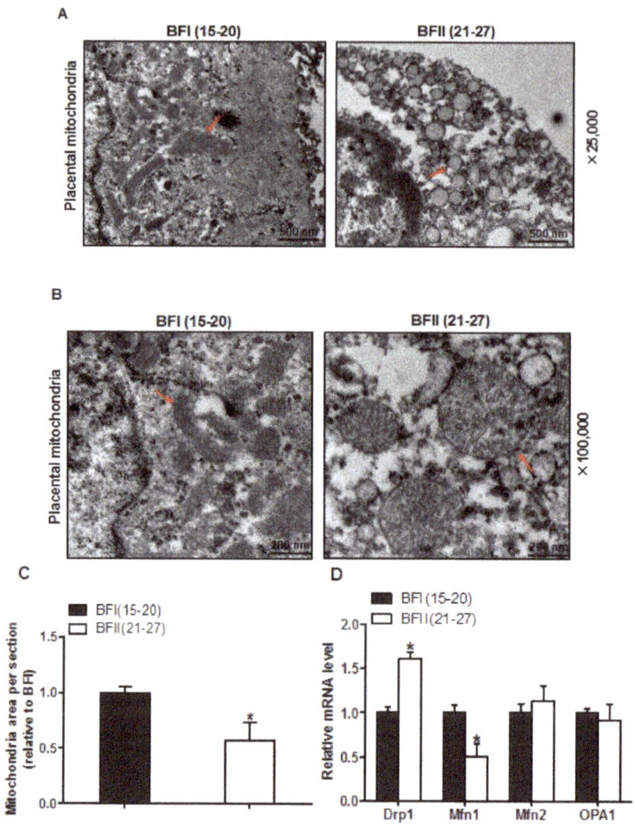

Figure 4. Alterations in the mitochondrial structure of placenta of sow challenged with excessive back-fat. (**a** and **b**) Transmission electron microscopy images at original magnifications of 25,000× (**a**) and 100,000× (**b**) in placental mitochondria from BFI and BFII sows; (**c**) Quantification of mitochondria area in the placental villi sections from 7-BFI and 7-BFII sows (analysis of 10 random images per placenta). Results were normalized by the mean value for the BFI sow set to 1 unit; (**d**) mRNA levels of mitochondrial fission- and fusion-related regulators, determined by quantitative RT-PCR, in placentas from BFI and BFII sows (n = 10/group). Values are expressed as mean ± SEM. * $p < 0.05$ compared with the BFI group. BF: back-fat thickness; Red arrow: mitochondria.

3.6. ROS Induce Mitochondrial Alterations and Dysfunction in Cultured Cytotrophoblasts

Since exposure to lipid oversupply leads to increased mitochondrial ROS production in a variety of peripheral tissues, we examined the effects of high lipid level on ROS production and mitochondrial density and functions in pig placental trophoblasts. ROS production was markedly induced by fatty acid (400 μM) treatment for 48 h (DCF: 48.5% in NEFA vs. 25.14% in BSA), and the addition of VE (2 mM) blocked this effect ($p < 0.05$, Figure 5A). Incubation with fatty acid also decreased ATP content (Figure 5B), CS activity (Figure 5C), and mtDNA level (Figure 5D). Furthermore, the mRNA levels of Mfn1, PGC1α, and genes implicated in mtDNA replication and repair, including gamma DNA polymerase subunit 1 (POLG1) and 2 (POLG2) and single-strand DNA binding protein 1 (SSBP1), were decreased in cytotrophoblasts treated with fatty acid for 48 h (Figure 5E). Concurrently, JC-1 fluorescent staining analysis showed mitochondrial membrane potential, which represented the oxidative respiration level, was reduced in fatty acid treatment (UR: 56.59% in NEFA vs. 89.89% in control, Figure S1A). The addition of VE counteracted all these effects, which indicated that ROS contributed to the observed mitochondrial alterations and dysfunction in cultured cytotrophoblasts (Figure 5B–E and Figure S1A). Transmission electron microscopy study nicely illustrated that the addition of fatty acid for 48 h negatively altered mitochondrial structure in cytotrophoblasts compared with control cells (Figure S1B).

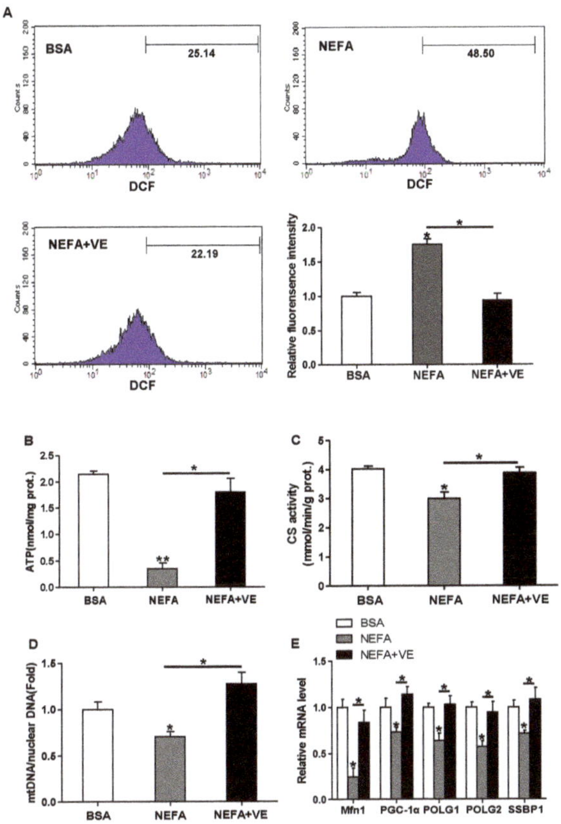

Figure 5. ROS induced mitochondrial alterations in pig primary trophoblast cells challenged with 400 μM NEFA. (**a**) ROS generation, measured by DCF production using flow cytometry or a spectrophotometer (n = 3), respectively, in cytotrophoblasts isolated from placenta of BFI sows. Before assessment of ROS production, cells were incubated with 400 μM NEFA in the presence or absence of 2 mM VE for 48 h. Value in scale bar depicts the percentage of DCF-positive cells; (**b**) Mitochondrial ATP production in cytotrophoblasts incubated with 400 μM NEFA in the presence or absence of 2 mM VE for 48 h (n = 3). ATP levels were normalized to total protein level of whole cell lysates extract; (**c** and **d**) Mitochondrial biogenesis was assessed by both citrate synthase activity (**c**) and mtDNA copy number (**d**) in cytotrophoblasts incubated with 400 μM NEFA in the presence or absence of 2 mM VE for 48 h (n = 3); (**e**) Relative mRNA expression of genes implicated in mitochondrial biogenesis and mtDNA replication (n = 3). Results were expressed as fold change relative to the values of untreated cells (BSA treatment) set to 1 unit. Values are expressed as mean ± SEM. * $p < 0.05$; ** $p < 0.01$ compared with the control group. ROS: reactive oxygen species; DCF: dichlorofluorescein; VE: Vitamin E; mtDNA: mitochondrial DNA; BSA: bovine serum albumin; NEFA: non-esterified fatty acid; Control: BSA group.

In order to further reveal the mechanisms of ROS induced mitochondrial alterations in cytotrophoblasts, western blot analysis of key regulators of mitochondrial function was performed.

As shown in Figure S1, microtubule affinity regulating kinase 4 (Mark4), phos-Mark4 (Thr 214), and phos- NF-κB (Ser536) were up-regulated ($p < 0.05$) in cytotrophoblasts incubated with fatty acid for 48 h, compared to control cells (Figure S1C). In contrast, the expression of phos-AMPK (Thr 172) was decreased in fatty acid treatment compared with control group ($p < 0.05$, Figure S1D). Moreover, the addition of VE alleviated these effects (Figure S1D). The protein content of AMPK and NF-κB was not affected by the addition of fatty acid (Figure S1C,D).

4. Discussion

As a mitochondrial rich organ, the placenta is essential for the normal growth and development of fetus. Thus, abnormalities in mitochondria may have detrimental consequences on placental function and therefore fetal growth and development. Indeed, cumulative evidence strongly suggests that alterations in mitochondrial respiration function and enzymatic activity of mitochondrial complexes in the human placenta are associated with pregnancy loss complicated by maternal obesity [14,31]. However, whether these changes are correlated with excessive back-fat of sows during pregnancy is not clear. Here, we investigated the influences of excessive back-fat on the amount, structure, and function of mitochondria in the porcine placenta. Our data indicate that mitochondrial defects were evident in term placenta from BFII sows. Furthermore, we found that increased oxidative stress in pig placenta due to excessive back-fat is probably one of the major determinants of mitochondrial alterations. This is supported by data showing that (a) an increase in ROS production occurred specifically in the placenta of BFII sows; (b) excessive ROS production was also associated with mitochondrial abnormalities in placenta from BFII sows; (c) incubation of cultured trophoblast cells with high lipid concentration induced ROS production and altered mitochondrial density and functions; and (d) these effects were alleviated by antioxidant treatment. Together, these findings suggest excessive back-fat aggravates a lipotoxic placental milieu (summarized in Figure 6) that induces oxidative stress and mitochondrial dysfunction in the full-term pig placenta.

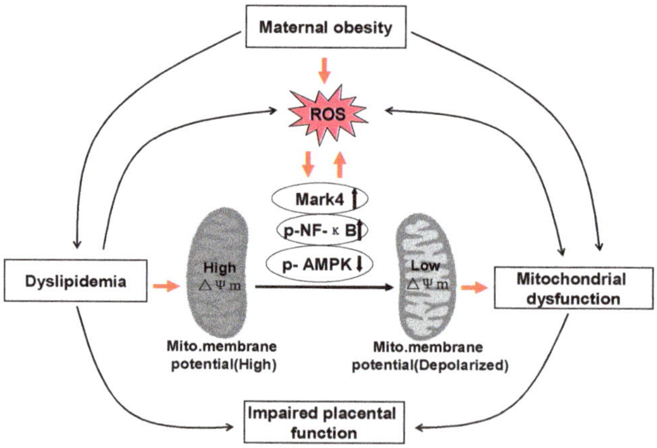

Figure 6. A schematic diagram depicts a causative role of obesity-induced oxidative stress in mitochondrial injury and placental dysfunction. Excessive back-fat aggravates dyslipidemia (lipotoxicity), which induces oxidative stress and mitochondrial dysfunction in the full-term pig placenta. Increased back-fat promotes mitochondrial oxidative injury by activating Mark4 and NF-κB and reducing AMPK activation. ↑: up-regulation of gene expression. ↓: down-regulation of gene expression. Arrows indicates a positive regulation. Interactions depicted are based on studies performed in various tissues (in some cases placenta) and have been previously published. ROS: reactive oxygen species.

In this study, excessive back-fat of sows was associated with increased plasma lipid and leptin levels, which indicated that the BFII sows had the trend for a higher adiposity than BFI sows. These metabolic alterations were associated with increased systemic (elevated plasma H_2O_2 level) and placenta (high levels of placental protein carbonylation and GSSG) oxidative stress from BFII sows compared to BFI sows, probably because of an increase in placental ROS production and a reduction in antioxidant defenses (decreased activities of SOD, GSH-PX, and CAT) in placenta of BFII sows. Furthermore, these disturbances were associated with mitochondrial changes in term placenta of sows with excessive back-fat. There was a significant decrease in placental ATP content associated with reduced expression levels of subunits encoding the complexes of mitochondrial electron transport chain (complexes I, II, III, and V), indicating a mitochondrial dysfunction in the placenta from BFII sows compared with BFI group. This observation agrees with previous reports in obese placenta of human model [14,15]. Since placenta is an extremely metabolically active tissue with rich energy-producing mitochondria, it is conceivable that mitochondrial abnormalities in ATP production may be not able to support placental substantially large energy requirements, therefore negatively affecting placental function and fetal development [14]. Consistent with this notion, a reduction in litter weight and placental efficiency was observed in BFII sows, which suggested that the placentas from BFII sows failed to support proper fetal development. Although a significant difference in birth weight between BFII and BFI groups was not observed, it is noted that a sexually dimorphic adaptation of the placenta may contribute to the differences in fetal growth and survival responding to the adverse intrauterine environment [32]. Recently, it has been demonstrated that pro-inflammatory cytokine production in placenta and the effect of inflammation on placental mitochondrial function exist a sexually dimorphic response [33]. Hence, the possibility that a sexual dimorphism may contribute to the lack of difference in the data cannot be completely ruled out. Further studies to assess the consequence of sexual dimorphic responses in placental mitochondrial energetics and function during porcine pregnancy associated with excessive back-fat are warranted.

To further explore the effect of excessive back-fat on placental mitochondrial respiration, we utilized an in vitro model of pig trophoblast cell culture. It is important that isolated trophoblasts can retain their in vivo phenotype in culture without the influence of the other different cells types (e.g., immunocytes and vascular cells) in whole placenta on the results of mitochondrial energy metabolism. Consistent with previous studies in human placenta and other tissues [5,15], we found a reduction in mitochondrial oxidative respiration with increased back-fat. The decreased mitochondrial maximum respiration and spare respiratory capacity reveals that cytotrophoblasts from placenta of BFII sows have a compromised cellular ability to meet energetic needs, and such a decrease in energetic metabolism may render them more susceptible to cellular stressors like lipids and ROS [34]. Measurement of activities of the mitochondrial respiratory complexes is commonly used as markers of mitochondrial function [35]. Consistent with reduced mitochondrial respiration, we observed that there was a significant decrease in mitochondrial complex I activity in term placenta of BFII sows compared to BFI group. As complex I of the mitochondrial respiratory chain is particularly prone to damage by oxidative stress [36], its reduction in activity associated with increased back-fat may be explained by a significant increase in ROS production that was also seen in the placenta of BFII sows. Indeed, increased local concentrations of ROS near the electron transport chain components are believed to be potentially high, making the mitochondrial respiratory complexes themselves likely targets of ROS impairment [13]. However, the activity of coupled complexes II and III was not significantly different between BFII and BFI sows. In contrast to our findings, Hastie et al. (2014) found a significant increase in the activity of placental combined complexes II and III associated with maternal obesity [14]. Differences in placental mitochondria collection may explain the discrepancy in our results. Of note, the mitochondrial inner membrane contains significantly higher complex I than any of the other complexes [36]. Thus, damage to mitochondrial complex II and III would have lesser impact on mitochondrial respiration than damage to complex I.

There is currently growing amounts of evidence showing that, as well as mitochondrial function, mitochondrial content is also altered in the human term placenta associated with maternal obesity [14,15]. Consistent with our expectation, we found that there was a significant decrease in mitochondrial number associated with a reduction in mtDNA content and CS activity (quantitative indexes of mitochondrial content) in placenta from BFII sows compared with BFI sows. These findings were further confirmed by reduced expression levels of mitochondria-encoded genes (ATP6, ND1, COX3, and CYTB) rather than increased mRNA contents of nuclear-encoded mitochondrial genes (ACADM and CS), which suggested that the control of mitochondrial biogenesis is altered in the placenta of BFII sows. Mitochondrial biogenesis is controlled by multiple transcription factors including PGC-1α, PGC-1β, NRF1, TFAM, TFB1M, and TFB2M [30]. Several studies have linked mitochondrial biogenesis transcription factors to placental mitochondrial content or pathology [37,38]. In agreement, we found that PGC-1α, PGC-1β, and downstream targets of PGC-1α, such as NRF1 and TFAM were down-regulated in term placenta of BFII sows compared to BFI group. Furthermore, a striking phenotype of placenta in BFII sows resided in the structural abnormalities of the mitochondria, as revealed by electron microscopy. A number of mitochondria appeared swollen, with fewer cristae, and the inner or outer membranes were sometimes disrupted in the placenta of sows with excessive back-fat. It should be noted that the regulation of mitochondrial dynamic network (maintaining normal mitochondrial structure) occurs through mitochondrial biogenesis and continuous cycles of fission and fusion [39]. In this trial, we observed, however, there was a significant decrease in the mitochondrial fusion regulator (Mfn1) associated with increased expression of Drp1, a protein participating in the mitochondrial fission, which suggested that aggravated mitochondrial fission as a mechanism may contribute to decreased placental mitochondrial content in BFII sows, potentially by disruption of mitochondrial structures [40]. Although our data pointed out that excessive back-fat was associated with mitochondrial alterations in the porcine placenta, further experiments are needed to determine the mediators of placental mitochondrial dysfunction.

To confirm the implication of oxidative stress in placental mitochondrial alterations associated with excessive back-fat, our in vitro data in cultured placental trophoblast cells demonstrated that treatment with high fatty acid concentration induced ROS production and mitochondrial damage in cytotrophoblasts, in agreement with previous reports that elevated lipids can induce cellular dysfunction through the activation of inflammation and the overproduction of ROS causing mitochondrial dysfunction in various models [6,41]. Although we were not able to determine whether ROS production is the only factor contributing to mitochondrial dysfunction, it is noticed that oxidative stress induced by lipid resulted in a decrease in the amount of mtDNA and CS activity in cultured cytotrophoblasts, mitochondrial abnormalities in function (reduced ATP production and mitochondrial membrane potential) and structure (mitochondrial swelling and disruption), and a concomitant reduction in expression of genes associated with mitochondrial biogenesis (PGC-1α and Mfn1) and mtDNA replication (POLG1, POLG2, and SSBP1), and these effects were reversed by antioxidant treatment, supporting a critical role of ROS in mediating mitochondrial alterations in placenta of sows with excessive back-fat. In addition, obese pregnancy has been previously linked to a heightened inflammatory state in placenta of human beings and animals such as pigs, with significant increases in pro-inflammatory cytokines such as TNFα, IL-6, and IL-8, which may impair mitochondrial function through activation of NF-κB signaling pathway [33]. To further estimate the mechanism of ROS-induced mitochondrial dysfunction in the term placenta of BFII sows, we initially investigated the status of activation of inflammatory NF-κB signaling in in vitro model of pig trophoblast cells challenged with high lipid. Consistently, we found that elevated lipid promoted the activation of NF-κB associated with increased ROS and impaired mitochondrial function and structure in cytotrophoblasts. Moreover, our previous studies showed that excessive back-fat is associated with increased activation of Mark4 and reduced activity and expression of AMPK in pig term placenta, indicating a potential mechanism for increased placental inflammation and oxidative stress [8,17]. In this trial, increased activation of microtubule affinity-regulating kinase 4 (Mark4) and reduced activation of AMPK were also observed in cultured trophoblast cells treated with high fatty acid, and addition of vitamin E alleviated these effects, suggesting that Mark4 and AMPK may be involved in regulation of ROS-mediated mitochondrial abnormalities in pig placenta induced by increased back-fat. In agreement with this conclusion, it has been shown that Mark4 promotes mitochondrial oxidative injury and adipose inflammation via activating NF-κB signal pathway [23]. Furthermore, studies have identified that AMPK activation could prevent inflammatory signaling pathways [1] and increases the expression of genes involved in mitochondrial function in human or mice skeletal muscle [42,43]. However, the precise molecular mechanisms by which these regulators improve or impair mitochondrial function in porcine placenta associated with excessive back-fat need to be further studied.

5. Conclusions

In summary, the present study demonstrates that excessive back-fat incites placental mitochondrial alterations that correspond with reduced ATP content and mitochondrial dysfunction in the pig full-term placenta. Moreover, augmented mitochondrial abnormalities may be attributed to increased placental oxidative stress in conditions associated with elevated maternal circulating lipids such as excessive back-fat during pregnancy of sows, and increased NF-κB signaling may contribute to ROS-mediated mitochondrial dysfunction in pig placenta induced by increased back-fat. Thus, our results suggest that excessive back-fat-induced oxidative stress and subsequent mitochondrial dysfunction has potential as a causal mechanism to impact pig placental function and therefore impair fetal growth and development.

Supplementary Materials: The following are available online at http://www.mdpi.com/2076-2615/10/2/360/s1. Figure S1: Effects of ROS on mitochondrial structure and function in pig primary trophoblast cells. Table S1: Primer sets used for real-time PCR.

Author Contributions: All the authors contributed to this manuscript. Conceptualization, P.Y.; formal analysis, L.T. and J.H.; funding acquisition, L.T.; investigation, L.T. and J.H.; resources, A.W.; supervision, L.T.; writing—original draft, L.T. All authors have read and agreed to the published version of the manuscript.

Funding: This research was surported by National Nature Science Foundation of China (No. 31702120), Fundamental Research Funds for the Central Universities of China (No. KJQN201831) and grants from the Nature Science Foundation of Jiangsu Province of China (No. BK20150672).

Acknowledgments: We sincerely acknowledge the staff at Research Farm and National Experimental Teaching Demonstration Center of Animal Science of Nan Jing Agricultural University for their helpful assistance in sample collection and for offering technical platform.

Conflicts of Interest: The authors declare no conflict of interest.

References

1. Saben, J.; Lindsey, F.; Zhong, Y.; Thakali, K.; Badger, T.M.; Andres, A.; Gomez-Acevedo, H.; Shankar, K. Maternal obesity is associated with a lipotoxic placental environment. *Placenta* **2014**, *35*, 171–177. [CrossRef] [PubMed]
2. Aye, I.L.; Lager, S.; Ramirez, V.I.; Gaccioli, F.; Dudley, D.J.; Jansson, T.; Powell, T.L. Increasing maternal body mass index is associated with systemic inflammation in the mother and the activation of distinct placental inflammatory pathways. *Biol. Reprod.* **2014**, *90*, 129. [CrossRef] [PubMed]
3. Brewer, C.J.; Balen, A.H. The adverse effects of obesity on conception and implantation. *Reproduction* **2010**, *140*, 347–364. [CrossRef]
4. Li, J.W.; Hu, J.; Wei, M.; Guo, Y.Y.; Yan, P.S. The Effects of Maternal Obesity on Porcine Placental Efficiency and Proteome. *Animals* **2019**, *9*, 546. [CrossRef]
5. Bonnard, C.; Durand, A.; Peyrol, S.; Chanseaume, E.; Chauvin, M.A.; Morio, B.; Vidal, H.; Rieusset, J. Mitochondrial dysfunction results from oxidative stress in the skeletal muscle of diet-induced insulin-resistant mice. *J. Clin. Invest.* **2008**, *118*, 789–800. [CrossRef]
6. Anderson, E.J.; Lustig, M.E.; Boyle, K.E.; Woodlief, T.L.; Kane, D.A.; Lin, C.T.; Price, J.W., 3rd; Kang, L.; Rabinovitch, P.S.; Szeto, H.H.; et al. Mitochondrial H2O2 emission and cellular redox state link excess fat intake to insulin resistance in both rodents and humans. *J. Clin. Invest.* **2009**, *119*, 573–581. [CrossRef]

7. Liang, T.; Jinglong, X.; Shusheng, D.; Aiyou, W. Maternal obesity stimulates lipotoxicity and up-regulates inflammatory signaling pathways in the full-term swine placenta. *Anim. Sci. J.* **2018**, *89*, 1310–1322. [CrossRef]
8. Jarvie, E.; Hauguel-de-Mouzon, S.; Nelson, S.M.; Sattar, N.; Catalano, P.M.; Freeman, D.J. Lipotoxicity in obese pregnancy and its potential role in adverse pregnancy outcome and obesity in the offspring. *Clin. Sci.* **2010**, *119*, 123–129. [CrossRef]
9. Dube, E.; Gravel, A.; Martin, C.; Desparois, G.; Moussa, I.; Ethier-Chiasson, M.; Forest, J.C.; Giguere, Y.; Masse, A.; Lafond, J. Modulation of fatty acid transport and metabolism by maternal obesity in the human full-term placenta. *Biol. Reprod.* **2012**, *87*, 1–11. [CrossRef] [PubMed]
10. Tian, L.; Dong, S.S.; Hu, J.; Yao, J.J.; Yan, P.S. The effect of maternal obesity on fatty acid transporter expression and lipid metabolism in the full-term placenta of lean breed swine. *J. Anim. Physiol. Anim. Nutr. (Berl.)* **2018**, *102*, e242–e253. [CrossRef] [PubMed]
11. Malti, N.; Merzouk, H.; Merzouk, S.A.; Loukidi, B.; Karaouzene, N.; Malti, A.; Narce, M. Oxidative stress and maternal obesity: Feto-placental unit interaction. *Placenta* **2014**, *35*, 411–416. [CrossRef] [PubMed]
12. Turrens, J.F. Mitochondrial formation of reactive oxygen species. *J. Physiol.* **2003**, *552*, 335–344. [CrossRef] [PubMed]
13. Sena, L.A.; Chandel, N.S. Physiological roles of mitochondrial reactive oxygen species. *Mol. Cell* **2012**, *48*, 158–167. [CrossRef] [PubMed]
14. Hastie, R.; Lappas, M. The effect of pre-existing maternal obesity and diabetes on placental mitochondrial content and electron transport chain activity. *Placenta* **2014**, *35*, 673–683. [CrossRef]
15. Mele, J.; Muralimanoharan, S.; Maloyan, A.; Myatt, L. Impaired mitochondrial function in human placenta with increased maternal adiposity. *Am. J. Physiol. Endocrinol. Metab.* **2014**, *307*, E419–E425. [CrossRef] [PubMed]
16. Zhou, Y.; Xu, T.; Cai, A.; Wu, Y.; Wei, H.; Jiang, S.; Peng, J. Excessive backfat of sows at 109 d of gestation induces lipotoxic placental environment and is associated with declining reproductive performance. *J. Anim. Sci.* **2018**, *96*, 250–257. [CrossRef]
17. Tian, L.; Wen, A.Y.; Dong, S.S.; Xiao, K.Y.; Li, H.; Yan, P.S. Excessive backfat of sows at mating promotes oxidative stress and up-regulates mitochondrial-mediated apoptotic pathway in the full-term placenta. *Livestock Sci.* **2019**, *222*, 71–82. [CrossRef]
18. Amaral Filha, W.S.; Bernardi, M.L.; Wentz, I.; Bortolozzo, F.P. Reproductive performance of gilts according to growth rate and backfat thickness at mating. *Anim. Reprod. Sci.* **2010**, *121*, 139–144. [CrossRef]
19. Tian, L.; Wen, A.; Dong, S.; Yan, P. Molecular Characterization of Microtubule Affinity-Regulating Kinase4 from Sus scrofa and Promotion of Lipogenesis in Primary Porcine Placental Trophoblasts. *Int. J. Mole. Sci.* **2019**, *20*, 5. [CrossRef]
20. Zhang, H.; Huang, Y.; Wang, L.; Yu, T.; Wang, Z.; Chang, L.; Zhao, X.; Luo, X.; Zhang, L.; Tong, D. Immortalization of porcine placental trophoblast cells through reconstitution of telomerase activity. *Theriogenology* **2016**, *85*, 1446–1456. [CrossRef]
21. Strakovsky, R.S.; Pan, Y.X. A decrease in DKK1, a WNT inhibitor, contributes to placental lipid accumulation in an obesity-prone rat model. *Biol. Reprod.* **2012**, *86*, 81. [CrossRef] [PubMed]
22. Muralimanoharan, S.; Maloyan, A.; Mele, J.; Guo, C.; Myatt, L.G.; Myatt, L. MIR-210 modulates mitochondrial respiration in placenta with preeclampsia. *Placenta* **2012**, *33*, 816–823. [CrossRef] [PubMed]
23. Liu, Z.; Gan, L.; Chen, Y.; Luo, D.; Zhang, Z.; Cao, W.; Zhou, Z.; Lin, X.; Sun, C. Mark4 promotes oxidative stress and inflammation via binding to PPARgamma and activating NF-kappaB pathway in mice adipocytes. *Sci. Rep.* **2016**, *6*, 21382. [CrossRef] [PubMed]
24. Wan, J.; Zhang, J.; Chen, D.; Yu, B.; He, J. Effects of alginate oligosaccharide on the growth performance, antioxidant capacity and intestinal digestion-absorption function in weaned pigs. *Anim. Feed Sci. Technol.* **2017**, *234*, 118–127. [CrossRef]
25. Shi, X.; Burkart, A.; Nicoloro, S.M.; Czech, M.P.; Straubhaar, J.; Corvera, S. Paradoxical effect of mitochondrial respiratory chain impairment on insulin signaling and glucose transport in adipose cells. *J. Biol. Chem.* **2008**, *283*, 30658–30667. [CrossRef]
26. Li, J.; Romestaing, C.; Han, X.; Li, Y.; Hao, X.; Wu, Y.; Sun, C.; Liu, X.; Jefferson, L.S.; Xiong, J.; et al. Cardiolipin remodeling by ALCAT1 links oxidative stress and mitochondrial dysfunction to obesity. *Cell Metab.* **2010**, *12*, 154–165. [CrossRef]

27. Livak, K.J.; Schmittgen, T.D. Analysis of relative gene expression data using real-time quantitative PCR and the 2-ΔΔCt method. *Methods Cell Sci.* **2001**, *25*, 402–408.
28. Vandesompele, J.; De Preter, K.; Pattyn, F.; Poppe, B.; Van Roy, N.; De Paepe, A.; Speleman, F. Accurate normalization of real-time quantitative RT-PCR data by geometric averaging of multiple internal control genes. *Genome Biol.* **2002**, *3*, Research0034. [CrossRef]
29. Lager, S.; Jansson, N.; Olsson, A.L.; Wennergren, M.; Jansson, T.; Powell, T.L. Effect of IL-6 and TNF-alpha on fatty acid uptake in cultured human primary trophoblast cells. *Placenta* **2011**, *32*, 121–127. [CrossRef]
30. Holland, O.; Dekker Nitert, M.; Gallo, L.A.; Vejzovic, M.; Fisher, J.J.; Perkins, A.V. Review: Placental mitochondrial function and structure in gestational disorders. *Placenta* **2017**, *54*, 2–9. [CrossRef]
31. Boyle, K.E.; Newsom, S.A.; Janssen, R.C.; Lappas, M.; Friedman, J.E. Skeletal muscle MnSOD, mitochondrial complex II, and SIRT3 enzyme activities are decreased in maternal obesity during human pregnancy and gestational diabetes mellitus. *J. Clin. Endocrinol. Metab.* **2013**, *98*, E1601–E1609. [CrossRef] [PubMed]
32. Sood, R.; Zehnder, J.L.; Druzin, M.L.; Brown, P.O. Gene expression patterns in human placenta. *Proc. Natl. Acad. Sci. USA* **2006**, *103*, 5478–5483. [CrossRef] [PubMed]
33. Muralimanoharan, S.; Guo, C.; Myatt, L.; Maloyan, A. Sexual dimorphism in miR-210 expression and mitochondrial dysfunction in the placenta with maternal obesity. *Int. J. Obes.* **2015**, *39*, 1274–1281. [CrossRef] [PubMed]
34. Brand, M.D.; Nicholls, D.G. Assessing mitochondrial dysfunction in cells. *Biochem. J.* **2011**, *435*, 297–312. [CrossRef] [PubMed]
35. Frazier, A.E.; Thorburn, D.R. Biochemical analyses of the electron transport chain complexes by spectrophotometry. *Methods Mol. Biol.* **2012**, *837*, 49–62.
36. Musatov, A.; Robinson, N.C. Susceptibility of mitochondrial electron-transport complexes to oxidative damage. Focus on cytochrome c oxidase. *Free Radic. Res.* **2012**, *46*, 1313–1326. [CrossRef]
37. Mando, C.; De Palma, C.; Stampalija, T.; Anelli, G.M.; Figus, M.; Novielli, C.; Parisi, F.; Clementi, E.; Ferrazzi, E.; Cetin, I. Placental mitochondrial content and function in intrauterine growth restriction and preeclampsia. *Am. J. Physiol. Endocrinol. Metab.* **2014**, *306*, E404–E413. [CrossRef]
38. Poidatz, D.; Dos Santos, E.; Duval, F.; Moindjie, H.; Serazin, V.; Vialard, F.; De Mazancourt, P.; Dieudonne, M.N. Involvement of estrogen-related receptor-gamma and mitochondrial content in intrauterine growth restriction and preeclampsia. *Fertil. Steril.* **2015**, *104*, 483–490. [CrossRef]
39. Twig, G.; Hyde, B.; Shirihai, O.S. Mitochondrial fusion, fission and autophagy as a quality control axis: The bioenergetic view. *Biochim. Biophys. Acta* **2008**, *1777*, 1092–1097. [CrossRef]
40. Vishnyakova, P.A.; Volodina, M.A.; Tarasova, N.V.; Marey, M.V.; Tsvirkun, D.V.; Vavina, O.V.; Khodzhaeva, Z.S.; Kan, N.E.; Menon, R.; Vysokikh, M.Y.; et al. Mitochondrial role in adaptive response to stress conditions in preeclampsia. *Sci. Rep.* **2016**, *6*, 32410. [CrossRef]
41. Nair, U.; Bartsch, H.; Nair, J. Lipid peroxidation-induced DNA damage in cancer-prone inflammatory diseases. A review of published adduct types and levels in humans. *Free Radic. Biol. Med.* **2007**, *43*, 1109–1120. [CrossRef] [PubMed]
42. Coletta, D.K.; Sriwijitkamol, A.; Wajcberg, E.; Tantiwong, P.; Li, M.; Prentki, M.; Madiraju, M.; Jenkinson, C.P.; Cersosimo, E.; Musi, N.; et al. Pioglitazone stimulates AMP-activated protein kinase signalling and increases the expression of genes involved in adiponectin signalling, mitochondrial function and fat oxidation in human skeletal muscle in vivo: A randomised trial. *Diabetologia* **2009**, *52*, 723–732. [CrossRef] [PubMed]
43. Smith, B.K.; Mukai, K.; Lally, J.S.; Maher, A.C.; Gurd, B.J.; Heigenhauser, G.J.; Spriet, L.L.; Holloway, G.P. AMP-activated protein kinase is required for exercise-induced peroxisome proliferator-activated receptor co-activator 1 translocation to subsarcolemmal mitochondria in skeletal muscle. *J. Physiol.* **2013**, *591*, 1551–1561. [CrossRef] [PubMed]

animals

MDPI

Article

Endometritis Changes the Neurochemical Characteristics of the Caudal Mesenteric Ganglion Neurons Supplying the Gilt Uterus

Barbara Jana [1,*] and Jarosław Całka [2]

1 Division of Reproductive Biology, Institute of Animal Reproduction and Food Research of the Polish Academy of Sciences, Tuwima 10, 10-748 Olsztyn, Poland

2 Department of Clinical Physiology, Faculty of Veterinary Medicine, University of Warmia and Mazury, Oczapowskiego 14, 11-041 Olsztyn, Poland; calkaj@uwm.edu.pl

* Correspondence: b.jana@pan.olsztyn.pl; +48-89-539-31-37

Received: 27 April 2020; Accepted: 19 May 2020; Published: 20 May 2020

Simple Summary: Uterine inflammation is a very frequent pathology in domestic animals leading to disturbances in reproductive processes and causing significant economic losses. The uterus possesses nerves from either the autonomic or sensory part of the peripheral nervous system. Most of the uterus-projecting neurons are localized in the caudal mesenteric ganglion. These neurons synthesize and release numerous biologically active substances in the uterus, which regulate uterine functions. The effect of inflammation on uterine innervation is poorly recognized. This study showed that *Escherichia coli*-induced uterine inflammation in pig led to a reduction in the total population of uterine neurons in the caudal mesenteric ganglion, and in the populations of these cells in the dorsal and central areas of this ganglion. In the caudal mesenteric ganglion of gilts after intrauterine bacterial injection, the population of uterine neurons presenting positive staining for dopamine-β-hydroxylase (an enzyme participating in noradrenaline synthesis) and negative staining for galanin, as well as the population of uterine neurons presenting negative staining for dopamine-β-hydroxylase but positive staining for neuropeptide Y, were decreased. In these gilts, there were increased numbers of uterine neurons which, besides dopamine-β-hydroxylase, also expressed neuropeptide Y, galanin and vasoactive intestinal peptide. The above changes suggest that inflammation of the gilt uterus may affect the function(s) of this organ by its action on the neurons of the caudal mesenteric ganglion.

Abstract: This study analyzed the influence of uterine inflammation on the neurochemical characteristics of the gilt caudal mesenteric ganglion (CaMG) uterus-supplying neurons. The horns of uteri were injected with retrograde tracer Fast Blue on day 17 of the first studied estrous cycle. Twenty-eight days later (the expected day 3 of the third studied estrous cycle), either saline or *Escherichia coli* suspension were administered into each uterine horn. Only the laparotomy was done in the control gilts. After 8 days, the CaMGs and uteri were harvested. The infected gilts presented a severe acute endometritis. In the CaMGs, the populations of uterine perikarya possessing dopamine-β-hydroxylase (DβH) and/or neuropeptide Y (NPY), somatostatin (SOM), galanin (GAL) and vasoactive intestinal polypeptide (VIP) were analyzed using the double immunofluorescence method. In the CaMG, bacterial injection decreased the total number of the perikarya (Fast Blue-positive), the small and large perikarya populations in the dorsal and central regions, and the small and large perikarya populations coded DβH+/GAL- and DβH-/NPY+. After bacterial treatment, there was an increase in the numbers of small and large perikarya coded DβH+/NPY+, small perikarya coded DβH+/GAL+ and DβH+/SOM- and large perikarya coded DβH+/VIP+. To summarize, uterine inflammation influences the neurochemical characteristics of the CaMG uterus-supplying neurons, which may be important for pathologically changed organ functions.

Keywords: uterus; inflammation; sympathetic innervation; caudal mesenteric ganglion; chemical plasticity; gilt

1. Introduction

Uterine inflammatory process is a very common condition in domestic animals, which leads to economic losses. Although this disorder may occur in females which have not yet given birth (after natural mating, insemination), uterine inflammation takes place predominantly after parturition as a result of disturbances in uterine involution and/or immunological reaction. Metritis/endometritis is evoked for the most part by bacteria [1–4]. Situations such as difficult labor, fetal membrane retention and uterine contractility disturbances are conducive to the origin of this pathology. Mild uterine inflammation does not cause any serious deviations in the estrous cycle course. In turn, in more advanced cases of inflammation, an inflammatory exudate (mucopurulent) is present in the uterine cavity. The myometrium of animals, especially cows, loses its ability to contract [5]. During uterine inflammation, the production and release of cytokines [6–9], prostaglandin (PG)$F_{2\alpha}$, PGE_2 [2,10], PGI_2 [11], leukotrienes (LT)B_4 and LTC_4 [12], as well as nitric oxide [13] are markedly increased.

The uterus in pigs is innervated by nerves from either the autonomic or sensory part of the peripheral nervous system. In the pig, most of the uterus-projecting neurons (further referred to as uterine neurons or uterine perikarya), indicated by fluorescent retrograde neuronal tracer Fast Blue (FB), exist in the caudal mesenteric ganglion (CaMG), where they form the "uterine region". The vast majority of all uterine perikarya in the CaMG are noradrenergic. These structures possess enzymes that synthetize noradrenaline (NA): tyrosine hydroxylase (TH) and dopamine-β-hydroxylase (DβH). However, other neurotransmitters such as galanin (GAL), neuropeptide Y (NPY), vasoactive intestinal polypeptide (VIP) and somatostatin (SOM) are also expressed in those perikarya. Moreover, in the sow, these neurotransmitters are present in the nerve fibers in the area of CaMG [14,15] and in the uterus, under luminal epithelium, near the blood vessels, endometrial glands and myocytes of the myometrium [16,17]. It is known that NA, acting through α- and/or β-adrenergic receptors, controls the contractile activity of the uterus [18,19] and PGs production in endometrial and myometrial cells under physiological conditions [20,21]. NPY, via its receptor subtype 1, stimulates the contractility of myometrium [22] and uterine arteries [23,24]. Similarly, GAL [25] and VIP [26] affect uterine contractility, while SOM regulates endometrial cell proliferation and motility [27].

Data are scarce on the relationship between uterine inflammation and morphology, as well as on the immunochemical characteristics of neurons projecting to the uterus. In rats, the inflammation of the uterus evoked abnormal behavior associated with visceral pain [28] and upregulated the uterine substance P (SP)-positive perikarya numbers in the dorsal root ganglia (DRGs) [29]. Due to its embryological, anatomical and physiological similarities to humans, the pig is a valuable species in biomedical research, also for the study of uterine function [30,31]. One published study conducted on pigs, describing the influence of uterine inflammation on the innervation of the uterus, indicated that in gilts, *Escherichia coli* (*E. coli*)-induced inflammatory state changes the morphological and neurochemical characteristics of the DRG neurons supplying the uterus [32]. In addition, it was found that there was a reduction in the total population of nerve fibers in the inflamed porcine uterus, including DβH-positive around inflamed uterine structures [33]. Based on these findings, it is hypothesized that the uterine inflammatory state affects the neurochemical characteristics of the uterine neurons in CaMG. To test this hypothesis, the effect of uterine inflammation was examined on (1) the total number of uterine perikarya and their size and localization, and on (2) the uterine perikaryal cell count containing DβH and/or NPY, SOM, GAL or VIP in the gilt CaMG.

2. Materials and Methods

2.1. Animals

The experiment was performed taking into account the principles of animal care (National Institute of Health publication No. 86–23, revised in 1985) and the specifics of national law concerning animal protection. The Local Ethics Committee of the University of Warmia and Mazury in Olsztyn approved all procedures and granted consent (no. 65/2015).

This research was carried out on 11 gilts (crossbred Large White × Landrace) at the age of 7–8 months and body weight (BW) of approximately 90–120 kg. By using a tester boar, behavioral estrus was estimated. In all studied gilts, disruption in reproductive state did not occur, vaginal discharges were not revealed, and the second estrous cycle was regular. The gilts were kept in laboratory conditions (Faculty of Veterinary Medicine, University of Warmia and Mazury, Olsztyn, Poland). They were maintained in individual pens with an area of about 5 m^2 under the following conditions: natural daylight—14.5 ± 1.5 h, night—9.5 ± 1.5 h and temperature 18 ± 2 °C. The pigs were fed with commercial diets according to their nutritional requirements and had access to water. After transport, the pigs were divided (randomly) into three groups: *Escherichia coli* (*E. coli* group, n = 4), the saline (SAL group, n = 3)-treated gilts and control (CON control, n = 4) gilts—subjected to "sham" operation (details are below). The study was started after three days (adaptive period). During the experiment, the animals were not medically treated.

2.2. Experimental Procedures

The experimental procedure was previously described [32]. On day 17 of the first studied estrous cycle (day 0 of the study), before surgery, the gilts were pre-medicated with atropine (0.05 mg/kg, administered intramuscularly (i.m.); Atropinum sulf. WZF, Warszawskie Zakłady Farmaceutyczne Polfa S.A., Warsaw, Poland), azaperone (2 mg/kg BW, administered i.m. Stresnil, Janssen Pharmaceutica, Beerse, Belgium) and ketamine hydrochloride (10 mg/kg BW, administered intravenously (i.v.); Ketamina, Biowet, Puławy, Poland). General anesthesia was reached with ketamine hydrochloride and prolonged by the application of supplementary doses of this medicine (1 mg/kg BW every 5 min, administered i.v.). After laparotomy, the uterine horns were injected with Fast Blue (FB, 5% aqua solution, EMS-CHEMIE, GmbH, Gross-Umstadt, Germany) to indicate the cell bodies of neurons projecting to the uterus. FB was administered using a Hamilton syringe with a 26-gauge needle into the wall of each uterine horn in paracervical, middle and paraoviductal portions. In each part (ring about 2 cm wide), 13 FB injections were done (volume of each injection—2 μL, total volume per place—26 μL). The needle of the Hamilton syringe was kept in each place for 1 min following injection to limit the leakage of FB outside the uterine tissue. Next, the place of injection was rinsed using isotonic saline and wiped with gauze.

Twenty-eight days later (the necessary period for FB to achieve the external sources of innervation of the uterus in pigs), on the expected day 3 of the third studied estrous cycle, the gilts were anaesthetized (as explained above). In gilts, after laparotomy was done, either 50 mL of *E. coli* suspension (*E. coli* group; 1 mL of suspension containing 10^9 colony-forming units, strain O25:K23/a/:H1; National Veterinary Research Institute, Department of Microbiology, Puławy, Poland), or 50 mL of saline solution (SAL group) were administered into both uterine horns. In the gilts of the CON group, only laparotomy was carried out. After 8 days (the expected day 11 of the third studied estrous cycle), euthanasia of gilts was performed using an overdose of ketamine hydrochloride (administered i.v.) and the gilts were transcardially perfused via the ascending aorta with 4% buffered paraformaldehyde (pH 7.4). Next, the bilateral CaMGs were obtained from gilts of all groups. The ganglia were post-fixed by immersion in the same fixative for 10 min, then washed with 0.1 M PB (pH 7.4) for two days and stored at 4 °C in an 18% buffered sucrose solution (pH 7.4), with natrium azide (0.001%). Later, the CaMGs were kept at −80 °C until further examination. For the microscopic study, the fragments of uterine horns were fixed in 4% paraformaldehyde solution (pH 7.4) for 24 h, and the tissues were then washed in 0.1 M phosphate-buffered saline (PBS, pH 7.4) and embedded in paraffin. The findings of the histological assessment of uteri were published previously [32].

2.3. Immunohistochemical Analysis

Serial cryostat sections (with a thickness of 10 μm, Frigocut, Reichert-Jung, Nussloch, Germany) of the CaMG were placed on chrome alum-coated slides. The presence of FB-positive perikarya was checked in serial sections of the bilateral CaMGs using an Olympus BX51 microscope (Olympus, Warsaw, Poland), equipped for epi-illumination fluorescence microscopy (V1 module: excitation filter 330–385 nm, barrier filter 420 nm). Sections with FB-positive perikarya were used for double-labeling immunofluorescence to determine DβH and/or NPY, SOM, GAL and VIP immunoreactivity [34]. The sections were dried (at 32 °C, 45 min), rinsed in a phosphate buffer with 0.8% sodium chloride and 0.02% potassium chloride (PBS, 3 × 10 min) and incubated in 10% normal goat serum in PBS with 0.3% Triton X-100 (Sigma, Saint Louis, MO, USA) and 1% bovine serum albumin (BSA; Sigma, USA) for 20 min. Next, the sections were incubated overnight (at 4 °C) with primary antibodies diluted in PBS containing 0.3% Triton X-100 and 1% BSA, raised against DβH (rabbit polyclonal, Cat. # AB1585, Merck Millipore, Kenilworth, NJ, USA, 1:500) and/or NPY (mouse monoclonal, Cat. # ABS 028-08-02, ThermoFisher Scientific, Waltham, MA, USA, 1:1000), SOM (rat monoclonal, Cat # 8330-0009, AbD Serotec, Kidlington, UK, 1:60), GAL (guinea pig polyclonal, Cat. # T-50-36, Penisula, San Carlos, CA, USA, 1:800) or VIP (mouse polyclonal, Cat # 9535-0504, BioGene Ltd., Huntingdon, UK, 1:2000). On the next day, the sections were rinsed (PBS, 5 × 15 min) and incubated with secondary antibodies (in PBS with 0.25% BSA and 0.1% Triton X-100) for 4 h (Alexa Fluor 488 nm donkey anti-rabbit, Cat # A21206, Alexa Fluor 546 nm donkey anti-mouse, Cat # A10036, Alexa Fluor 546 nm donkey anti-rat, Cat # A11081 and Alexa Fluor 546 nm donkey anti-guinea pig, Cat # A11074, all from ThermoFisher Scientific, Waltham, MA, USA, and diluted 1:1000) to visualize the following antibody combinations: DβH/NPY, DβH/SOM, DβH/GAL and DβH/VIP. After that, the sections were rinsed (PBS, 3 × 5 min) and covered with a polyethylene glycol/glycerin solution with 1,4-diazabicyclo[2.2.2]octane (DABCO, Sigma, USA). To control for immunofluorescence specificity, standard tests (pre-absorption for the applied antisera with the respective antigen at a content of 20–50 μg antigen/mL diluted antiserum, exclusion of primary or secondary antisera and replacement by non-immune sera of all the primary antisera used) were conducted. The specificity of retrograde tracing was examined by the use of the various tests presented previously [32]. FB-positive and double-immunostained perikarya were investigated and photographed with the appropriate filter sets for fluorescein isothiocyanate (FITC, B1 module, excitation filter 450–480 nm, barrier filter 515 nm) and CY3 (G1 module excitation filter 510–550 nm, barrier filter 590 nm). DβH-, NPY-, SOM-, GAL-, VIP-immunoreactive and/or all FB-positive perikarya were calculated in every fourth section of the bilateral CaMGs. CaMG uterine perikarya profiles with a visible nucleus were only scored to avoid double counting. The perikarya distribution was determined for particular regions of CaMG, taking into account the method reported previously [34]. In brief, the cranial region of the CaMG constitutes an area corresponding to I splanchnic lumbar nerve and intermesenteric nerve, the dorsal region forms the area corresponding to II and III splanchnic nerves, the caudal region composes the area corresponding to IV splanchnic lumbar nerve and hypogastric nerve and the ventral region constitutes the area corresponding to caudal colonic nerves. In addition, the perikarya were classified as small (with a diameter of 23 ± 10 μm) or large (with a diameter of 51 ± 17 μm) [34], using Cell F Imaging Software (Olympus, PL). The perikarya sizes were estimated by measurements of their long and short axis. The average value of the perikaryon diameter was assigned by the use of the equation $d = \sqrt{l} \times k$, where d equals the diameter of a circle with a surface area which is the most similar to the surface area of an ellipsoidal figure with a long axis (l) and a short axis (k) [35]. By adding the small and large perikaryal cell count from all areas of the CaMG, the total population of perikarya for each studied group was determined. The images were captured by a digital camera connected to a PC and analyzed with AnalySIS software (version 3.02, Soft Imaging System, Münster, Germany).

2.4. Statistical Analysis

Data obtained from the CaMGs of gilts from the CON, SAL and *E. coli* groups were averaged per one CaMG, region, small perikarya, large perikarya and perikarya, with individual neurochemical characteristics for each group. Data concerning neurochemical characteristics were expressed as percentages of the total population of small or large uterine perikarya stained for two substances in each group, accepted as 100%. Statistical significances (mean ± standard error of the mean (SEM)) were estimated by a one-way analysis of variance (ANOVA) followed by the Bonferroni test (Statistica 13 software, StatSoft Inc., Tulsa, OK, USA). Differences were evaluated as significant if the probability was $p < 0.05$. Before the experiment, a statistical power calculation was not performed. The number of gilts in the studied group was based on the earlier studies, in which 3 to 4 animals were used for neuro-immunofluorescence experiments.

3. Results

3.1. The Population and Localization of the CaMG Uterine Perikarya

In the *E. coli* group, the total number of FB-positive uterine perikarya (small and large perikarya combined) was lower ($p < 0.01$) compared to the CON and SAL groups (70 ± 24.2 versus 228 ± 45.3, 245 ± 35.1, respectively). In the *E. coli* group, a reduction ($p < 0.01$) in the number of small perikarya was found in relation to the CON and SAL groups (45.5 ± 14.3 versus 137.9 ± 18.4, 146 ± 21.4, respectively). In the gilts, after intrauterine infusion of bacteria, the number of large perikarya was also lower ($p < 0.05$) than in the CON and SAL groups (30.8 ± 6.5 versus 95 ± 25.1, 102 ± 16.8, respectively).

In relation to the CON and SAL groups, in the *E. coli* group, the numbers of small perikarya was lower in the cranial ($p < 0.01$), caudal ($p < 0.001$), dorsal (CON—$p < 0.05$, SAL—$p < 0.01$) and central ($p < 0.01$) ganglional areas (Figure 1A). The populations of large perikarya in the dorsal ($p < 0.05$), ventral ($p < 0.01$) and central ($p < 0.01$) CaMG areas of the bacterial-treated gilts were reduced compared to other groups (Figure 1B).

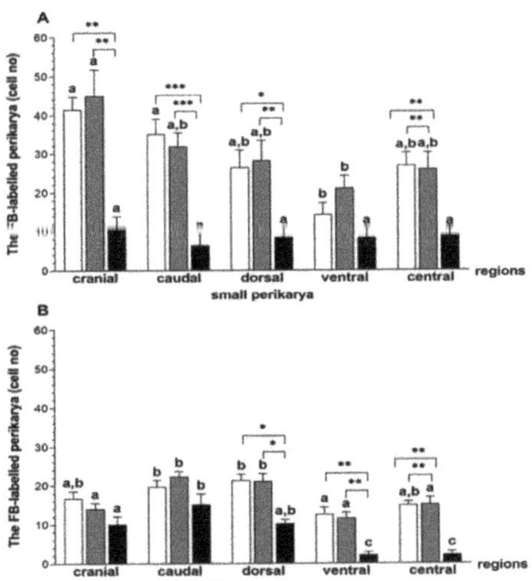

Figure 1. Small (**A**) and large (**B**) perikaryal cell count (mean ± standard error of the mean (SEM)) in the different regions of the caudal mesenteric ganglion (CaMG) projecting to the uterus of gilts from the control (CON) (white bars), saline (SAL) (grey bars) and *E. coli* (black bars) groups. Different letters (a, b, c) show differences ($p < 0.05$–0.001) among the particular regions within the CON, SAL and *E. coli* groups; * $p < 0.05$, ** $p < 0.01$ and *** $p < 0.001$ show differences between all groups in the same ganglion region.

3.2. The Populations of the CaMG Uterine Perikarya Expressing DβH, NPY, SOM, GAL and VIP

In the *E. coli* group, the numbers of small (CON—$p < 0.05$, SAL—$p < 0.001$) and large ($p < 0.001$) perikarya presenting positive staining for DβH and NPY increased, while the numbers of these perikarya presenting negative staining for DβH and positive staining for NPY were reduced ($p < 0.05$) in relation to other groups. The population of large perikarya presenting negative staining for DβH and NPY after intrauterine bacterial infusion was also lower ($p < 0.05$) than in the CON group (Figure 2A, Figure 3A–H). The number of small perikarya expressing DβH but not SOM increased in the *E. coli* group in relation to the CON ($p < 0.01$) and SAL ($p < 0.001$) groups. The numbers of small and large uterine perikarya presenting negative staining for DβH and SOM were reduced compared to the CON ($p < 0.01$) and SAL ($p < 0.001$) groups (Figure 2B, Figure 3I–P). Intrauterine injections of *E. coli* increased ($p < 0.001$) the number of small perikarya expressing DβH and GAL compared to other groups. In turn, in the *E. coli* group, the numbers of small (CON—$p < 0.05$, SAL—$p < 0.01$) and large ($p < 0.05$) perikarya presenting positive staining for DβH and negative for GAL and the number of small perikarya with negative staining for DβH and GAL ($p < 0.05$) were reduced compared to other groups (Figure 2C, Figure 4A–H). The population of large perikarya presenting positive staining for DβH and VIP was higher after intrauterine bacterial treatment than in the CON ($p < 0.05$) and SAL ($p < 0.01$) groups. The small (CON—$p < 0.001$, SAL—$p < 0.01$) and large (CON—$p < 0.01$, SAL—$p < 0.001$) perikarya populations without positive staining for DβH and VIP were reduced compared to other groups (Figure 2D, Figure 4I–P). The numbers of uterine perikarya expressing DβH and/or NPY, SOM, GAL or VIP as well as those without these substances within the gilt CaMG from the CON, SAL and *E. coli* groups are given in Figure 2A–D.

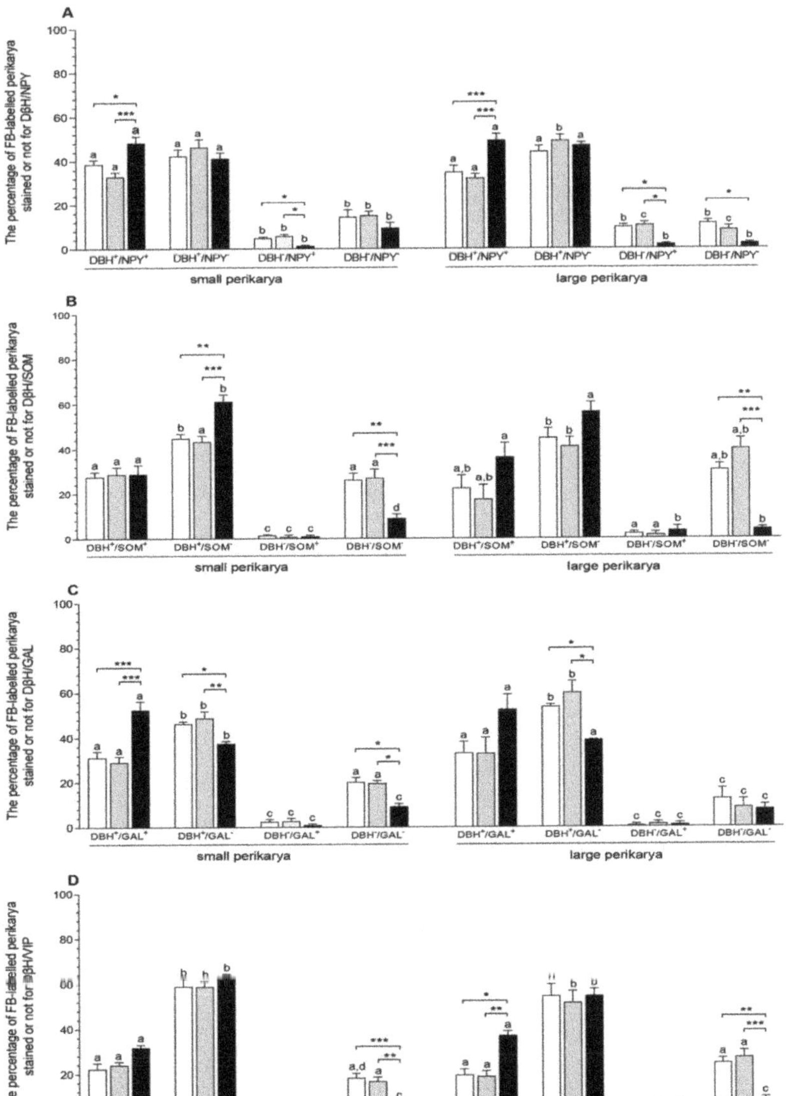

Figure 2. The populations (expressed as percentages, mean ± SEM) of small and large uterine perikarya expressing dopamine-β-hydroxylase (DβH) and/or neuropeptide Y (NPY) (**A**), DβH and/or somatostatin (SOM) (**B**), DβH and/or galanin (GAL) (**C**) and DβH and/or vasoactive intestinal polypeptide (VIP) (**D**) as well as those without these substances in the CaMG of gilts from the CON (white bars), SAL (grey bars) and *E. coli* (black bars) groups. Data are expressed as percentages of the total population of small or large uterine perikarya stained for two substances in each group, accepted as 100%. Different letters (a, b, c, d) show differences ($p < 0.01$, $p < 0.001$) among the particular populations of uterine perikarya within the CON, SAL and *E. coli* groups; * $p < 0.05$, ** $p < 0.01$ and *** $p < 0.001$ show differences between all groups for the same population of uterine perikarya.

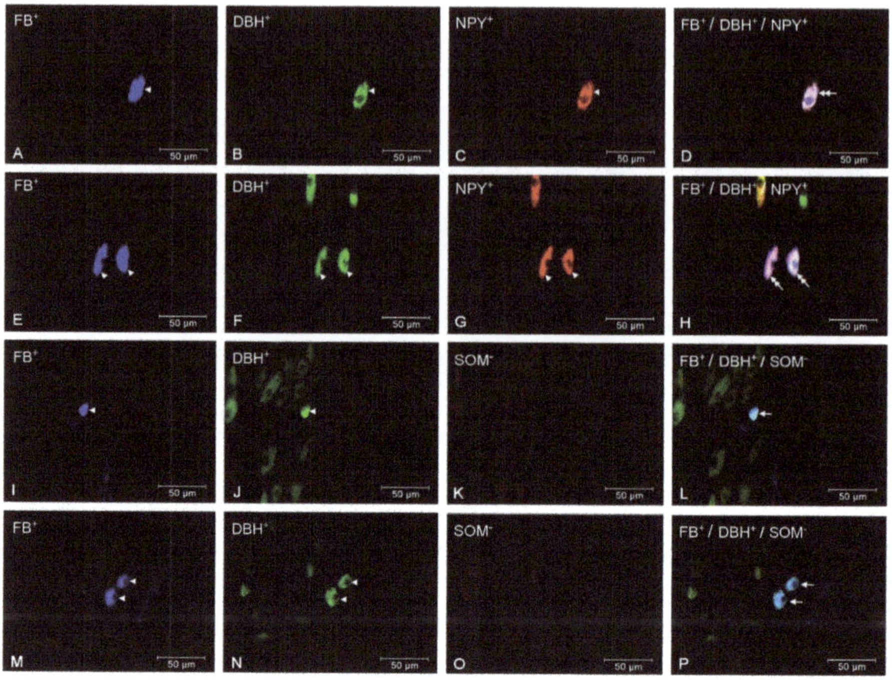

Figure 3. Micrographs demonstrating the presence of DβH (B, F, J, N), NPY (C, G) and SOM (K, O) in the CaMG uterine perikarya of gilts from the CON (**A–D**), SAL (**I–L**) and *E. coli* (**E–H**, **M–P**) groups. The arrowhead indicates: Fast Blue (FB)-positive perikaryon, perikaryon presenting positive staining for DβH and NPY and perikaryon presenting positive staining for DβH. The double arrow indicates FB-positive uterine perikaryon expressing DβH and NPY. The arrow indicates FB-positive uterine perikaryon expressing DβH. The photographs (D, H, L, P) were made by digital superimposition of three color channels: FB-positive (blue), DβH-positive (green) and NPY- or SOM-positive (red). One large uterine perikaryon DβH and NPY immunoreactive is visible in the gilt of the CON group (**A–D**). In the CaMG of *E. coli* group, an elevation in the population of large perikarya containing these substances is observed (**E–H**). One small perikaryon expressing DβH, but not SOM, is present in the ganglion of the SAL group (**I–L**). In the *E. coli* group, two small perikarya expressing DβH, but not SOM, are observed in the CaMG (**M–P**).

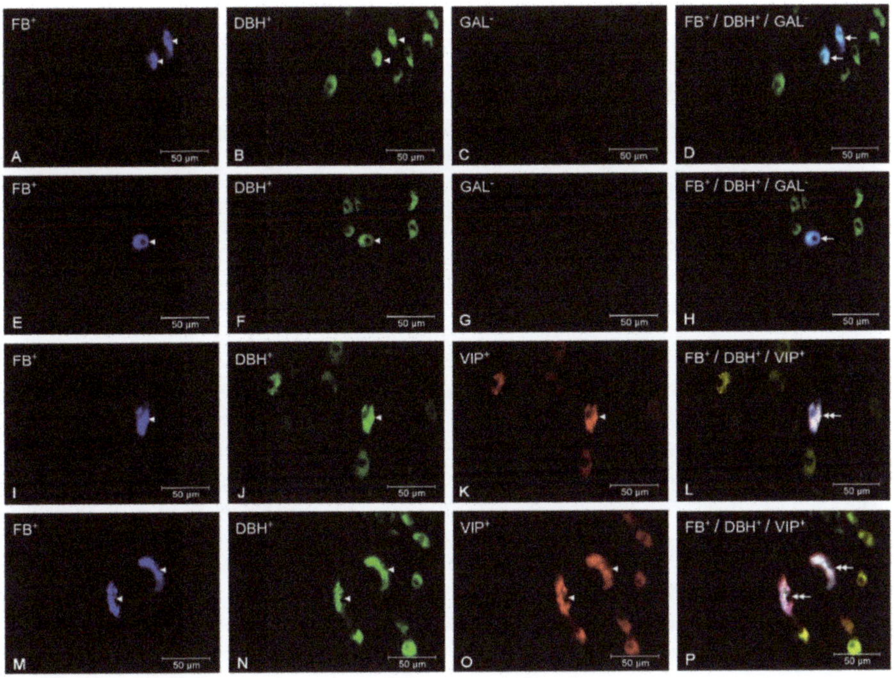

Figure 4. Micrographs demonstrating the presence DβH (B, F, J, N), GAL (C, G) and VIP (K, O) in the CaMG uterine perikarya of gilts from the CON (**A–D**), SAL (**I–L**) and *E. coli* (**E–H**, **M–P**) groups. The arrowhead indicates: FB-positive perikaryon, perikaryon presenting positive staining for DβH and perikaryon presenting positive staining for DβH and VIP. The arrow indicates FB-positive uterine perikaryon expressing DβH. The double arrow indicates FB-positive uterine perikaryon expressing DβH and VIP. The photographs (D, H, L P) were made by digital superimposition of three color channels: FB-positive (blue), DβH-positive (green) and GAL- or VIP-positive (red). In the CaMG from the CON group, two small uterine perikarya containing DβH but not GAL are visible (**A–D**). Following intrauterine bacterial injection, a drop in the number of small perikarya expressing DβH but immunonegative to GAL is present (**E–H**). In the CaMG of gilt from the SAL group, one large perikaryon reactive to DβH and VIP is visible (**I–L**). In the gilt of the *E. coli* group, an augmentation in the number of large perikarya occurs (**M–P**).

4. Discussion

This study, for the first time, showed how the inflammation of the uterus alters the neurochemical characteristics of the CaMG's uterine perikarya. The results of the histopathological assessment of pig uteri used in the current experiment were presented previously. In the *E. coli*-treated uteri, severe acute endometritis was revealed [32]. In the CaMG of gilts from the CON and SAL groups, the total number, size and localization of uterine perikarya, and the numbers of these structures expressing DβH and/or NPY, SOM, GAL or VIP were similar. This shows that administration of saline into the uterus did not significantly change the neurochemical features of the examined perikarya.

The current study revealed that after intrauterine infusion of bacteria in the gilt, the total population of uterine perikarya in CaMG was reduced. Uterine inflammation also led to a decrease in the total population of uterine perikarya in the DRGs of animals examined in the current experiment [32]. Although in the present study the steroid levels in peripheral blood of gilts were not determined, it is possible that they may influence the uterine perikarya populations in the *E. coli* group. In pigs and cows with an inflamed uterus, the level of 17β-estradiol (E_2) was decreased, and in pigs, the androstenedione level was also augmented [10,36–38]. A drop in the total number of CaMG uterine perikarya in pigs with uterine inflammation may be a consequence of a decreased level of E_2 and an elevated level of androgens. The neuroprotective properties of estrogen are connected with their ability to activate various membrane-associated intracellular signaling pathways and nuclear receptors for estrogen (ERs) and to induce growth factors' production, as well as their antioxidant actions [39,40]. Moreover, ERs are expressed in rat uterine perikarya in the DRGs [41] and porcine ovary perikarya in the CaMG [42]. The drop in CaMG of ovarian neurons and an augmentation in the set of these cells with androgen receptors (ARs) after long-term testosterone (T) administration in gilts was reported [43]. In addition, androgens have the ability to decrease growth factor production in different tissues [44,45]. It is possible that the reduced survival of uterine perikarya in the CaMG revealed in the present study may also be a consequence of increased synthesis and the release of inflammatory factors in the course of uterine inflammation. This is supported by reports that show elevated levels of pro-inflammatory cytokines (tumor necrosis factor α (TNF-α), interleukin 1 β), as well as $PGF_{2\alpha}$, LTB_4 and LTC_4 in peripheral blood of females with an inflamed uterus [7,9,10,12] and the presence of TNF-α, LTB_4 and LTC_4 receptors in the perikarya within DRG [46–48]. It is known that pro-inflammatory cytokines participate in neuron death by toxic free radical generation [49]. The loss of both small and large perikarya in the CaMG after intrauterine *E. coli* administration may be associated with a similar content of the ERs and/or ARs [42,43], as well as receptors for inflammatory mediators [46–48]. However, the exact mechanism of steroid hormone- and inflammatory factor-dependent drop in the neuronal CaMG number needs further studies.

The current study found a decrease in the gilt CaMG populations of small and large uterine perikarya following intrauterine injection of bacteria. In addition, in the DRGs of gilts used in the current study, uterine inflammation led to a drop in the population of small uterine perikarya [32]. Although in the present experiment the effect of the inflammatory process on the dynamic changes in perikarya size has not been determined, the revealed changes in the numbers of small and large uterine perikarya may result from their atrophy or hypertrophy, although this hypothesis needs confirmation. The small and large uterine perikarya were found in all examined areas of the CaMG of gilts from the CON, SAL and *E. coli* groups. In the CaMG of gilts after intrauterine bacterial injection, the numbers of small and large perikarya were reduced in the dorsal and central regions. In addition, uterine inflammation decreased the numbers of small perikarya in the cranial and caudal regions and the number of large perikarya in the ventral region. Previously, it was mentioned that the changes in populations of uterine perikarya inside the particular DRGs of gilts with uterine inflammation were not related to distinct regions [32]. Partly compatible to the current findings is another study, which reported a drop in ovarian perikarya (small, large) in particular areas of the CaMG of E_2-injected gilts [42] and a reduction of small ovarian perikarya in these ganglia of T-injected gilts [43].

In the CaMG of gilts after intrauterine bacterial treatment, a rise was found in the number of small and large uterine perikarya expressing DβH and NPY, the number of small uterine perikarya presenting positive staining for DβH and GAL, the number of small uterine perikarya immunoreactive to DβH but immunonegative to SOM and the number of large uterine perikarya with positive staining for DβH and VIP. In contrast, a decrease was noted in the number of small and large uterine perikarya expressing DβH but not GAL, in the number of small and large uterine perikarya immunonegative to DβH, but immunoreactive to NPY, as well as in the number of uterine perikarya presenting negative staining for the studied substances (except for small uterine perikarya immunonegative to DβH and NPY and large uterine perikarya immunonegative to DβH and GAL). Similarly, expression of GAL was increased in the uterine perikarya in the Th10-S4 DRGs of gilts with uterine inflammation used in the present study [32] and in the colon-projecting neurons in the CaMG of pigs with colitis [50]. A drop in the populations of ovarian perikarya possessing DβH, but not GAL, also took place in the porcine CaMG after long-term exposure to E_2 [42] and T [43]. Contrary to the current study, a drop in the TH, NPY and VIP expression in the colon-projecting neuronal cells in CaMG-led colitis [50] and a reduction in the numbers of ovarian perikarya with positive staining for DβH and NPY and immunoreactive to DβH, but not to SOM, was revealed in the CaMG after E_2 [42] and T [43] administration. The alterations in neurochemical characteristics of the CaMG neurons in gilts after intrauterine *E. coli* injection may be connected with the hormonal state and inflammatory factors having pro-/anti-inflammatory effects. It was reported that receptors for steroid hormones are present in ganglia neurons supplying the reproductive organs in females [41–43,51]. The localization of multiple receptors for PGE_2 [52], LTs [46] and TNF-α [47] was found in the DRG neurons. Although co-expression of steroid hormones and inflammatory mediators' receptors with particular neurotransmitters in the sympathetic ganglia perikarya have not yet been specified, it cannot be excluded that the differences found depend on the neurochemical characteristics of neurons and their size. This may result from varied density and cellular distribution of receptors for steroid hormones and inflammatory factors in the examined populations of neurons. E_2 acting by particular ERs-type in cellular line PC12, elevated or reduced transcriptional activity of the TH gene—catecholamine synthesizing enzyme [53]. Moreover, increased synthesis of SP and calcitonin gene-related peptide in rat uterine cervix-related DRG neurons is under the influence of the E_2-ERs system [54]. Further experiments should be carried out to recognize the co-localization of receptors for steroids and inflammatory mediators with particular neurotransmitters in the CaMG uterine perikarya as well as receptor bases of these substances' influence on neurochemical characteristics.

In the *E. coli* group, as found in the current experiment, there was a rise in the number of the CaMG uterine perikarya, which expressed NA, NPY, GAL or VIP, but not SOM. In contrast, a reduction in the populations of noradrenergic and non-noradrenergic perikarya occurred. It is proposed that these alterations may lead to disruptions in a diversity of uterine-sympathetic and non-sympathetic activities. It was revealed that NA exerts an effect on the contractility in healthy [18,19] and inflamed [11] porcine uterus. Similarly, under physiological conditions, NPY, GAL and VIP regulate this uterine activity [22,25,26]. Thus, it is possible that the above-mentioned neurotransmitters may be involved in the elimination of the exudate from the uterus with inflammation. Additionally, NA is able to affect uterine PGs synthesis [20,21], while NPY modulates the uterine blood supply [23,24]. Neuroprotective GAL function in the inflamed uterus may also be possible [55]. Moreover, this study revealed that the inflammatory uterine process led to changes in the uterine small and large perikaryal cell populations, which were partially dependent on their neurochemical features. This may suggest a different action of small and large uterine perikarya on the function of an inflamed uterus.

5. Conclusions

It was found that *E. coli*-induced inflammatory state of the uterus in the pig causes changes in spatial and neurochemical organization patterns of the CaMG neurons innervating the uterus. Alterations in the noradrenergic and non-noradrenergic uterine neuronal populations suggest that inflammation of the uterus may affect the function(s) of this organ by acting on the CaMG uterine neurons. However, further studies are needed to identify the mechanism of the effect of inflammation on uterine innervation. These findings confirm the use of the pig as an appropriate model for the research of the inflammatory states of the reproductive system as well as the importance of the inflammatory process as modulators of neuronal plasticity. Finally, the alterations in the expression of neurotransmitters in the CaMG uterine neurons during uterine inflammation can be used to develop neurotransmitter analogues that restore normal uterine function.

Author Contributions: Conceptualization, B.J.; methodology, B.J. and J.C.; formal analysis, B.J. and J.C.; investigation, B.J. and J.C.; data curation, B.J.; writing—original draft preparation, B.J.; writing—review and editing, B.J. and J.C.; supervision, B.J.; project administration, B.J.; funding acquisition, B.J. All authors have read and agreed to the published version of the manuscript.

Funding: The research was funded by the National Science Center, Poland (grant No. 2014/15/B/NZ5/O3572), as well as the Polish Academy of Sciences. Project financially co-supported by Minister of Science and Higher Education in the range of the program entitled "Regional Initiative of Excellence" for the years 2019-2022, Project No. 010/RID/2018/19, amount of funding 12.000.000 PLN.

Acknowledgments: We thank Marta Czajkowska, Joanna Kalinowska and Andrzej Pobiedziński for their technical assistance.

Conflicts of Interest: The authors declare no conflict of interest.

References

1. De Winter, P.J.J.; Verdonck, M.; De Kruif, A.; Devriese, L.A.; Haesebrouck, R. Bacterial endometritis and vaginal discharge in the sow: Prevalence of different bacterial species and experimental reproduction of the syndrome. *Anim. Reprod. Sci.* **1995**, *37*, 325–335. [CrossRef]
2. Mateus, L.; Lopes da Costa, L.; Diniz, P.; Ziecik, A.J. Relationship between endotoxin and prostaglandin (PGE2 and PGFM) concentrations and ovarian function in dairy cows with puerperal endometritis. *Anim. Reprod. Sci.* **2003**, *76*, 143–154. [CrossRef]
3. Roberson, J.; Moll, D.; Saunders, G. Chronic Staphylococcus aureus endometritis in a virgin gilt. *Vet. Rec.* **2007**, *161*, 821–822. [PubMed]
4. Tummaruk, P.; Kesdangsakonwut, S.; Prapasarakul, N.; Kaeoket, K. Endometritis in gilts: Reproductive data, bacterial culture, histopathology, and infiltration of immune cells in the endometrium. *Comp. Clin. Pathol.* **2010**, *19*, 575–584. [CrossRef]
5. Shynlova, O.; Lee, Y.H.; Srikhajon, K.; Lye, S.J. Physiologic uterine inflammation and labor onset: Integration of endocrine and mechanical signals. *Reprod. Sci.* **2013**, *20*, 154–167. [CrossRef]
6. Herath, S.; Fischer, D.P.; Werling, D.; Williams, E.J.; Lilly, S.T.; Dobson, H.; Bryant, C.E.; Sheldon, I.M. Expression and function of Toll-like receptor 4 in the endometrial cells of the uterus. *Endocrinology* **2006**, *147*, 562–570. [CrossRef]
7. Gabler, C.; Drillich, M.; Fischer, C.; Holder, C.; Heuwieser, W.; Einspanier, R. Endometrial expression of selected transcripts involved in prostaglandin synthesis in cows with endometritis. *Theriogenology* **2009**, *71*, 993–1004. [CrossRef]
8. Christoffersen, M.; Woodward, E.M.; Bojesen, A.M.; Petersen, M.R.; Squires, E.L.; Lehn-Jensen, H.; Troedsson, M.H. Effect of immunomodulatory therapy on the endometrial inflammatory response to induced infectious endometritis in susceptible mares. *Theriogenology* **2012**, *78*, 991–1004. [CrossRef]
9. Shao, C.Y.; Wang, H.; Meng, X.; Zhu, J.Q.; Wu, Y.Q.; Li, J.J. Characterization of the innate immune response in goats after intrauterine infusion of E. coli using histopathological, cytologic and molecular analyses. *Theriogenology* **2012**, *78*, 593–604. [CrossRef]
10. Jana, B.; Kucharski, J.; Dzienis, A.; Deptuła, K. Changes in prostaglandin production and ovarian function in gilts during endometritis induced by Escherichia coli infection. *Anim. Reprod. Sci.* **2007**, *97*, 137–150. [CrossRef]
11. Jana, B.; Jaroszewski, J.J.; Czarzasta, J.; Włodarczyk, M.; Markiewicz, W. Synthesis of prostacyclin and its effect on the contractile activity of the inflamed porcine uterus. *Theriogenology* **2013**, *79*, 470–485. [CrossRef] [PubMed]
12. Jana, B.; Czarzasta, J.; Jaroszewski, J. Synthesis of leukotrienes in porcine uteri with endometritis induced by infection with Escherichia coli. *Reprod. Fertil. Dev.* **2014**, *26*, 1007–1016. [CrossRef] [PubMed]
13. Jana, B.; Andronowska, A.; Kucharski, J. Nitric oxide mediates an inflammatory effect of Escherichia coli in the porcine uterus. *Polish J. Vet. Sci.* **2000**, *4*, 207–212.
14. Wasowicz, K.; Majewski, M.; Lakomy, M. Distribution of neurons innervating the uterus of the pig. *J. Auton. Nerv. Syst.* **1998**, *74*, 13–22. [CrossRef]
15. Wasowicz, K. Uterus-innervating neurons in porcine inferior mesenteric ganglion: An immunohistochemical characteristic. *Anat. Histol. Embryol.* **2003**, *32*, 154–160. [CrossRef]
16. Majewski, M.; Sienkiewicz, W.; Kaleczyc, J.; Mayer, B.; Czaja, K.; Lakomy, M. The distribution and co-localization of immunoreactivity to nitric oxide synthase, vasoactive intestinal polypeptide and substance P within nerve fibres supplying bovine and porcine female genital organs. *Cell Tissue Res.* **1995**, *281*, 445–464. [CrossRef]
17. Rytel, L.; Gonkowski, S.; Janowski, T.; Wojtkiewicz, J.; Pomianowski, A. The neurochemical characterization of parasympathetic nerve fibres in the porcine wall under physiological conditons and after exposure to bisphenol A (BPA). *Neurotox. Res.* **2019**, *35*, 867–882.
18. Kitazawa, T.; Nakagoshi, K.; Teraoka, H.; Taneike, T. 5-HT(7) receptor and beta(2)-adrenoceptor share in the inhibition of porcine uterine contractility in a muscle layer-dependent manner. *Eur. J. Pharmacol.* **2001**, *433*, 187–197. [CrossRef]

19. Markiewicz, W.; Jaroszewski, J.J. β2- and β3-adrenergic receptors stimulation relaxes porcine; myometrium in the peri-implantation period. *J. Anim. Sci.* **2016**, *94*, 4611–4618. [CrossRef]
20. Quaas, L.; Zahradnik, P. The effects of alpha- and beta-adrenergic stimulation on contractility and prostaglandin (prostaglandins E2 and F2a and 6-keto-prostaglandin F1 alpha) production of pregnant human myometrial strips. *Am. J. Obstet. Gynecol.* **1985**, *152*, 852–856. [CrossRef]
21. Skarzynski, D.J.; Uenoyama, Y.; Kotwica, J.; Okuda, K. Noradrenaline stimulates the production of prostaglandin f2alpha in cultured bovine endometrial cells. *Biol. Reprod.* **1999**, *60*, 277–282. [CrossRef] [PubMed]
22. Atke, A.; Henriksen, J.S.; Jacobsen, H.S.; Vilhardt, H. Characterization of the rat myometrial contractile response to neuropeptide Y. *Recept. Signal Transduct. Res.* **1996**, *16*, 25–38. [CrossRef]
23. Morris, J.L. Neuropeptide Y inhibits relaxations of the guinea pig uterine artery produced by VIP. *Peptides* **1990**, *11*, 381–386. [CrossRef]
24. Jovanovic, S.; Grbovic, L.; Jovanovic, A. Pregnancy is associated with altered response to neuropeptide Y in uterine artery. *Mol. Hum. Reprod.* **2000**, *6*, 352–360. [CrossRef]
25. Niiro, N.; Nishimura, J.; Hirano, K.; Nakano, H.; Kanaide, H. Mechanisms of galanin-induced contraction in the rat myometrium. *Br. J. Pharmacol.* **1998**, *124*, 1623–1632. [CrossRef] [PubMed]
26. Bryman, I.; Norström, A.; Lindblom, B.; Dahlström, A. Histochemical localization of vasoactive intestinal polypeptide and its influence on contractile activity in the non-pregnant and pregnant human cervix. *Gynecol. Obstet. Investig.* **1989**, *28*, 57–61. [CrossRef]
27. Annunziata, M.; Luque, R.M.; Durán-Prado, M.; Baragli, A.; Grande, C.; Volante, M.; Gahete, M.D.; Deltetto, F.; Camanni, M.; Ghigo, E.; et al. Somatostatin and somatostatin analogues reduce PDGF-induced endometrial cell proliferation and motility. *Hum. Reprod.* **2012**, *27*, 2117–2129. [CrossRef]
28. Wesselmann, U.; Czakanski, P.P.; Affaitati, G.; Giamberardino, M.A. Uterine inflammation as a noxious visceral stimulus: Behavioral characterization in the rat. *Neurosci. Lett.* **1998**, *246*, 73–76. [CrossRef]
29. Li, J.; Micevych, P.; McDonald, J.; Rapkin, A.; Chaban, V. Inflammation in the uterus induces phosphorylated extracellular signal-regulated kinase and substance P immunoreactivity in dorsal root ganglia neurons innervating both uterus and colon in rats. *J. Neurosci. Res.* **2008**, *86*, 2746–2752. [CrossRef]
30. Verma, N.; Rettenmeier, A.W.; Schmitz-Spanke, S. Recent advances in the use of *Sus scrofa* (pig) as a model system for proteomic studies. *Proteomics* **2011**, *11*, 776–793. [CrossRef]
31. Swindle, M.M.; Makin, A.; Herron, A.J.; Clubb, F.J., Jr.; Frazier, K.S. Swine as models in biomedical research and toxicology testing. *Proc. Natl. Acad. Sci. USA* **2012**, *109*, 16612–16617. [CrossRef] [PubMed]
32. Bulc, M.; Całka, J.; Meller, K.; Jana, B. Endometritis affects chemical coding of the dorsal root ganglia neurons supplying uterus in the sexually mature gilts. *Res. Vet. Sci.* **2019**, *124*, 417–425. [CrossRef] [PubMed]
33. Meller, K.; Całka, J.; Palus, K.; Jana, B. Inflammation-induced changes in expression of DβH, SP and GAL in the porcine uterine nerve fibres. In Proceedings of the 4th Winter Workshop of the Society for Biology of Reproduction "Central and local Regulations of Reproductive Processes", Zakopane, Poland, 3–4 February 2016.
34. Majewski, M.; Heym, C. The origin of ovarian neuropeptide Y (NPY)-immunoreactive nerve fibres from the inferior mesenteric ganglion in the pig. *Cell Tissue Res.* **1991**, *266*, 591–596. [CrossRef] [PubMed]
35. Neuhuber, W.L.; Sandoz, P.A.; Fryscak, T. The central projections of primary afferent neurons of greater splanchnic and intercostal nerves in the rat. A horseradish peroxidase study. *Anat. Embryol.* **1986**, *174*, 123–144. [CrossRef] [PubMed]
36. Peter, A.T.; Bosu, W.T.K. Effects of intrauterine infection on the function of the corpora lutea formed after first postpartum ovulations in dairy cows. *Theriogenology* **1987**, *127*, 593–609. [CrossRef]
37. Opsomer, G.; Gröhn, Y.T.; Hertl, J.; Coryn, M.; Deluyker, H.; de Kruif, A. Risk factors for post-partum ovarian dysfunction in high producing dairy cows in Belgium: A field study. *Theriogenology* **2000**, *53*, 841–857. [CrossRef]
38. Jana, B.; Kucharski, J.; Ziecik, A.J. Effect of intrauterine infusion of Escherichia coli on hormonal patterns in gilts during the oestrous cycle. *Reprod. Nutr. Dev.* **2004**, *44*, 37–48. [CrossRef]
39. Green, P.S.; Bishop, J.; Simpkins, J.W. 17 alpha-estradiol exerts neuroprotective effects on SK-N-SH cells. *J. Neurosci.* **1997**, *17*, 511–515.

40. Cardona-Gómez, G.P.; Mendez, P.; DonCarlos, L.L.; Azcoitia, I.; Garcia-Segura, L.M. Interactions of estrogens and insulin-like growth factor-I in the brain: Implications for neuroprotection. *Brain Res. Brain Res. Rev.* **2001**, *37*, 320–334. [CrossRef]
41. Papka, R.E.; Workley, M.; Usip, S.; Mowa, C.N.; Fahrenkrug, J. Expression of pituitary adenylate cyclase activating peptide in the uterine cervix, lumbosacral dorsal root ganglia and spinal cord of rats during pregnancy. *Peptides* **2006**, *27*, 743–752. [CrossRef]
42. Koszykowska, M.; Całka, J.; Szwajca, P.; Jana, B. Long-term estradiol-17β administration decreases the number of neurons in the caudal mesenteric ganglion innervating the ovary in sexually mature gilts. *J. Reprod. Dev.* **2011**, *57*, 62–71. [CrossRef] [PubMed]
43. Jana, B.; Rytel, L.; Czarzasta, J.; Całka, J. Reduction of the number of neurones in the caudal mesenteric ganglion innervating the ovary in sexually mature gilts following testosterone administration. *J. Neuroendocrinol.* **2013**, *25*, 826–838. [CrossRef] [PubMed]
44. Pieretti, S.; Mastriota, M.; Tucci, P.; Battaglia, G.; Trabace, L.; Nicoletti, F.; Scaccianoce, S. Brain nerve growth factor unbalance induced by anabolic androgenic steroids in rats. *Med. Sci. Sports Exerc.* **2013**, *45*, 29–35. [CrossRef] [PubMed]
45. Zhao, J.; Harada, N.; Okajima, K. Dihydrotestosterone inhibits hair growth in mice by inhibiting insulin-like growth factor-I production in dermal papillae. *Growth Horm. IGF Res.* **2011**, *21*, 260–267. [CrossRef]
46. Andoh, T.; Kuraishi, Y. Expression of BLT1 leukotriene B4 receptor on the dorsal root ganglion neurons in mice. *Brain Res. Mol. Brain Res.* **2005**, *137*, 263–266. [CrossRef] [PubMed]
47. Okubo, M.; Yamanaka, H.; Kobayashi, K.; Fukuoka, T.; Dai, Y.; Noguchi, K. Expression of leukotriene receptors in the rat dorsal root ganglion and the effects on pain behaviors. *Mol. Pain* **2010**, *17*, 57. [CrossRef] [PubMed]
48. Fischer, B.D.; Ho, C.; Kuzin, I.; Bottaro, A.; O'Leary, M.E. Chronic exposure to tumor necrosis factor in vivo induces hyperalgesia, upregulates sodium channel gene expression and alters the cellular electrophysiology of dorsal root ganglion neurons. *Neurosci. Lett.* **2017**, *13*, 195–201. [CrossRef] [PubMed]
49. Hu, S.; Peterson, P.K.; Chao, C.C. Cytokine-mediated neuronal apoptosis. *Neurochem. Int.* **1997**, *30*, 427–431. [CrossRef]
50. Wojtkiewicz, J.; Równiak, M.; Crayton, R.; Barczewska, M.; Bladowski, M.; Robak, A.; Pidsudko, Z.; Majewski, M. Inflammation-induced changes in the chemical coding pattern of colon-projecting neurons in the inferior mesenteric ganglia of the pig. *J. Mol. Neurosci.* **2012**, *46*, 450–458. [CrossRef]
51. Zoubina, E.V.; Smith, P.G. Distributions of estrogen receptors alpha and beta in sympathetic neurons of female rats: Enriched expression by uterine innervation. *J. Neurobiol.* **2002**, *52*, 14–23. [CrossRef]
52. Wise, H. Lack of interaction between prostaglandin E2 receptor subtypes in regulating adenylyl cyclase activity in cultured rat dorsal root ganglion cells. *Eur. J. Pharmacol.* **2006**, *27*, 69–77. [CrossRef] [PubMed]
53. Sabban, E.L.; Maharjan, S.; Nostramo, R.; Serova, L.I. Divergent effects of estradiol on gene expression of catecholamine biosynthetic enzymes. *Physiol. Behav.* **2010**, *9*, 163–168. [CrossRef] [PubMed]
54. Mowa, C.N.; Papka, R.E. The role of sensory neurons in cervical ripening: Effects of estrogen and neuropeptides. *J. Histochem. Cytochem.* **2004**, *52*, 1249–1258. [CrossRef] [PubMed]
55. Theodorsson, A.; Holm, L.; Theodorsson, E. Hypothermia-induced increase in galanin concentrations and ischemic neuroprotection in the rat brain. *Neuropeptides* **2008**, *1*, 79–87. [CrossRef]

animals

MDPI

Communication

Epithelial to Mesenchymal Transition (EMT) in a Laryngeal Squamous Cell Carcinoma of a Horse: Future Perspectives

Federico Armando [1], Francesco Godizzi [2], Elisabetta Razzuoli [3,*], Fabio Leonardi [4], Mario Angelone [4], Attilio Corradi [1], Daniela Meloni [3], Luca Ferrari [1] and Benedetta Passeri [1]

1 Pathology Unit, Department of Veterinary Science, University of Parma, Strada del Taglio 10, 43126 Parma, Italy; federico.armando@unipr.it (F.A.); attilio.corradi@unipr.it (A.C.); luca.ferrari@unipr.it (L.F.); benedetta.passeri@unipr.it (B.P.)

2 Department of Veterinary Science (DIMEVET), University of Milan, Via dell'Università 6, 26900 Lodi, Italy; francesco.godizzi@unimi.it

3 National Reference Center of Veterinary and Comparative Oncology (CEROVEC), Piazza Borgo Pila 39/24, 16129 Genoa, Italy; daniela.meloni@izsto.it

4 Department of Veterinary Science, Strada del Taglio 10, 43126 Parma, Italy; fabio.leonardi@unipr.it (F.L.); marioangelonevet@libero.it (M.A.)

* Correspondence: elisabetta.razzuoli@izsto.it; Tel.: +39-010-542274; Fax: +39-010-566654

Received: 13 November 2020; Accepted: 4 December 2020; Published: 7 December 2020

Simple Summary: Squamous cell carcinoma (SCC) is one of the most common cancers in horses, and it can arise at any site on the skin and mucosae. Recent studies associated equine papillomavirus type 2 (EcPV2) infections with this type of cancers of the oral tract and genitals. Larynx and pharynx are frequently recognized as sites of SCC. In humans, squamous cell carcinoma of the larynx (SCCL) is a common cancer associated with papilloma virus (PV) infection and epithelial to mesenchymal transition (EMT). EMT can occur under different biological conditions, upon the same programmed changes: embryogenesis and organ development fibrosis, wound healing, and cancer metastases. This work reports for the first time in a SCCL of a horse a wide immunohistochemical EMT characterization, by analyzing main epithelial markers (E-cadherin, β-catenin, and pan-cytokeratin AE3/AE1), main mesenchymal markers (N-cadherin and vimentin), and the main EMT-related transcription factors (TWIST-1, ZEB-1, and HIF-1α). This work illustrates an example of tumor cell adaptation during the metastatic process in the equine SCCL, taking also into consideration the potential influence provided by EcPV2 oncoproteins on the EMT process.

Abstract: Squamous cell carcinoma (SCC) is one of the most frequent tumors of skin and muco-cutaneous junctions in the horse. Equine papillomavirus type 2 (EcPV2) has been detected in equine SCC of the oral tract and genitals, and recently also in the larynx. As human squamous cell carcinoma of the larynx (SCCL), it is strongly etiologically associated with high-risk papillomavirus (h-HPV) infection. This study focuses on tumor cells behavior in a naturally occurring tumor that can undergo the so-called epithelial to mesenchymal transition (EMT). A SCCL in a horse was investigated by immunohistochemistry using antibodies against E-cadherin, pan-cytokeratin AE3/AE1, β-catenin, N-cadherin, vimentin, ZEB-1, TWIST, and HIF-1α. EcPV2 DNA detection and expression of oncogenes in SCC were investigated. A cadherin switch and an intermediate filaments rearrangement within primary site tumor cells together with the expression of the EMT-related transcription factors TWIST-1, ZEB-1, and HIF-1α were observed. DNA obtained from the tumor showed EcPV2 positivity, with E2 gene disruption and E6 gene dysregulation. The results suggest that equine SCCL might be a valuable model for studying EMT and the potential interactions between EcPV2 oncoproteins and the EMT process in SCCL.

Keywords: horse; squamous cell carcinoma of the larynx (SCCL); epithelial to mesenchymal transition (EMT); equine papillomavirus type 2 (EcPV2)

---

1. Introduction

In horses and other equids, squamous cell carcinoma (SCC) is the most common malignant skin neoplasia, and accounts for 7–37% of equine skin lesions [1]. SCC can arise at any site on the skin and mucosa. However, non-pigmented skin and muco-cutaneous junctions have been reported to be preferential SCC sites [2,3]. Larynx and pharynx are reported to be sites of SCC development in horses. Equine oral SCC is invasive and lymph node metastases were reported in 30% of horses with oral SCC [4]. The longest survival time reported in 11 horses with laryngeal SCC was 4 months [4]. Squamous cell carcinoma of the larynx (SCCL) is the sixth most common human cancer and has strong etiologic association with smoking [5] and also with high-risk human papillomavirus (h-HPV) infection [6]. Indeed, about 26.6% of patients with SCCL are h-HPV positive. Particularly, h-HPV16 is the most common type (19.8%). Interestingly, equine papillomavirus type 2 (EcPV2) also has been recently associated with SCC of the penis, vulva, and stomach [7–10], as well as it has been recently detected in equine SCCL [11]. These data suggest that the horse is a good spontaneous animal model for studying the pathogenic mechanisms of this tumor in humans.

The transition of epithelial cells to a mesenchymal phenotype, the so-called epithelial to mesenchymal transition (EMT), can occur under different biological conditions, following the same programmed changes: embryogenesis and organ development [12], fibrosis [13], wound healing [14], and cancer metastases [15]. This process is characterized by decreased expression of the typical epithelial proteins (i.e., E-cadherin and cytokeratin) and, due to a prompt activation of the so-called "master genes regulators" (i.e., TWIST, ZEB, SNAIL1, SLUG), a mesenchymal phenotype is gradually acquired [16]. Epithelial cells loosen cell-to-cell adhesion structures, modulate their polarity, and rearrange their cytoskeleton, with intermediate filaments typically switching from cytokeratins to vimentin [17]. Moreover, in cells undergoing EMT, a cadherin switch takes place and is characterized by decreased E-cadherin and transiently increased N-cadherin expression [18–20]. Extensive studies focused on this event and proposed to use it for diagnostic, prognostic, and even therapeutic approaches [21]. To the authors' knowledge, the current work reports, for the first time, a wide immunohistochemical EMT characterization in a SCCL of a horse, by analyzing main epithelial markers (E-cadherin, β-catenin, and cytokeratin), main mesenchymal markers (N-cadherin and vimentin), and the main EMT-related transcription factors (TWIST-1, ZEB-1, and HIF-1α). Therefore, naturally occurring tumors in domestic and farm animals represent a unique opportunity to study cancer in vivo [14]. The authors present these results from an equine SCCL in order to: (1) provide preliminary results for setting future studies on the equine species as suitable model to study EMT in human cancers, not only for penile carcinomas [14], but also for other mucosal carcinoma such as oral carcinoma, thus opening new perspectives for future studies in this field; (2) provide hints for future studies in horses regarding the potential interaction between EcPV oncoproteins and the EMT process.

2. Materials and Methods

2.1. Post-Mortem Analysis

The horse was euthanized due to poor clinical conditions and necropsy was immediately performed and followed by proper tissue sampling for histopathological evaluation.

2.2. *Histolopathological Evaluation*

Collected organ specimens were formalin-fixed (10% *v*/*v*, pH 7.4) and paraffin-embedded (Bio-Plast 56–58 °C, Bio-Optica, Milano, Italy). Samples were routinely processed, and 5-μm thick sections were stained using Mayer's hematoxylin and eosin (HE).

2.3. *Immunohistochemistry*

The primary antibodies were titrated according to the manufacturer's recommendations. Among the antibodies employed in this study, only vimentin and TWIST-1 are reported by the manufacturer as validated against equine tissues. The other antibodies are validated against either canine (E-cadherin, β-catenin, ZEB-1, HIF-1α, vimentin, and TWIST-1) or bovine (N-cadherin, pan-cytokeratin AE1/AE3) tissues. Therefore, different tissues (i.e., digestive tract, heart, kidney, and skin) were sampled from another horse which was unrelated to the study and did not present significant histopathological findings. In addition, heart and skin were sampled from a cow without significant histopathological findings, and digestive tract, kidney, and skin from a dog without significant histopathological findings. The canine and bovine tissues were examined together with the equine tissues, in order to obtain a preliminary evaluation of the antibody cross-reactivity on the equine species. Only those primary antibodies exhibiting cross-reactivity with the expected target tissue from the control canine or bovine tissues, compared to the equine tissues, were employed against the horse neoplastic tissue, namely the antibodies against E-cadherin, β-catenin, ZEB-1, HIF-1α, vimentin, TWIST-1, N-cadherin, and pan-cytokeratin AE1/AE3. Once that cross-reactivity was confirmed, the tumor samples from the horse were always examined together with the appropriate controls (canine or bovine tissues and the equine counterpart). Immunohistochemistry was performed as previously described [22] with minor modifications. Briefly, after dewaxing–rehydration, tissue sections were exposed to antigen retrieval; then, sections were cooled at room temperature for 20 min before being soaked into 3% H_2O_2 for 12 min. Slides were rinsed twice in PBS, pH 7.4, followed by serum blocking with normal goat or rabbit serum. Incubation with primary antibody was carried out overnight at 4 °C. After being washed twice in PBS, pH 7.4, the slides were incubated for 30 min with a biotinylated goat anti-rabbit, or a goat anti-mouse, or a rabbit anti-goat IgG antibody. Afterwards, an avidin–biotin complex (ABC) peroxidase kit (Vectastain, Elite, ABC-Kit PK-6100, Vector Labs, Burlingame, CA, USA) and 3′3′-diaminobenzidine (DAB) system (DAB-Kit-SK4100, Vector Labs) were used for the detection of antigen–antibody reactions. Nuclei were counterstained with Mayer's hematoxylin. For negative controls, the primary antibodies were replaced by rabbit or goat serum, or Balb/c ascitic fluid at corresponding concentrations. Antibody details and positive controls are reported in Table 1. EMT markers expression was qualitatively analyzed in each whole section.

Table 1. Details of the antibodies used for immunostaining, including primary antibody, host species, clonality, epitope retrieval method, blocking serum, dilution of primary antibody, secondary antibody, and positive control.

Target Antigen	Antibody Details/Clone	Blocking Serum	Heat Induced Epitope Retrieval (HIER)	Primary Antibody Dilution	Secondary Antibody (1:200)	Positive Control
E-cadherin	Monoclonal mouse anti-human, IgG2a, clone 36/E-Cadherin BD 610181 (BD transduction laboratories)	Goat	Microwave 400 W, 3 cycles, 5 min. each, sodium citrate buffer, pH 6.0	1:100	biotinylated goat anti-mouse IgG (BA-1000-Vector Labs)	Horse, skin Dog, skin
Pan-ckytokeratin AE3/AE1	Monoclonal mouse anti-human IgG1 SC-81714 (Santa Cruz Biotechnology)	Goat	Microwave 400 W, 3 cycles, 5 min. each, sodium citrate buffer, pH 6.0	1:100	biotinylated goat anti-mouse IgG (BA-1000-Vector Labs)	Horse, skin Cow, skin

Table 1. *Cont.*

Target Antigen	Antibody Details/Clone	Blocking Serum	Heat Induced Epitope Retrieval (HIER)	Primary Antibody Dilution	Secondary Antibody (1:200)	Positive Control
β-catenin	Polyclonal goat anti-human, IgG, AB0095-200 (Sicgen)	Rabbit	Microwave 400 W, 3 cycles, 5 min. each, sodium citrate buffer, pH 6.0	1:3000	biotinylated rabbit anti-goat IgG (BA-1000-Vector Labs)	Horse, intestine Dog, intestine
N-cadherin	Polyclonal rabbit anti-human, IgG, 22018-1-AP (Proteintech)	Goat	Microwave 400 W, 3 cycles, 5 min. each, sodium citrate buffer, pH 6.0	1:3000	biotinylated goat anti-rabbit IgG (BA-1000-Vector Labs)	Horse, heart Cow, heart
Vimentin	Monoclonal mouse anti-human IgG1, Clone V9 (Dako)	Goat	Microwave 400 W, 3 cycles, 5 min. each, sodium citrate buffer, pH 6.0	1:100	biotinylated goat anti-mouse IgG (BA-1000-Vector Labs)	Horse, heart (vessels)
ZEB-1	Polyclonal rabbit anti-human, IgG, LS-C31478 (LSBio)	Goat	Microwave 400 W, 3 cycles, 5 min. each, sodium citrate buffer, pH 6.0	1:200	biotinylated goat anti-rabbit IgG (BA-1000-Vector Labs)	Horse, kidney Dog, kidney
TWIST-1	Polyclonal rabbit anti-human, IgG, Orb329955 (Biorbyt)	Goat	Microwave 400 W, 3 cycles, 5 min. each, sodium citrate buffer, pH 6.0	1:800	biotinylated goat anti-rabbit IgG (BA-1000-Vector Labs)	Horse, kidney
HIF-1α	Polyclonal rabbit anti-human, IgG, NB100-449 (NovusBio)	Goat	Microwave 400 W, 3 cycles, 5 min. each, sodium citrate buffer, pH 6.0	1:1000	biotinylated goat anti-rabbit IgG (BA-1000-Vector Labs)	Horse, kidney Dog, kidney

2.4. Detection of EcPV2 and Evaluation of Oncogene Expression

DNA and RNA extraction was performed to check EcPV2 presence and expression of oncogenes in SCCL. One sample of kidney obtained from a healthy horse was used as negative control. To prevent cross-contamination, a different blade was used for each sample. Four sections (5 μm) were obtained from FFPE samples for total acid nucleic extraction, which was performed using an AllPrep DNA FFPE Kit (Qiagen, Hilden, Germany) and an AllPrep RNA FFPE Kit (Qiagen, Hilden) in accordance with the manufacturer's instructions.

During extraction, samples were treated with DNase (RNase-Free DNase set, Qiagen). To evaluate DNA amplifiability, β-actin gene amplification was performed [7]; then, EcPV2-L1 DNA presence was tested using a Real-Time protocol previously described [7]. Briefly, TaqMan® probe-based real-time PCR (CustomProbe 2× qPCR Master Mix, Canvax, Cordoba, Spain, cat. E0339) was performed using a CFX96™ Real-Time System (Bio-Rad, Rome, Italy). The reaction mix had a final concentration of 1× TaqMan® master mix, 200 nM probe, 100 nM each primer combination, and 200 ng of DNA. Reverse transcription (RT) was performed using a SuperScript™ IV VILO™ Master Mix with ezDNase™ Enzyme (Invitrogen, ThermoFisher Scientific (Waltham, MA, USA), cat. 11766050) according to the manufacturer's instructions; cDNA was used to check gene expression by TaqMan® probe-based real-time PCR (CustomProbe 2× qPCR Master Mix, Canvax, Spain, cat. E0339) performed using a CFX96™ Real-Time System (Bio-Rad, Paris, France). The PCR mix (25 μL) had a final concentration of 1× TaqMan® master mix, 200 nM probe, 100 nM each primer combination, and 5 μL of template (cDNA or DNA and RNA used as negative control to exclude possible contamination by genomic DNA). The thermal profile used for amplification was the following: 95 °C for 10 min, followed by 40 cycles at 95 °C for 15 s, and 60 °C for 60 s for annealing/extension and detection of the fluorescence signal. The fluorescence threshold limit was set automatically. Moreover, the expression of E2, E6, and L1 genes was evaluated by RT-real-time PCR using primers and probes (Table 2) and a protocol described by Porcellato and co-workers [7]. Briefly, RT was performed using a SuperScript™ IV VILO™ Master Mix (Invitrogen, ThermoFisher Scientific) according to the manufacturer's instructions. The cDNA was used to check gene expression: 5 μL of template were added to 20 μL of PCR mix at a final concentration of 1× master mix (iTaq Universal Probes Supermix, Bio-Rad, Italy), 200 nM probe, and 100 nM each primer combination. RNA was used as control to exclude possible contamination by EcPV2 genomic DNA. Nuclease-free water was used as negative control. Each sample was tested in triplicate.

Table 2. Primer set and probes for real-time PCR and RT-real-time PCR.

Gene	Primers	Accession Number	Amplicon (Base Pairs)
Ec-PV2-E2	For: 5′-AAAAGGGAGGGTACGTTGTC-3′ Rev: 5′-CCTGGTAGTAGACATGCTGC-3′	NC_012123.1	90
Ec-PV2-E6	For: 5′-CGTTGGCCTTCTTTGCATCT-3′ Rev: 5′-AGGTTCAGGTCTGCTGTGTT-3′	NC_012123.1	81
Ec-PV2-L1	For: 5′-TTGTCCAGGAGAGGGGTTAG-3′ Rev: 5′-TGCCTTCCTTTTCTTGGTGG-3′	NC_012123.1	81
pEc-PV2-E2	FAM-GCCAAGACAGCCACGACGCCAT-TAMRA	NC_012123.1	22
pEc-PV2-E6	FAM-CCGTGTGGCTATGCTGATGACATTTGG-TAMRA	NC_012123.1	27
pEc-PV2-L1	FAM-CGTCCAGCACCTTCGACCACCA-TAMRA	NC_012123.1	22

3. Results

3.1. Post-Mortem and Microscopical Analysis

A 17-year-old female, 550 Kg, Maremmano Horse was referred with severe clinical signs due to a dorso-cranial dislocation of the epiglottis which caused the reduction of 70% of the laryngeal lumen, diagnosed by an endoscopic exam, performed under sedation with butorphanol combined with detomidine [23]. The animal was humanely euthanized due to the poor condition. Necropsy revealed a locally expansive, multilobular, white-yellowish, firm mass localized at the base of the larynx. The lesion was partially ulcerated, with irregular margins and central necrotic areas (Figure 1). The neoplasia expanded to the nearby tissues and regional lymph nodes were markedly enlarged. Microscopically, the laryngeal tumor was completely effacing the submucosa, non-encapsulated, densely cellular, and poorly demarcated. Neoplastic cells were variably arranged in anastomosing bands and chords occasionally forming lobules embedded in a moderate amount of fibrous collagen stroma. Neoplastic cells were large, variably from polygonal to spindle-shaped with indistinct cell borders and an intermediate to high nuclear/cytoplasmic ratio (Figure 2). The cytoplasm was moderate and eosinophilic nuclei were large, round to oval, with vesicular chromatin and 1 or 2 round, basophilic nucleoli. Anisocytosis and anisokaryosis were high, and mitoses ranged from 0–2 per HPF (400×). Multifocally, there were wide areas of coagulative necrosis within the tumor. A moderate amount of small mature lymphocytes and plasma cells were found within and surrounding the tumor area. Remarkably, scattered tumor cells displayed a more prominent spindled shape. Interestingly, the regional lymph node was 80% effaced by necrotic and metastatic events (Figure 2). Numerous neoplastic cells with a morphology and histological pattern similar to those in the laryngeal tumor were found among resident lymphoid cells. Given this aggressive behavior of the neoplasia and this particular phenotypical change in morphology of the neoplastic cells which were variably from polygonal to elongated, an immunohistochemical panel for the EMT phenomenon was performed.

3.2. Cadherin Switch and Intermediate Filaments Rearrangements Suggest an EMT Phenomenon in Equine Laryngeal Squamous Cell Carcinomas

Immunohistochemical analysis of primary site tumor cells revealed an increased number of cells with cytoplasmic E-cadherin expression rather than membranous, together with a gradual overall loss of cells expressing this adhesion molecule moving towards the invasive front of the tumor (Figure 3). Interestingly, the number of cells expressing N-cadherin resulted to be increased within primary site tumor cells, even though it was expressed at nuclear level instead of being membranous (Figure 3). Interestingly, moving from the tumor center towards the invasive front, a decreased number of cells expressing cytokeratin was observed, while scattered neoplastic cells with a prominent mesenchymal morpholgy acquired cytoplasmic vimentin expression (Figure 3). Noteworthy, numerous neoplastic cells were scattered throughout the tumor area and expressed the EMT-related transcription factors TWIST-1, ZEB-1, and HIF-1α (Figure 4). In addition, β-catenin was found to be frequently expressed at a nuclear level, or (less frequently) cytoplasmically, rather than being membraneous (Figure 3). On the other hand, immunohistochemical analysis of the regional lymph node (medial retropharingeal lymph node) revealed a multifocal strong cytokeratin expression together with a more frequent membranous E-cadherin immunolabeling (Figure 5). Neoplastic cells did not express vimentin within the lymph nodes, but still expressed N-cadherin at nuclear level (Figure 5). Moreover, we detected a slight decrease of the number of neoplastic cells expressing the EMT-related transcription factors (TWIST-1, ZEB-1, and HIF-1α) and nuclear β-catenin (Figures 5 and 6). These findings demonstrate that the morphological changes and the aggressive behavior are most likely due to the EMT process activated within these tumor cells. Considering that the current literature suggests an emerging role for EcPV2 in several equine squamous cell carcinomas [11] and given that a correlation between SSCL and h-HPV infection [6] was demonstrated also in humans, we investigated whether, also in this case, the papillomavirus infection played a role, making the equine species a promising model for this type of tumor.

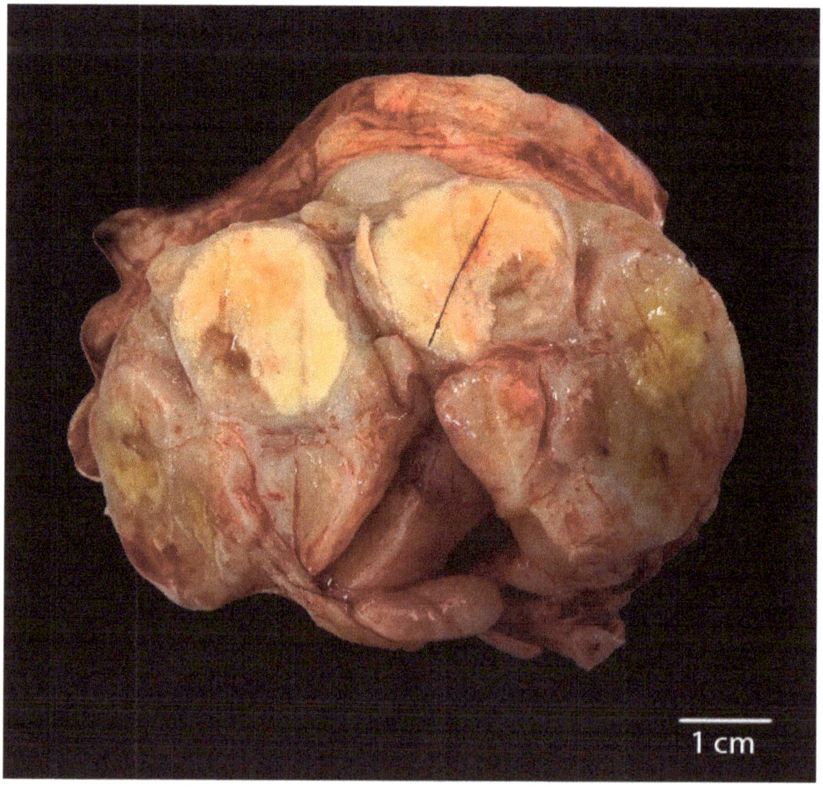

Figure 1. Cutting surface from the laryngeal region of the horse tumor sampled during the post-mortem analysis. A white to yellowish multilobular mass, located at the basis of the larynx, is partially ulcerated and shows multifocal central necrotic areas.

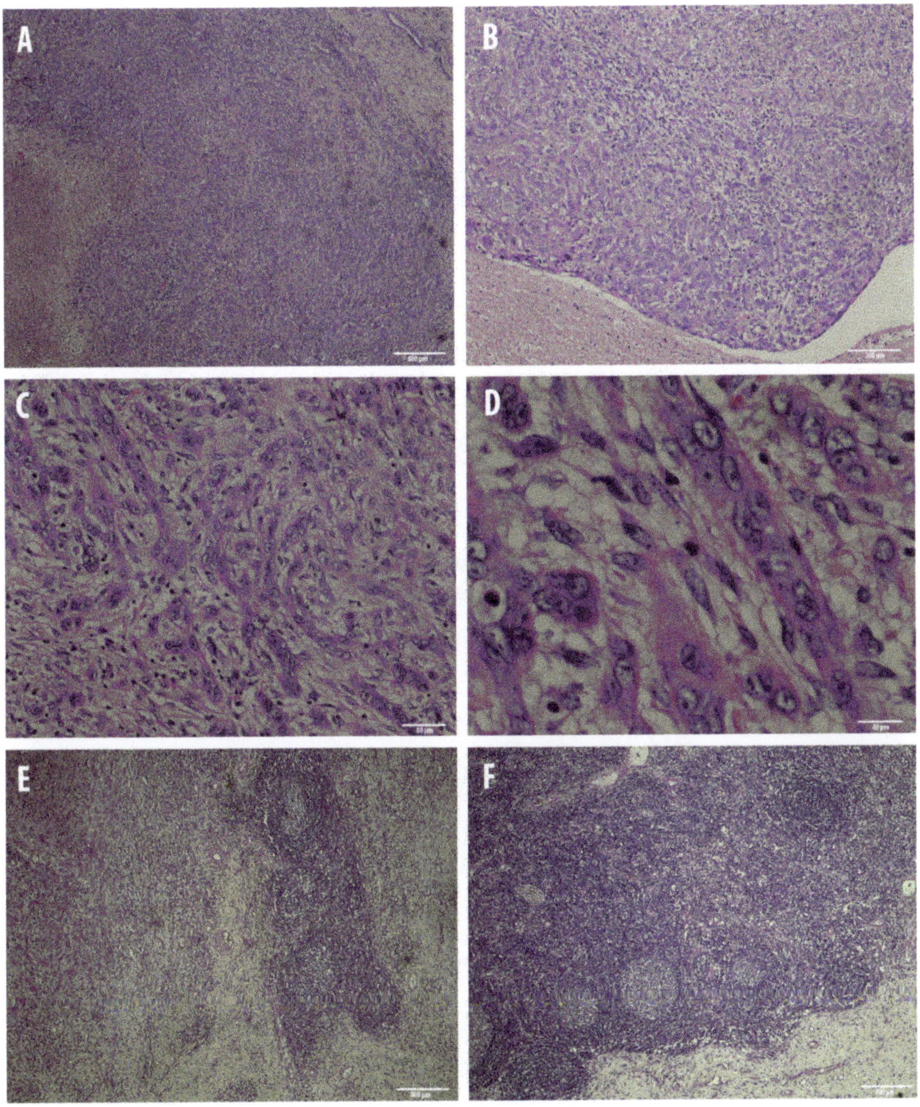

Figure 2. Overview of the histological findings of the squamous cell carcinoma of the larynx (**A**–**D**) and lymph node metastasis (**E**,**F**). (**A**): Neoplastic cells embedded in a moderate amount of fibrous collagen stroma are completely effacing the submucosa; of note, the multifocal necrotic areas within the tumor (4×, haematoxilyn–eosin, H&E). (**B**): neoplastic cells are variably arranged in anastomosing bands and chords occasionally forming lobules (10×, haematoxilyn–eosin, H&E). (**C**): Arborizing chords of variably differentiated squamous cells filling the submucosa and invading the nearby tissue (20×, haematoxilyn–eosin, H&E). (**D**): highly pleomorphic cell population with elongated, spindled cells often present in the invasive front of the tumor (40×, haematoxilyn–eosin, H&E). (**E**): few hyperplastic lymphoid follicles on the right with numerous invasive metastatic cells on the left (4×, haematoxilyn–eosin, H&E). (**F**): normal activated lymph node architecture scattered within the metastatic organ (4×, haematoxilyn–eosin, H&E).

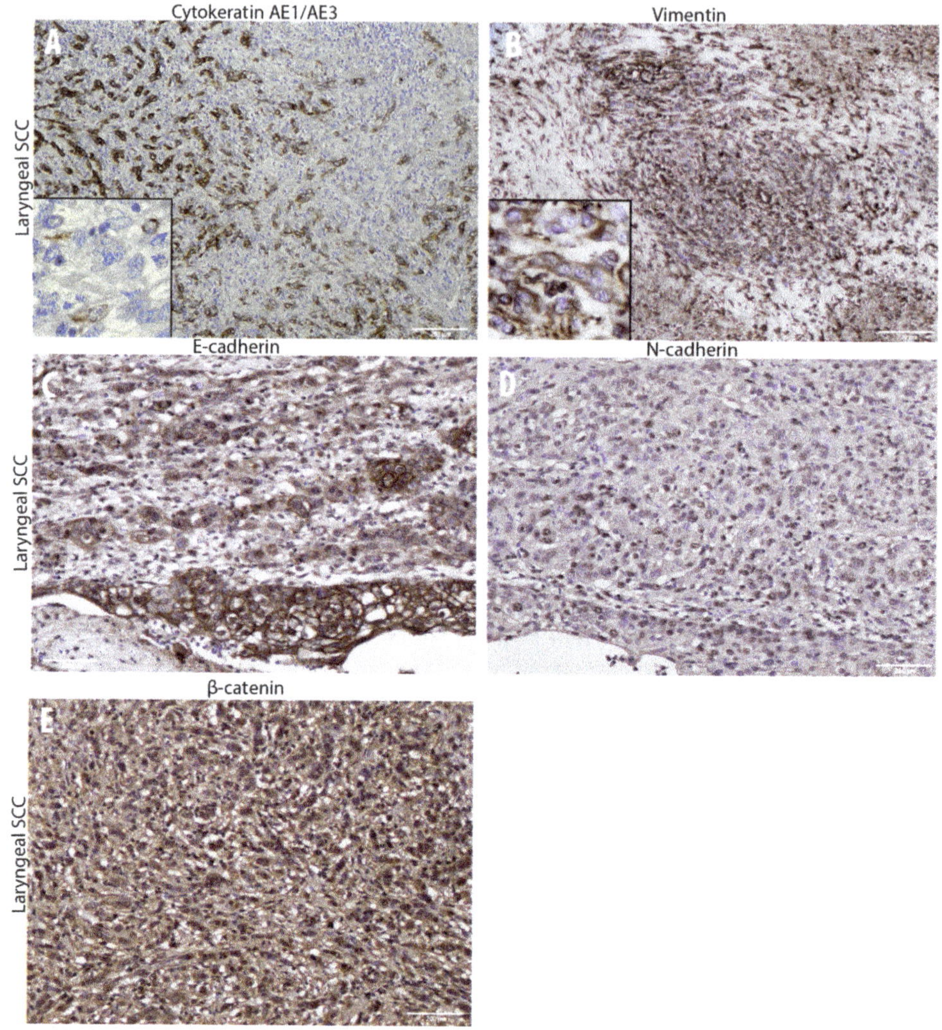

Figure 3. Overview of the immunohistochemical characterization of structural and adhesion molecules (**A**–**E**) of the laryngeal squamous cell carcinoma that supports an Epithelial to Mesenchymal transition (EMT) within the tumor. (**A**) immunohistochemistry for pan-cytokeratin AE1/AE3 displays a gradual cytoplasmic loss (insert) of this epithelial marker within neoplastic cells (10×). (**B**): Immunohistochemistry for vimentin reveals multifocal neoplastic cells characterized by elongated shape with a diffuse cytoplasmic expression (insert) of this mesenchymal marker (10×). (**C**): Immunohistochemistry displays a severe loss of membranous E-cadherin expression comparing the normal/dysplastic epithelium with the tumor cells invading the underlying tissue (10×). (**D**): A high number of nuclear N-cadherin immunolabeled neoplastic cells are present within the tumor (10×). (**E**): Immunohistochemistry reveals a moderate number of β-catenin cytoplasmic immunolabeled neoplastic cells together with numerous neoplastic cells also showing nuclear immunolabeling (10×).

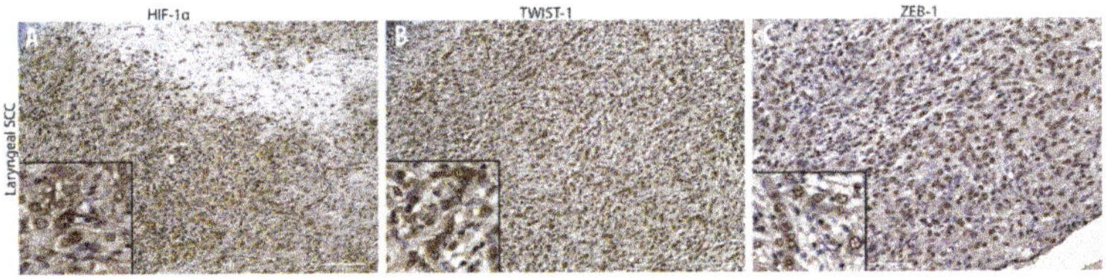

Figure 4. Overview of the immunohistochemical characterization of transcription factors (**A**–**C**) of the laryngeal squamous cell carcinoma that supports an Epithelial to Mesenchymal Transition (EMT) within the tumor. Immunohistochemical analysis reveals numerous nuclear immunolabeled neoplastic cells expressing HIF-1α (**A**, insert, 10×), TWIST-1 (**B**, insert, 10×), and ZEB-1 (**C**, insert, 10×) within the laryngeal squamous cell carcinoma.

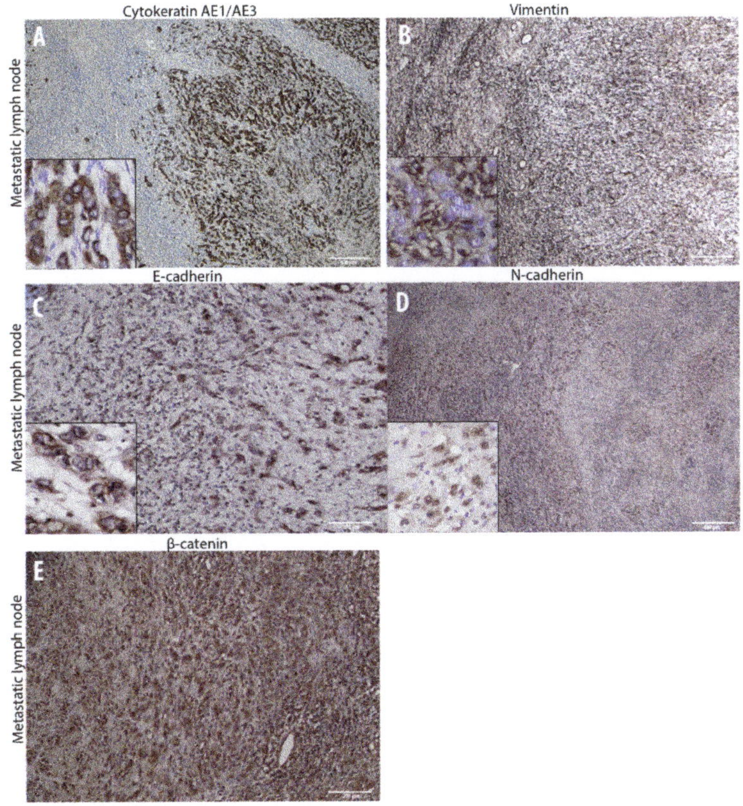

Figure 5. Overview of the immunohistochemical characterization of structural and adhesion molecules (**A**–**E**) of the metastatic lymph node that supports a Mesenchymal to Epithelial Transition (MET) within the tumor cells. (**A**) immunohistochemistry for pan-cytokeratin AE1/AE3 within the metastatic lymph node reveals a diffuse cytoplasmic (insert) immunolabeling in numerous metastatic neoplastic cells surrounded by resident lymphoid cells (4×). (**B**): Immunohistochemistry for vimentin reveals no positive immunolabeled metastatic neoplastic cells (insert) surrounded by positive stromal resident cells (4×). (**C**): Immunohistochemistry displays a gradual re-acquisition of membranous E-cadherin expression (insert) in metastatic neoplastic cells (10×). (**D**): A moderate number of nuclear N-cadherin immunolabeled neoplastic cells are present in the metastatic lymph node (4×). (**E**): Immunohistochemistry reveals a moderate number of β-catenin immunolabeled metastatic neoplastic cells showing either a cytoplasmic or a nuclear expression (10×).

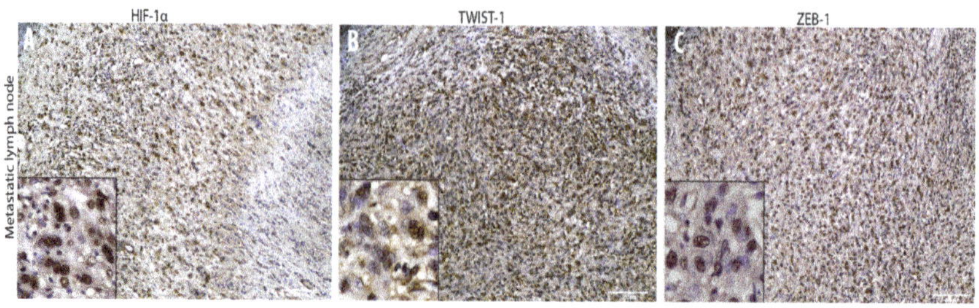

Figure 6. Overview of the immunohistochemical characterization of transcription factors (**A–C**) of the metastatic lymph node that supports a Mesenchymal to Epithelial Transition (MET) within the tumor cells. Immunohistochemical analysis reveals a moderate number of nuclearly immunolabeled metastatic neoplastic cells expressing HIF-1α (**A**, insert, 10×), TWIST-1 (**B**, insert, 10×), and ZEB-1 (**C**, insert, 10×) within the metastatic lymph node.

3.3. EcPV2 Detection Suggests a Potential Role in Equine Laryngeal Squamous Cell Carcinomas

DNA obtained from the tumor sample was amplifiable (Table 2) and it showed positivity for EcPV2 L1 (Table 2), with a mean Cq of 22.5 ± 0.17, obtained as mean of three replicates. The negative sample was amplifiable, but negative for EcPV2. Regarding gene expression, only the E6 gene was expressed (Table 3), with a mean Cq of 30.4 ± 0.24.

Table 3. Results of equine papillomavirus type 2 (EcPV2) detection and expression.

Sample ID	Type of Sample	RT-PCR		RT-qPCR	
		B2M	L1	E2	E6
1	Epiglottis tumor	21.1 ± 0.62	26.5 ± 0.17	>40	30.4 ± 0.24
2	Kidney negative sample	21.6 ± 0.27	-	>40	>40

Data are expressed as mean Cq (quantitation cycle) ± SD (standard deviation) of 3 replicates; B2M = beta-2-microglobulin.

4. Discussion

Based on the current findings, a laryngeal squamous cell carcinoma with a regional lymph node metastasis was diagnosed. The wide immunohistochemical panel used in the current study allowed us to detect a gradual decreased number of cells expressing epithelial markers within the primary tumor site invasive front, together with a gradual increased number of cells expressing mesenchymal markers and key transcription factors for the EMT process. On the other hand, lymph node metastasis revealed a moderate switch in markers expression compared to the primary carcinoma site that might suggest a partial/incomplete mesenchymal to epithelial transition (MET) process. Specifically, lowered numbers of cells expressing E-cadherin and cytokeratin were detected among primary site cancer cells together with aberrant N-cadherin expression and a moderate number of cells immunolabeled for vimentin. These findings were further supported by numerous cells displaying nuclear staining for TWIST-1, ZEB-1, β-catenin, and HIF-1α. In contrast, the metastatic lymph node displayed a partial reversion of the aforementioned epithelial and mesenchymal markers, thus suggesting an MET process. It is well reported that TWIST-1 is considered as the main regulator of EMT [24,25] and is up-regulated in a large number of malignant tumors determining the onset of the metastatic process, via promoting invasiveness in both spontaneous and experimental models [25,26]. Noteworthy, the tumor in the present study had both a high number of cells expressing TWIST-1 and a very low number of cells expressing membranous E-cadherin or, interestingly, the cells exhibited cytoplasmic internalization of the protein. This aberrant cytoplasmic expression has been recently related in equine penile carcinoma to a more aggressive behavior due to AKT/MAPK pathway activation [27,28]. Interestingly, the decreased number of cells expressing E-cadherin (E-cadherin loss) was replaced by the increased number of cells expressing nuclear N-cadherin. The aforementioned results might be in line with a study on human nasopharyngeal carcinoma by Luo and colleagues, reporting a correlation between nuclear N-cadherin and a poorer prognosis [29]. Another aberrant localization found in the current case is represented by β-catenin in both primary and metastatic sites. Normally, the membranous E-cadherin/β-catenin complex maintains the intercellular tight junction and minor free β-catenin cytoplasmic molecules are controlled by multiprotein complexes, while on the other hand, nuclear localization of β-catenin is essential for the progression of various human cancers, such as nasopharyngeal carcinoma, via transcriptional upregulation of downstream genes [30]. These findings are in line with the low number of cells expressing β-catenin at a cytoplasmic level, compared to the higher number of cells expressing β-catenin at a nuclear level found in this study. It is well documented that solid tumors generally have a hypoxic microenvironment [31]. Previous studies suggested that moderate hypoxic conditions might trigger an EMT process via HIF-1α, leading different human cancer cells to significantly increase their invasiveness [32]. Noteworthy, Yang and colleagues demonstrated that HIF-1α directly binds to the hypoxia response element (HRE) in the TWIST promoter, regulating the expression of this transcription factor [33]. These results seem to be in agreement with the HIF-1α expression found in this study. However, all the results regarding the EMT process obtained at a protein level in the present study by immunohistochemistry will need to be further confirmed also at a genetic level in future studies aimed at increasing the knowledge on these aspects by focusing also on the possible role of some microRNA (miRNA) families. The molecular analysis which demonstrated the presence of EcPV2 and its oncogene expression is in agreement with previous studies [11,34,35]. In particular, our data showed E6 but not E2 expression. The lack of E2 expression suggests virus genome integration and loss or disruption of the E2 gene. This can cause deregulation of E6 expression and, in turn, the increase of this event triggers cancer progression [36]. Moreover, a previous study in humans demonstrated a role for E6 in innate immune gene repression [36]. It is also important to consider the potential role of papillomavirus oncoproteins in triggering the EMT process. Interestingly, Liu and colleagues reported that E6 and E7 oncoproteins enhance the expression of HIF-1α, as well as of ZEB-1, SNAIL-1, SLUG, and TWIST-1 in non-small cell lung cancer (NSCLC) cells, thus promoting the EMT process [37]. Recently, a study using human lung samples confirmed the E7 oncoprotein role in promoting EMT in human lung cancers, reporting correlations with E7 and E-cadherin, N-cadherin, and TGF-β expression [38]. According to the authors, the findings of the present work about EcPV oncoproteins and the EMT-related transcription factors and structural/adhesion proteins are in line with the literature and might represent a promising starting point to be further investigated.

5. Conclusions

In conclusion, this interesting case of equine metastatic SCCL provides an example of tumor cell adaptation during the metastatic process in the equine species, taking also into account the possible influence of EcPV2 oncoproteins on the EMT process. This is an opportunity to propose the equine species in future studies for evaluation of the potential interactions between EcPV2 oncoproteins and the EMT process both in human and animal cancers, thus opening new study perspectives in this field.

Author Contributions: Conceptualization: F.A., E.R., and B.P.; Data curation: F.A., F.G., E.R., and B.P.; Formal Analysis: F.A., F.G., E.R., M.A., and F.L.; Investigation: F.A., E.R., and B.P.; Methodology: F.A. and E.R.; Project administration: E.R. and B.P.; Supervision: A.C. and B.P.; Visualization: F.A., E.R., L.F., and B.P.; Writing—original draft: F.A., F.G., E.R., A.C., L.F., B.P., M.A., and F.L.; Writing—review & editing: F.A., F.G., E.R., A.C., L.F., B.P., M.A., F.L., and D.M. All authors have read and agreed to the published version of the manuscript.

Funding: The study was funded by the Italian Ministry of Health (Grant IZS PLV 15/18 RC).

Acknowledgments: The authors are grateful to Laura Pecorari and to Francesca Altomare for the precious collaboration.

Conflicts of Interest: The authors declare no conflict of interest.

References

1. Sykora, S.; Brandt, S. Papillomavirus infection and squamous cell carcinoma in horses. *Vet. J.* **2017**, *223*, 48–54. [CrossRef]

2. Scott, D.W.; Miller, W.H.J. Squamous cell carcinoma. In *Equine Dermatology*, 1st ed.; W.B. Saunders: St. Louis, MO, USA, 2003; pp. 707–712.
3. Van den Top, J.G.; de Heer, N.; Klein, W.R.; Ensink, J.M. Penile and preputial squamous cell carcinoma in the horse: A retrospective study of treatment of 77 affected horses. *Equine Vet. J.* **2008**, *40*, 533–537. [PubMed]
4. Jones, D.L. Squamous cell carcinoma of the larynx and pharynx in horses. *Cornell Vet.* **1994**, *84*, 15–24.
5. Schultz, P. Vocal fold cancer. *Eur. Ann. Otorhinolaryngol. Head Neck Dis.* **2011**, *128*, 301–308. [CrossRef] [PubMed]
6. Li, X.; Gao, L.; Li, H.; Gao, J.; Yang, Y.; Zhou, F.; Gao, C.; Li, M.; Jin, Q. Human papillomavirus infection and laryngeal cancer risk: A systematic review and meta-analysis. *J. Infect. Dis.* **2013**, *207*, 479–488. [CrossRef] [PubMed]
7. Porcellato, I.; Modesto, P.; Cappelli, K.; Varello, K.; Peletto, S.; Brachelente, C.; Martini, I.; Mechelli, L.; Ferrari, A.; Ghelardi, A.; et al. Equus caballus papillomavirus type 2 (EcPV2) in co-occurring vulvar and gastric lesions of a pony. *Res. Vet. Sci.* **2020**, *132*, 137–161. [CrossRef] [PubMed]
8. Porcellato, I.; Mecocci, S.; Mechelli, L.; Cappelli, K.; Brachelente, C.; Pepe, M.; Orlandi, M.; Gialletti, R.; Passeri, B.; Ferrari, A.; et al. Equine Penile Squamous Cell Carcinomas as a Model for Human Disease: A Preliminary Investigation on Tumor Immune Microenvironment. *Cells* **2020**, *9*, 2364. [CrossRef] [PubMed]
9. Alloway, E.; Linder, K.; May, S.; Rose, T.; DeLay, J.; Bender, S.; Tucker, A.; Luff, J. A Subset of Equine Gastric Squamous Cell Carcinomas Is Associated With Equus Caballus Papillomavirus–2 Infection. *Vet. Path.* **2020**, *57*, 427–431. [CrossRef]
10. Torres, S.M.F.; Koch, S.N. Papillomavirus-associated diseases. *Vet. Clin. N. Am. Equine Pract.* **2013**, *29*, 643–655. [CrossRef]
11. Hibi, H.; Hatama, S.; Obata, A.; Shibahara, T.; Kadota, K. Laryngeal squamous cell carcinoma and papilloma associated with Equus caballus papillomavirus 2 in a horse. *J. Vet. Med. Sci.* **2019**, *81*, 1029–1033. [CrossRef]
12. Micalizzi, D.S.; Farabaugh, S.M.; Ford, H.L. Epithelial-mesenchymal transition in cancer: Parallels between normal development and tumor progression. *J. Mammary Gland Biol. Neoplasia* **2010**, *15*, 117–134. [CrossRef] [PubMed]
13. Thiery, J.P.; Acloque, H.; Huang, R.Y.J.; Nieto, M.A. Review article: Epithelial-mesenchymal transitions in development and disease. *Cell* **2009**, *139*, 871–890. [CrossRef] [PubMed]
14. Bonnet, A.S.; Willis, C.; Pittaway, R.; Smith, K.; Mair, T.; Priestnall, S.L. Molecular carcinogenesis in equine penile cancer: A potential animal model for human penile cancer. In *Urologic Oncology: Seminars and Original Investigations*; Elsevier: Amsterdam, The Netherlands, 2018; Volume 36, pp. 532.e9–532.e18.
15. Lamouille, S.; Xu, J.; Derynck, R. Molecular mechanisms of epithelial–mesenchymal transition. *Nat. Rev. Mol. Cell. Biol.* **2014**, *15*, 178–196. [CrossRef] [PubMed]
16. Chang, H.; Liu, Y.; Xue, M.; Liu, H.; Du, S.; Zhang, L.; Wang, P. Synergistic action of master transcription factors controls epithelial-to-mesenchymal transition. *Nucleic Acids Res.* **2016**, *44*, 2514–2527. [CrossRef] [PubMed]
17. Savagner, P. The epithelial–mesenchymal transition (EMT) phenomenon. *Ann. Oncol.* **2010**, *21*, 89–92. [CrossRef]
18. Mani, S.A.; Guo, W.; Liao, M.J.; Eaton, E.N.; Ayyanan, A.; Zhou, A.Y.; Brooks, M.; Reinhard, F.; Zhang, C.C.; Shipitsin, M.; et al. The epithelial-mesenchymal transition generates cells with properties of stem cells. *Cell* **2008**, *133*, 704–715. [CrossRef]
19. Wheelock, M.J.; Shintani, Y.; Maeda, M.; Fukumoto, Y.; Johnson, K.R. Cadherin switching. *J. Cell. Sci.* **2008**, *121*, 727–735. [CrossRef]
20. O'Connor, J.W.; Gomez, E.W. Biomechanics of TGFβ-induced epithelial-mesenchymal transition: Implications for fibrosis and cancer. *Clin. Transl. Med.* **2014**, *3*, 23. [CrossRef]
21. Santamaria, P.G.; Moreno-Bueno, G.; Portillo, F.; Cano, A. EMT: Present and future in clinical oncology. *Mol. Oncol.* **2017**, *11*, 718–738. [CrossRef]
22. Armando, F.; Gambini, M.; Corradi, A.; Becker, K.; Marek, K.; Pfankuche, V.M.; Mergani, A.E.; Brogden, G.; de Buhr, N.; von Köckritz-Blickwede, M.; et al. Mesenchymal to epithelial transition driven by canine distemper virus infection of canine histiocytic sarcoma cells contributes to a reduced cell motility in vitro. *J. Cell. Mol. Med.* **2020**, *24*, 9332–9348. [CrossRef]
23. Leonardi, F.; Costa, G.L.; Dubau, M.; Sabbioni, A.; Simonazzi, B.; Angelone, M. Effects of intravenous romifidine, detomidine, detomidine combined with butorphanol, and xylazine on tear production in horses. *Equine Vet. Educ.* **2018**, *32*, 53–57. [CrossRef]

24. Kalluri, R.; Weinberg, R.A. The basics of epithelial-mesenchymal transition. *J. Clin. Investig.* **2009**, *119*, 1420–1428. [CrossRef] [PubMed]
25. Grzegrzolka, J.; Biala, M.; Wojtyra, P.; Kobierzycki, C.; Olbromski, M.; Gomulkiewicz, A.; Piotrowska, A.; Rys, J.; Podhorska-Okolow, M.; Dziegiel, P. Expression of EMT markers SLUG and TWIST in breast cancer. *Anticancer Res.* **2015**, *35*, 3961–3968. [PubMed]
26. Armando, F.; Ferrari, L.; Arcari, M.L.; Azzali, G.; Dallatana, D.; Ferrari, M.; Lombardi, G.; Zanfabro, M.; Di Lecce, R.; Lunghi, P.; et al. Endocanalicular transendothelial crossing (ETC): A novel intravasation mode used by HEK-EBNA293-VEGF-D cells during the metastatic process in a xenograft model. *PLoS ONE* **2020**, *15*, e0239932. [CrossRef]
27. Rodriguez, F.J.; Lewis-Tuffin, L.J.; Anastasiadis, P.Z. E-cadherin's dark side: Possible role in tumor progression. *Biochim. Biophys. Acta* **2012**, *1826*, 23–31. [CrossRef]
28. Kourtidis, A.; Lu, R.; Pence, L.J.; Anastasiadis, P.Z. A central role for cadherin signaling in cancer. *Exp. Cell Res.* **2017**, *358*, 78–85. [CrossRef]
29. Luo, W.R.; Wu, A.B.; Fang, W.Y.; Li, S.Y.; Yao, K.T. Nuclear expression of N-cadherin correlates with poor prognosis of nasopharyngeal carcinoma. *Histopathology* **2012**, *61*, 237–246. [CrossRef]
30. Wang, W.; Wen, Q.; Luo, J.; Chu, S.; Chen, L.; Xu, L.; Zang, H.; Alnemah, M.M.; Li, J.; Zhou, J.; et al. Suppression of β-catenin nuclear translocation by CGP57380 decelerates poor progression and potentiates radiation-induced apoptosis in nasopharyngeal carcinoma. *Theranostics* **2017**, *7*, 2134–2149. [CrossRef]
31. Marie-Egyptienne, D.T.; Lohse, I.; Hill, R.P. Cancer stem cells, the epithelial to mesenchymal transition (EMT) and radioresistance: Potential role of hypoxia. *Cancer Lett.* **2013**, *341*, 63–72. [CrossRef]
32. Cannito, S.; Novo, E.; Compagnone, A.; Valfrè di Bonzo, L.; Busletta, C.; Zamara, E.; Paternostro, C.; Povero, D.; Bandino, A.; Bozzo, F.; et al. Redox mechanisms switch on hypoxia-dependent epithelial-mesenchymal transition in cancer cells. *Carcinogenesis* **2008**, *29*, 2267–2278. [CrossRef]
33. Yang, M.H.; Wu, M.Z.; Chiou, S.H.; Chen, P.M.; Chang, S.Y.; Liu, C.J.; Teng, S.C.; Wu, K.J. Direct regulation of TWIST by HIF-1alpha promotes metastasis. *Nat. Cell. Biol.* **2008**, *10*, 295–305. [CrossRef] [PubMed]
34. Greenwood, S.; Chow-Lockerbie, B.; Epp, T.; Knight, C.; Wachoski-Dark, G.; MacDonald-Dickinson, V.; Wobeser, B. Prevalence and Prognostic Impact of Equus caballus Papillomavirus Type 2 Infection in Equine Squamous Cell Carcinomas in Western Canadian Horses. *Vet. Path.* **2020**, *57*, 623–631. [CrossRef]
35. Knight, C.G.; Munday, J.S.; Rosa, B.V.; Kiupel, M. Persistent, widespread papilloma formation on the penis of a horse: A novel presentation of equine papillomavirus type 2 infection. *Vet. Dermatol.* **2011**, *22*, 570–574. [CrossRef] [PubMed]
36. Evans, M.R.; James, C.D.; Bristol, M.L.; Nulton, T.J.; Wang, X.; Kaur, N.; White, E.A.; Windle, B.; Morgan, I.M. Human papillomavirus 16 E2 regulates keratinocyte gene expression relevant to cancer and the viral life cycle. *J. Virol.* **2019**, *93*, e01918–e01941. [CrossRef] [PubMed]
37. Liu, J.; Huang, B.; Xiu, Z.; Zhou, Z.; Liu, J.; Li, X.; Tang, X. PI3K/Akt/HIF-1α signaling pathway mediates HPV-16 oncoprotein-induced expression of EMT-related transcription factors in non-small cell lung cancer cells. *J. Cancer* **2018**, *9*, 3456–3466. [CrossRef] [PubMed]
38. Rezaei, M.; Mostafaei, S.; Aghaei, A.; Hosseini, N.; Darabi, H.; Nouri, M.; Etemadi, A.; Neill, A.O.; Nahand, J.S.; Mirzaei, H.; et al. The association between HPV gene expression, inflammatory agents and cellular genes involved in EMT in lung cancer tissue. *BMC Cancer* **2020**, *20*, 916. [CrossRef]

Publisher's Note: MDPI stays neutral with regard to jurisdictional claims in published maps and institutional affiliations.

animals

MDPI

Article

Productive and Physiological Response of Male Rabbits to Dietary Supplementation with Thyme Essential Oil

Ahmed A. A. Abdel-Wareth [1] and Abdallah E. Metwally [2,*]

[1] Department of Animal and Poultry Production, Faculty of Agriculture, South Valley University, Qena 83523, Egypt; a_bkr1@yahoo.com

[2] Department of Nutrition and Clinical Nutrition, Faculty of Veterinary Medicine, Zagazig University, Zagazig 44511, Egypt

* Correspondence: abdallah.metwally75@gmail.com; Tel.: +201009599211

Received: 16 September 2020; Accepted: 8 October 2020; Published: 10 October 2020

Simple Summary: The present study was carried out to compare the potential effects of the levels of thyme essential oil on the productive performance and serum metabolic profile of male rabbits. Rabbits were assigned to five dietary treatments including a basal diet as a negative control, a basal diet supplemented with an antibiotic as a positive control, and a basal diet supplemented with 60, 120, or 180 mg/kg of thyme essential oil. The main results showed that the levels of thyme essential oil contributed to the improvement of productive and the physiological response compared with the negative and positive control of male rabbits.

Abstract: The present study aimed at assessing the efficiency of thyme essential oil (TEO) as an alternative to antibiotics for improving the productive performance and serum metabolic profile of male rabbits. A total of one hundred and fifty 70-day-old male Californian rabbits were assigned to five dietary treatments, including a basal diet as a negative control (NC), a basal diet supplemented with an antibiotic as a positive control (PC), and a basal diet supplemented with 60, 120, or 180 mg/kg of TEO. The experiment period lasted for 60 days. Supplementation of TEO levels significantly ($p < 0.01$) increased daily body weight gain and improved feed conversion ratio of male Californian rabbits compared to NC and PC groups. Similarly, the TEO remarkably enhanced the semen characteristics of rabbits compared to NC and PC groups. Supplementation of TEO significantly decreased aspartate transaminase, alanine transaminase, urea, and creatinine compared with NC and PC groups. Supplementation with TEO increased serum testosterone concentration compared to NC and PC treatments. Our data demonstrate that TEO levels up to 180 mg/kg can play a major role as an alternative to dietary antibiotics, in improving the productive performance, semen quality, testosterone levels, and the kidney and liver functions in California male rabbits.

Keywords: blood biochemistry; performances; rabbits; semen quality; thyme essential oil

1. Introduction

Recently, the rabbit industry has played a major role in meeting the increasing requirements of animal protein for human consumption and is becoming an important source of the national economies in Egypt [1,2]. Rabbit meat is characterized by a high protein and low fat and cholesterol content [1]. These nutritional qualities are of great value to the meat industry and consumers. A ban on antibiotic growth promoters by several countries and the menace of antibiotic-resistant bacteria have forced researchers to look for alternatives for improving efficiency in animal production [2,3]. Due to this ban, much research has been conducted in order to explore the use of phytogenic substances as alternate feed additives in animal nutrition [4,5]. Phytogenic substances are generally considered safe and are frequently utilized in the food and feed industries [6].

The effects the phytogenic substances have on the intestinal, antioxidant status, and antimicrobial activity are considered essential for the biological activities. Moreover, the thyme essential oil (TEO), as a phytogenic feed additive, can influence the rabbit performance and welfare positively [7]. Thyme (*Thymus vulgaris*) is an aromatic plant that belongs to the Lamiaceae family, and great attention has been paid to its pharmaceutical and therapeutic effects. Furthermore, supplementation of dietary thyme extract at 0.5 g/kg significantly improved gut integrity and antioxidant status of rabbits [8] mainly because of the thyme active components. However, other studies have not observed any effect of the dietary inclusion of thyme on growth development [9] or the carcass of young rabbits [10]. The main components of TEO are thymol, carvacrol, γ-terpinene, p-cymene, β-myrcene, linalool, and terpinen-4-ol [1,11]. These active components had the best oxidative status [12] and may reduce serum cholesterol [13], which could improve the reproductive performance of rabbits. El-Ratel et al. [14] reported that oral administration of phytobiotics improved the liver function of rabbits compared to the control. Similarly, supplementation of 2.5% of thyme leaves to rabbit diets remarkably enhanced kidney function by decreasing urea and creatinine levels [15]. However, no information is available regarding the effects of TEO, as alternatives to dietary antibiotic growth promoters, on the performance, semen quality, kidney and liver functions, and testosterone levels in male rabbits. To explore the effects of TEO on the productive and physiological status of male rabbits, we investigated the efficiency of TEO as an alternative to antibiotics for improving the productive performance, testosterone levels, and liver and kidney functions of male rabbits.

2. Materials and Methods

2.1. Animals and Housing

A total of one hundred and fifty, 70-day-old male Californian rabbits, weighing 1250 ± 30 g, were utilized in this study. The rabbits were acquired from a Breeding Agricultural Research Centre, Faculty of Agriculture, South Valley University, Egypt. The Institutional Animal Care and Use Committee of University of South Valley University approved the experimental protocol used in this study in accordance with the guidelines of the Egyptian Research Ethics Committee and the guidelines in the Guide for the Care and Use of Laboratory Animals (2011). All procedures in the current study were in accordance with the European Union Directive 2010/63/EU regarding the protection of animals utilized in experimentation. Rabbits were housed separately in cages (one animal per cage) of galvanized wire net (width × length × height: 60 cm × 60 cm × 40 cm), equipped with an automatic drinker and a manual feeder. Farm temperature was maintained at 23 °C, and the rabbits were exposed to a cycle of 16 h of light and 8 h of darkness during the experimental period. Fresh tap water was available ad libitum via stainless steel nipples located inside each cage. The study was performed following the ARRIVE guidelines for the reporting of animal experiments [16]. Rabbits were housed under the same managerial, hygienic, and housing conditions during the whole experimental period. The health condition of all the rabbits was closely monitored through daily health checks. After the experimentation, all the remaining animals were released.

2.2. Experimental Diets and Growth Performance

The male Californian rabbits were assigned to five dietary treatment groups of 30 rabbits each. Dietary treatments included a basal diet as a negative control (NC), a basal diet supplemented with a 500 mg/kg oxytetracycline antibiotic as a positive control (PC), and a basal diet supplemented with 60, 120, or 180 mg/kg of TEO. The experiment period lasted for 60 days. Table 1 presents the experimental ration ingredients and chemical composition. The diets were formulated to contain adequate levels of nutrients for rabbits as per the National Research Council (NRC) [17]. Rabbits were individually identified, and their body weight (BW) values were recorded at day 0 (70 days old), mid-experiment (100 days old), and end of the experiment (130 days old). Daily body weight gain (DBWG) was calculated for each period per pen. Additionally, the feed consumption for each pen between weighing was determined through the measurement of feed residue on the same day as the rabbits were weighed. Feed conversion ratio (FCR) was calculated as feed per gain based on the weight of feed consumed divided by DBWG per pen. Besides, mortality was observed daily during the entire experimental period.

Table 1. Ingredients and chemical compositions (as-fed basis) of the control diet fed to rabbits throughout the experiment period.

Ingredients	%	Chemical Analysis	%
Yellow maize grain	32.00	Dry matter	91.40
Wheat bran	20.00	Ash	9.80
Soybean meal (44% CP)	18.00	Crude protein	17.00
Wheat straw	12.00	Crude fiber	12.60
Lucerne hay	5.00	Fat-ether extract	2.90
Rice bran	5.00	Digestible energy (MJ/kg)	9.42
Linseed straw	2.80	Calcium	1.30
Sunflower meal	2.50	Phosphorus	0.86
Limestone	2.00	Lysine	0.60
Sodium chloride	0.30	Methionine	0.41
Vitamin–mineral premix [1]	0.30		
dl-Methionine	0.10		

[1] Per kg of ration: vitamin A, 10.000 IU; vitamin D_3, 900 IU; vitamin E, 50.0 mg; vitamin K, 2.0 mg; vitamin B_1, 2.0 mg; folic acid, 5.0 mg; pantothenic acid, 20.0 mg; vitamin B_6, 2.0 mg; choline, 1200 mg; vitamin B_{12}, 0.01 mg; niacin, 50 mg; biotin, 0.2 mg; Cu, 0.1 mg; Fe, 75.0 mg; Mn, 8.5 mg; Zn, 70 mg.

2.3. Chemical Analysis

The feed was analyzed for moisture by oven drying (method no. 930.15), ash by incineration (method no. 942.05), protein by Kjeldahl (method no. 984.13), ether extract by Soxhlet fat analysis (method no. 920.39), and calcium and phosphorus (Ca and P; method no. 999.10) as described by the AOAC International [18]. Lysine and, after performic acid oxidation in 6 M HCl for 24 h at 100 °C under an N atmosphere, and methionine were determined as methionine sulfone after cold performic acid oxidation overnight and hydrolyzing with 7.5 N HCl at 110 °C (procedure 4.1.11; alternative 1; AOAC, 2000) [18] for 24 h, followed by analysis using an amino acid analyzer (Hitachi L-8800, Tokyo, Japan).

The gross energy (GE) contents of the diet and feces were measured using an adiabatic bomb calorimeter (Parr Instrument Company, Moline, IL, USA). A digestibility trial was conducted using thirty 100-day-old male Californian rabbits to determine the dry matter (DM) and digestible energy (DE) of the basal diet according to Perez et al. [19]. Rabbits were housed in individual metabolism cages (measuring width × length × height to be 50 cm × 60 cm × 40 cm). A 10-day adaptation period was followed by continuous feces collection for 5 consecutive days (collection period). Samples of daily feces of each rabbit were taken and oven dried at 60 °C for 48 h, then was ground and stored for proximate chemical analysis. Chemical analyses for the digestibility trial (DM and GE of diets and feces) were conducted at the Animal Nutrition Laboratory, Faculty of Agriculture, South Valley University, Egypt. The DE refers to GE intake minus energy lost in feces according to Hall et al. [20].

Analysis of the chemical composition of hydrodistilled TEO (Table 2) was conducted using a gas chromatography–mass spectrometry (GC/MS) system as per Abozid and Asker [21]. The TEO was purchased from El Hawag Natural Oils Company, Cairo, Egypt. The TEO was extracted by hydrodistillation in a Clevenger-type apparatus for three hours. The TEO was analyzed at the Department of Medicinal and Aromatic Plants Research, National Research Centre, Egypt, by gas chromatography (Delsi 121C gas chromatograph). Constituents were identified by GC/MS, using a Sigma 300 apparatus attached to an HP 5970 300 mass spectrometer.

Table 2. Major chemical compounds of hydrodistilled thyme essential oil (TEO) as detected by gas chromatography–mass spectrophotometer (GC-MS).

Chemical Compounds	Rt.	%	Mol. Weight (gm/mol)	Chemical Formula
p-Cymene	6.99	23.59	134.218	$C_{10}H_{14}$
B-linalool	9.61	0.74	154.25	$C_{10}H_{18}$
Carvone (carvacrol)	15.70	9.80	150.22	$C_{10}H_{14}$
Anethole	17.49	2.50	148.2	$C_{10}H_{12}$
Thymol	17.70	39.45	150.22	$C_{10}H_{14}$
Carvacrol	18.09	2.07	150.217	$C_{10}H_{14}$
trans-Caryophyllene	22.46	0.98	204.36	$C_{15}H_{24}$
γ-Terpinene	25.14	12.49	136.23	$C_{10}H_{16}$
Aromadenrene	34.84	2.12	204.35	$C_{15}H_{24}$
Ledol	48.66	2.24	222.358	$C_{15}H_{26}$

2.4. Semen Collection and Assay

Fifteen male rabbits from each group were selected for semen characterization, including volume of each ejaculate, sperm livability, sperm motility, abnormal sperm, and sperm forward motility of sexually mature rabbits (130 days of age), which were assessed as described previously by Abdel-Wareth et al. [4] and El-Desoky et al. [22].

Assessments of live, dead, and abnormal sperms were performed by counting 200 sperm cells using an Eosin–Nigrosin staining mixture. Complete or partial, purple-stained sperm cells were considered nonviable, whereas nonstained sperm cells were considered viable. Percentages of motile sperms on a warm stage showing progressive forward movement were visually calculated in several microscopic fields under 100× magnification using light microscope.

2.5. Blood Biochemical Assay

At the end of the experimental period, rabbits were fasted 12 h prior to blood sampling. The treated animals were anesthetized by intramuscular injection of ketamine and xylazine, and then 5 mL of blood was withdrawn from one of the external ear veins. Blood samples were centrifuged at 3000 rpm for 15 min, where the serum was collected and stored at −20 °C until assayed for biochemical analysis. Serum testosterone concentrations were measured by immunoassay using commercial kits (Monobind, Inc., Lake Forest, CA 92630, USA). Liver enzymes, alanine transaminase (ALT) and aspartate transaminase (AST); the kidney function markers, creatinine and urea, concentrations, were also measured using standard diagnostic kits (Monobind, Inc., Lake Forest, CA 92630, USA).

2.6. Statistical Analysis

The statistical analysis was conducted by ANOVA followed by Duncan's test using SAS software [23]. The cage was the experimental unit for each parameter. The significance effects were declared at $p < 0.05$. Orthogonal polynomial contrasts were also utilized for determining the linear and quadratic effects of levels of TEO inclusion considering only negative control (0 mg/kg TEO) as a control, whereas positive control was not included in this analysis. Significance was declared at $p \leq 0.05$.

3. Results

3.1. Growth Performance

Table 3 presents the effects of TEO levels, as an alternative to dietary antibiotics, on growth performance. The BW significantly increased ($p < 0.001$) with the increasing levels of TEO supplementations at 100 and 130 days of age. There was no significant difference on initial BW at 70 days of age among treatments. Furthermore, the TEO supplementations at 60, 120, 180 mg/kg increased DBWG and improved FCR of male Californian rabbits compared to the PC and NC groups during the periods of 70–100, 100–130, and 70–130 days of age. The highest DBWG and most efficient FCR were found in the 180 mg, 120 mg, and 60 mg TEO groups, then the PC and NC. The weakest performance was recorded for the NC group compared with PC and TEO groups. In terms of the daily feed intake (DFI), the highest DFI was observed ($p < 0.001$) in 180 and 120 mg TEO supplementations compared with 60 mg TEO, NC, and PC groups during the periods of 70–100 and 70–130 days of age.

Table 3. Effects of thyme essential oil on the productive performance of male rabbits.

Items	Treatments (T)					SEM [3]	*p* Value		
	NC [1]	PC [2]	60 mg	120 mg	180 mg		T [4]	Lin [5]	Quad [6]
Body weight, g									
70 d	1255	1259	1258	1249	1253	6	0.614	0.882	0.263
100 d	1700 [d]	1751 [c]	1807 [b]	1819 [a]	1835 [a]	14	<0.001	0.003	0.303
130 d	2288 [d]	2354 [c]	2428 [b]	2441 [a]	2469 [a]	21	<0.001	<0.001	0.053
Daily body weight gain, g									
70–100 d	14.85 [d]	16.40 [c]	18.33 [b]	19.02 [ab]	19.41 [a]	0.22	<0.001	<0.001	< 0.001
100–130 d	19.59	20.10	20.67	20.71	21.14	0.38	0.076	0.006	0.640
70–130 d	17.22 [b]	18.25 [b]	19.50 [a]	19.87 [a]	20.28 [a]	0.25	<0.001	<0.001	0.033
Daily feed intake, g									
70–100 d	54.46 [c]	55.43 [c]	57.64 [b]	58.61 [ab]	59.65 [a]	0.51	<0.001	<0.001	0.574
100–130 d	65.20	62.77	63.04	64.48	64.71	0.64	0.053	0.719	0.013
70–130 d	59.83 [c]	59.10 [c]	60.34 [bc]	61.54 [ab]	62.18 [a]	0.43	<0.001	<0.001	0.110
Feed conversion ratio									
70–100 d	3.67 [a]	3.38 [b]	3.15 [c]	3.08 [c]	3.07 [c]	0.048	0.001	<0.001	<0.001
100–130 d	3.33 [a]	3.13 [b]	3.05 [b]	3.12 [b]	3.06 [b]	0.056	0.014	0.006	0.046
70–130 d	3.48 [a]	3.24 [b]	3.09 [c]	3.09 [c]	3.07 [c]	0.042	0.001	<0.001	0.002

[a–d] Means within a row with different superscripts differ significantly at $p < 0.05$. [1] NC: basal diet as a negative control. [2] PC: a basal diet was supplemented with an antibiotic as a positive control. [3] SEM: standard error of means. [4] T: NC, PC, and thyme essential oil treatments. [5,6] Lin and Quad: linear and quadratic responses, respectively, to thyme essential oil inclusion level considering NC (0 mg/kg thyme essential oil) as a control; PC was not included in this analysis.

On the other hand, the supplemented TEO groups did not affect DFI ($p \geq 0.05$) compared to control groups during the period of 70–100 days of age.

3.2. Semen Characteristics

Table 4 displays the effects of TEO on semen characteristics. Results demonstrated that TEO increased the sperm livability, sperm motility, and ejaculate volume compared with PC and NC groups at the end of treatments. Abnormal sperm was reduced ($p < 0.01$) as TEO increased, compared to PC and NC groups. Moreover, supplementation of TEO up to 180 mg/kg to male rabbit diets remarkably increased ($p < 0.001$) sperm forward motility% and sperm livability% in comparison with control groups. However, no significant difference was observed in the semen pH values between treatments. Overall, the PC (oxytetracycline) exhibited a significant increase in semen quality compared to NC.

3.3. Blood Biochemical Constituents

The effects of dietary supplemental TEO on blood serum constituents of male rabbits are shown in Figures 1–3. Rabbits fed the diets supplemented with TEO at 60, 120, and 180 mg/kg had significantly ($p < 0.001$) reduced serum urea and creatinine compared to PC and NC groups. Moreover, supplementation of TEO to diets of male rabbits significantly ($p < 0.05$) reduced the activity of serum ALT and AST in comparison with PC and NC groups. Furthermore, male rabbits fed diets supplemented with the TEO up to 180 mg/kg exhibited the highest improvement in serum testosterone concentrations ($p < 0.05$) compared to PC and NC groups.

Table 4. Effects of thyme essential oil on semen quality of male rabbits at 130 days of age.

Items	Treatments (T)					SEM [1]	*p* Value		
	NC [1]	PC [2]	60 mg	120 mg	180 mg		T [4]	Lin [5]	Quad [6]
				Semen quality					
Semen volume, ml	0.52 [d]	0.56 [c]	0.61 [b]	0.64 [a]	0.65 [a]	0.09	0.001	0.001	0.004
Abnormal sperm, %	16.7 [a]	15.4 [ab]	12.6 [bc]	11.9 [cd]	11.2 [d]	0.41	0.001	0.001	0.432
Live sperm, %	75.0 [d]	75.4 [d]	79.2 [c]	81.6 [b]	83.6 [a]	0.63	0.001	0.001	0.458
Semen pH, value	7.11	7.09	7.10	7.11	7.10	0.03	0.990	0.965	0.853
Forward motility, %	55.1 [c]	55.3 [c]	65.7 [b]	69.5 [a]	70.8 [a]	0.57	0.001	0.001	0.056

[a–d] Means within a row with different superscripts differ significantly at $p < 0.05$. [1] NC: basal diet as a negative control. [2] PC: a basal diet supplemented with an antibiotic as a positive control. [3] SEM: standard error of means. [4] T: NC, PC, and thyme essential oil treatments. [5,6] Lin and Quad: linear and quadratic responses, respectively, to thyme essential oil inclusion level considering NC (0 mg/kg thyme essential oil) as a control; PC was not included in this analysis.

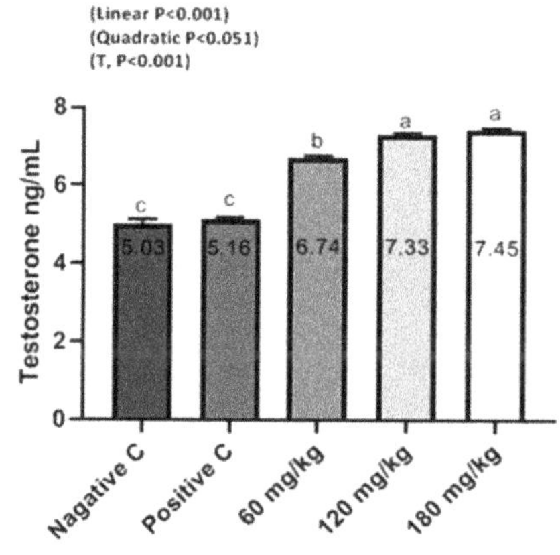

Figure 1. Testosterone ng/mL of male rabbits in response to dietary treatments including a basal diet (NC), a basal diet supplemented with a 500 mg/kg oxytetracycline antibiotic (PC), and a basal diet supplemented with 60, 120, or 180 mg/kg of TEO at 130 days of age. [a–c] Means within a row with different superscripts differ significantly at $p < 0.05$.

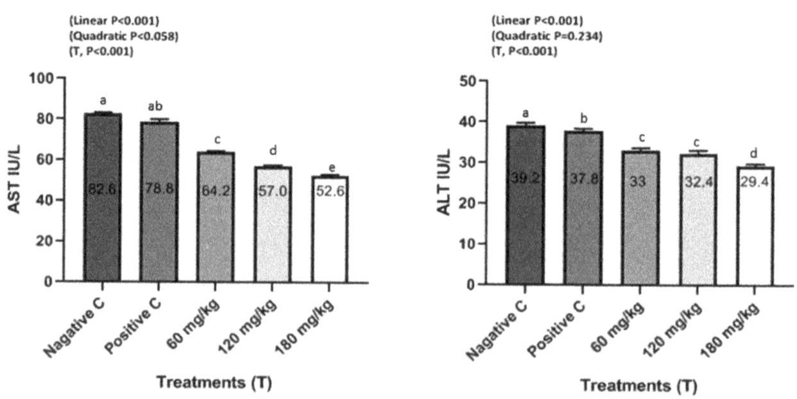

Figure 2. Liver enzyme (alanine transaminase (ALT) and aspartate transaminase (AST)) levels of male rabbits in response to dietary treatments including a basal diet (NC), a basal diet supplemented with a 500 mg/kg oxytetracycline antibiotic (PC), and a basal diet supplemented with 60, 120, or 180 mg/kg of TEO at 130 days of age. [a–e] Means within a row with different superscripts differ significantly at $p < 0.05$.

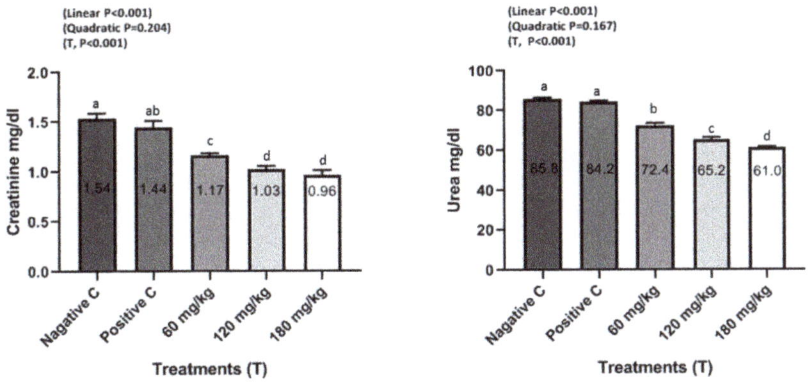

Figure 3. Serum creatinine mg/dl and urea mg/dl of male rabbits in response to dietary treatments including a basal diet (NC), a basal diet supplemented with a 500 mg/kg oxytetracycline antibiotic (PC), and a basal diet supplemented with 60, 120, or 180 mg/kg of TEO at 130 days of age. $^{a\text{–}d}$ Means within a row with different superscripts differ significantly at $p < 0.05$.

4. Discussion

It is extremely difficult to compare the studies that applied different essential oils because the results will be based on numerous factors, including essential composition, doses, extract methods and application, animal age, and housing conditions. The TEO feed additive, as an alternative to dietary antibiotics, is highly safe and could be utilized for improving the productive and reproductive performance of rabbits. The effects the phytogenic substances have on the intestinal health, antioxidant status, and antimicrobial activity are considered essential for the biological activities; however, studies lack the determination of active components, and their mechanisms are still not clear [24–26]. Therefore, more studies under standardization are required to explore the mechanisms of TEO on rabbit production and reproduction. In the present study, the main active component of TEO was thymol which constitutes 40% of its analyzed composition. The composition of TEO used in the present study was consistent with that reported in the literature [1,11]. TEO levels caused the highest DFI and the lowest FCR which could be associated with the ability of TEO to improve nutrient absorption. In the current study, it is observed that dietary TEO enhanced the productive performance associated with the present active component, thymol. The current study demonstrated that DBWG and FCR were improved in rabbits fed a control diet supplemented with TEO up to 180 mg/kg compared to PC and NC groups. These improvements may be because the active compounds (thymol and carvacrol) stimulated digestive enzymes, thus leading to the improvement in nutrient digestibility [5,27]. Thymol stimulates the appetite, and the secretion of endogenous digestive enzymes and nutrient absorption mainly because thymol could protect the microvilli [28]. Supplementation of TEO with olive oil enhanced the growth performance of male Californian rabbits under high-temperature environments [1]. Similarly, oral administration of aqueous thyme extract with 50 mg/kg BW improved ($p < 0.001$) BWG, FI, and FCR of rabbits compared to the control [29].

The TEO levels significantly increased the reproductive performance of male rabbits at the end of the treatments period (Table 4). Interestingly, the improvements in the reproductive performance of male rabbits were in parallel with those of testosterone concentration (Figure 1). Ruiz-Olvera et al. [30] demonstrated that semen volume and sperm motility are associated with serum testosterone concentrations. Supplementation of thyme aqueous extracts significantly increased the semen quality of rabbits [31]. In addition, semen volume, sperm motility, sperm concentration, and sperm livability significantly increased ($p < 0.05$) in rabbits received aqueous thyme extract with 50 mg/kg compared to the control [29].

Blood biochemical analysis reflects internal organ integrity. In this direction, this study indicated that TEO did not cause obvious damage to any of the internal organs, including the kidney, liver, and heart. These obtained results can be confirmed by liver and kidney function. In the current study, supplementation of TEO to male rabbits significantly reduced serum urea, creatinine, and ALT and AST activities and increased serum testosterone levels compared to PC and NC groups, but the values were within the normal physiological range [14,16,29–31]. These results agree with those of El-Ratel et al. [14], where it is reported that the activity of AST and ALT of rabbits was remarkably enhanced by oral administration of phytogenic additives (5 or 10 mg allicin/BW) compared to the control. Supplementation of 2.5% of thyme leaves to rabbit diets significantly decreased urea and creatinine [15]. Similarly, Abdel-Gabbar et al. [31] demonstrated that thyme extracts at 100 mg/kg induced significant ($p < 0.001$) reduction in creatinine and urea as well as ALT and AST of rabbits compared to the control group. However, Dieumou et al. [32] reported that the administration of garlic and ginger oils via stomach tube in four doses, 0 (control), 10 mg/kg/day, 20 mg/kg/day, and 40 mg/kg/day, showed no clear variations recorded in ALT, AST and serum creatinine features due to ginger treatments. Shanoon et al. [33] indicated that there were no obvious differences in serum ALT, AST and blood creatinine features, indicating that none of the examined three dosages of the oil administered to birds was harmful. In addition, Herve et al. [34] found that the serum features of AST or ALT clearly decreased with 100 or 150 mg/kg BW of ginger roots oil when compared to control data, without any harmful outcomes on feed uptake and BW profits.

Therefore, the effect of TEO on the productive and reproductive performance of rabbits is better known. It increased both productive and reproductive performance and the liver and kidney functions, as observed in the current study.

5. Conclusions

Considering the above findings, the levels of TEO up to 180 mg/kg can play a major role in enhancing productive performance, semen quality, testosterone levels, and kidney and liver functions in California male rabbits.

Author Contributions: Conceptualization, A.A.A.A.-W.; Data curation, A.A.A.A.-W. and A.E.M.; Formal analysis, A.E.M.; Methodology, A.A.A.A.-W.; Writing—original draft, A.A.A.A.-W.; Writing—review and editing, A.E.M. All authors have read and agreed to the published version of the manuscript.

Funding: This research received no external funding.

Conflicts of Interest: The authors declare that they have no conflict of interests.

Abbreviations

TEO: thyme essential oil; NC: negative control; PC: positive control; BW: body weight; BWG: body weight gain; FCR: feed conversion ratio; AST: aspartate transaminase; ALT: alanine transaminase.

References

1. Sheiha, A.M.; Abdelnour, S.A.; El-Hack, M.E.A.; Khafaga, A.F.; Metwally, K.A.; Ajarem, J.S.; Maodaa, S.N.; Allam, A.A.; El-Saadony, M.T. Effects of Dietary Biological or Chemical-Synthesized Nano-Selenium Supplementation on Growing Rabbits Exposed to Thermal Stress. *Animal* **2020**, *10*, 430. [CrossRef]
2. Abdel-Wareth, A.A.A.; Taha, E.M.; Südekum, K.-H.; Lohakare, J. Thyme oil inclusion levels in a rabbit ration: Evaluation of productive performance, carcass criteria and meat quality under hot environmental conditions. *Anim. Nutr.* **2018**, *4*, 410–416. [CrossRef]
3. Zotte, A.D.; Szendrő, Z. The role of rabbit meat as functional food. *Meat Sci.* **2011**, *88*, 319–331. [CrossRef] [PubMed]
4. Abdel-Wareth, A.A.A.; Ahmed, A.; Hassan, H.; El-Sadek, M.A.; Ghazalah, A.; Lohakare, J. Nutritional impact of nano-selenium, garlic oil, and their combination on growth and reproductive performance of male Californian rabbits. *Anim. Feed. Sci. Technol.* **2019**, *249*, 37–45. [CrossRef]

5. Abouelezz, K.F.M.; Abou-Hadied, M.; Yuan, J.; Elokil, A.A.; Wang, G.; Wang, S.; Wang, J.; Bian, G. Nutritional impacts of dietary oregano and Enviva essential oils on the performance, gut microbiota and blood biochemicals of growing ducks. *Animal* **2019**, *13*, 2216–2222. [CrossRef] [PubMed]
6. Rašković, A.; Pavlović, N.; Kvrgić, M.; Sudji, J.; Mitić, G.; Čapo, I.; Mikov, M. Effects of pharmaceutical formulations containing thyme on carbon tetrachloride-induced liver injury in rats. *BMC Complement. Altern. Med.* **2015**, *15*, 442. [CrossRef]
7. Attia, Y.A.; Bakhashwain, A.A.; Bertu, N.K. Thyme oil (Thyme vulgaris L.) as a natural growth promoter for broiler chickens reared under hot climate. *Ital. J. Anim. Sci.* **2016**, *16*, 275–282. [CrossRef]
8. Placha, I.; Chrastinova, L.; Laukova, A.; Čobanová, K.; Takáčová, J.; Strompfová, V.; Chrenková, M.; Formelová, Z. Faix, Štefan Effect of thyme oil on small intestine integrity and antioxidant status, phagocytic activity and gastrointestinal microbiota in rabbits. *Acta Vet. Hung.* **2013**, *61*, 197–208. [CrossRef]
9. Gerencsér, Z.; Szendrő, Z.; Matics, Z.; Radnai, I.; Kovács, M.; Nagy, I.; Cullere, M.; Bosco, A.D.; Zotte, A.D. Effect of dietary supplementation of spirulina (Arthrospira platensis) and thyme (Thymus vulgaris) on apparent digestibility and productive performance of growing rabbits. *World Rabbit. Sci.* **2014**, *22*, 1. [CrossRef]
10. Zotte, A.D.; Cullere, M.; Sartori, A.; Bosco, A.D.; Gerencsér, Z.; Matics, Z.; Kovács, M.; Szendrő, Z. Effect of dietary supplementation of spirulina (Arthrospira platensis) and thyme (Thymus vulgaris) on carcass composition, meat physical traits, and vitamin B12 content on growing rabbits. *World Rabbit. Sci.* **2014**, *22*, 11. [CrossRef]
11. Lee, S.-J.; Umano, K.; Shibamoto, T.; Lee, K.-G. Identification of volatile components in basil (*Ocimum basilicum* L.) and thyme leaves (Thymus vulgaris L.) and their antioxidant properties. *Food Chem.* **2005**, *91*, 131–137. [CrossRef]
12. Rota, M.C.; Herrera, A.; Martinez, R.M.; Sotomayor, J.A.; Jordán, M. Antimicrobial activity and chemical composition of Thymus vulgaris, Thymus zygis and Thymus hyemalis essential oils. *Food Control* **2008**, *19*, 681–687. [CrossRef]
13. Abdulkarimi, R.; Daneshyar, M.; Aghazadeh, A. Thyme (Thymus vulgaris) extract consumption darkens liver, lowers blood cholesterol, proportional liver and abdominal fat weights in broiler chickens. *Ital. J. Anim. Sci.* **2011**, *10*, e20. [CrossRef]
14. El-Ratel, I.T.; Abdel-Khalek, A.-K.E.; Gabr, S.A.; Hammad, M.E.; El-Morsy, H.I. Influence of allicin administration on reproductive efficiency, immunity and lipid peroxidation of rabbit does under high ambient temperature. *J. Anim. Physiol. Anim. Nutr.* **2020**, *104*, 539–548. [CrossRef]
15. Salem, A.A. Effect of Feeding on Olive Oil and Thyme on Pregnancy and Lactation Periods. *Int. J. Nutr. Food Sci.* **2015**, *4*, 19. [CrossRef]
16. Kilkenny, C.; Browne, W.J.; Cuthill, I.C.; Emerson, M.; Altman, D.G. Improving bioscience research reporting: The ARRIVE guidelines for reporting animal research. *PLoS Biol.* **2010**, *8*, e1000412. [CrossRef]
17. NRC. *Nutrient Requirements of Rabbits*; National Academies Press: Washington, DC, USA, 1977.
18. AOAC. *Official Methods of Analysis of AOAC International*, 17th ed.; Association Office Analysis Chemistry: Gaithersburg, MD, USA, 2000.
19. Perez, J.M.; Lebas, F.; Gidenne, T.; Maertens, L.; Xiccato, G.; Parigi-Bini, R.; Dalle Zotte, A.; Cossu, M.E.; Carazzolo, A.; Villamide, M.J.; et al. European Reference Method For In Vivo Determination of Diet Digestibility in Rabbits. *World Rabbit. Sci.* **2010**, *3*, 41–43. [CrossRef]
20. Hall, J.; Melendez, L.D.; Jewell, D.E. Using Gross Energy Improves Metabolizable Energy Predictive Equations for Pet Foods Whereas Undigested Protein and Fiber Content Predict Stool Quality. *PLoS ONE* **2013**, *8*, e54405. [CrossRef]
21. Abozid, M.M.; Asker, M.M. Chemical composition, antioxidant and antimicrobial activity of the essential oil of the thyme and rosemary. *Int. J. Acad. Res.* **2013**, *5*, 186–195. [CrossRef]
22. El-Desoky, N.I.; Hashem, N.M.; Elkomy, A.; Abo-Elezz, Z.R. Physiological response and semen quality of rabbit bucks supplemented with Moringa leaves ethanolic extract during summer season. *Animal* **2017**, *11*, 1549–1557. [CrossRef]
23. Bulsara, M.K. Book review: Alex Dmitrienko, Geert Molenberghs, Christy Chuang-Stein, Walter Offen 2005: Analysis of clinical trials using SAS: A practical guide. Cary, NC: SAS Institute Inc. 420 pp. ISBN 1 59047 504 6. *Stat. Methods Med. Res.* **2007**, *16*, 375–376. [CrossRef]

24. Bozkurt, M.; Hippenstiel, F.; Abdel-Wareth, A.A.A.; Kehraus, S.; Küçükyilmaz, K.; Südekum, K.-H. Effects of selected herbs and essential oils on performance, egg quality and some metabolic activities in laying hens-A review. *Euro. Poult. Sci.* **2014**, *78*, 15. [CrossRef]
25. Hafeez, A.; Männer, K.; Schieder, C.; Zentek, J. Effect of supplementation of phytogenic feed additives (powdered vs. encapsulated) on performance and nutrient digestibility in broiler chickens. *Poult. Sci.* **2016**, *95*, 622–629. [CrossRef] [PubMed]
26. Hippenstiel, F.; Abdel-Wareth, A.A.A.; Kehraus, S.; Südekum, K.-H. Effects of selected herbs and essential oils; and their active components on feed intake and performance of broilers-A review. *Arch. Geflugelk.* **2011**, *75*, 226–234.
27. El-Hack, M.E.A.; Alagawany, M.; Farag, M.R.; Tiwari, R.; Karthik, K.; Dhama, K.; Zorriehzahra, J.; Adel, M.; Zorriehzahra, M.J. Beneficial impacts of thymol essential oil on health and production of animals, fish and poultry: A review. *J. Essent. Oil Res.* **2016**, *28*, 365–382. [CrossRef]
28. Shanoon, A.; Jassim, M.S. Effects of Thymus vulgaris and Zingiber officinale Aqueous on Semen Parameters, Testes Weight and Histology Measurements of Broiler Breeder Male. *Int. J. Poult. Sci.* **2012**, *11*, 594–598. [CrossRef]
29. Kandeil, M.A.; Mohamed, A.E.H.; Gabbar, M.A.; Ahmed, R.R.; Ali, S.M. Ameliorative effects of oral ginger and/or thyme aqueous extracts on productive and reproductive performance of V-line male rabbits. *J. Anim. Physiol. Anim. Nutr.* **2019**, *103*, 1437–1446. [CrossRef]
30. Ruiz-Olvera, S.F.; Rajmil, L.; Sanchez-Curbelo, J.-R.; Vinay, J.; Rodriguez-Espinosa, J.; Ruiz-Castañé, E. Association of serum testosterone levels and testicular volume in adult patients. *Andrologia* **2017**, *50*, e12933. [CrossRef]
31. Gabbar, M.A.; Ahmed, R.R.; Kandeil, M.A.; Mohamed, A.E.H.; Ali, S.M. Administration of ginger and/or thyme has ameliorative effects on liver and kidney functions of V-line rabbits: Histological and biochemical studies. *J. Anim. Physiol. Anim. Nutr.* **2019**, *103*, 1758–1767. [CrossRef]
32. Dieumou, F.E.; Teguia, A.; Kuite, J.R.; Tamokou, J.D.; Fonge, B.N.; Dongmo, M.C. Effects of ginger (*Z. officinale*)and garlic (*Allium sativum*) essential oils on growth performance and gut microbial population of broiler chickens. *Livestock Res. Rural Dev.* **2009**, *21*, 23–32.
33. Shanoon, A.K.; Jassim, M.S.; Amin, Q.H.; Ezaddin, I.N. Eects of Ginger (*Zingiber officinale*) Oil on Growth Performance and Microbial Population of Broiler Ross 308. *Int. J. Poult. Sci.* **2012**, *11*, 589–593. [CrossRef]
34. Tchoffo, H.; Raphaël, K.J.; Ferdinand, N.; Herman, N.V.; Marvel, N.M.W.; D'Alex, T.C.; Vitrice, F.T.L. Effects of Ginger (Zingiber officinale, Roscoe) Essential Oil on Growth and Laying Performances, Serum Metabolites, and Egg Yolk Antioxidant and Cholesterol Status in Laying Japanese Quail. *J. Vet. Med.* **2019**, *2019*, 7857504. [CrossRef]

animals

MDPI

Communication

Severe Acute Respiratory Syndrome Coronavirus 2 (SARS-CoV-2) Exhibits High Predicted Binding Affinity to ACE2 from Lagomorphs (Rabbits and Pikas)

Silvia Preziuso

School of Biosciences and Veterinary Medicine, University of Camerino, 62032 Camerino, Italy; silvia.preziuso@unicam.it

Received: 18 June 2020; Accepted: 17 August 2020; Published: 20 August 2020

Simple Summary: *Severe acute respiratory syndrome coronavirus 2* (SARS-CoV-2) is responsible for the pandemic COVID-19. The virus infects human cells by binding of the virus spike to the cell receptor ACE2. Some studies suggest that dogs, cats and other animal species could be infected by SARS-CoV-2, while very limited data are available on lagomorphs. There are several occasions where rabbits and other lagomorphs are in close contact with humans. To investigate the interaction between SARS-CoV-2 spikes and ACE2 of lagomorphs, predictive computer-based models were used in this study. The structure of ACE2 of lagomorphs was obtained on the basis of the amino acid sequences computationally. The interaction with the model of SARS-CoV-2 spikes was studied and described in depth on the basis of the complex human ACE2-SARS-CoV-2 published before. The interaction among SARS-CoV-2 spikes and ACE2 of other companion or laboratory animals is also described for comparative purposes. The results predict that ACE2 of lagomorphs are likely to bind SARS-CoV-2 spikes and suggest that further studies would be justified to confirm these results and to evaluate the risks to humans being in close contact with lagomorphs, such as veterinarians, farmers, slaughterhouse workers, butchers or pet owners.

Abstract: *Severe acute respiratory syndrome coronavirus 2* (SARS-CoV-2) is responsible for the pandemic COVID-19. The virus infects human cells by binding of the virus spike to the cell receptor ACE2. The crystal structure of SARS-CoV-2 spikes in complex with human ACE2 has recently been solved, and the main amino acid residues involved in the virus–receptor complex have been detected. To investigate the affinity of ACE2 of lagomorphs to the SARS-CoV-2 spike, ACE2 sequences from rabbits and American pikas were compared with human ACE2 and with ACE2 from mammals with different susceptibility to the virus. Models of the complex formed by SARS-CoV-2 spike and ACE2 from lagomorphs and from other mammals were created for comparative studies. ACE2 of lagomorphs showed fewer substitutions than human ACE2 in residues involved in the ACE2-SARS-CoV-2 spike complex, similar to cats. Analysis of the binding interface of the simulated complexes ACE2-SARS-CoV-2 spike showed high affinity of the ACE2 of lagomorphs to the viral spike protein. These findings suggest that the spike of SARS-CoV-2 could bind the ACE2 receptor of lagomorphs, and future studies should investigate the role of lagomorphs in SARS-CoV-2 epidemiology. Furthermore, the risks to humans coming into close contacts with these animals should be evaluated.

Keywords: SARS-CoV-2; ACE2; lagomorphs; virus–receptor complex; *in silico*

1. Introduction

Since its discovery in China, *Severe acute respiratory syndrome coronavirus 2* (SARS-CoV-2) has been spreading rapidly worldwide, causing the Coronavirus Disease 2019 (COVID-19) pandemic [1].

SARS-CoV-2 belongs to the genus *Betacoronavirus* and shows very high sequence similarity to bat *Rhinolophus affinis* coronavirus RaTG13 in the whole genome (93.7% amino acid similarity) [2] and to the Guangdong pangolin coronaviruses in the receptor-binding domain (RBD) [3,4]. RBD is part of the C-terminal domain of the S1 subunit in the SARS-CoV-2 spike protein (S), which interacts with the human angiotensin-converting enzyme 2 (hACE2) receptor [5,6]. The structure of SARS-CoV-2 spike has been investigated by cryo-electron microscopy, and high affinity for hACE2 was found [7]. Priming of coronavirus spike proteins by host cell proteases is essential for viral entry into cells. For example, the cellular serine protease TMPRSS2 is employed by SARS-CoV-2 for the priming of the spike protein [8]. Host protease processing acts as a species barrier, and the addition of exogenous protease to cell cultures may facilitate coronaviruses cell entry. However, the viral spike does not enter the cells, regardless of protease addition, if the receptor is absent or incompatible [9]. Despite other entry mechanisms and potential receptors are under investigations [10,11], the angiotensin-converting enzyme 2 (ACE2) is considered the main receptor for SARS-CoV2 so far. ACE2 protein is abundantly expressed not only in different cells of the human respiratory tract, but also in many other cells in several organs [12]. hACE2 is also the receptor of the other *Betacoronavirus* SARS-CoV, which caused the severe acute respiratory syndrome (SARS) epidemic in 26 countries in 2002–2003 [13].

The crystal structure of SARS-CoV-2 RBD in complex with its receptor hACE2 has been solved (PBD ID 6LZG, PDB 6M17) [6,14]. Despite the overall similarity between the complexes SARS-CoV-RBD and SARS-CoV-2-RBD, several sequence variations and conformational deviations were found in their respective interfaces with ACE2 [14]. Furthermore, the SARS-CoV-2-RBD binding interface has more residues than SARS-RBD (21 versus 17) that directly interact with hACE2, forming more Van der Waals contacts and H-bonds and consequently forming a stronger binding [6]. The specific binding with its receptor determines the host range of a virus. Investigating animal species susceptibility to SARS-CoV-2 is helpful for evaluating the risks of transmission between humans and animals and to prevent further spread of the epidemic, although entry and post-entry barriers may limit the virus replication. Studies based on experimental infections in animals require high-level biosafety facilities and pose ethical issues and legislative concerns. Preliminary screening of potential susceptible animal species is useful before planning in vitro and in vivo experiments. The animal host range of SARS-CoV-2 has been predicted by sequence analysis and structure simulation studies [15,16]; however, not all key residues at the SARS-CoV-2-ACE2 interface, but rather only a limited number of them, have been considered in these studies. Evidence of SARS-CoV-2 infection in animals was reviewed while this article was under revision [17,18]. Lagomorphs may act as an important wildlife reservoir for zoonotic viruses. In vitro and in vivo studies on interaction between SARS-CoV-2 and ACE2 of lagomorphs are lacking, despite the common use of these animals in laboratories, as pets, for food production or for hunting.

In this study, the ACE2 sequences of lagomorphs were investigated and the binding interfaces of the complexes SARS-CoV-2-ACE2 of lagomorphs were predicted in comparison with humans. In addition, comparison was carried out with SARS-CoV-2-ACE2 models of some mammals which showed different degrees of susceptibility to the virus in recent studies.

2. Materials and Methods

2.1. Sequence Analysis

ACE2 sequences from lagomorphs available in databases (rabbit *Oryctolagus cuniculus* XP_002719891 and American pika *Ochotona princeps* XP_004597549.2) were downloaded. These sequences were compared with ACE2 sequences from other companion mammals showing different degrees of susceptibility to SARS-CoV-2 as dogs (*Canis lupus familiaris* NP_001158732.1), cats (*Felis catus* XP_023104564.1), ferrets (*Mustela putorius furo* NP_001297119.1), Chinese hamsters (*Cricetulus griseus* XP_003503283.1), golden Syrian hamsters (*Mesocricetus auratus* XP_005074266.1) [16–23]. Sequences of mice (*Mus musculus* NP_081562.2) and brown rats (*Rattus norvegicus* NP_001012006.1) were also considered because recent structural studies suggest that mouse and rat ACE2 are likely poor receptors for SARS-CoV-2 [5,16]. Sequence of the human ACE2 was used as reference sequence (NP_001358344.1).

Pairwise alignment of amino acid sequences was achieved using computer program MUSCLE (https://www.ebi.ac.uk/Tools/msa/muscle/). Sequences were edited with BioEdit software version 7.2 [24]. ACE2 phylogenetic tree was inferred with the software MEGA version 10.1 [25]. The estimated best-fitting substitution model was Jones-Taylor-Thornton with gamma-distributed rates among sites. Then the maximum likelihood method with bootstrap values based on 1000 repetitions was used. Pairwise distances between the sequences were calculated with the software MEGA version 10.1.

2.2. Structure Simulation of the Complex SARS-CoV-2-RBD-ACE2

The 3D structure of the SARS-CoV-2-RBD-ACE2 complex for each animal species was simulated by homology modelling using SWISS-MODEL [26]. The structure of SARS-CoV-2 spike RBD complexed with its receptor ACE2 was used as template (PDB ID: 6LZG) [6]. The obtained structures were analyzed by Chimera software version 1.14 [27].

Prediction of binding affinity and different types of intermolecular interactions was carried out with the web server PRODIGY (PROtein binDIng enerGY prediction) (https://bianca.science.uu.nl/prodigy/) [28,29]. This web application predicts the binding affinity (BA, expressed as Gibbs free energy ΔG; the lower the value is, the stronger the predicted BA is), counts the number of interfacial contacts (ICs) made at the interface of a protein–protein complex within a 5.5 Å distance threshold, and classifies them according to the polar/nonpolar/charged character of the interacting amino acids. This information is then combined with properties on the Non-Interacting Surface (NIS), which can influence the BA [28,30].

3. Results

3.1. Sequence Analysis

Amino acid sequence alignment of ACE2 molecules from humans, dogs, cats, rabbits, American pikas, Chinese hamsters, golden Syrian hamsters, mice, brown rat, and ferret showed changes in many residues all along the 807 aa sequences. Phylogenetic and pair distance analysis showed that whole ACE2 sequences of rabbits and cats were respectively the first and the second sequences more related with hACE2, while American pikas ACE2 was the most dissimilar to hACE2 (Figure 1 and Table 1). However, not all ACE2 amino acids are involved in the complex with SARS-CoV-2-RBD, being only 22 residues involved in the binding with SARS-CoV-2-RBD [6]. The positions of the key residues involved in the binding with SARS-CoV-2-RBD are reported in the first row of Table 2. A few changes were found among these 22 ACE2 residues in different animal species (Table 2). When only these residues were considered, Chinese hamster and golden Syrian hamster ACE2 were the closest to hACE2 sequence, followed by cats, rabbits and pikas, while mouse and rat ACE2 were the most dissimilar to hACE2 (Table 2, Supplementary Table S1). Among the 22 residues involved in the binding with SARS-CoV-2-RBD [6], ACE2 of Chinese hamsters and golden Syrian hamsters showed only 2 changes (H34Q and M82N). ACE2 of rabbits, American pikas, cats and dogs showed 4 changes, ACE2 of ferrets showed seven changes, and ACE2 of mice and rats showed eight changes. Most changes among sequences from different animal species were found at ACE residues Q24, D30 H34 and M82 (Table 2). The effect of a change in the sequence can be favorable, unfavorable or neutral. The consequences of changes in key residues of the ACE2 sequences were predicted by structure simulations.

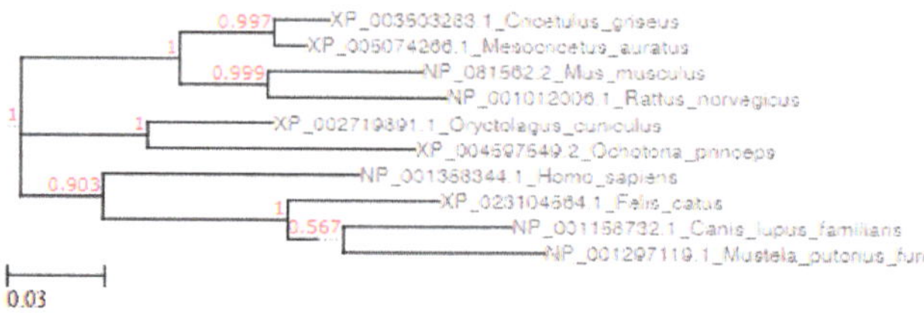

Figure 1. Phylogenetic analysis of the ACE2 sequences of lagomorphs and of other mammals. Human ACE2 sequence was used as reference sequence. The phylogenetic analysis was performed with a maximum likelihood (ML) method using the Jones-Taylor-Thornton model with a gamma distribution and with bootstrap values based on 1000. The tree scale is reported on the bottom left. The tree was visualized by the online tool ETE Toolkit (www.etetoolkit.org).

Table 1. Estimates of evolutionary divergence between sequences. The whole 807 amino acid long sequences of ACE2were considered. The number of amino acid substitutions per site between sequences are shown. Analyses were conducted using the JTT matrix-based model. The rate variation among sites was modeled with a gamma distribution (shape parameter = 5).

	ACE2 Sequence	1	2	3	4	5	6	7	8	9
1	NP_001358344.1_Homo_sapiens	-								
2	XP_023104564.1_Felis_catus	0.170	-							
3	NP_001158732.1_Canis_lupus_familiaris	0.188	0.105	-						
4	XP_002719891.1_Oryctolagus_cuniculus	0.164	0.179	0.187	-					
5	XP_004597549.2_Ochotona_princeps	0.219	0.224	0.220	0.126	-				
6	XP_003503283.1_Cricetulus_griseus	0.174	0.195	0.198	0.149	0.178	-			
7	XP_005074266.1_Mesocricetus_auratus	0.172	0.200	0.204	0.147	0.175	0.029	-		
8	NP_081562.2_Mus_musculus	0.204	0.213	0.219	0.183	0.205	0.114	0.109	-	
9	NP_001012006.1_Rattus_norvegicus	0.201	0.219	0.229	0.183	0.210	0.118	0.112	0.103	-
10	NP_001297119.1_Mustela_putorius_furo	0.199	0.115	0.112	0.195	0.232	0.201	0.206	0.216	0.235

Table 2. Alignment of ACE2 sequences and comparison among residues involved in the contact with SARS-CoV-2 RBD. Points mean that the amino acid is identical to the corresponding amino acid in the hACE2 sequence.

ACE2	Amino Acid Position										
	19	24	27	28	30	31	34	35	37	38	41
Homo sapiens	S	Q	T	F	D	K	H	E	E	D	Y
Felis catus	.	L	.	.	E	.	.	.	.	E	.
Canis lupus familiaris	.	L	.	.	E	.	Y	.	.	E	.
Oryctolagus cuniculus	.	L	.	.	E	.	Q	.	.	.	.
Ochotona princeps	.	L	.	.	.	.	Q	.	.	.	.
Cricetulus griseus	.	.	.	.	.	.	Q	.	.	.	.
Mesocricetus auratus	.	.	.	.	.	.	Q	.	.	.	.
Mus musculus	.	N	.	.	N	N	Q	.	.	.	.
Rattus norvegicus	.	K	S	.	N	.	Q	.	.	.	.
Mustela putorius furio	.	L	.		E	.	Y	.	.	E	.
ACE2	**Amino Acid Position**										
	42	45	79	82	83	330	353	354	355	357	393
Homo sapiens	Q	L	L	M	Y	N	K	G	D	R	R
Felis catus	.	.	.	T	.	.	.	.	.	.	.
Canis lupus familiaris	.	.	.	T	.	.	.	.	.	.	.
Oryctolagus cuniculus	.	.	.	T	.	.	.	.	.	.	.
Ochotona princeps	.	.	.	T	.	.	.	D	.	.	.
Cricetulus griseus	.	.	.	N	.	.	.	.	.	.	.
Mesocricetus auratus	.	.	.	N	.	.	.	.	.	.	.
Mus musculus	.	.	T	S	F	.	H	.	.	.	.
Rattus norvegicus	.	.	I	N	F	.	H	.	.	.	.
Mustela putorius furio	.	.	H	T	.	.	.	R	.	.	.

3.2. Structure Simulation of the Complex SARS-CoV-2-RBD-ACE2

The structure of the protein complex between SARS-CoV-2-RBD and ACE2 of cats, dogs, rabbits, American pikas, Chinese hamsters, golden Syrian hamsters, mice, rats and ferrets were predicted by using the whole sequence of ACE2 (Figures 2–4 and Figure S1).

The positions of the ACE2 residues involved in the complex with SARS-CoV-2-RBD in animals were the same as the ACE2 residues involved in the complex with SARS-CoV-2-RBD in humans (listed in the first column of Table S2). The number of contacts predicted by both Chimera and PRODIGY between ACE2 and SARS-CoV-2-RBD was higher in all animals, apart from mice and rats, than in humans (Figure 2). In particular, the total number of the three strongest types of contacts charged-charged, charged-polar and charged-nonpolar in animal ACE2 complexes with spike SARS-CoV-2-RBD was similar to humans except for mice, which showed the lowest number (n = 18), and ferrets, which showed the highest number (n = 43). Thus, the number of contacts predicted between SARS-CoV-2-RBD and ACE2 of animal species proved to be more susceptible to the virus infection was higher than in animals proved to be less susceptible to the virus infection. This difference is evident in particular when only the three strongest type of contacts are considered. Predicted interactions between ACE2 and spike SARS-CoV-2-RBD showed that BAs expressed as Gibbs free energy values in rabbits and American pikas were similar to humans (Figure 2).

All the animal species included in the study showed similar BAs calculated by the PRODIGY method, suggesting that the spike protein of SARS-CoV-2 could have similar affinity for ACE2 of mammals. However, the lowest BA value was obtained in mice, which is considered a species with low susceptibility to SARS-CoV-2 infection [31]. Further investigations are required to evaluate whether the BAs values obtained by the PRODIGY method could be useful for predicting the interaction between spike SARS-CoV-2-RBD and ACE2 of different animals species.

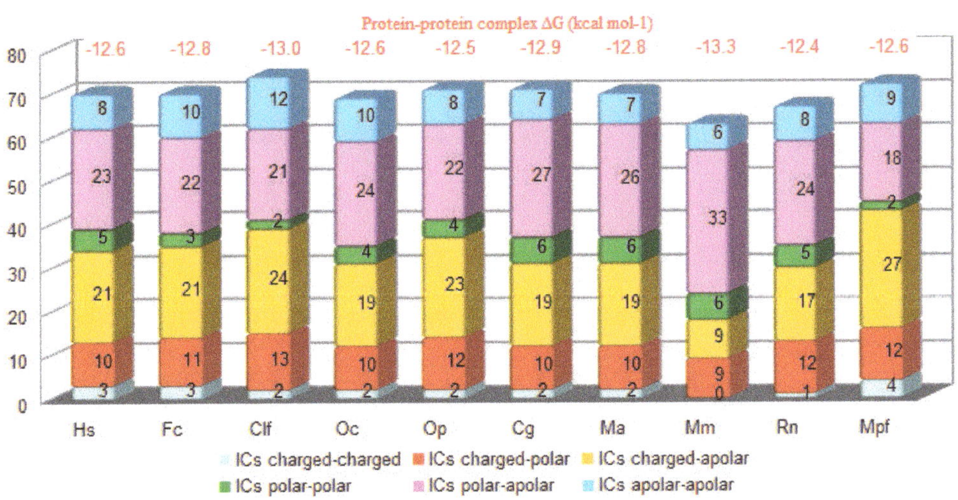

Figure 2. Interfacial contacts (ICs) (axis Y) observed in the ACE2-SARS-CoV-2 predicted complex in humans (Hs), cats (Fc), dogs (Clf), rabbit (Oc), pika (Op), Chinese hamster (Cg), golden Syrian hamster (Ma), ferrets (Mpf), mouse (Mm), rat (Rn) (axis X) by PRODIGY. Predicted protein–protein complex binding affinity expressed as Gibbs free energy (ΔG; kcal mol-1) for each complex is reported in red on the top of each column.

The predicted key residues involved in ACE2-SARS-CoV-2-RBD complex formation in animals and the predicted distances are reported in the Supplementary Table S2. In our models, Q24 hACE was changed in all but Chinese and golden Syrian hamster ACE2 and all Q24L, Q24N and Q24K changes corresponded to a longer predicted distance with viral N487 (Table 2 and Supplementary

Table S2). Similarly, D30N changes corresponded to lack of interaction with viral K417 in mice and rats, and in general D30E changes corresponded to a longer distance with viral K417 in cats, dogs and ferrets, but a shorter distance in rabbits. Shorter distance between ACE2 and spike SARS-CoV-2-RBD residues was found when H34Q substitution was present, while the change H34Y detected in dogs and ferrets corresponded to a longer distance. A longer distance was found also between Y449 and the changed residue D38E in cats, dogs and ferrets. Residue M82 was changed in all animals examined. The distance between viral F486 and T82 was shorter than M82, while the distance between viral F486 and N82 or S82 was longer than M82 (Supplementary Table S2).

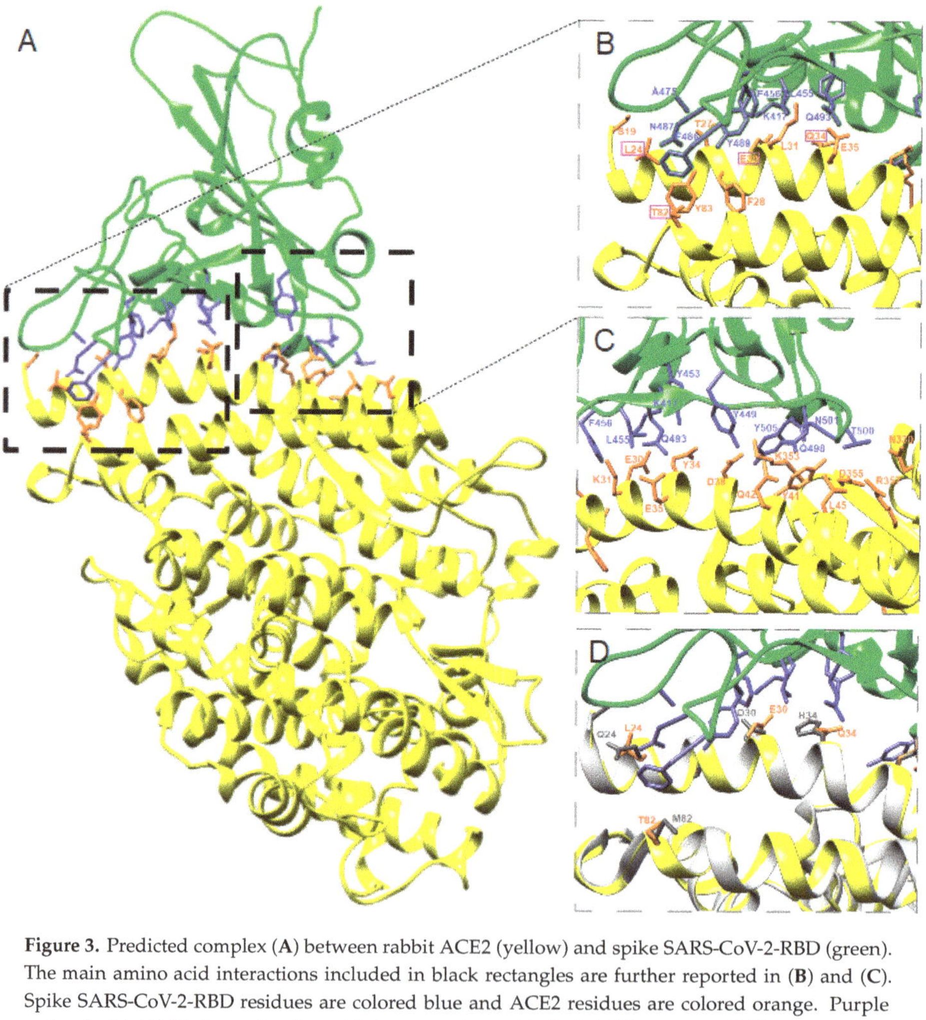

Figure 3. Predicted complex (**A**) between rabbit ACE2 (yellow) and spike SARS-CoV-2-RBD (green). The main amino acid interactions included in black rectangles are further reported in (**B**) and (**C**). Spike SARS-CoV-2-RBD residues are colored blue and ACE2 residues are colored orange. Purple rectangles highlight changed amino acids in comparison with the hACE2 sequence. Differences among hACE and rabbit ACE2 residues involved in the complex with SARS-CoV-2-RBD are shown in (**D**). hACE is colored white and its residues are colored grey, rabbit ACE2 is colored yellow and its residues are colored orange, SARS-CoV-2-RBD is colored green and its residues are colored blue.

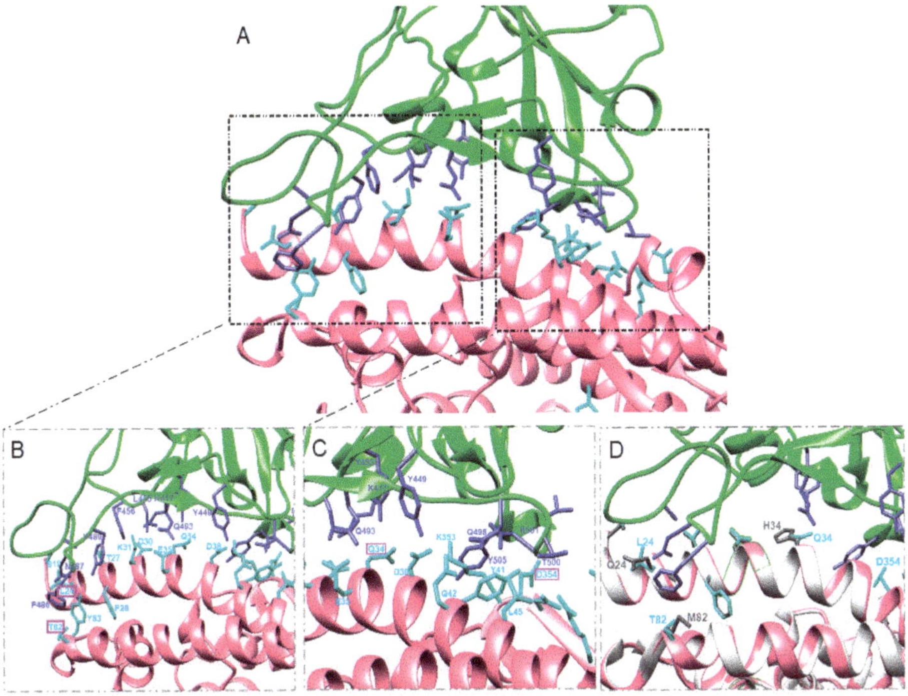

Figure 4. Predicted complex (**A**) between American pika ACE2 (pink) and SARS-CoV-2-RBD (green). The main amino acid interactions included in black rectangles are further reported in (**B**) and (**C**). SARS-CoV-2 residues are colored blue and ACE2 residues are colored cyan. Purple rectangles highlight changed amino acids in comparison with the hACE2 sequence. Differences among hACE and American pika ACE2 residues involved in the complex with SARS-CoV-2-RBD are shown in (**D**). hACE is colored white and its residues are colored grey, American pika ACE2 is colored pink and its residues are colored cyan, SARS-CoV-2 is colored green and its residues are colored blue.

The predicted structures of the protein complexes ACE2-SARS-CoV-2-RBD in rabbits and pikas showed that the four changes present in the respective ACE2 sequences do not affect significantly the binding between amino acids in ACE2 and in SARS-CoV-2-RBD in comparison with humans (Figures 3 and 4). Indeed, some contacts between SARS-CoV-2-RBD and ACE2 of rabbits or American pikas were closer than those observed between SARS-CoV-2-RBD and hACE2, suggesting that changes in the ACE2 sequences could result in closer binding. Sequence studies should not be limited at detecting changes in the ACE2 sequences, but the consequences of these changes on the virus-receptor complex should be evaluated.

4. Discussion

SARS-CoV-2 is responsible for a pandemic and, in the absence of a vaccine, the only measures for prevention of COVID-19 mostly consist of avoiding direct and indirect contact between infected and susceptible humans. SARS-CoV-2 originated from bats, and pangolins have been proposed as intermediate hosts [3,4]. To assess the susceptibility of different animal species to SARS-CoV-2 is important not only for epidemiological purposes, but also for risk evaluation and management. Different animals are kept as pets and live in close contact with humans. Since from the first cases of COVID-19, pet owners have been asking scientists whether animals can spread the virus to people. The risk of human-to-animal transmission of SARS-CoV-2 and then the risk of animal-to-human transmission need to be investigated. In vivo studies by experimental infection of candidate animal hosts have many limitations due to ethical, legal and safety issues. Furthermore, experimental infections to assess the susceptibility of animals to SARS-CoV-2 are limited by the limited availability of high-level biosafety facilities and trained workers. Predictive models are helpful to identify the animal species that are more likely infectable by the virus. Furthermore, interpretation of data obtained by predictive models together with results obtained by in vitro and in vivo studies can suggest the key amino acid residues that play a role in the virus–receptor interaction. In this study, the most recent knowledge about the complex SARS-CoV-2-RBD-hACE2 [6] was used to predict the susceptibility of lagomorph animals in comparison with other mammals showing different degree of susceptibility to the virus by *in silico*, in vitro or in vivo investigations [16–23,32].

The complex structure between spike SARS-CoV-2-RBD and hACE2 includes strong polar contacts among key residues [6]. SARS-CoV-2-RBD residue A475 interacts with hACE2 residue S19, N487 with Q24, E484 with K31, and Y453 with H34. Residue K417 contributes ionic interactions with hACE2 D30. Several residues (G446, Y449, G496, Q498, T500, and G502) in the viral bulged loops are in close proximity with hACE2 amino acids D38, Y41, Q42, K353, and D355, forming a concentration of H-bonds. Further virus–receptor contacts include SARS-CoV-2-RBD Y489 and F486 packing against hACE2 residues F28, L79, M82, and Y83, forming a small patch of hydrophobic interactions at the interface [6]. Changes in these residues could interfere with the optimal binding between viral RBD and ACE2.

Despite the genetic variability of animal ACE2 sequences, the spike SARS-CoV-2-RBD is predicted to bind very conserved residues among animals. A lower number of changes in the ACE2 residues involved in the receptor-virus complex was detected in Chinese hamsters and in golden Syrian hamsters in comparison with humans, as reported recently [32]. The only two changes were M82N, corresponding to a longer distance with the viral residue F486, and H34Q, corresponding to a closer contact with the viral amino acid Y453 (Supplementary Table S2). Recent experiments have shown that golden Syrian hamsters are susceptible to SARS-CoV-2 infection, and they developed clinical signs and lesions and infected naïve contact hamsters housed in the same cage in a 1:1 ratio with infected hamsters [19]. High viral titer was found in the lungs, and the virus was also found in the intestine. These findings suggest that SARS-CoV-2 infection in golden Syrian hamsters occurs despite the M82N and H34Q changes.

American pikas showed two changes at the same positions as golden Syrian hamsters in ACE2. In addition to the same H34Q substitution of golden Syrian hamsters, American pikas have a change at M82T, which shows closer contact with F486 than humans and hamsters (Supplementary Table S2). The same M82T substitution is also present in cats, dogs and ferrets, but not in rats and mice. American pika ACE2 also has the favorable change G354D, and thus a total of three out of four changes result in closer contacts with their viral residues. The fourth substitution in American pika ACE2 is Q24L, which together with M82T is present also in cats, dogs and ferrets, but not in rats and mice.

Recent SARS-CoV-2 experimental infections in cats showed that the virus can replicate efficiently, younger cats are more permissive, and airborne viral transmission between cats may occur [20]. Cats and rabbits share three out of four changes in the SARS-CoV-2 spike-contacting regions of ACE2. The fourth change in rabbit is H34Q, which corresponds to a shorter distance between Q34 and Y453 in all animals having this substitution. In comparison, the fourth change in cats is D38E, corresponding to a longer distance between E38 and the viral amino acid Y449. Furthermore, the predicted protein–protein complex ΔG in rabbits is the same as in humans. These findings suggest that SARS-CoV-2-RBD could bind ACE2 receptor of rabbits and American pikas similarly to in humans, although these findings need to be confirmed.

Experimental studies also showed that SARS-CoV-2 can replicate in the upper respiratory tract of ferrets for up to eight days, but not in the lungs or in other organs. Dogs were found to have low susceptibility to SARS-CoV-2 because the virus was found only sporadically in rectal swabs and not in the respiratory tract [20]. Dogs and cats have the same amino acid differences in the SARS-CoV-2 spike-contacting regions of ACE2, but dogs have the extra change H34Y, which is absent in cats. This substitution corresponds to a longer distance between Y34 and viral Y453 and could interfere with efficient virus binding. H34Y change is also present in ferrets, which showed less efficient viral replication than cats [20].

As observed about SARS-CoV [33] and SARS-CoV-2 [31], mice and rats are not ideal model animals for studying SARS-CoV-2. Although based on the analysis of a small number of residues, some studies have suggested that mice and efficiently rats could not bind SARS-CoV-2 RBD [5,15,16]. In particular, mouse or rat ACE2 contains an asparagine at the 30 position and a histidine at the 353 position, which does not fit into the virus–receptor interaction as well as aspartate or a lysine do, respectively. Furthermore, three changes are present at positions L79, M82 and Y83, which together with F28 form a small patch of hydrophobic interactions at the interface. These findings, in addition to the low number of the interfacial contacts and the number of charged-charged, charged-polar and charged-nonpolar contacts found in this study, support the hypothesis that ACE2 of mice and rats are not optimal receptors for SARS-CoV-2. Experimental infection of wild type mice and of hACE2 transgenic mice confirmed that human ACE2 is required to establish an effective infection [31].

5. Conclusions

In conclusion, the simulated complexes ACE2-SARS-CoV-2-RBD showed high affinity of ACE2 of lagomorphs to the viral spike protein, suggesting that lagomorphs could be susceptible to SARS-CoV-2. There are several occasions where rabbits can be in close contact with humans. For example, rabbits are reared and used for food in many countries, and thus farmers, slaughterhouse workers, veterinarians, butchers, etc. could be exposed to the virus. Furthermore, different species of rabbits are kept as pet animals and the risk for veterinarians should be evaluated when they examinate rabbits living with symptomatic or asymptomatic SARS-CoV-2-infected rabbit owners. The recent finding of the potential for pangolins as the zoonotic reservoir of SARS-CoV-2-like coronaviruses raises the question of the role of wildlife in the epidemiology of the virus. Pikas distribution varies greatly based on species. Two species live in North America, the rest range throughout Asia. Pikas are wild animals, they are not usually kept as pets but are housed in zoos. Some pika species are at risk of extinction due to attempts to capture them and to breed them in captivity. Other lagomorphs are wild animals. Unfortunately, only two sequences of ACE2 of lagomorphs (*Oryctolagus cuniculus* and *Ochotona princeps*) are available in GenBank so far. Future studies should be aimed at sequencing ACE2 protein from different species of lagomorphs. In silico predictions are helpful to detect the animal species having receptors with highest affinity for the virus and to predict favorable and unfavorable virus–receptor interactions. Predicted positive interactions should be confirmed by in vitro and in vivo studies because replication of viruses in species other than their natural hosts can be limited by entry and post-entry barriers.

Supplementary Materials: The following are available online at http://www.mdpi.com/2076-2615/10/9/1460/s1, Table S1: Estimates of Evolutionary Divergence between Sequences. Only 22 residues corresponding to the sites of contact between SARS-CoV-2 RBD and ACE2 were considered. The number of amino acid substitutions per site from between sequences are shown. Analyses were conducted using the JTT matrix-based model. The rate variation among sites was modeled with a gamma distribution (shape parameter = 5); Table S2: Predicted distances among residues involved in contacts between ACE2 and SARS-CoV-2 calculated by Chimera software. Data are reported only for residues showing substitutions. The lowest distance is reported when an ACE2 residue shows more than one contact with the same SARS-CoV-2 residue; Figure S1: Structure simulation of the complex SARS CoV-2-ACE2 of cats (A), dogs (B), Chinese hamsters (C), golden Syrian hamsters (D), mice (E), rats (F) and ferrets (G).

Author Contributions: Study design, acquisition, analysis, and interpretation of data, and article writing were carried out by S.P. The author has read and agreed to the published version of the manuscript.

Funding: This research received no external funding.

Acknowledgments: Molecular graphics and analyses performed with UCSF Chimera, developed by the Resource for Biocomputing, Visualization, and Informatics at the University of California, San Francisco. A special thanks to Andrea Camporeale for his support and technical assistance. Thanks to Delio Camporeale for the information about pikas.

Conflicts of Interest: The authors declare no conflict of interest.

References

1. Zhu, N.; Zhang, D.; Wang, W.; Li, X.; Yang, B.; Song, J.; Zhao, X.; Huang, B.; Shi, W.; Lu, R.; et al. A Novel Coronavirus from Patients with Pneumonia in China, 2019. *N. Engl. J. Med.* **2020**, *382*, 727–733. [CrossRef] [PubMed]
2. Li, C.; Yang, Y.; Ren, L. Genetic evolution analysis of 2019 novel coronavirus and coronavirus from other species. *Infect. Genet. Evol.* **2020**, *82*, 104285. [CrossRef] [PubMed]
3. Lam, T.T.Y.; Shum, M.H.H.; Zhu, H.C.; Tong, Y.G.; Ni, X.B.; Liao, Y.S.; Wei, W.; Cheung, W.Y.M.; Li, W.J.; Li, L.F.; et al. Identifying SARS-CoV-2 related coronaviruses in Malayan pangolins. *Nature* **2020**, *583*, 282–285. [CrossRef] [PubMed]
4. Xiao, K.; Zhai, J.; Feng, Y.; Zhou, N.; Zhang, X.; Zou, J.J.; Li, N.; Guo, Y.; Li, X.; Shen, X.; et al. Isolation of SARS-CoV-2-related coronavirus from Malayan pangolins. *Nature* **2020**, *583*, 286–289. [CrossRef]
5. Wan, Y.; Shang, J.; Graham, R.; Baric, R.S.; Li, F. Receptor Recognition by the Novel Coronavirus from Wuhan: An Analysis Based on Decade-Long Structural Studies of SARS Coronavirus. *J. Virol.* **2020**, *94*. [CrossRef]
6. Wang, Q.; Zhang, Y.; Wu, L.; Niu, S.; Song, C.; Zhang, Z.; Lu, G.; Qiao, C.; Hu, Y.; Yuen, K.Y.; et al. Structural and Functional Basis of SARS-CoV-2 Entry by Using Human ACE2. *Cell* **2020**, *181*, 894–904. [CrossRef]
7. Wrapp, D.; Wang, N.; Corbett, K.S.; Goldsmith, J.A.; Hsieh, C.L.; Abiona, O.; Graham, B.S.; Jason, S.; McLellan, J.S. Cryo-EM structure of the 2019-nCoV spike in the prefusion conformation. *Science* **2020**, *367*, 1260–1263. [CrossRef]
8. Hoffmann, M.; Kleine-Weber, H.; Schroeder, S.; Krüger, N.; Herrler, T.; Erichsen, S.; Schiergens, T.S.; Herrler, G.; Wu, N.H.; Nitsche, A.; et al. SARS-CoV-2 Cell Entry Depends on ACE2 and TMPRSS2 and Is Blocked by a Clinically Proven Protease Inhibitor. *Cell* **2020**, *181*, 271–280. [CrossRef]
9. Letko, M.; Marzi, A.; Munster, V. Functional assessment of cell entry and receptor usage for SARS-CoV-2 and other lineage B betacoronaviruses. *Nat. Microbiol.* **2020**, *5*, 562–569. [CrossRef]
10. Daly, J.L.; Simonetti, B.; Antón-Plágaro, C.; Kavanagh Williamson, M.; Shoemark, D.K.; Simón-Gracia, L.; Klein, K.; Bauer, M.; Hollandi, R.; Greber, U.F.; et al. Neuropilin-1 is a host factor for SARS-CoV-2 infection. *bioRxiv* **2020**. [CrossRef]
11. Cantuti-Castelvetri, L.; Ojha, R.D.; Pedro, L.D.; Djannatian, M.; Franz, J.; Kuivanen, S.; Kallio, K.; Kaya, T.; Anastasina, M.; Smura, T.; et al. Neuropilin-1 facilitates SARS-CoV-2 cell entry and provides a possible pathway into the central nervous system. *bioRxiv* **2020**. [CrossRef]
12. Hamming, I.; Timens, W.; Bulthuis, M.L.C.; Lely, A.T.; Navis, G.J.; van Goor, H. Tissue distribution of ACE2 protein, the functional receptor for SARS coronavirus. A fifirst step in understanding SARS pathogenesis. *J. Pathol.* **2004**, *203*, 631–637. [CrossRef] [PubMed]
13. Li, F.; Li, W.; Farzan, M.; Harrison, S.C. Structure of SARS coronavirus spike receptor-binding domain complexed with receptor. *Science* **2005**, *309*, 1864–1868. [CrossRef] [PubMed]
14. Yan, R.; Zhang, Y.; Li, Y.; Xia, L.; Guo, Y.; Zhou, Q. Structural basis for the recognition of SARS-CoV-2 by full-length human ACE2. *Science* **2020**, *367*, 1444–1448. [CrossRef] [PubMed]
15. Liu, Z.; Xiao, X.; Wei, X.; Li, J.; Yang, J.; Tan, H.; Zhu, J.; Zhang, Q.; Wu, J.; Liu, L. Composition and divergence of coronavirus spike proteins and host ACE2 receptors predict potential intermediate hosts of SARS-CoV-2. *J. Med. Virol.* **2020**, *92*, 595–601. [CrossRef]
16. Luan, J.; Lu, Y.; Jin, X.; Zhang, L. Spike protein recognition of mammalian ACE2 predicts the host range and an optimized ACE2 for SARS-CoV-2 infection. *Biochem. Biophys. Res. Commun.* **2020**, *526*, 165–169. [CrossRef]
17. Abdel-Moneim, A.S.; Abdelwhab, E.M. Evidence for SARS-CoV-2 Infection of Animal Hosts. *Pathogens* **2020**, *9*, 529. [CrossRef]
18. Hossain, M.G.; Javed, A.; Akter, S.; Saha, S. SARS-CoV-2 host diversity: An update of natural infections and experimental evidence. *Microbiol. Immunol. Infect.* **2020**. [CrossRef]
19. Chan, J.F.; Zhang, A.J.; Yuan, S.; Poon, V.K.; Chan, C.C.; Lee, A.C.; Chan, W.M.; Fan, Z.; Tsoi, H.W.; Wen, L.; et al. Simulation of the clinical and pathological manifestations of Coronavirus Disease 2019 (COVID-19) in golden Syrian hamster model: Implications for disease pathogenesis and transmissibility. *Clin. Infect. Dis.* **2020**, ciaa325. [CrossRef]

20. Shi, J.; Wen, Z.; Zhong, G.; Yang, H.; Wang, C.; Huang, B.; Liu, R.; He, X.; Shuai, L.; Sun, Z.; et al. Susceptibility of ferrets, cats, dogs, and other domesticated animals to SARS-coronavirus 2. *Science* **2020**, *368*, 1016–1020. [CrossRef]
21. Sit, T.H.C.; Brackman, C.J.; Ip, S.M.; Tam, K.W.S.; Law, P.Y.T.; To, E.M.W.; Yu, V.Y.T.; Sims, L.D.; Tsang, D.N.C.; Chu, D.K.W.; et al. Infection of dogs with SARS-CoV-2. *Nature* **2020**. [CrossRef] [PubMed]
22. Sailleau, C.; Dumarest, M.; Vanhomwegen, J.; Delaplace, M.; Caro, V.; Kwasiborski, A.; Hourdel, V.; Chevaillier, P.; Barbarino, A.; Comtet, L.; et al. First detection and genome sequencing of SARS-CoV-2 in an infected cat in France. *Transbound. Emerg. Dis.* **2020**. [CrossRef]
23. Sia, S.F.; Yan, L.M.; Chin, A.W.H.; Fung, K.; Choy, K.T.; Wong, A.Y.L.; Kaewpreedee, P.; Perera, R.A.P.M.; Poon, L.L.M.; Nicholls, J.M.; et al. Pathogenesis and transmission of SARS-CoV-2 in golden hamsters. *Nature* **2020**, *583*, 834–838. [CrossRef] [PubMed]
24. Hall, T.A. BioEdit: A user-friendly biological sequence alignment editor and analysis program for Windows 95/98/NT. *Nucl. Acids Symp. Ser.* **1999**, *41*, 95–98.
25. Kumar, S.; Stecher, G.; Li, M.; Knyaz, C.; Tamura, K. MEGA X: Molecular Evolutionary Genetics Analysis across Computing Platforms. *Mol. Biol. Evol.* **2018**, *35*, 1547–1549. [CrossRef] [PubMed]
26. Waterhouse, A.; Bertoni, M.; Bienert, S.; Studer, G.; Tauriello, G.; Gumienny, R.; Heer, F.T.; de Beer, T.A.P.; Rempfer, C.; Bordoli, L.; et al. SWISS-MODEL: Homology modelling of protein structures and complexes. *Nucleic Acids Res.* **2018**, *46*, W296–W303. [CrossRef]
27. Pettersen, E.F.; Goddard, T.D.; Huang, C.C.; Couch, G.S.; Greenblatt, D.M.; Meng, E.C.; Ferrin, T.E. UCSF Chimera—A visualization system for exploratory research and analysis. *J. Comput. Chem.* **2004**, *25*, 1605–1612. [CrossRef]
28. Vangone, A.; Bonvin, A.M. Contacts-based prediction of binding affinity in protein–protein complexes. *eLife* **2015**, *4*, e07454. [CrossRef]
29. Xue, L.C.; Rodrigues, J.P.; Kastritis, P.L.; Bonvin, A.M.; Vangone, A. PRODIGY: A web server for predicting the binding affinity of protein-protein complexes. *Bioinformatics* **2016**, *32*, 3676–3678. [CrossRef]
30. Kastritis, P.L.; Moal, I.H.; Hwang, H.; Weng, Z.; Bates, P.A.; Bonvin, A.M.; Janin, J. A structure-based benchmark for protein-protein binding affinity. *Protein. Sci.* **2011**, *20*, 482–491. [CrossRef]
31. Bao, L.; Deng, W.; Huang, B.; Gao, H.; Liu, J.; Ren, L.; Wei, Q.; Yu, P.; Xu, Y.; Qi, F.; et al. The pathogenicity of SARS-CoV-2 in hACE2 transgenic mice. *Nature* **2020**. [CrossRef]
32. Luan, J.; Jin, X.; Lu, Y.; Zhang, L. SARS-CoV-2 spike protein favors ACE2 from Bovidae and Cricetidae. *J. Med. Virol.* **2020**, *92*, 1649–1656. [CrossRef] [PubMed]
33. Roberts, A.; Wood, J.; Subbarao, K.; Ferguson, M.; Wood, D.; Cherian, T. Animal models and antibody assays for evaluating candidate SARS vaccines: Summary of a technical meeting 25–26 August 2005, London, UK. *Vaccine* **2006**, *24*, 7056–7065. [CrossRef] [PubMed]

animals

MDPI

Article

Diurnal Rhythm of Plasma Melatonin Concentration in the Domestic Turkey and Its Regulation by Light and Endogenous Oscillators

Magdalena Prusik * and Bogdan Lewczuk

Department of Histology and Embryology, Faculty of Veterinary Medicine, University of Warmia and Mazury in Olsztyn, Oczapowskiego Str. 13, 10-719 Olsztyn, Poland; lewczukb@uwm.edu.pl

* Correspondence: mprusik@uwm.edu.pl; Tel.: +48-89-523-39-49

Received: 9 March 2020; Accepted: 10 April 2020; Published: 13 April 2020

Simple Summary: Environmental light regulates a wide range of phenomena in almost all organisms on Earth. Daily and seasonal changes in the photoperiod duration are the most important factors controlling the secretion of melatonin (MLT), a pineal hormone that affects many physiological processes in birds. The results of previous studies on the effect of MLT on the productivity and health of poultry have been promising. However, there are very few studies on the daily profiles of plasma MLT concentrations in domestic birds; therefore, we decided to examine plasma MLT levels in 10-week-old domestic turkeys exposed to different light conditions. The results demonstrated that plasma MLT concentration in turkeys kept under a 12 h light: 12 h dark cycle showed a prominent diurnal rhythm. Night-time light exposure caused a rapid decrease in plasma MLT concentrations. The housing of turkeys in continuous dim red light revealed endogenously generated diurnal rhythm of MLT secretion. The rhythm of the plasma MLT level in a reversed cycle of 12 h dark: 12 h light adapted quickly to the new lighting condition.

Abstract: The aim of this study was to characterize the diurnal rhythm of plasma melatonin (MLT) concentration and its regulation by light and endogenous oscillators in 10-week-old domestic turkeys. Three experiments were performed to examine (i) the course of daily changes in plasma MLT concentration in turkeys kept under a 12 h light: 12 h dark (12L:12D) cycle; (ii) the influence of night-time light exposure lasting 0.5, 1, 2, or 3 h on the plasma MLT level; and (iii) the occurrence of circadian fluctuations in plasma MLT levels in birds kept under continuous dim red light and the ability of turkeys to adapt their pineal secretory activity to a reversed light-dark cycle (12D:12L). The plasma MLT concentration was measured with a direct radioimmunoassay. The plasma MLT concentration in turkeys kept under a 12L:12D cycle changed significantly in a daily rhythm. It was low during the photophase and increased stepwise after the onset of darkness to achieve the maximal level in the middle of the scotophase. Next, it decreased during the second half of the night. The difference between the lowest level of MLT and the highest level was approximately 18-fold. The exposure of turkeys to light during the scotophase caused a rapid, large decrease in plasma MLT concentration. The plasma MLT concentration decreased approximately 3- and 10-fold after 0.5 and 1 h of light exposure, respectively, and reached the day-time level after 2 h of exposure. In turkeys kept under continuous darkness, the plasma MLT level was approximately 2.5-fold higher at 02:00 h than at 14:00 h. In birds kept under 12D:12L, the plasma MLT level was significantly higher at 14:00 h than at 02:00 h. The results showed that plasma MLT concentrations in 10-week-old turkeys have a prominent diurnal rhythm, which is endogenously generated and strongly influenced by environmental light.

Keywords: domestic turkey; pineal; melatonin; plasma; rhythm

---

1. Introduction

Environmental light regulates a wide range of physiological processes in almost all organisms on Earth. The daily and seasonal changes in the photoperiod duration are the most important factors controlling melatonin (MLT) secretion by the pineal gland. Day-night rhythmic MLT secretion regulates in birds, among others, the diurnal rhythm of locomotor activity and feed intake [1–4]; body weight [2]; reproductive system functions [2,5–8]; immune system activity [9–12]; seasonal singing, migration, and spatial orientation during flight [8,13]; thermal homeostasis [14–16]; and energy metabolism [1,17]. MLT is also an important component of the antioxidative defense system [18]. The influence of the pineal organ and its hormone on the productivity and health of poultry has been previously studied, mainly for chickens and turkeys, with promising results. MLT supplementation and/or properly set lighting schedules resulted in a reduction in the incidence of sudden death syndrome [19], activity-related heat production [16,17], and isolation distress [20], in chickens. The better use of feed [2], the influence on growth of the ovaries and egg production [2,5], the improvement of male reproduction [21], and the acceleration of development of the cellular and humoral immune responses [10,11] have been described as the effects of such treatment in turkeys.

In the face of a growing number of studies proving the significant role of the pineal gland and its chief hormone in poultry farming, there are very few studies focusing on the profiles of plasma MLT concentration in domestic birds [18,22–28]. In most cases, the published data involve very young birds aged 2–4 weeks. Liou et al. [24] described significant differences in the nocturnal patterns of plasma MLT between chicks and laying hens. The peak of plasma MLT in laying chicken was broader than that in chicks, which was considered to be essential for the regulation of oviposition.

The aim of our study was to determine in 10-week-old turkeys: 1) the course of the diurnal rhythm of plasma MLT concentration under a 12 h light: 12 h dark (12L:12D) cycle; 2) the influence of night-time light exposure on the plasma MLT level; 3) the occurrence of circadian fluctuations in plasma MLT levels in birds kept under a continuous dim red light; and 4) the ability of turkeys to adapt their pineal secretory activity to a reversed light-dark cycle. The data presented in this article show that the domestic turkey distinguishes among poultry species by the very high-amplitude of the diurnal rhythm of plasma MLT level.

2. Materials and Methods

2.1. Chemicals

Antimelatonin antibody R/R/19540-16876 was provided by Dr. Jean-Pierre Ravault (Institut National De La Recherche Agronomique, France). 2-[^{125}I]-iodomelatonin was purchased from Perkin Elmer (USA), gelatin was obtained from Merck (Germany), and the other reagents were obtained from Sigma (USA).

2.2. Animals and Materials

Female turkeys (*Meleagris gallopavo gallopavo var. domesticus*) were kept under a cycle of 12-h photophase (from 07:00 h to 19:00 h) and 12-h scotophase, starting from the 6th week of their postembryonic life. During the photophase, full-spectrum fluorescent lamps provided light with an intensity of 100 lx at the floor level and during the scotophase the turkeys were kept in red light with an intensity of 3 lx. Dim red light was used as an alternative for darkness, because it enables one to perform animal euthanasia and blood sample collection during scotophase without changes in light condition. Moreover, dim red light was necessary for maintenance of animals kept in continuous darkness in Experiment III. The animals had free access to standard food and water.

2.3. Experiments

Three experiments were performed to determine the diurnal profile of the plasma MLT concentration in turkeys kept under a 12L:12D cycle (Experiment I), changes in the plasma MLT level in response to light exposure at night (Experiment II), and the circadian or diurnal variations in the hormone levels in the plasma of turkeys kept in continuous darkness or under a reversed dark-light cycle (12D:12L) (Experiment III). All experimental procedures on animals were performed in accordance with Polish and European Union laws (Approval of the Local Ethics Committee in Olsztyn for project KBN 6 PO6K 023 21).

2.3.1. Experiment I

The experiment was performed in two replicates. Forty-two turkeys reared under a 12L:12D cycle between the ages of 6 and 10 weeks were used in each replicate. The birds were euthanized at 08:00, 10:00, 12:00, 14:00, 16:00, and 18:00 h (three individuals per time-point, n = 6 from two replicates) and at 20:00, 22:00, 24:00, 02:00, 04:00, and 06:00 h (four individuals per time-point, n = 8 from two replicates). Blood samples were collected to measure the plasma MLT concentrations. Plasma samples were frozen at −20 °C until subsequent MLT assay.

2.3.2. Experiment II

The experiment was performed in two replicates on 24 turkeys in each replicate. Ten-week old birds were divided into two equal groups: control and experimental group. At 01:00 h, the birds from the experimental group were exposed to fluorescent light with an intensity of 100 lx, while the control animals were kept under dim red light. Turkeys in both groups were killed at 01:30, 02:00, 03:00, and 04:00 h (three individuals per time-point from each group, n = 6 from two replicates), and blood samples were collected for MLT assay.

2.3.3. Experiment III

Three groups of birds, housed between 6 and 9 weeks of age under the 12L:12D cycle, were placed in separate rooms. During the next week, they were kept as follows: group I, in a normal light-dark cycle (12L:12D); group II, in a continuous dim red light with intensity below 3 lux (0L:24D); and group III, in a reversed dark-light cycle with a photophase from 19:00 h to 07:00 h and a scotophase from 07:00 h to 19:00h. After 7 days, the turkeys from each group were killed at 14:00 h and 02:00 h (five individuals per time-point from each group) and blood samples were collected.

2.4. Melatonin Radioimmunoassay

MLT concentration in the plasma samples was measured by a direct radioimmunoassay [25–29] with rabbit R/R/19540-16876 antiserum [30] and ^{125}I-melatonin used as a tracer. Antiserum R/R/19540-16876 (200 μL), diluted 1:15000 in assay buffer (tricine, 0.1 M; sodium chloride, 9 g/L; gelatin 1 g/L), was added to a 100 μL sample or standard (0–1000 pg/mL prepared in charcoal-stripped turkey plasma). The mixture was incubated at room temperature for 30 min and then 100 μL of ^{125}I-melatonin, diluted in the assay buffer to approximately 10000 cpm in 0.1 mL, was added. After overnight incubation at 4 °C, the antibody-bound melatonin was separated from the free fraction by incubation with 250 μL dextran-coated charcoal (1.2 g Norit A and 60 mg dextran in 100 mL of assay buffer) for 15 min at 4 °C. After centrifugation (3000× *g*, 20 min at 4 °C), the radioactivity of 350 μL of supernatant was measured using the liquid scintillation method. The concentration of MLT in samples was determined using ImmunoFit EIA/RIA ver. 3.0A software (Beckman, Pasadena, CA, USA). The sensitivity of the assay was 4 pg/mL. Intra- and interassay coefficients of variation were below 10%.

The assay was validated by running the samples containing different amounts of exogenous MLT and the night-time samples diluted with charcoal-stripped turkey plasma. MLT added to samples of turkey plasma was quantitatively recovered (97–105%, R = 0.998–0.999). The samples diluted with charcoal-stripped plasma gave a displacement parallel to that of the standard curve.

2.5. Statistical Analysis

The data were analyzed using a one-way analysis of variance followed by Duncan's test (Experiment I) or by t-test (Experiments II and III) using the Statistica 10.0 (StatSoft, Tulsa, OK, USA) software program. A value of $p \leq 0.05$ was considered significant.

3. Results

3.1. Experiment I

The plasma MLT concentration changed during a diurnal cycle and was significantly higher between 20:00 h and 06:00 h than between 08:00 h and 18:00 h (Figure 1). The lowest level of MLT (7.9 ± 1.9 pg/mL) was measured at 14:00 h; however, there were no significant differences between the investigated time-points of the photophase. The MLT concentration increased stepwise during the first half of the scotophase, reaching a maximum value of 145.1 ± 8.4 pg/mL at 02:00 h, and then decreased. At 06:00 h, the mean MLT level was approximately 3-fold lower than that at 04:00 h.

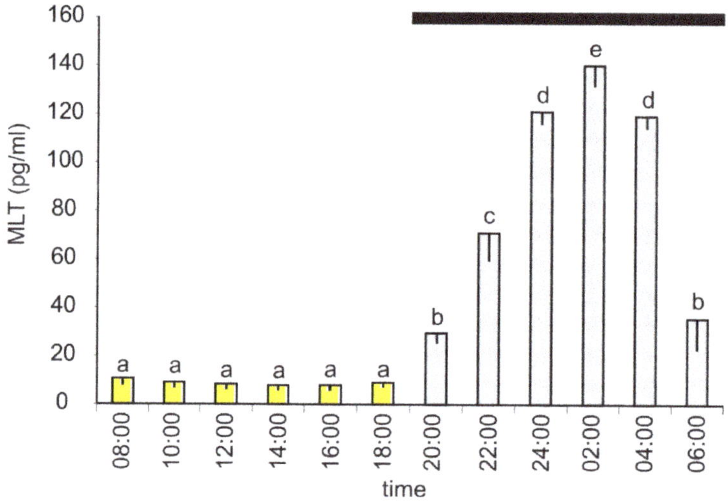

Figure 1. Experiment I. Concentration (mean, and standard error of the mean, SEM) of melatonin in plasma of 10-week-old turkeys kept under a 12L:12D cycle. Values flagged with different letters are significantly different. Black horizontal bar: period of darkness.

3.2. Experiment II

The mean plasma MLT concentrations in turkeys kept in darkness (group I) at 01:30, 02:00, 03:00, and 04:00 h varied between 120 and 150 pg/mL. The light exposure (group II) caused a rapid decrease in plasma MLT concentration (Figure 2). The MLT concentration was more than 3-fold lower after 30-min of light exposure (at 01:30 h) and approximately 10-fold lower after 60-min of light exposure (at 02:00 h) compared to the corresponding controls. The plasma MLT concentration after 2 and 3 h of light exposure decreased to approximately 10 pg/mL.

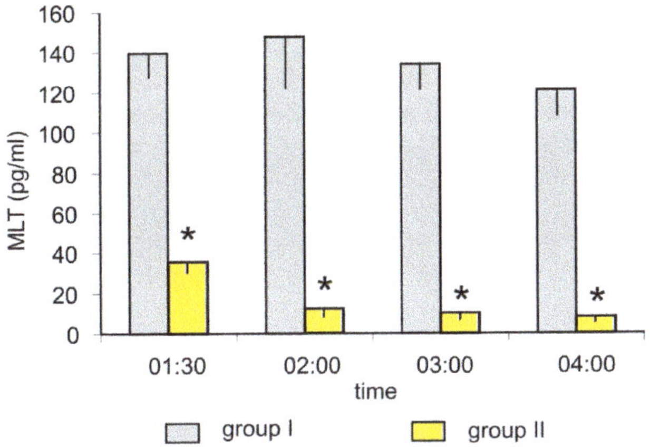

Figure 2. Experiment II. Concentration (mean and SEM) of melatonin in blood plasma in control (group I) and light-exposed (group II) turkeys at 01:30, 02:00. 03:00, and 04:00 h. The exposition of birds from group II started at 01:00 h. Values flagged with "*" differed significantly from the corresponding controls.

3.3. *Experiment III*

The mean plasma MLT concentration in turkeys kept under a 12L:12D cycle (group I) was significantly higher at 02:00 h than at 14:00 h (Figure 3). In contrast, the hormone level was significantly higher at 14:00 h than at 02:00 h in the birds kept under a reversed 12D:12L cycle (group II). The mean plasma MLT concentration of turkeys kept in the continuous dim red light (group III) was significantly higher (approximately 2.5-fold) at 02:00 h than at 14:00 h.

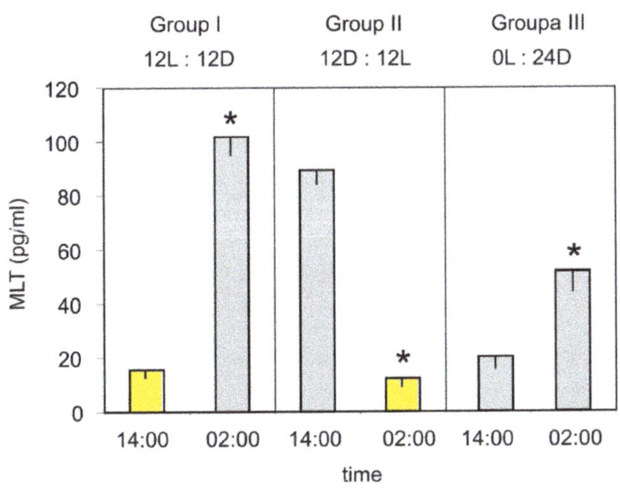

Figure 3. Experiment III. Concentration (mean and SEM) of melatonin in the blood plasma in 10-week-old turkeys kept under 12L:12D cycle (group I), 12D:12D cycle (group II), and continuous dim red light (group III) measured at 14:00 h and 02:00 h. Values significantly different between samples taken at 14:00 h and 02:00 h within each group were flagged with "*".

4. Discussion

The obtained results showed that the plasma MLT concentration in 10-week-old turkeys, housed under a 12L:12D cycle, changed in a distinct diurnal rhythm. The MLT concentration was low during the photophase and increased stepwise after the onset of darkness to achieve a maximal level in the middle of the scotophase. Then, it decreased during the second half of the night. The profile of plasma MLT concentration in turkeys resembles type B of Reiter's classification based on studies performed in mammals [31]. The maximal concentration of MLT, noted at 02:00 h, was approximately 18-fold higher than the minimal concentration at 14:00 h (Experiment I). The amplitude of day-night changes in plasma MLT concentration in 10-week-old turkeys was markedly higher than that in other domestic birds, especially when compared to chickens [18,22,24] and ducks [25]. Plasma MLT in chickens was 3–5 fold higher at night than during the day [18,22,24]. It is worth to note that both species—the turkey and the chicken—belong to the same avian family. The amplitude of day-night changes in plasma MLT in ducks reported by Zawilska et al. [25] was 11-fold.

Our results differ slightly from those reported by Zawilska et al. [27,28] in younger turkeys, in which MLT concentrations measured during both photophase and scotophase were higher and the ratio between the minimum and maximum MLT concentrations was lower than in the present study. These differences suggest that the diurnal profile of plasma MLT concentration undergoes changes during the postembryonic development of the turkey.

The daily profile of plasma MLT concentration described in the present study was similar to the diurnal cycle of MLT secretion in previous in vitro studies that were performed on the pineal organs of turkeys at a similar age [32,33]. However, the ratio between the maximum and minimum concentrations of MLT in the plasma was significantly lower than the amplitude of the MLT rhythm reported in the superfusion culture (18-fold in vivo vs. 40-fold in vitro). This difference may have several sources. First, the MLT level in the plasma is not only a result of the secretory activity of the pineal organ but is also, to some extent, a result of the absorption of MLT into the blood stream from the gastrointestinal tract [34,35] or other sites including the retina [36]. A significant share of the pool of the hormone circulating in the blood during the day-time may have MLT adsorbed from the seeds of cereals that are the main component of the feed for turkeys [37,38]. Studies performed on chickens and pigs revealed that MLT from the feed significantly affected the plasma levels of the hormone during the day [34,35]. In turn, the increase in secretory activity of the pineal organ at night makes the percentage of MLT originating from nonpineal sources very small. Second, plasma MLT concentrations are influenced by the intensity of the metabolism of this hormone in the liver [39,40]. Third, the pineal organ in culture is deprived of neuronal and hormonal regulatory factors and its activity may also be affected by artificial in vitro conditions.

The exposure of turkeys to light during the scotophase caused a rapid decrease in plasma MLT concentration. Our study revealed that 30 and 60 min after light exposure, the plasma MLT concentrations were less than 4 and 1.5 times higher than the level occurring during the day, respectively. Two-hour exposure to light caused a decrease in the plasma MLT concentration close to the day-time level. The light-evoked decrease in vivo was significantly higher than that observed in similar in vitro experiments [32,33]. Results of the in vitro studies showed that even 3-h exposure to light did not decrease the MLT secretion to the level characteristic of the light phase of the cycle. The obtained data strongly suggested that light affects the pineal gland in two ways: directly on photosensitive pinealocytes and indirectly through the retina and adrenergic innervation. The stronger inhibitory effect of light in the in vivo experiments than in the in vitro experiments was probably the result of the inhibitory effect of norepinephrine released from the sympathetic fibers during light exposure. This explanation is supported by the data from in vitro experiments showing a stronger effect of norepinephrine than light on MLT secretion during the scotophase [32].

The experiment that housed the birds under continuous dim red light showed that the rhythm of plasma MLT concentration in the domestic turkey was generated endogenously. After 7 days in such conditions, the concentration of this hormone was still approximately 2.5-fold higher during the subjective night than during the subjective day. The lower amplitude of circadian fluctuations of plasma MLT concentrations resulted both from higher levels during the day-time and lower levels during the night-time. Similar data were obtained from younger turkeys, in which the circadian rhythm of MLT secretion persisted for 7 days under continuous dim red light and had an amplitude that was 50–80% lower than the rhythm in the 12L:12D cycle [27,41]. The circadian rhythm of MLT secretion with gradually decreasing amplitude was also observed in the superfusion culture of the turkey pineal organs for 5 days in continuous darkness [33]. However, the amplitude of this rhythm was very low starting from the third day of culture. The longer persistence of circadian MLT rhythm in vivo than in vitro suggests that the circadian oscillators that are located outside the pineal organ, that is, in the suprachiasmatic nucleus and retina, retain their activity without external photic time cues for a longer period than the pineal oscillator and influence pineal activity. In hens kept under continuous darkness, the circadian rhythm of plasma MLT persisted for 3 days without a decrease in the amplitude [22]. However, it decreased quickly in birds, in which sympathetic innervation of the pineal gland was interrupted [22,42]. In the superfusion culture, the circadian rhythm of MLT secretion persisted longer in the experiments with the turkey pineal organs than the chicken pineal organ [22,32,33,43]. In contrast, the circadian rhythm of MLT secretion from the pineal glands of Japanese quail cultured in constant darkness was very weak or completely abolished [44]. However, the Japanese quail is the only known exception among birds investigated to date in which the pineal oscillator does not play a significant role in the circadian pacemaking system [45]. It is worth noting that circadian rhythms in birds are controlled not only by light but also by other factors acting especially on peripheral oscillators located in other sites than pineal gland, the retina, and the suprachiasmatic nucleus. For example, it has been proven that different feeding regimes significantly affected mRNA expression of circadian clock genes in the liver, jejunum, and kidney [46–48].

After 7 days of housing of 10-week-old turkeys under a reversed dark-light cycle (12D:12L), the diurnal rhythm of the plasma MLT concentration was adapted to the new lighting conditions. In an in vitro experiment, turkey pinealocytes entrained their secretory activity to such changes on the second day of culture [32,33]. The adaptation of the chicken pinealocytes to the 12D:12L cycle was observed on the third day of culture [43].

Our results showed that the turkey pineal organ not only secrets MLT in a daily rhythm with very high amplitude but also responds quickly and precisely to changes in light conditions. These findings have important practical aspects, considering the significance of the domestic turkey as meat-producing animal. Schwean-Lardner et al. [49] exposed turkeys to different cycles with varying periods of darkness and studied body weight, feed efficiency, skeletal disorders, mortality, mobility, and ocular measures, in birds. Most of these factors achieved their best results in cycles with the shortest days. Only feed efficiency was higher in cycles with longer days. MLT extends many variable effects on the organism, which are related to the wide distribution of its receptors in the brain and other organs: the heart, arteries, adrenal gland, kidney, lung, liver, gallbladder, small intestine, adipocytes, ovaries, uterus, breast, prostate, and skin [50]. The pineal hormone may also act intracellularly by binding to calmodulin and Z retinoid nuclear receptors. MLT is one of the most powerful natural antioxidants acting by direct chelation of oxygen and nitrogen reactive species and by mobilization of the intracellular antioxidant enzymatic system [50].

5. Conclusions

In summary, in the present study, we found that 10-week-old turkeys manifest the prominent diurnal rhythm of the plasma MLT concentration, which is endogenously generated and strongly controlled by light. Turkeys are thus a very good avian model for further in vivo studies on the mechanisms regulating MLT secretion from the pineal gland.

Author Contributions: Conceptualization, B.L. and M.P.; methodology, B.L.; validation, B.L. and M.P.; investigation, M.P and B.L.; writing—original draft preparation, M.P.; writing—review and editing, B.L.;

visualization, M.P.; supervision, B.L.; project administration, B.L. All authors have read and agreed to the published version of the manuscript.

Funding: The publication costs were supported by Minister of Science and Higher Education in the range of the program entitled "Regional Initiative of Excellence" for the years 2019-2022, Project No. 010/RID/2018/19, amount of funding 12.000.000 PLN

Acknowledgments: The authors thank Jean-Pierre Ravault from Institut National De La Recherche Agronomique, Nouzilly, France, for gifting the rabbit antiserum R/R/19540-16876.

Conflicts of Interest: The authors declare no conflicts of interest.

References

1. Bermudez, F.F.; Forbes, J.M.; Injidi, M.H. Involvement of melatonin and thyroid hormones in the control of sleep, food intake and energy metabolism in the domestic fowl. *J. Physiol.* **1983**, *337*, 19–27. [CrossRef] [PubMed]
2. Siopes, T.D. Pineal gland and ocular influences on turkey breeder hens. 2. Body weight, feed intake, and egg characteristics. *Poult. Sci.* **1987**, *66*, 528–534. [CrossRef] [PubMed]
3. Chabot, C.C.; Menaker, M. Circadian feeding and locomotor rhythms in pigeons and house sparrows. *J. Biol. Rhythms* **1992**, *7*, 287–299. [CrossRef]
4. Kumar, V.; Gwinner, E. Pinealectomy shortens resynchronisation times of house sparrow (*Passer domesticus*) circadian rhythms. *Naturwissenschaften* **2005**, *92*, 419–422. [CrossRef]
5. Siopes, T.D. Initiation of egg production by turkey breeder hens: Sexual maturation and age at lighting. *Poult. Sci.* **2010**, *89*, 1490–1496. [CrossRef]
6. Siopes, T.D.; Underwood, H.A. Pineal gland and ocular influences on turkey breeder hens. 1. Reproductive performance. *Poult. Sci.* **1987**, *66*, 521–527. [CrossRef]
7. Sharp, P.J. Photoperiodic regulation of seasonal breeding in birds. *Ann. N. Y. Acad. Sci.* **2005**, *1040*, 189–199. [CrossRef]
8. Kumar, V.; Wingfield, J.C.; Dawson, A.; Ramenofsky, M.; Rani, S.; Bartell, P. Biological clocks and regulation of seasonal reproduction and migration in birds. *Physiol. Biochem. Zool.* **2010**, *83*, 827–835. [CrossRef]
9. Moore, C.B.; Siopes, T.D. Effects of lighting conditions and melatonin supplementation on the cellular and humoral immune responses in Japanese quail *Coturnix coturnix japonica*. *Gen. Comp. Endocrinol.* **2000**, *119*, 95–104. [CrossRef]
10. Moore, C.B.; Siopes, T.D. Effect of melatonin supplementation on the ontogeny of immunity in the Large White turkey poult. *Poult. Sci.* **2002**, *81*, 1898–1903. [CrossRef]
11. Moore, C.B.; Siopes, T.D. Enhancement of cellular and humoral immunity following embryonic exposure to melatonin in turkeys (*Meleagris gallopavo*). *Gen. Comp. Endocrinol.* **2005**, *143*, 178–183. [CrossRef] [PubMed]
12. Markowska, M.; Majewski, P.M.; Skwarło-Sońta, K. Avian biological clock-immune system relationship. *Dev. Comp.* **2017**, *66*, 130–138. [CrossRef] [PubMed]
13. Wang, G.; Harpole, C.E.; Paulose, J.; Cassone, V.M. The role of the pineal gland in the photoperiodic control of bird song frequency and repertoire in the house sparrow, *Passer domesticus*. *Horm. Behav.* **2014**, *65*, 372–379. [CrossRef] [PubMed]
14. Underwood, H. The circadian rhythm of thermoregulation in Japanese quail: I. Role of the eyes and pineal. *J. Comp. Physiol. A* **1994**, *175*, 639–653. [CrossRef]
15. Underwood, H.; Edmonds, K. The circadian rhythm of thermoregulation in Japanese quail: II. Multioscillator control. *J. Biol. Rhythms* **1995**, *10*, 234–247. [CrossRef]
16. Rozenboim, I.; Miara, L.; Wolfenson, D. The thermoregulatory mechanism of melatonin-induced hypothermia in chicken. *Am. J. Physiol.* **1998**, *274*, R232–R236. [CrossRef]
17. Apeldoorn, E.J.; Schrama, J.W.; Mashaly, M.M.; Parmentier, H.K. Effect of melatonin and lighting schedule on energy metabolism in broiler chickens. *Poult. Sci.* **1999**, *78*, 223–229. [CrossRef]
18. Albarrán, M.T.; Lopez-Burillo, S.; Pablos, M.I.; Reiter, R.J.; Agapito, M.T. Endogenous rhythms of melatonin, total antioxidant status and superoxide dismutase activity in several tissues of chick and their inhibition by light. *J. Pineal Res.* **2001**, *30*, 227–233. [CrossRef]
19. Clark, W.D.; Classen, H.L. The effects of continuously or diurnally fed melatonin on broiler performance and health. *Poult. Sci.* **1995**, *74*, 1900–1904. [CrossRef]

20. Nelson, E.; Panksepp, J.; Ikemoto, S. The effects of melatonin on isolation distress in chickens. *Pharmacol. Biochem. Behav.* **1994**, *49*, 327–333. [CrossRef]
21. Saleh, Y.S.; Attia, K.A.; Sawiress, F.A.R. Effect of Melatonin Supplement on Some Fertility Parameters of Turkey Toms. In Proceedings of the 5th International Symposium on Turkey Diseases, Berlin, Germany, 16–19 June 2004; pp. 40–50.
22. Cassone, V.M.; Menaker, M. Sympathetic regulation of chicken pineal rhythms. *Brain Res.* **1983**, *272*, 311–317. [CrossRef]
23. Underwood, H.; Binkley, S.; Siopes, T.; Mosher, K. Melatonin rhythms in the eyes, pineal bodies, and blood of Japanese quail (*Coturnix coturnix japonica*). *Gen. Comp. Endocrinol.* **1984**, *56*, 70–81. [CrossRef]
24. Liou, S.S.; Cogburn, L.A.; Biellier, H.V. Photoperiodic regulation of plasma melatonin levels in the laying chicken (*Gallus domesticus*). *Gen. Comp. Endocrinol.* **1987**, *67*, 221–226. [CrossRef]
25. Zawilska, J.B.; Rosiak, J.; Vivien-Roels, B.; Skene, D.J.; Pévet, P.; Nowak, J.Z. Daily variation in the concentration of 5-methoxytryptophol and melatonin in the duck pineal gland and plasma. *J. Pineal Res.* **2002**, *32*, 214–218. [CrossRef] [PubMed]
26. Zawilska, J.B.; Berezińska, M.; Rosiak, J.; Vivien-Roels, B.; Skene, D.J.; Pévet, P.; Nowak, J.Z. Daily variation in the concentration of melatonin and 5-methoxytryptophol in the goose pineal gland, retina, and plasma. *Gen. Comp. Endocrinol.* **2003**, *134*, 296–302. [CrossRef]
27. Zawilska, J.B.; Lorenc, A.; Berezińska, M.; Vivien-Roels, B.; Pévet, P.; Skene, D.J. Diurnal and circadian rhythms in melatonin synthesis in the turkey pineal gland and retina. *Gen. Comp. Endocrinol.* **2006**, *145*, 162–168. [CrossRef]
28. Zawilska, J.B.; Lorenc, A.; Berezińska, M.; Vivien-Roels, B.; Pévet, P.; Skene, D.J. Photoperiod-dependent changes in melatonin synthesis in the turkey pineal gland and retina. *Poult. Sci.* **2007**, *86*, 1397–1405. [CrossRef]
29. Fraser, S.; Cowen, P.; Franklin, M.; Franey, C.; Arendt, J. Direct radioimmunoassay for melatonin in plasma. *Clin. Chem.* **1983**, *29*, 396–397. [CrossRef]
30. Ravault, J.P.; Arendt, J.; Tobler, I.; Chesneau, D.; Maulin, O. Entrainment of melatonin rhythms in rams by symmetrical light-dark cycles of different period length. *Chronobiol. Int.* **1989**, *6*, 329–339. [CrossRef]
31. Reiter, R.J. Pineal melatonin: Cell biology of its synthesis and of its physiological interactions. *Endocr. Rev.* **1991**, *12*, 151–180. [CrossRef]
32. Prusik, M. Mechanisms Regulating Melatonin Secretion in the Turkey Pineal Gland. Ph.D. Thesis, University of Warmia and Mazury, Olsztyn, Poland, 2005.
33. Prusik, M.; Lewczuk, B. Roles of direct photoreception and the internal circadian oscillator in the regulation of melatonin secretion in the pineal organ of the domestic turkey: A novel in vitro clock and calendar model. *Int. J. Mol. Sci.* **2019**, *20*, 4022. [CrossRef] [PubMed]
34. Hattori, A.; Migitaka, H.; Iigo, M.; Itoh, M.; Yamamoto, K.; Ohtani-Kaneko, R.; Hara, M.; Suzuki, T.; Reiter, R.J. Identification of melatonin in plants and its effect on plasma melatonin levels and binding to melatonin receptors in vertebrates. *Biochem. Mol. Biol. Int.* **1995**, *35*, 627–634. [PubMed]
35. Bubenik, G.A.; Pang, S.F.; Cockshut, J.R.; Smith, P.S.; Grovum, L.W.; Friendship, R.M.; Hacker, R.R. Circadian variation of portal, arterial and venous blood levels of melatonin in pigs and its relationship to food intake and sleep. *J. Pineal Res.* **2000**, *28*, 9–15. [CrossRef] [PubMed]
36. Reppert, S.M.; Sagar, S.M. Characterization of the day-night variation of retinal melatonin content in the chick. *Investig. Ophthalmol. Vis. Sci.* **1983**, *24*, 294–300.
37. Zieliński, H.; Kozłowska, H.; Lewczuk, B. Melatonin in cereal grains as a potential cancer prevention agent. In *Dietary Anticarcinogenes and Antimutagens. Chemical and Biological Aspects*; Johnson, I.T., Fenwick, G.R., Eds.; Royal Society of Chemistry: Cambridge, UK, 2000; pp. 266–273.
38. Zieliński, H.; Kozłowska, H.; Lewczuk, B. Bioactive compunds in the cereal grains before and after hydrotermal processsing. *Innov. Food Sci. Emerg. Technol.* **2001**, *2*, 159–169. [CrossRef]
39. Zawilska, J.B.; Skene, D.J.; Arendt, J. Physiology and pharmacology of melatonin in relation to biological rhythms. *Pharmacol. Rep.* **2009**, *6*, 383–410. [CrossRef]
40. Amaral, F.G.D.; Cipolla-Neto, J. A brief review about melatonin, a pineal hormone. *Arch. Endocrinol. Metab.* **2018**, *62*, 472–479. [CrossRef]
41. Lorenc-Duda, A.; Berezińska, M.; Bothorel, B.; Pévet, P.; Zawilska, J.B. Turkey retina and pineal gland differentially respond to constant environment. *J. Comp. Physiol. A* **2008**, *194*, 907–913. [CrossRef]

42. Cassone, V.M.; Takahashi, J.S.; Blaha, C.D.; Lane, R.F.; Menaker, M. Dynamics of noradrenergic circadian input to the chicken pineal gland. *Brain Res.* **1986**, *384*, 334–341. [CrossRef]
43. Csernus, V.; Ghosh, M.; Mess, B. Development and control of the circadian pacemaker for melatonin release in the chicken pineal gland. *Gen. Comp. Endocrinol.* **1998**, *110*, 19–28. [CrossRef]
44. Murakami, N.; Nakamura, H.; Nishi, R.; Marumoto, N.; Nasu, T. Comparison of circadian oscillation of melatonin release in pineal cells of house sparrow, pigeons and Japanese quail, using cell perfusion system. *Brain Res.* **1994**, *651*, 209–214. [CrossRef]
45. Steele, C.T.; Zivkovic, B.D.; Siopes, T.; Underwood, H. Ocular clocks are tightly coupled and act as pacemakers in the circadian system of Japanese quail. *Am. J. Physiol. Regul. Integr. Comp. Physiol.* **2003**, *284*, R208–R218. [CrossRef] [PubMed]
46. Lin, X.; Liu, Y.; Meng, T.; Xie, C.; Wu, X.; Yin, Y. Circadian calcium feeding regime in laying hens related to zinc concentration, gene expression of circadian clock, calcium transporters and oxidative status. *J. Trace Elem. Med. Biol.* **2018**, *50*, 518–526. [CrossRef] [PubMed]
47. Liu, Y.; Lin, X.; Zhou, X.; Wan, D.; Wang, Z.; Wu, X.; Yin, Y. Effects of dynamic feeding low and high methionine diets on egg quality traits in laying hens. *Poult. Sci.* **2017**, *96*, 1459–1465. [CrossRef] [PubMed]
48. Liu, Y.; Wan, D.; Zhou, X.; Ruan, Z.; Zhang, T.; Wu, X.; Yin, Y. Effects of dynamic feeding low- and high-methionine diets on the variation of glucose and lipid metabolism-related genes in the liver of laying hens. *Poult. Sci.* **2019**, *98*, 2231–2240. [CrossRef]
49. Schwean-Lardner, K.; Vermette, C.; Leis, M.; Classen, H.L. Basing Turkey Lighting Programs on Broiler Research: A Good Idea? A Comparison of 18 Daylength Effects on Broiler and Turkey Welfare. *Animals (Basel)* **2016**, *6*, 27. [CrossRef]
50. Tordjman, S.; Chokron, S.; Delorme, R.; Charrier, A.; Bellissant, E.; Jaafari, N.; Fougerou, C. Melatonin: Pharmacology, Functions and Therapeutic Benefits. *Curr. Neuropharmacol.* **2017**, *15*, 434–443. [CrossRef]

animals

MDPI

Article

Productive Results, Oxidative Stress and Contaminant Markers in European Sea Bass: Conventional vs. Organic Feeding

Antonio Carminato [1], Francesco Pascoli [1], Angela Trocino [2], Lisa Locatello [3], Lisa Maccatrozzo [2], Renato Palazzi [4], Giuseppe Radaelli [2,*], Cristina Ballarin [2], Martina Bortoletti [2] and Daniela Bertotto [2]

1 Italian Health Authority and Research Organization for Animal Health and Food Safety, Viale dell'Università 14, 35020 Padova, Italy; acarminato@izsvenezie.it (A.C.); fpascoli@izsvenezie.it (F.P.)

2 Department of Comparative Biomedicine and Food Science (BCA), University of Padova, Viale dell'Università 16, 35020 Padova, Italy; angela.trocino@unipd.it (A.T.); lisa.maccatrozzo@unipd.it (L.M.); cristina.ballarin@unipd.it (C.B.); martina.bortoletti@phd.unipd.it (M.B.); daniela.bertotto@unipd.it (D.B.)

3 Department of Biology, University of Padova, Via U. Bassi 58/b, 35121 Padova, Italy; lisa.locatello@unipd.it

4 Veneto Agricoltura, Innovation and Development Section, Viale dell'Università 14, 35020 Padova, Italy; renato.palazzi@venetoagricoltura.org

* Correspondence: giuseppe.radaelli@unipd.it; Tel.: +39-049-8272593

Received: 10 June 2020; Accepted: 15 July 2020; Published: 18 July 2020

Simple Summary: Over the years, aquaculture moved to organic production given the rising interest of consumers towards healthy and ecologically friendly food. Among the cultured species, European sea bass (*Dicentrarchus labrax*) is one of the leading farmed fish products in the Mediterranean area and thus one of the most economically important. For these reasons, further investigations on the effects of organic feeding on this species are of primary interest. In the present study, European sea bass were fed two different diets, organic and conventional, and growing performances, oxidative stress, and contaminant markers were determined. Although conventional diet gave the best results in terms of production, groups fed with the organic one also showed a positive growth trend and importantly no negative effects on fish welfare were observed, demonstrating the feasibility of this diet. This work represents an insight into the emerging aquaculture organic production.

Abstract: In the present study European sea bass (*Dicentrarchus labrax*) subjected to two different diets (organic vs. conventional) were evaluated in terms of growing performances, oxidative stress, and contaminant markers. Growing performances were evaluated using biometric measures and condition factor (K), whereas insulin-like growth factor (IGF-I and IGF-II) levels were assessed trough Real-Time PCR analysis. For oxidative stress, immunohistochemical staining for 8-hydroxy-2′-deoxyguanosine (8-OHdG) and 4-hydroxy-2-nonenal (HNE) was performed, whereas total glutathione (GSH) in blood serum was determined by an enzymatic method adapted. Cytochrome P4501A (CYP1A) and melanomacrophage centers (MMCs) were evaluated as contaminant markers trough immunohistochemical and histochemical approaches, respectively. The growing performances showed a positive trend in both groups but a greater productivity in conventional fed fish compared to the organic ones. A significant higher expression of MMCs was observed in organic vs. conventional diet fed fish. Fillet analysis showed a higher MUFA content and a lower PUFAs n-6 content in organically fed sea bass indicating that diets with a content in fatty acids closer to that of wild fish will definitely affect the fatty acid profile of the fish flesh. On the other hand, the diet composition did not seem to affect neither the oxidative stress parameters (GSH, 8-OHdG, HNE) nor the CYP1A expression.

Keywords: immunohistochemical analysis; RT-PCR; IGFs; GSH; CYP1A; *Dicentrarchus labrax*; organic aquaculture

---

1. Introduction

In recent years, the development of organic production responds to a growing consumer request for a high degree of food safety, high nutritional value, sustainable production and eco-environmental attention [1,2]. One of the main topic of the European Commission (EC) organic farming regulation (N.889/2008) [3] is the feed formulations that aim to ensure fish and consumer health and high quality final product with a low environmental impact. Several studies showed the importance of the diet for the optimum growing performances of fishes [4–6].

Growth in vertebrates depends on a regulatory network in which the growth hormone (GH)-insulin-like growth factor (IGF)-I axis plays a key role in growth regulation together with insulin, thyroid hormones and sex steroids [7]. While IGF-I mRNA is expressed mainly in liver of adult fish, as in mammals and non-mammalian vertebrates [8–10], IGF-II is proved to be ubiquitously expressed, working essentially as a growth factor [11–14].

Food composition is also considered an important factor in preserving welfare, growth, development, and reproduction of the fish. An important role among all the nutrients is certainly played by the Highly Unsaturated Fatty Acids (HUFAs) [15–17], the deficiency of which causes a decreased growth rate and a weakening of immune function. Under farming conditions, the fish are often subjected to unavoidable stressors such as manipulation, size grading, stocking density, fasting, transport, conditions of pre-slaughter and slaughter techniques that could affect health status [18–20].

Oxidative stress is considered one of the major upstream components of the signaling cascade involved in many cellular functions [21]. Under conditions of oxidative stress, many effects of cellular dysfunction such as oxidation of proteins, polyunsaturated fatty acids (PUFAs) and DNA are mediated. 4-hydroxy-2-nonenal (HNE) is the most abundant and toxic α,β-unsaturated aldehyde, which originates from the β-cleavage of hydroperoxides from ω-6 PUFAs. It is mainly involved in the inhibition of protein and DNA synthesis and is also considered a potent mutagenic agent [22]. DNA damage may be due to modification of bases, such as the oxidation of deoxyguanosine (dG) to form 8-hydroxy-2′-deoxyguanosine (8-OHdG). If not repaired, this damage can lead to an incorrect pairing between adenine (A) and 8-OHdG, rather than cytosine (C), causing a G:C to T:A transverse mutation [23,24].

High dietary levels of lipids and vitamins likely influence the oxidative status as pointed out by several studies that show a protective effect of such diets [25,26]. In order to prevent oxidation-induced lesions and mortalities, there must be effective antioxidant systems involving compounds such as glutathione (GSH). Total GSH is used as an indicator of oxidative stress due to its action against ROS or molecules such as benzoates and others.

Due to the presence of fish meal and fish oil, commercially marine feed could represent a source of persistent organic pollutants and heavy metals such as mercury, cadmium and arsenic [27,28]. The assessment of the Cytochrome P4501A (CYP1A) in fish represents an environmental biomarker since CYP1A is involved in the biotransformation of a variety of aquatic contaminants such as oil compounds, dioxins, PCBs, and PAHs [29–31]. Beyond the liver, which represents the main site of CYP1A expression in fish, the epithelia of organs in direct contact with the environment (gills, intestine and kidney) and the vascular endothelia exhibit a remarkable CYP1A expression as well. Nowadays, melanomacrophage centers' (MMCs) evaluation is often used as a reliable indicator of pollutant exposure, in particular heavy metals [32–35] indicating sub-lethal effects [36]. MMCs are groups of pigments containing cells located within the tissues of cold-blooded vertebrates. In fish, MMCs located primarily in the spleen, kidney, and liver [37] can increase after chemical pollutant exposure but also as a result of diseases [38,39] such as chronic inflammatory lesions, ovarian atresia, and changes induced by starvation [40].

The present study was conducted in a commercial farm during the productive season in order to evaluate the potential of organic feeding in European sea bass aquaculture in terms of growth, welfare and product quality. Growth and welfare have been investigated by measure of biometric parameters, growth factor expression, and oxidative stress indicators while fillet composition and contaminant markers have been evaluated to test the effect on the product quality. Tissue microarray technology has been applied to optimize the histological analysis. Although skin coloration might allow to distinguish organically vs. conventionally-produced sea bass [41], this study gives new insights into organic sea bass aquaculture.

2. Materials and Methods

2.1. Rearing Conditions

European sea bass (initial body weight 62.7 ± 4.0 g) were reared at the "Centro Ittico Valle Bonello" (Porto Tolle, RO, Italy) in a period of the year ranging from April to November into two separated 300 m^3 outdoor ponds (50 × 6 × 1 m), one for organic and one for conventional feeding trial. All fish had been previously fed a conventional diet and at the start of the trial, fish were moved from the pre-fattening tanks and randomly distributed to the fattening ponds. One of the pond received the conventional diet, the other one was fed with the organic diet. During the trial, fish in the two ponds were fed two times per day by an automatic system and at a rate in the range of 1–2% live weight which was adjusted according to changes in water temperature Figure 1.

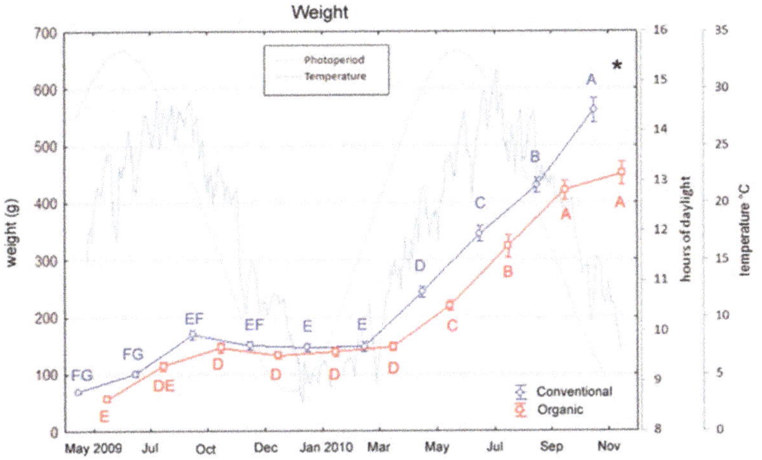

Figure 1. Variations in weight of sea bass over an 18-month period (mean ± SE) reared under conventional and organic aquaculture. Different letters indicate significant differences in samplings within the same feeding system (organic or conventional; $p < 0.05$). Asterisk indicates significant differences between feeding systems (organic vs. conventional) within the same sampling ($p < 0.05$).

In all the ponds, stocking density was the same (initial stocking density: 2 kg/m^3; 10,000 fish per pond), as all the other parameters (water temperature and photoperiod, both environmental). The dissolved oxygen was kept at normoxic conditions for European sea bass (>7.5 mg/L) as recommended by EFSA (European Food Safety Authority) [40] and ranged from 8 to 12 mg/L in both ponds. Fish of the two groups were fed with different types of diet, one certified organic and one conventional Table 1. Overall mortality at the end of the trial in both the ponds was 15% and final densities were 12 and 14 kg/m^3 for organic and conventional groups, respectively.

Table 1. Proximate composition (% as-fed) and fatty acid profile (% total fatty acids methyl esters) of the diets.

Rearing System (R)	Organic	Conventional
Proximate composition		
Water (%)	7.10	6.32
Ether extract (%)	20.3	17.2
Crude protein (%)	40.7	44.5
Crude fibre (%)	0.73	1.18
Ash (%)	12.3	7.73
Fatty acid profile		
C14:0	5.03	3.05
C15:0	0.37	0.23
C16:0	13.1	12.3
C17:0	0.35	0.32
C18:0	2.50	3.82
C20:0	0.21	0.28
Other SFAs	0.77	0.53
Total SFAs	22.4	20.6
C16:1n-7	3.90	3.03
C18:1n-7	3.93	3.99
C18:1n-9	18.5	21.3
C20:1n-9	2.76	0.84
C22:1n-9	0.71	0.13
C22:1n-11	8.08	0.62
C24:1n-9	0.46	0.19
Other MUFAs	0.81	0.35
Total MUFAs	39.0	30.4
C18:3n-3	2.92	4.96
C18:4n-3	5.82	0.79
C20:5n-3	4.47	5.96
C22:5n-3	0.91	0.74
C22:6n-3	5.90	3.10
PUFAs n-3	20.0	15.6
C18: 2n-6	13.3	29.3
C20:4n-6	0.24	0.31
PUFAs n-6	13.5	29.6
Ratio of n-3 to n-6 PUFAs	1.48	0.53
Other PUFAs	1.00	0.67
Total PUFAs	34.5	45.8
Unknown FAs	4.15	3.21

FAs: fatty acids; SFAs: Saturated FAs; MUFAs: monounsaturated FAs; PUFAs: polyunsaturated FAs.

2.2. Animal Sampling and Samples Preparation

Fish were monitored for 18 months until they reached the commercial size. Bimonthly, 20 fishes per pond were rapidly netted and immediately after capture, put into iced brackish water and brought to the near laboratory facilities. A big net was used to guarantee a representative sampling and all the sampled animals were subjected to the same procedure that had taken no more than 10 min. Animals were then euthanized with an overdose of anesthetic (MS222, 500 ppm), immediately bled from the caudal vein, and the following organs were collected: liver, spleen, gut, and head kidney. Samples for histology were immediately fixed in 4% paraformaldehyde prepared in phosphate-buffered saline while liver tissue sampled also for molecular analyses, was immediately fixed in trizol and processed as described below. Serum was isolated by allowing the blood to clot overnight at +4 °C and standard centrifugation then conserved at −20 °C until analysis. During each sampling, biometric measures (weight, total and standard lengths) were recorded and Fulton's condition factor (K) was calculated (K = fish weight/fish total length ^{3}) [42]. All animals were treated as requested by EFSA guidelines [43].

2.3. Histochemistry

Tissue samples were washed in phosphate-buffered saline (PBS), dehydrated through a graded series of ethanol and embedded in paraffin. Serial sections were cut at a thickness of 4 μm. Haematoxylin and eosin (H & E) staining and Schmorl's reaction, used to evaluate MMCs for count (see below), were performed as described in Bancroft and Gamble [44].

2.4. Tissue Microarray

The array was built using paraffin liver and gut donor blocks of the 200 sampled fishes. Trained technologists using the Beecher TMA instrument (Beecher Instruments, Sun Prairie, WI) removed 1 core of 0.6 mm per donor blocks and transferred it into a recipient block. Cores were arranged in sectors, each containing 4 rows with 10 cores per row following a previously drawn, preset unequivocal plot. Positive and negative control samples were included in each array. Serial 4 μm thick sections were cut.

2.5. Immunohistochemistry

Immunohistochemical staining for 8-OHdG, HNE, and CYP1A was performed using the automated immunostainer Benchmark Ultra Ventana as described in Pascoli et al. [34]. Immunoreactive scoring for 8-OHdG and HNE was done by counting positive nuclei in 100 cells. Samples stained with CYP1A were evaluated for the presence and distribution of immunopositivity, and a grade from negative (−) to strong (++) was assigned to the intensity of the reaction.

2.6. Melanomacrophage Centers (MMCs) Count

Serial sections of spleen were stained with H & E sequential stain to ascertain structural details, and with Schmorl's reaction (as described above) to detect the melanomacrophage centers (MMCs). Microscopic quantitative assessment of MMCs was made through a computerized image analyzer system (Olympus CellB, Japan) on sections of spleen since it is the organ which exhibited the highest number of MMCs, as also reported in literature [45]. This quantitative assessment proceeded as follows: (1) Each haul was represented by 3 sections from each spleen. (2) Three fields from each spleen section were analyzed and the amount of MMCs was recorded.

2.7. Glutathione (GSH)

Total GSH in blood serum was determined by an enzymatic recycling method adapted for microtitre plate reader [46]. Initially, a standard curve is prepared by diluting GSH Standard stock solution with TF-E (0.1 M phosphate buffer, 0.6 mM EDTA); then in decreasing concentrations of standard solution (SS) thus obtained, are added precise amount of TF-E. Subsequently, two reagents are prepared, the Reaction Solution (RS) and the Reductase. Respectively, for the RS 8.3 mg of NADPH are dissolved in 1 mL of distilled water, to which are then added 0.04 TF-E and 600 μL 5.5′-ditiobis-2-nitrobenzoic acid (DTNB). For Reductase preparation, 15 μL of the commercial reductase solution (Glutathione Reductase, 205 units/mg protein, Sigma-Aldrich) is diluted with 53.4 μL of ammonium sulphate 3.6 M (Sigma a-4915, MW132.1), then 65 μL of the obtained solution is diluted with 3835 μL of TF-E. A 96 multiwell plate is loaded with 30 μL of TF-E (blank), the standard curve (30 μL of the various points, in descending order) and samples, 15 μL of each in duplicate. The RS is then added and the plate is read at 405 nm (Microplate Photometer Spectracount, Packard Instrument, Meriden, CT, USA) for about ten minutes, until no differences in absorbances were recorded, so the reductase activity is over. At this point, 25 μL of reductase are added to all wells and read on for 20 min. Samples results are compared to standard and expressed in nmol/mL.

2.8. Qualitative Reverse Transcription/PCR and Quantitative Real-Time PCR

Qualitative reverse transcription/PCR and quantitative real-time PCR were performed by following the methods detailed in Bertotto et al. [47]. RNAs were extracted from the liver tissue (50 mg sample) of 12 individuals (6 conventional, 6 organic) per sampling period (4 periods: May, December; May and November of the following year) using TRIZOL Reagent (Gibco-BRL, Gaithersburg, MD, USA).

Total RNA (1.5 µg) was retrotranscribed into cDNA. First-strand cDNAs were synthesized by using Superscript II RNase reverse transcriptase protocols (Invitrogen, Life technologies, UK) and a mixture of random hexamers as primer (synthetized by MWG-Biotech, Ebersberg, Germany). The obtained cDNAs were used as templates for PCR expression analysis. We refer the reader to Bertotto et al. [47] for details on IGF-I primer design and efficiency. The primer for IGF-II was designed by using Primer Express 3.0 software (Applied Biosystems) (forward, 5′-AGTGTTGTTTCCGTAGCTGTGA-3′, reverse 5′-ATCCTGAGGGCCAAAAAGTATCG-3′) and its specificity checked by PCRs. Data were normalized to the housekeeping gene β-actin.

Quantification assays to detect the relative expression of IGF-I and IGF-II mRNA were carried out by using the ABI 7500 Real-Time PCR System (Applied Biosystems) as described by Bertotto et al. [47]. Data from SYBR Green I PCR amplicons were collected with ABI 7500 System SDS Software. The ΔΔCt method was used for relative quantification (comparative method) using a calibrator sample as basis for comparative results (see Chemistry Guide, Applied Biosystem, 2003). Dissociation melting curves confirmed the specific amplification of the cDNA target and the absence of nonspecific products.

2.9. Proximate Composition and FA Analysis

For proximate composition and fatty acids (FA) analysis, a total of 16 sea bass (8 specimens per rearing system) were collected in January 2011. All fish were slaughtered by immersion in ice slurry and immediately transported to the laboratory in thermally insulated boxes and stored on ice in a refrigerated room (2 °C) for subsequent analysis on the day following collection. Fresh minced fillets were analyzed for FA composition as detailed by Trocino et al. [48].

2.10. Statistical Analyses

Statistical analyses were carried out with STATISTICA 9 (StatSoft) and SAS (ver. 9.1) software. All data are reported as mean ± SEM. Data were checked for normality using a Kolmogorov-Smirnov test and in the case of IGF-I and IGF-II expression they were log-transformed to meet the assumption. The effect of feeding condition (organic vs. conventional) and sampling period (10 sampling periods from May to November) on growing performances (weight, length, condition index), IGF-I and IGF-II expression levels, MMCs count and GSH levels were analyzed by means of univariate two-way factorial ANOVAs (GLM). Feeding condition and sampling periods were included in the model as independent fixed factors; growing and oxidative stress parameters as dependent variables. HSD-Tukey's post-hoc tests were performed when identifying a significant effect. In all analyses a $p < 0.05$ value was considered as statistically significant.

The fillet data collected in the study were analyzed with the GLM procedure of SAS. The diet was used as the experimental factor. Published data [49] were also included for comparison.

3. Results

3.1. Growth

Fish growth was evaluated using biometric measures and condition factor (K). Since standard and total length were highly correlated (r = 0.99; $p < 0.001$), total length (TL) is considered hereinafter.

The two-way ANOVA evidenced a significant effect of both the groups and the sampling period on weight (diet: $F_{1,9}$= 16.78, $p < 0.001$; period: $F_{1,9}$ =331.36, $p < 0.001$) and TL (diet: $F_{1,9}$ = 8.99, $p < 0.001$; period: $F_{1,9}$ = 258.01, $p < 0.001$). The interaction between the two factors (diet × period) was also significant for both variables (weight: $F_{1,9}$ = 4.25, $p < 0.001$; TL: $F_{1,9}$ = 2.20, p = 0.02). Indeed, conventional fed fish show generally higher weight and length than organic ones Figures 1 and 2. Moreover, in both groups, fish exhibited a clear seasonal trend with an interruption of growth during cold months and a recovery of growth from May to the end of the sampling period (Tukey's tests: March significantly differed from the following months, all $p < 0.05$. Figures 1 and 2). The interaction between factors (diet x period) was evident at the end of the feeding period, when conventional fishes showed a higher increase in weight and TL than organic ones. Indeed, at the end of the trial weight and TL of conventional fishes significantly differed from those of organic ones (Tukey's test: weight, $p < 0.001$; TL, $p = 0.038$).

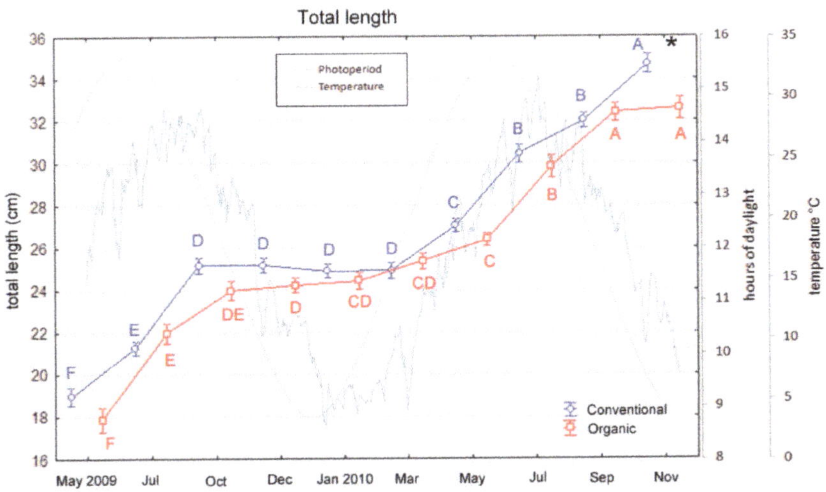

Figure 2. Variations in total length of sea bass over an 18-month period (mean ± SE) reared under conventional and organic aquaculture. Different letters indicate significant differences in samplings within the same feeding system (organic or conventional; ($p < 0.05$). Asterisk indicates significant differences between feeding systems (organic vs. conventional) within the same sampling ($p < 0.05$).

Regarding the condition factor "K", the analyses evidenced a significant effect of both the groups and the sampling period (diet: $F_{1,9} = 5.44$, $p = 0.02$; period: $F_{1,9} = 84.58$, $p < 0.001$), but no interaction between the two factors (sampling × period). Indeed, in both groups, as observed for growth and TL, fish exhibited a seasonal trend in K values that were lower during cold months Figure 3 and significantly increased from May to the end of the feeding period (Tukey's tests: March significantly differed from the following months, all $p < 0.001$. Figure 3). However, this trend was similar for conventional and organic fed fishes and no difference between the two groups was observed at the end of the feeding period.

3.2. *Immunohistochemistry*

8-OHdG and HNE immunostaining were evaluated as DNA damage and oxidative stress markers, respectively, whereas CYP1A as environmental biomarker. Counting stained nuclei for 8-OHdG and CYP1A performed in liver and gut did not show any statistically difference between the two groups (ANOVA, $p > 0.05$; Figure 4). The anti-HNE staining was detected in the spleen, head kidney, and liver mostly in the MMCs and spare macrophages Figure 5. Immunopositivity was found both in organic and conventional samples with no differences in intensity between the two groups.

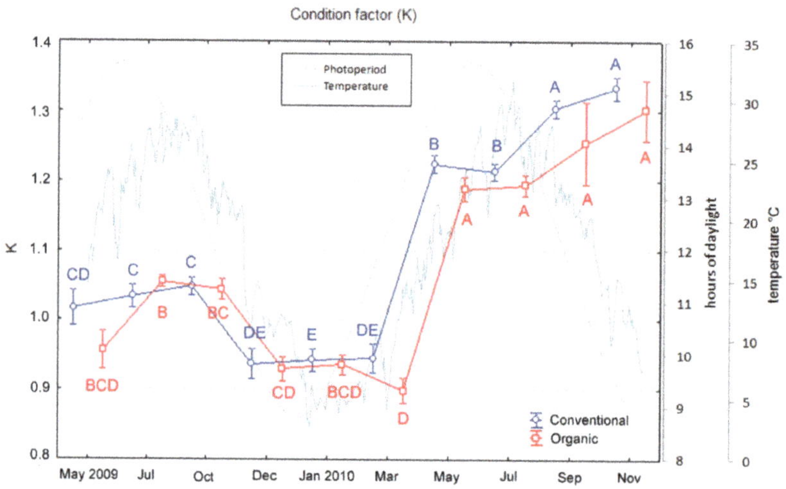

Figure 3. Variations in condition factor (K) of sea bass over an 18-month period (mean ± SE) reared under conventional and organic aquaculture. Different letters indicate significant differences in samplings within the same feeding system (organic or conventional; $p < 0.05$).

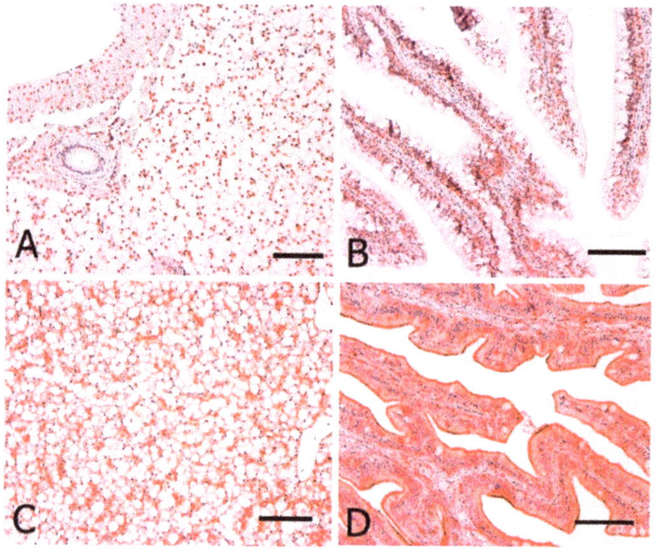

Figure 4. Immunohistochemical localization of CYP1A and 8-OHdG in sea bass. All panels are counterstained with Harryr's hematoxylin. Scale bar: 50 μm. (**A**) Liver, strong and diffuse nuclear immunoreactivity of hepatocytes to anti-8-OHDG antibody; (**B**) Gut, nuclear immunoreactivity of most enterocytes to anti-8-OHDG antibody; (**C**) Liver, endothelial, and sinusoidal immunoreactivity of hepatocytes to anti-CYP-1A; (**D**) Gut, diffuse strong immunoreactivity of the intestinal mucosa to anti-CYP-1A.

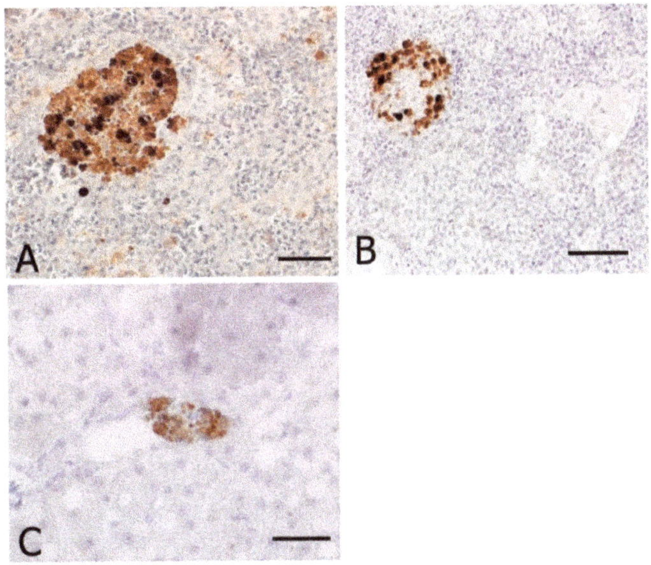

Figure 5. Immunohistochemical localization of HNE in sea bass. All panels are counterstained with Harryr's hematoxylin. Scale bar: 50 μm. HNE-immunostaining is present in several melanomacrophage centers located in the parenchyma of (**A**) spleen; (**B**) kidney; (**C**) liver.

3.3. MMCs Count

MMCs counts, as markers of pollutant exposure, were performed on spleen sections and revealed several differences between the two groups Figure 6. In most samples, organic fed fishes exhibited a higher number of MMCs respect to conventional ones, except for March and May of the second year, where no significant differences were found (Tukey's test, $p < 0.01$).

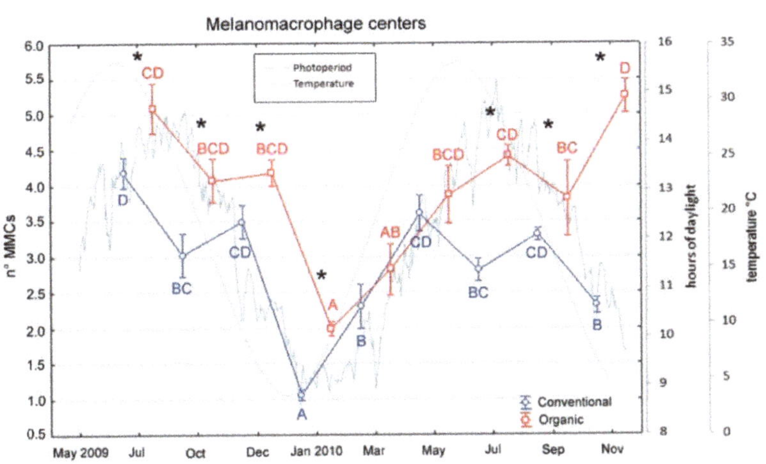

Figure 6. Variations in melanomacrophage centers number of sea bass over an 18-month period (mean ± SE) reared under conventional and organic aquaculture. Different letters indicate significant differences in samplings within the same feeding system (organic or conventional; $p < 0.05$). Asterisks indicate significant differences between feeding systems within the same sampling ($p < 0.05$).

3.4. Total Glutathione (GSH)

Total glutathione (GSH) was evaluated as oxidative stress marker. Total GSH assay showed significant differences between the two groups only in July and November of the second year (Tukey's tests, $p < 0.05$; Figure 7). In both samplings, conventional fed fishes showed higher values than organic ones (35.9 ± 0.9 vs. 34.0 ± 1.8 and 36.2 ± 0.5 vs. 29.0 ± 2.2 nmol/mL).

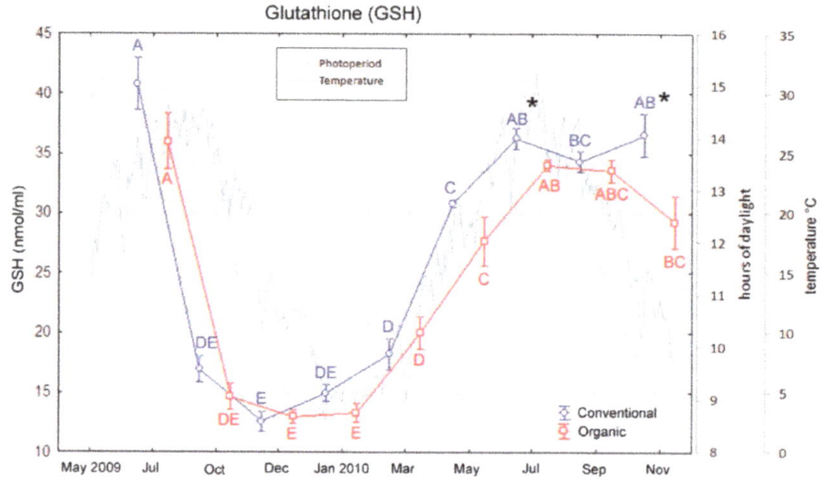

Figure 7. Variations in glutathione of sea bass over an 18-month period (mean ± SE) reared under conventional and organic aquaculture. Different letters indicate significant differences in samplings within the same feeding system (organic or conventional; $p < 0.05$). Asterisks indicate significant differences between systems within the same sampling ($p < 0.05$).

3.5. IGF-I and IGF-II expression

Due to their role in growth regulation, IGF-I and IGF-II expression has been assessed. Neither the feeding condition nor the sampling period had a significant effect on IGF-I and IGF-II expression, and no significant interaction between the two factors was observed (IGF-I: diet: $F_{1,3} = 0.13$, $p = 0.72$; period: $F_{1,3} = 0.27$, $p = 0.85$; diet × period: $F_{1,3} = 0.20$, $p = 0.89$. IGF-II: diet: $F_{1,3} = 0.004$, $p = 0.95$; period: $F_{1,3} = 0.78$, $p = 0.51$; diet × period: $F_{1,3} = 0.30$, $p = 0.82$).

3.6. Proximate Composition and FA Analysis

The chemical composition of the experimental diets fed during the last period of growth of sea bass differed among treatments. In detail, the conventional diet showed lower ether extracts and ash compared to the organic one (17.2% vs. 20.3% and 7.73% vs. 12.3%, respectively), whereas an opposite trend was recorded for crude protein (44.5% vs. 40.7%) Table 1.

The proximate composition and fatty acid profile of sea bass fillets are reported in Table 2. Despite the starvation period prior to slaughtering, the ether extract content tended to be higher in sea bass that were fed with the diet containing the highest ether extract ($p = 0.06$), i.e., the conventional diet, in the last period of feeding compared to those fed the organic diet. Some differences among treatments were found in the fatty acid profile of sea bass. In fact, MUFAs content was higher in fillets of sea bass fed organic diet than conventional one (37.4% vs. 33.8%; $p = 0.02$), especially due to the higher content of C20:1n-9 and C22:1n-11. Besides, the PUFAs n-6 content was lower in the organic than in conventional fillets (12.0% vs. 15.8%; $p = 0.04$) due to the lower content of C18: 2n-6 (11.0% vs. 14.3%; $p = 0.04$).

Table 2. Proximate composition (% as-fed) and fatty acid profile (% of total fatty acid methyl esters) of sea bass fillets.

Rearing System (R)	Organic	Conventional	Probability	RSD
Live weight (g)	432	473	0.18	57
Proximate composition				
Water (%)	71.3	69.3	0.08	0.60
Ether extract (%)	7.93	9.72	0.06	0.47
Crude protein (%)	19.5	19.5	0.92	0.37
Ash (%)	1.14	1.16	0.41	0.01
Fatty acid profile				
C14:0	4.32	3.30	0.10	0.34
C15:0	0.53	0.40	0.15	0.06
C16:0	16.6	17.1	0.66	0.94
C17:0	0.54	0.49	0.32	0.04
C18:0	7.45	7.00	0.84	2.10
Other SFAs	4.18	2.90	0.22	0.72
Total SFAs	33.6	31.2	0.62	4.16
C16:1n-7	4.50	4.52	0.59	0.03
C18:1n-9	19.6	22.8	0.05	0.77
C18:1n-7	2.24	2.44	0.04	0.04
C20:1n-9	4.24	1.75	< 0.001	6.15
C22:1n-11	3.41	0.69	< 0.001	7.40
Other MUFAs	3.41	1.54	0.03	0.34
Total MUFAs	37.4	33.8	0.02	0.53
C18:3n-3	1.71	1.77	0.75	0.00
C18:4n-3	1.04	0.73	0.09	0.10
C20:5n-3	2.72	4.41	0.10	0.56
C22:5n-3	0.70	0.96	0.29	0.18
C22:6n-3	3.36	3.67	0.79	1.06
PUFAs n-3	10.1	12.0	0.45	2.10
C18: 2n-6	11.0	14.3	0.04	0.69
C20:4n-6	0.34	0.46	0.08	0.04
PUFAs n-6	12.0	15.8	0.04	0.79
Ratio of n-3 to n-6 PUFAs	0.83	0.76	0.64	0.12
Other PUFAs	3.03	3.63	0.10	0.20
Total PUFAs	23.8	29.9	0.18	2.96
Unknown FAs	5.14	5.13	0.99	0.00

FAs: fatty acids; SFAs: Saturated FAs; MUFAs: monounsaturated FAs; PUFAs: polyunsaturated FAs.

4. Discussion

The present study investigated the productivity, the oxidative stress status and contaminant response in European sea bass fed an organic feed vs. a conventional one. As regards growth, conventional fed fishes showed a significant increase in growth at the end of the trial, suggesting that conventional feeding leads to a greater productivity. Nevertheless, the expression of IGF-I and IGF-II mRNA was similar throughout the whole experimental period confirming their roles on growth. As expected, the feed intake was affected by temperature, since both groups of fish experienced reduction of food intake during the cold season recording a temporary growth arrest [50]. In a different study on the effects of feed restriction in sea bass, Escobar-Aguirre et al. observed a decrease in plasma IGF-I levels of treated fish [51], thus confirming that the lower levels found in colder months could be related to the stop of feed intake due to the winter starvation period.

The diet composition did not seem to affect neither the oxidative stress parameters (GSH, 8-OHdG, HNE) nor the CYP1A expression, although several studies demonstrated the anti-oxidative effect of lipid and vitamin-rich diets [25,26]. It is likely that a more pronounced variation in the composition of such nutrients would need to affect antioxidant defenses. Nevertheless, total GSH exhibited some seasonal effects: Lowest values were found during cold months (October–March), whereas higher levels were seen during summer months. This seasonal tendency is in agreement with previous studies on the impact of temperature on fish GSH levels in various organs [52,53]. Moreover, it is well known that antioxidant levels are modified by food availability as Pascual et al. pointed out in their study in which they observed an increase of antioxidant activities and a disruption of GSH redox status in gilthead sea bream (*Sparus aurata*) kept under a food deprivation condition [54]. In this work, immunopositivity for HNE was detected in spleen, kidney, and liver, but with no differences in intensity between organic and conventionally fed fish. However, Fiocchi et al. in their study, aimed to evaluate the stress oxidative biomarkers in sea bass, also observed immunoprecipitates for HNE in the same organs [55], validating the use of this substance as stress biomarker.

Conversely, a significant higher expression of MMCs was observed in organic vs. conventional diet fed fish. However, this preliminary result, to be confirmed by a greater number of data, is discordant with Montero et al. [56] that reported an increased number of MMCs in *Sparus aurata* juveniles fed with a diet, low in EPA and DHA but could be related to a more efficient system of detoxification in the organic fishes or to the source of proteins and lipids within the organic diet. Organic farming condition could have developed a more reactive system of detoxification in fishes e.g., a higher MMCs rate leading to a better ability to respond versus potential stressors. Since few studies on MMCs centers in organic farmed sea bass are available, this result could be compared with what Magrone et al. [57] and Arciuli et al. [58] found in sea bass fed a diet enriched with polyphenols extracted from red grape. In fact, they also observed an increase of splenic and kidney MMCs number, area and activity in fish fed polyphenols, demonstrating how alternative feeding could indeed enhance the action of the immune and oxidative system. Otherwise, piscivorous fish represent an important accumulator of heavy metals and/or other pollutants since fish meal and oil used in organic aqua-feed are notoriously a source of contaminants [27]. As reported in salmon feeds [59,60], the fish oils obtained from feral pelagic fish species are considered the main source of persistent organic pollutants (POPs). In order to contain this problem, decontamination techniques have recently been developed to effectively remove persistent organic contaminants from fish oils [61,62]. Moreover, wild foods are notoriously highly nutritious but restricted in quantity and the use of fish-derived meal and oil, if on the one hand is closer to the natural life of the fish, on the other hand it does not match the principle of sustainability, causing overexploitation of fisheries and wild stocks [63] and should be reduced.

The chemical composition of the organic and conventional diets administered to the sea bass differed in terms of ether extract values, lower in the organic diet, and ash content and crude protein content higher compared with the conventional diet. The fillet FA profile is known to be strictly dependent on the diet. Indeed, the level of n-3 PUFAs largely depends on the dietary level and the types of supplemented fish meal and oils, especially vegetable oil [2,64–66]. Even if the precise source of fish oil used in the two diets was not declared by the producer, the high level of eicosenoic acid and cetoleic acid makes to presuppose that the organic diet was likely supplemented with a blend of fish oils containing herring oil. The organic aquaculture regulation fixes a maximum inclusion level of vegetable feeds at 60% of the diet and indirectly imposes a 40% level of fish meal and oil as a minimum and this should be reflected in the FA fillet composition of organic vs. conventional fed fishes as previously found by Trocino et al. [47].

5. Conclusions

In conclusion, this was a field study aimed to evaluate the growth performances as well as the expression of IGF-I and IGF-II mRNA, the oxidative stress and contaminant response in European sea bass fed with organic diet. The study was carried out as a pilot study in a small commercial plant to verify the feasibility of the production of sea bass using organic feed. These productive conditions impeded replications but give first results, although preliminary, on commercial production volumes. Our results highlighted a greater productivity in conventional fed fish comparing to the organic ones. The higher productivity was likely due to diet composition, since differences were significantly mitigated during starvation period. On the other hand, the considered oxidative stress and contaminant markers did not show any significant differences among groups. Feeding fish with diets with a content in fatty acids closer to that of wild fish definitely affect the nutritional value of the flesh in terms of the fatty acid profile. The consumption of the derived flesh could be considered more appropriate and healthier than the conventional one but due to the exploitation of some fisheries, wild-caught fish could not be the answer to the organic feed requirements. The findings and limitations raised in this study could stimulate a challenging debate in the field of organic fish nutrition in order to think a new concept of organic fish nutrition and consequently to better address required future researches in the same field.

Author Contributions: Conceptualization, D.B. and G.R.; methodology, formal analysis, investigation, C.B., M.B., A.C., L.L., L.M., F.P. and A.T.; resources, R.P.; writing—original draft preparation, A.C. and F.P.; writing—review and editing, D.B. and G.R.; supervision, D.B. and G.R.; funding acquisition, D.B. and F.P. All authors have read and agreed to the published version of the manuscript.

Funding: This research was supported by a grant of University of Padua to D. Bertotto (Progetto di Ateneo CPDA082999/08) and by European Social Fund grant to F. Pascoli (2105/101/1/722/2009).

Acknowledgments: Authors wish to thank Tommaso Brogin and Giovanni Caporale for their technical assistance.

Conflicts of Interest: The authors declare no conflict of interest. The funders had no role in the design of the study; in the collection, analyses, or interpretation of data; in the writing of the manuscript, or in the decision to publish the results.

References

1. Willer, H.; Schlatter, B.; Trávníček, J.; Kemper, L.; Lernoud, J. The world of organic agriculture 2020: Summary. In *World of Organic Agriculture. Statistics and Emerging Trends 2020*; Willer, H., Schlatter, B., Trávníček, J., Kemper, L., Lernoud, J., Eds.; Research Institute of Organic Agriculture (FiBL), Frick, and IFOAM–Organics International: Bonn, Germany, 2020; pp. 20–30.
2. Di Marco, P.; Petochi, T.; Mariño, G.; Priori, A.; Finoia, M.; Tomassetti, P.; Porrello, S.; Giorgi, G.; Lupi, P.; Bonelli, A.; et al. Insights into organic farming of European sea bass Dicentrarchus labrax and gilthead sea bream Sparus aurata through the assessment of environmental impact, growth performance, fish welfare and product quality. *Aquaculture* **2017**, *471*, 92–105. [CrossRef]
3. EUR-Lex, European Union Low. Organic Farming Regulation n.889/2008. Available online: http://eur-lex.europa.eu/legal-content/EN/TXT/?uri=CELEX%3A32008R0889/ (accessed on 25 March 2020).
4. Satoh, S.; Hernandez, A.; Tokoro, T.; Morishita, Y.; Kiron, V.; Watanabe, T. Comparison of phosphorus retention efficiency between rainbow trout (Oncorhynchus mykiss) fed a commercial diet and a low fish meal based diet. *Aquaculture* **2003**, *224*, 271–282. [CrossRef]
5. Wang, X.; Kim, K.-W.; Bai, S.C.; Huh, M.-D.; Cho, B.-Y. Effects of the different levels of dietary vitamin C on growth and tissue ascorbic acid changes in parrot fish (Oplegnathus fasciatus). *Aquaculture* **2003**, *215*, 203–211. [CrossRef]
6. Albrektsen, S.; Mundheim, H.; Aksnes, A. Growth, feed efficiency, digestibility and nutrient distribution in Atlantic cod (Gadus morhua) fed two different fish meal qualities at three dietary levels of vegetable protein sources. *Aquaculture* **2006**, *261*, 626–640. [CrossRef]
7. Jones, J.I.; Clemmons, D.R. Insulin-like growth factors and their binding proteins: Biological actions*. *Endocr. Rev.* **1995**, *16*, 3–34. [CrossRef] [PubMed]
8. Shamblott, M.J.; Chen, T.T. Age-related and tissue-specific levels of five forms of insulin-like growth factor mRNA in a teleost. *Mol. Mar. Biol. Biotechnol.* **1993**, *2*, 351–361. [PubMed]
9. Funkenstein, B.; Almuly, R.; Chan, S. Localization of IGF-I and IGF-I Receptor mRNA in Sparus aurata Larvae. *Gen. Comp. Endocrinol.* **1997**, *107*, 291–303. [CrossRef]
10. Patruno, M.; Sivieri, S.; Poltronieri, C.; Sacchetto, R.; Maccatrozzo, L.; Martinello, T.; Funkenstein, B.; Radaelli, G. Real-time polymerase chain reaction, in situ hybridization and immunohistochemical localization of insulin-like growth factor-I and myostatin during development of Dicentrarchus labrax (Pisces: Osteichthyes). *Cell Tissue Res.* **2007**, *331*, 643–658. [CrossRef]

11. Carnevali, O.; Cardinali, M.; Maradonna, F.; Parisi, M.; Olivotto, I.; Mosconi, G.; Funkenstein, B.; Polzonetti-Magni, A.M. Hormonal regulation of hepatic IGF-I and IGF-II gene expression in the Marine Teleost Sparus aurata. *Mol. Reprod. Dev.* **2005**, *71*, 12–18. [CrossRef]
12. Wood, A.W.; Duan, C.; Bern, H.A. Insulin-like growth factor signaling in fish. *Intern. Rev. Cytol.* **2005**, *243*, 215–285. [CrossRef]
13. Patruno, M.; Maccatrozzo, L.; Funkenstein, B.; Radaelli, G. Cloning and expression of insulin-like growth factors I and II in the shi drum (Umbrina cirrosa). *Comp. Biochem. Physiol. B* **2006**, *144*, 137–151. [CrossRef] [PubMed]
14. Radaelli, G.; Poltronieri, C.; Bertotto, D.; Funkenstein, B.; Simontacchi, C. Cellular localization of insulin-like growth factor-II protein in the sea bass (Dicentrarchus labrax) from hatching to adult. *Histol. Histopathol.* **2008**, *23*, 523–530. [PubMed]
15. Sargent, J.R.; Bell, J.G.; Bell, M.V.; Henderson, R.J.; Tocher, D.R. The metabolism of phospholipids and polyunsaturated fatty acids in fish. In *Coastal and Estuarine Studies*; American Geophysical Union (AGU): Washington, DC, USA, 1993; Volume 43, pp. 103–124.
16. Sargent, J.R.; Bell, J.G.; Bell, M.V.; Henderson, R.J.; Tocher, D.R. Requirement criteria for essential fatty acids. *J. Appl. Ichthyol.* **1995**, *11*, 183–198. [CrossRef]
17. Sargent, J.; McEvoy, L.; Bell, J. Requirements, presentation and sources of polyunsaturated fatty acids in marine fish larval feeds. *Aquaculture* **1997**, *155*, 117–127. [CrossRef]
18. Conte, F. Stress and the welfare of cultured fish. *Appl. Anim. Behav. Sci.* **2004**, *86*, 205–223. [CrossRef]
19. Ashley, P.J. Fish welfare: Current issues in aquaculture. *Appl. Anim. Behav. Sci.* **2007**, *104*, 199–235. [CrossRef]
20. Poli, B.M. Farmed fish welfare-suffering assessment and impact on product quality. *Ital. J. Anim. Sci.* **2009**, *8*, 139–160. [CrossRef]
21. Halliwell, B.; Gutteridge, J.M.C. *Free Radicals in Biology and Medicine*; Oxford University Press: Oxford, UK, 2015.
22. Aldini, G.; Dalle-Donne, I.; Facino, R.M.; Milzani, A.D.G.; Carini, M. Intervention strategies to inhibit protein carbonylation by lipoxidation-derived reactive carbonyls. *Med. Res. Rev.* **2007**, *27*, 817–868. [CrossRef]
23. Stegeman, J.J.; Hahn, M.E. Biochemistry and molecular biology of monooxygenases: Current perspective on forms, functions, and regulation of cytochrome P450 in aquatic species. In *Aquatic toxicology; Molecular, Biochemical and Cellular Perspectives*; Malins, D.C., Ostrander, G.K., Eds.; CRC Press: Boca Raton, FL, USA, 1994; pp. 87–203.
24. Ploch, S.A.; Lee, Y.P.; MacLean, E.; Di Giulio, R.T. Oxidative stress in liver of brown bullhead and channel catfish following exposure to tert-butyl hydroperoxide. *Aquat. Toxicol.* **1999**, *46*, 231–240. [CrossRef]
25. Mourente, G.; Díaz-Salvago, E.; Tocher, D.; Bell, J. Effects of dietary polyunsaturated fatty acid/vitamin E (PUFA/tocopherol ratio on antioxidant defence mechanisms of juvenile gilthead sea bream (Sparus aurata L., Osteichthyes, Sparidae). *Fish. Physiol. Biochem.* **2000**, *23*, 337–351. [CrossRef]
26. Mourente, G.; Díaz-Salvago, E.; Bell, J.; Tocher, D.R. Increased activities of hepatic antioxidant defence enzymes in juvenile gilthead sea bream (Sparus aurata L.) fed dietary oxidised oil: Attenuation by dietary vitamin E. *Aquaculture* **2002**, *214*, 343–361. [CrossRef]
27. Berntssen, M.; Julshamn, K.; Lundebye, A.K. Chemical contaminants in aquafeeds and Atlantic salmon (Salmo salar) following the use of traditional-versus alternative feed ingredients. *Chemosphere* **2010**, *78*, 637–646. [CrossRef] [PubMed]
28. Siano, F.; Bilotto, S.; Nazzaro, M.; Russo, G.; Di Stasio, M.; Volpe, M. Effects of conventional and organic feed on the mineral composition of cultured European sea bass (Dicentrarchus labrax). *Aquac. Nutr.* **2016**, *23*, 796–804. [CrossRef]
29. Nebert, D.W.; Gonzalez, F.J. P450 genes: Structure, evolution, and regulation. *Ann. Rev. Biochem.* **1987**, *56*, 945–993. [CrossRef] [PubMed]
30. Sarasquete, C.; Segner, H. Cytochrome P4501A (CYP1A) in teleostean fishes. A review of immunohistochemical studies. *Sci. Total. Environ.* **2000**, *247*, 313–332. [CrossRef]
31. Ribecco, C.; Hardiman, G.; Šášik, R.; Vittori, S.; Carnevali, O. Teleost fish (Solea solea): A novel model for ecotoxicological assay of contaminated sediments. *Aquatic Toxicol.* **2012**, *109*, 133–142. [CrossRef] [PubMed]
32. Meinelt, T.; Krüger, R.; Pietrock, M.; Osten, R.; Steinberg, C. Mercury pollution and macrophage centres in pike (Esox lucius) tissues. *Environ. Sci. Pollut. Res.* **1997**, *4*, 32–36. [CrossRef]

33. Suresh, N. Effect of cadmium chloride on liver, spleen and kidney melano macrophage centres in Tilapia mossambica. *J. Environ. Biol.* **2009**, *30*, 505–508.
34. Pascoli, F.; Negrato, E.; Di Giancamillo, A.; Bertotto, D.; Domeneghini, C.; Simontacchi, C.; Mutinelli, F.; Radaelli, G. Evaluation of oxidative stress biomarkers in Zosterisessor ophiocephalus from the Venice Lagoon, Italy. *Aquatic Toxicol.* **2011**, *101*, 512–520. [CrossRef]
35. Passantino, L.; Santamaria, N.; Zupa, R.; Pousis, C.; Garofalo, R.; Cianciotta, A.; Jirillo, E.; Acone, F.; Corriero, A. Liver melanomacrophage centres as indicators of Atlantic bluefin tuna, Thunnus thynnus L. well-being. *J. Fish. Dis.* **2013**, *37*, 241–250. [CrossRef]
36. Bucke, D.; Vethaak, A.; Lang, T. Quantitative assessment of melanomacrophage centres (MMCs) in dab Limanda limanda along a pollution transect in the German Bight. *Mar. Ecol. Prog. Ser.* **1992**, *91*, 193–196. [CrossRef]
37. Roberts, R.J. Melanin-containing cells of the teleost fish and their relation to disease. In *The Pathology of Fishes*; Ribelin, W.E., Migaki, G., Eds.; University of Wisconsin Press: Madison, WI, USA, 1975; pp. 399–428.
38. Agius, C. Preliminary studies on the ontogeny of the melano-macrophages of teleost haemopoietic tissues and age-related changes. *Dev. Comp. Immunol.* **1981**, *5*, 597–606. [CrossRef]
39. Kranz, H. Changes in splenic melano-macrophage centres of dab Limanda limanda during and after infection with ulcer disease. *Dis. Aquat. Org.* **1989**, *6*, 167–173. [CrossRef]
40. EFSA. Scientific Opinion of the panel on animal health and welfare on a request from the European commission on animal welfare aspects of husbandry systems for farmed European seabass and Gilthead seabream. *EFSA J.* **2008**, *844*, 1–21.
41. Costa, C.; Menesatti, P.; Rambaldi, E.; Argenti, L.; Bianchini, M.L. Preliminary evidences of colour differences in European sea bass reared under organic protocols. *Aquac Eng.* **2013**, *57*, 82–88. [CrossRef]
42. Fulton, T.W. Rate of growth of sea fishes. *Sci. Invest. Fish. Div. Scotl. Reprod.* **1902**, *20*, 1–22.
43. EFSA, European Food Safety Authority. Available online: http://www.efsa.europa.eu/ (accessed on 2 April 2013).
44. Bancroft, J.D.; Gamble, M. *Theory and Practice of Histological Techniques*, 6th ed.; Churchill Livingstone: Edinburgh, UK, 2008.
45. Agius, C.; Roberts, R.J. Melano-macrophage centres and their role in fish pathology. *J. Fish. Dis.* **2003**, *26*, 499–509. [CrossRef]
46. Baker, M.A.; Cerniglia, G.J.; Zaman, A. Microtiter plate assay for the measurement of glutathione and glutathione disulfide in large numbers of biological samples. *Anal. Biochem.* **1990**, *190*, 360–365. [CrossRef]
47. Bertotto, D.; Poltronieri, C.; Negrato, E.; Richard, J.; Pascoli, F.; Simontacchi, C.; Radaelli, G. Whole body cortisol and expression of HSP70, IGF-I and MSTN in early development of sea bass subjected to heat shock. *Gen. Comp. Endocrinol.* **2011**, *174*, 44–50. [CrossRef]
48. Trocino, A.; Xiccato, G.; Majolini, D.; Tazzoli, M.; Bertotto, D.; Pascoli, F.; Palazzi, R. Assessing the quality of organic and conventionally-farmed European sea bass (Dicentrarchus labrax). *Food Chem.* **2012**, *131*, 427–433. [CrossRef]
49. Pascoli, F.; Lanzano, G.; Negrato, E.; Poltronieri, C.; Trocino, A.; Radaelli, G.; Bertotto, D. Seasonal effects on hematological and innate immune parameters in sea bass Dicentrarchus labrax. *Fish Shellfish Immunol.* **2011**, *31*, 1081–1087. [CrossRef] [PubMed]
50. Pastoureaud, A. Influence of starvation at low temperatures on utilization of energy reserves, appetite recovery and growth character in sea bass, Dicentrarchus labrax. *Aquaculture* **1991**, *99*, 167–178. [CrossRef]
51. Escobar-Aguirre, S.; Felip, A.; Mazón, M.J.; Ballester-Lozano, G.; Pérez-Sánchez, J.; Björsson, B.T.; Zanuy, S.; Carrillo, M. Long-term feeding of a maintenance ration affects the release of Igf-1 and leptin, and delays maturation in a male teleost fish, Dicentrarchus labrax L. *Aquaculture* **2020**, *587*, 735467. [CrossRef]
52. Leggatt, R.A.; Brauner, C.J.; Iwama, G.K.; Devlin, R.H. The glutathione antioxidant system is enhanced in growth hormone transgenic coho salmon (Oncorhynchus kisutch). *J. Comp. Physiol. B* **2007**, *177*, 413–422. [CrossRef] [PubMed]
53. Pavlovic, S.; Mitić, S.S.B.; Radovanović, T.B.; Gavrilović, B.R.; Despotović, S.G.; Gavrić, J.P.; Saicic, Z. Seasonal variations of the activity of antioxidant defense enzymes in the Red Mullet (Mullus barbatus l.) from the Adriatic Sea. *Mar. Drugs* **2010**, *8*, 413–428. [CrossRef] [PubMed]
54. Pascual, P.; Pedrajas, J.R.; Toribio, F.; López-Barea, J.; Peinado, J. Effect of food deprivation on oxidative stress biomarkers in fish (Sparus aurata). *Chem. Biol. Interact.* **2003**, *145*, 191–199. [CrossRef]

55. Fiocchi, E.; Civettini, M.; Carbonara, P.; Zupa, W.; Lembo, G.; Manfrin, A. Development of molecular and histological methods to evaluate stress oxidative biomarkers in sea bass (Dicentrarchus labrax). *Fish. Physiol. Biochem.* **2020**, 1–12. [CrossRef]
56. Montero, D.; Izquierdo, M.; Tort, L.; Robaina, L.; Vergara, J. High stocking density produces crowding stress altering some physiological and biochemical parameters in gilthead seabream, Sparus aurata, juveniles. *Fish. Physiol. Biochem.* **1999**, *20*, 53–60. [CrossRef]
57. Magrone, T.; Fontana, S.; Laforgia, F.; Dragone, T.; Jirillo, E.; Passantino, L. Administration of a polyphenol-enriched feed to farmed sea bass (Dicentrarchus labraxL.) modulates intestinal and spleen immune responses. *Oxid. Med. Cell. Longev.* **2015**, *2016*, 1–11. [CrossRef]
58. Arciuli, M.; Fiocco, D.; Fontana, S.; Arena, M.P.; Frassanito, M.A.; Gallone, A. Administration of a polyphenol-enriched feed to farmed sea bass (Dicentrarchus labrax L.): Kidney melanomacrophages response. *Fish. Shellfish. Immunol.* **2017**, *68*, 404–410. [CrossRef]
59. Jacobs, M.; Covaci, A.; Schepens, P. Investigation of selected persistent organic pollutants in farmed Atlantic salmon (Salmo salar), salmon aquaculture feed, and fish oil components of the feed. *Environ. Sci. Technol.* **2002**, *36*, 2797–2805. [CrossRef]
60. Hites, R.A.A.; Foran, J.; Carpenter, D.O.; Hamilton, M.C.; Knuth, B.A.; Schwager, S.J. Global assessment of organic contaminants in farmed salmon. *Science* **2004**, *303*, 226–229. [CrossRef]
61. Berntssen, M.; Olsvik, P.; Torstensen, B.E.; Julshamn, K.; Midtun, T.; Goksøyr, A.; Johansen, J.; Sigholt, T.; Joerum, N.; Jakobsen, J.-V.; et al. Reducing persistent organic pollutants while maintaining long chain omega-3 fatty acid in farmed Atlantic salmon using decontaminated fish oils for an entire production cycle. *Chemosphere* **2010**, *81*, 242–252. [CrossRef] [PubMed]
62. Olli, J.J.; Breivik, H.; Mørkøre, T.; Ruyter, B.; Johansen, J.; Reynolds, P.; Thorstad, O.; Berge, G. Removal of persistent organic pollutants from Atlantic salmon (Salmo salar L.) diets: Influence on growth, feed utilization efficiency and product quality. *Aquaculture* **2010**, *310*, 145–155. [CrossRef]
63. Crampton, V.; Nanton, D.; Ruohonen, K.; Skjervold, P.O.; El-Mowafi, A. Demonstration of salmon farming as a net producer of fish protein and oil. *Aquac. Nutr.* **2010**, *16*, 437–446. [CrossRef]
64. Grigorakis, K. Compositional and organoleptic quality of farmed and wild gilthead sea bream (Sparus aurata) and sea bass (Dicentrarchus labrax) and factors affecting it: A review. *Aquaculture* **2007**, *272*, 55–75. [CrossRef]
65. Turchini, G.; Torstensen, B.E.; Ng, W.K. Fish oil replacement in finfish nutrition. *Rev. Aquac.* **2009**, *1*, 10–57. [CrossRef]
66. Benedito-Palos, L.; Bermejo-Nogales, A.; Karampatos, A.I.; Ballester-Lozano, G.F.; Navarro, J.C.; Diez, A.; Bautista, J.M.; Bell, J.G.; Tocher, D.R.; Obach, A.; et al. Modelling the predictable effects of dietary lipid sources on the fillet fatty acid composition of one-year-old gilthead sea bream (Sparus aurata L.). *Food Chem.* **2011**, *124*, 538–544. [CrossRef]

animals

MDPI

Article

A Detailed Study of Rainbow Trout (*Onchorhynchus mykiss*) Intestine Revealed That Digestive and Absorptive Functions Are Not Linearly Distributed along Its Length

Nicole Verdile [1,*], Rolando Pasquariello [1], Marco Scolari [2], Giulia Scirè [2], Tiziana A. L. Brevini [3] and Fulvio Gandolfi [1,*]

1 Department of Agricultural and Environmental Sciences, University of Milan, 20133 Milano, Italy; Rolando.pasquariello@unimi.it

2 Skretting Aquaculture Research Centre, 37100 Verona, Italy; Marco.Scolari@nutreco.com (M.S.); giulia.scire@skretting.com (G.S.)

3 Department of Health, Animal Science and Food Safety, University of Milan, 20133 Milano, Italy; tiziana.brevini@unimi.it

* Correspondence: nicole.verdile@unimi.it (N.V.); fulvio.gandolfi@unimi.it (F.G.); Tel.: +39-02-5031-6449 (N.V.); +39-02-5031-7990 (F.G.)

Received: 10 April 2020; Accepted: 20 April 2020; Published: 24 April 2020

Simple Summary: Aquaculture is the fastest growing food-producing sector due to the increase of fish intended for human consumption. However, aquaculture growth generates concerns, since carnivorous fish are extensively fed using fish-meal and fish-oil. This constitutes a severe limit to the aquaculture industry, questioning its sustainability. Consequently, alternative feeds are continuously searched through extensive in vivo feeding trials. Undoubtedly, to evaluate their impact on the gastrointestinal tract health, detailed knowledge of the intestine morphology and physiology is required. To date, extensive studies have been performed in several livestock species; however, available information on fish is limited nowadays, most importantly because their alimentary canal is able to easily adapt to external stimuli, and their intestinal morphology is affected by external factors. Therefore, it is essential to establish accurate reference values, especially along the productive cycle of animals raised in standardized conditions. Here, we performed a detailed characterization of the epithelial cells lining the intestinal mucosa in rainbow trout along the first year of development. We studied the absorptive and secretory activity as well as its ability to self-renewal. Our results indicate that, in this species, both digestive and absorptive functions are not linearly distributed along the intestinal length.

Abstract: To increase the sustainability of trout farming, the industry requires alternatives to fish-based meals that do not compromise animal health and growth performances. To develop new feeds, detailed knowledge of intestinal morphology and physiology is required. We performed histological, histochemical, immunohistochemical and morphometric analysis at typical time points of in vivo feeding trials (50, 150 and 500 g). Only minor changes occurred during growth whereas differences characterized two compartments, not linearly distributed along the intestine. The first included the pyloric caeca, the basal part of the complex folds and the villi of the distal intestine. This was characterized by a significantly smaller number of goblet cells with smaller mucus vacuoles, higher proliferation and higher apoptotic rate but a smaller extension of fully differentiated epithelial cells and by the presence of numerous pinocytotic vacuolization. The second compartment was formed by the proximal intestine and the apical part of the posterior intestine complex folds. Here we observed more abundant goblet cells with bigger vacuoles, low proliferation rate, few round apoptotic cells, a more extended area of fully differentiated cells and no pinocytotic vacuoles. Our results suggest that rainbow trout intestine is physiologically arranged to mingle digestive and absorptive functions along its length.

Keywords: intestine; epithelium; folds; renewal; rainbow trout

1. Introduction

Rainbow trout (*Oncorhynchus mykiss*) is one of the most widespread species in aquaculture due to its many merits that include adaptability to the farming environment, reproductive efficiency and disease resistance [1]. Nevertheless, the pressure to optimize the farming efficiency is constant and involves different aspects, including the continuous refinement of the diet [2]. Environmental sustainability and costs are the main drivers in the search for the optimal aquafeed, while progress depends on the accurate knowledge of intestine physiology in this and other related farmed species.

The gastrointestinal tract (GI) promotes the absorption of nutrients acting as a selective filter between the lumen and the circulatory system, but it simultaneously avoids the passage of harmful intraluminal xenobiotics [3]. Extensive and accurate studies on intestinal morphology have been performed in different livestock species [4], such as poultry, ruminants and pig [5], whereas in fish, a detailed morphological and functional characterization of the intestinal wall is still limited. In fact, fish being the largest group of vertebrates [6], are an extremely numerous and heterogeneous group. Several studies focused on the Teleostei intestinal features showed a wide variability among species due to diet, phylogeny and body shape [6]. Their GI can quickly and reversibly adapt to the environmental changes based on their physiological requirements [7]. The morphology of their alimentary canal is determined by different factors, such as taxonomy, feeding habitats, food type and frequency of food intake [8], making its detailed study and characterization complicated.

Current histological parameters used to determine the rainbow trout gut health in correspondence of dietary modifications are based on classical morphological changes. These include villus shortening, widening of lamina propria of villi, nuclear position disparity, intraepithelial lymphocyte infiltration, qualitative changes in mucus secretion [9] and massive enterocyte vacuolization. Parameters are accurately measured in the proximal intestine, whereas the analysis of the distal intestine is less detailed, possibly due to its complex morphology. This distal intestine is the major site of macromolecules absorption in most fish species, including salmonids [7]. In addition, only a few studies have been performed on rainbow trout intestine report quantitative data [10], while most of them are based on qualitative observations [6]. Finally, these studies are rarely carried out in animals raised in optimal and standardized conditions and do not consider the relationship between intestinal morphology and fish growth. Therefore, the aim of this work was to characterize the intestinal epithelial cells in rainbow trout to establish accurate reference values and to identify novel markers for increasing the sensitivity of earlier investigations. To this purpose, we studied individuals raised in standard conditions and ranging from 50 to 500 g in weight, the size range used for feeding trials.

2. Materials and Methods

2.1. Samples Collection

Five female rainbow trout (*Oncorhynchus mykiss*) for each age (7, 10, 12 months, weighing approximately 50, 150 and 500 g respectively) were euthanized under total anesthesia according to Annex IV EU guideline 2010/63, during non-experimental clinical veterinary practices. Fish were maintained in 600 L volume tanks with a water flow of 700 L/h and were reared under a photoperiod regimen of 24 h light. The water temperature was maintained constant at 15 °C.

These stages correspond to time points used in in vivo feeding trials. Moreover, in our experiment, these steps correspond to stepwise increase of digestible energy: 18 MJ/kg up to 50 g, 18.5 MJ/kg from 50 to 150 g and 19 MJ/kg up to 500 g (Optiline, Skretting, Verona, Italy). Animals included in this study were healthy and were raised in optimal conditions at Skretting Italia Spa (Verona, Italy).

Immediately after sacrifice, a longitudinal incision along the ventral line was performed and the whole gastrointestinal (GI) tract was removed.

At sampling time, each animal included in this study was weighed and its length measured. Once the whole intestinal tract was removed, the length of each tract was also defined. All the observed values are reported in Table 1 and are expressed as mean ± SD.

Table 1. Gross anatomy measurements in rainbow trout along the first year of development.

Trout Weight (g)	N.	Rainbow Trout Length (cm)	Whole Intestinal Length (cm)	Intestinal Segment Length (cm)	
				Proximal	Distal
50	5	16.8 ± 0.29	9.7 ± 0.62	6.0 ± 0.30	3.6 ± 0.28
150	5	22.7 ± 0.58	15.1 ± 0.80	9.0 ± 0.30	6.0 ± 0.50
500	5	32.4 ± 0.53	26.1 ± 0.76	18.2 ± 0.64	7.9 ± 0.11

Values are expressed as mean ± SD.

We collected segments of 2 cm of both anterior and posterior intestine. We defined as anterior intestine the tract comprised between the pyloric sphincter and the ileum-rectal valve. For consistency, we sampled only the segments with annexed pyloric caeca. We considered as distal intestine the tract downstream of the ileum-rectal valve characterized by larger diameter, darker pigmentation and circularly arranged blood vessels. For consistency, we collected all samples in the middle of this tract (Figure 1). Samples were immediately fixed in 10% neutral buffer formalin for 24 h at room temperature, dehydrated in graded alcohols, cleared with xylene and embedded in paraffin.

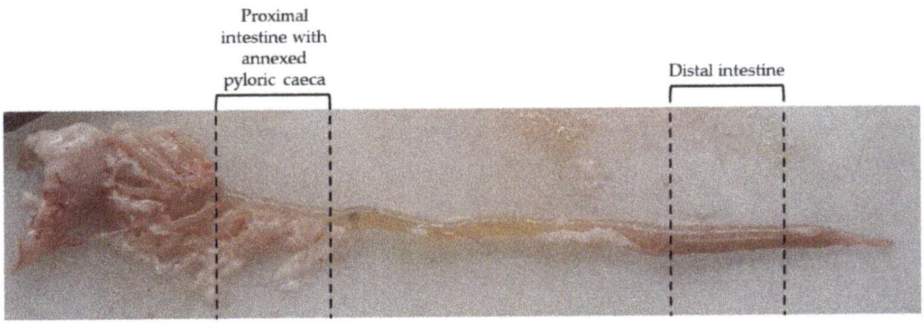

Figure 1. Selected regions of the samples collection.

2.2. Histology and Histochemistry

After dewaxing and re-hydration, 5 µm thick sections were stained with hematoxylin/eosin (HE) to evaluate the morphological features. Other sections were stained with Periodic acid–Schiff (PAS) or with Alcian Blue pH 2.5 (AB) for histochemical purposes. Subsequently, new slides, were stained with Periodic acid–Schiff (PAS)/Alcian Blue pH 2.5 staining kit (Bio Optica, 04-163802) to analyze the overall complex carbohydrates [11,12].

Lendrum's staining was used to identify acidophilic granules containing cells. Sections were incubated with Phloxine B to visualize acidophilic compounds and then with Tartrazine to remove the non-specific staining.

The alkaline phosphatase expression was used as a marker for the identification of fully differentiated epithelium. Briefly, slides were rehydrated and brought to distilled water, were then immersed in fresh Tris HCL (pH 9.5) solution to create the alkaline environment for 5 min. They were then incubated with BCIP/NBT substrate (5-bromo-4-chloro-3-indolyl phosphate/nitroblue tetrazolium, (Vector Laboratories, SK-4500 USA), which produces an indigo reaction product in the presence of alkaline phosphatase (AP) enzyme. Sections were then rinsed in tap water, counterstained using Mayer's hematoxylin, dehydrated and mounted.

2.3. Morphometric Evaluation

Images were acquired using a NanoZoomer S60 Digital slide scanner (Hamamatsu photonics, Hamamatsu city, Japan)) and observed at continuous magnifications between 20× and 80×. Villi morphometric evaluation was performed on HE-stained sections using the NDP.view software (Hamamatsu, Japan). For each fish, we measured 20 villi for each intestinal tract and 10 complex folds in the distal intestine. Villus height was calculated from the apex of the villus to the villus-intestinal folds junction; villus width was measured at the base, at the middle and at the apex and it was expressed as the mean of the three measurements. Fold height was measured from the compactum layer of the submucosa to the apex of the folds. Goblet cells were counted along 1 mm of epithelium for each intestinal compartment in each fish. The precise length of the epithelial lining was measured using the specific function of the NDP.view software. Volume of goblet cell mucus vacuoles was estimated using the same software. For each sample, 50 goblet cells were evaluated in all intestinal sections considered.

2.4. Quantitative Stereological Analysis

Systematic sampling was performed as described in detail previously [13]. Briefly, the estimation of the volume densities (Vv) of the separate layers of the intestine was based on the general principle of Delesse [14]. The volume of each layer was estimated from the fractional area of the structure of interest (e.g., the mucosa) and the total area of the reference compartment (e.g., the intestinal wall) was measured in histological cross-sections. Area-measurements were performed by point-counting [13], as specified below. For estimation of the volume densities of the different layers, the regions of interest containing the respective reference compartments were defined in the histological sections. Within these regions of interest, systematically randomly sampled areas (70% of the total sectional area of the region of interest) were photographed and superimposed with an adequately sized grid of equally distant points. The number of points hitting the interested structure and the respective reference compartment were counted and the fractional area of the structure of interest and the total area of the reference compartment were then calculated from the respective quotient of points hitting these structures. The magnification was chosen to allow the relevant portion of the intestinal wall to be contained in each field of vision. Vv were expressed as percentages and were calculated as follows:

$$Vv_{(\text{analyzed compartment, reference compartment})} = \left[\sum P_{(\text{analyzed compartment})} / \sum P_{(\text{reference compartment})}\right] \times 100$$

where $\sum P_{(\text{analyzed compartment})}$ is the number of points hitting the compartment under study, and $\sum P_{(\text{reference compartment})}$ is the number of points hitting the relevant structure.

2.5. Immunohistochemistry

Proliferating cell nuclear antigen (PCNA) localization was characterized by indirect immunohistochemistry using the Avidin Biotin Complex method (VECTASTAIN® Elite® ABC, Vector Laboratories, Burlingame, CA, USA) following manufacturer instructions. Briefly, slides were brought to boiling in 10 mM sodium citrate buffer, 0.05% Tween20 (pH 6) in a pressure cooker for 1 min for antigen retrieval. After cooling at room temperature for 20 min, sections were rinsed in PBS (phosphate-buffered saline, pH 7.4) and then were immersed in 3% H_2O_2 solution in methanol for 15 min to quench the endogenous peroxidase. To prevent aspecific binding, sections were then incubated in Normal Blocking Serum (Vectastain ABC Elite KIT, Burlingame, CA, USA) at room temperature for 30 min. Sections were incubated with Anti PCNA Mouse monoclonal antibody 1:1600 (Millipore Corporation, MAB424, Darmstadt, Germany) diluted in 4% BSA in PBS with 0.05% Tween20, for 60 min at room temperature in a humid chamber. Sections were then incubated with appropriate biotinylated secondary antibody for 30 min at room temperature in a humidified chamber followed by the avidin-biotinylated horseradish peroxidase (HRP) complex (Vectastain ABC Elite KIT, Burlingame, CA, USA) for another 30 min. Finally, sections were incubated with ImmPACT

NovaRED substrate (Vector Laboratories, SK-4105, Burlingame, CA, USA), which produces a red reaction product in the presence of peroxidase (HRP) enzyme. Sections were then briefly counterstained with Mayer's hematoxylin, dehydrated and permanently mounted with a mounting media (Bio-Optica, 05-BMHM100, Milano, Italy). Secondary antibodies controls were performed following the same staining protocol but omitting the primary antibody.

2.6. TUNEL Test

Peroxidase in situ detection Apoptosis kit (Millipore Corporation, S7100, Darmstadt, Germany) was used to detect cells undergoing apoptosis following the manufacturer instructions. In brief, sections were brought to boiling in 10 mM sodium citrate buffer, 0.05% Tween20 (pH 6) in a pressure cooker for 1 min to improve the access of the terminal deoxynucleotidyl transferase (TdT) to the fragmented DNA. Slides were then cooled at room temperature for 20 min, washed in PBS, and incubated in a humidified chamber for 15 min in 3% H_2O_2 in methanol to quench endogenous peroxidase. Sections were rinsed in distilled water, exposed to equilibration buffer (Millipore Corporation, S7100, Darmstadt, Germany) for 20 s and incubated with TdT enzyme digoxigenin-conjugated in a humidified chamber at 37 °C for 60 min. TdT enzyme was previously diluted in reaction buffer (1:32) (Millipore Corporation, S7100, Darmstadt, Germany). Reaction was stopped by immersing sections in fresh prepared stop/wash buffer in a Coplin jar for 10 min. Anti-Digoxigenin peroxidase conjugated antibody was applied to slides for 30 min at room temperature. Samples were then washed in PBS and were incubated with 3,3′-diaminobenzidine solution (ImmPACT® DAB, SK-4105 Vector Laboratoris, Burlingame, CA, USA), which, in the presence of a peroxidase (HRP) enzyme, produces a brown reaction product. Sections were then briefly counterstained with Mayer's hematoxylin, dehydrated and permanently mounted with a mounting media (Bio-Optica, 05-BMHM100, Milano, Italy).

2.7. Statistical Analysis

Quantitative data were expressed as mean ± SD. Results were analyzed by using One-way or Two-way analysis of variance (ANOVA) followed by all-pairwise multiple comparison test with the Holm–Sidak method. The two independent variables considered (animal weight and intestinal tract) were independent of each other, and there was no relationship between the observation in each group or between the groups themselves. Dependent variables considered (mucosa volume, goblet cells number and vacuoles volume and goblet cells producing acid, neutral and a mixed combination of them) were normally distributed according to the Shapiro–Wilk normality test. The homogeneity of variance was assessed through Bartlett's test. Differences were considered statistically significant if $p < 0.05$.

3. Results

3.1. Gross Anatomy

Macroscopically, the rainbow trout intestine corresponds to the general description of this organ in teleost fish [15]. It comprised a proximal intestine with blind diverticula called pyloric caeca annexed to its upper part and a distal intestine [15]. The latter is characterized by a larger diameter, dark pigmentation and circularly arranged blood vessels in agreement with a previous study performed in Brown trout [16]. Circular folds protruding from the distal intestinal wall towards the lumen were also evident even if this is not a typical teleost feature.

3.2. Microscopical Anatomy

Pyloric caeca, proximal and distal intestine are lined by a tunica mucosa constituted by epithelium and lamina propria forming villi along all tracts.

Villus length in pyloric caeca increased significantly in parallel with age (Table 2). Interestingly, in this region, at 500 gr we observed enterocytes supranuclear vacuolization (Figure 2).

Table 2. Evaluation of pyloric caeca histometry in rainbow trout along the first year of development.

Trout Weight	N.	Pyloric Caeca (μm)	Enterocytes Supranuclear Vacuolization (SNV)
50 g	5	380.91 ^{a} ± 35.85	−
150 g	5	719.50 ^{b} ± 32.45	−
500 g	5	1011.50 ^{c} ± 18.14	+

Values are expressed as mean ± SD. $^{a–c}$ Different superscripts in the same column indicate significant differences ($p < 0.05$) determined by one-way ANOVA (animal weight independent variable). The presence or the absence of enterocytes supranuclear vacuolization are indicated with "+" or "−" respectively.

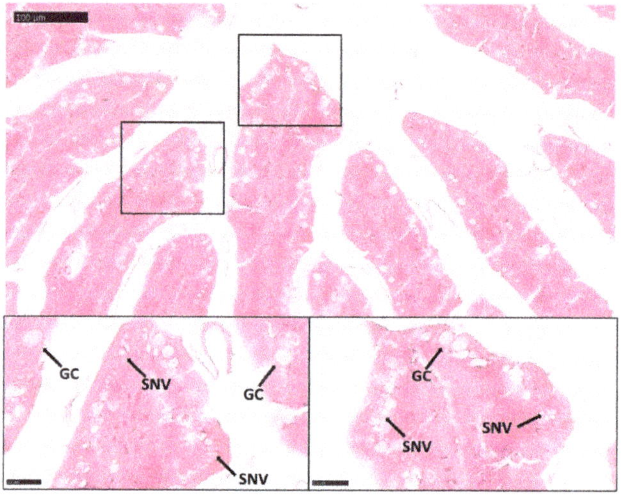

Figure 2. Hematoxylin/eosin (HE) stained section, showing the presence of enterocytes supranuclear vacuolization (SNV) and goblet cells (GC) in the pyloric caeca of 500 gr rainbow trout.

In the proximal intestine, we observed a wide variation of villus length. In order to reduce the wide standard deviation and making possible a meaningful statistical analysis, we divided them into two arbitrary groups: shorter and longer of 400 μm. Average short villi (below 400 μm) length remained constant during growth, whereas long villi (above 400 μm) increased their length significantly when animals reached the 500 gr size (Table 3). At the same time, villi in the larger animals became more branched (Figure 3) whereas short villi were rarer. No supranuclear vacuoles were observed in the proximal intestine enterocytes.

The large circular folds observed macroscopically in the distal tract, corresponded to complex folds formed by tall primary extroflexions of the muscular and mucosal layers. From their main axis, secondary extroflexions of epithelium and lamina propria formed villi like those present along the rest of the intestinal wall. Along the entire length of the posterior intestine, the complex folds were interspersed at regular intervals among normal villi. Also, in the distal intestine, villi were very heterogenous in length, so we divided them into shorter or longer than 400 μm. Their length, as well as the height of the complex folds, remained constant along development. Villi and folds in the distal tract had to be measured on longitudinally embedded samples to appreciate and to measure the typical folds structure. Unfortunately, due to the very small lumen of the 50 g samples, the opening procedure damaged the mucosa morphology to the point of compromising an accurate morphometric analysis of the height of the folds. Therefore, we decided to exclude this data from the analysis (Table 4).

Table 3. Evaluation of proximal intestine histometry in rainbow trout along the first year of development.

Trout Weight (g)	Proximal Intestine				
	N.	Short Villi Length (μm)	Long Villi Length (μm)	Villi Width (μm)	Villi Branching
50	5	251.0 ^{a} ± 13.5	470.4 ^{a} ± 64.6	96.0 ^{a} ± 11.4	−
150	5	310.3 ^{a} ± 44.5	555.8 ^{a} ± 64.2	122.9 ^{b} ± 10.1	+
500	5	250.3 ^{a} ± 42.1	657.8 ^{b} ± 49.1	116.9 ^{b} ± 14,4	++

Values are expressed as mean ± SD. $^{a–b}$ Different superscripts in the same column indicate significant differences ($p < 0.05$) determined by one-way ANOVA (animal weight independent variable). The absence or the occurrence of villi branching is indicated with "−" and "+" respectively. "++" indicate an increase in villus branching.

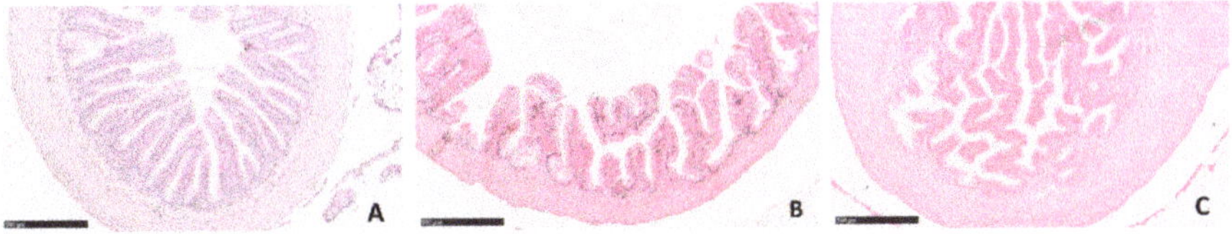

Figure 3. Branching of intestinal villi in the anterior intestine of rainbow trout during growth ((**A**) 50 g; (**B**) 150 g; (**C**) 500 g).

Table 4. Evaluation of distal intestine histometry in rainbow trout along the first year of development.

		Distal Intestine		
Trout Weight (g)	N.	Short Villi Length (μm)	Long Villi Length (μm)	Folds Height (μm)
50	5	−	−	−
150	5	211.1 ± 3.1	592.4 ± 19.3	1284.9 ± 51.0
500	5	202.4 ± 20.1	556.5 ± 37.0	1410.1 ± 201.7

Values within the same column indicate no significant differences ($p > 0.05$) determined by one-way ANOVA (animal weight independent variable).

In the distal intestine, villi protruding from the organ wall and those located at the basal part of the complex folds presented pinocytotic vacuoles, whereas those located at the folds apex did not. The latter, on the contrary, were characterized by numerous actively secreting goblet cells as observed in the villi of the proximal intestine (Figure 4).

Along all intestinal tracts, the tunica submucosa was composed of a thick compactum layer interposed between two thin granulosum layers, so-called because of the presence of the characteristic granular cells. The upper granulosum layer was always thinner than the bottom one. Notably, the tunica submucosa present in the complex folds axis was homogeneous with neither compactum nor granulosum layers.

The whole intestinal tract possessed a thick muscle layer composed of an inner circular and outer longitudinal layer. Circularly-arranged muscle cells were also clearly visible within the apical part of the complex folds of the distal intestine but not at their base.

No inflammation features were observed, such as villi shortening and nuclear positioning disparity. Only a few intraepithelial lymphocytes were found confirming the animals' healthy state.

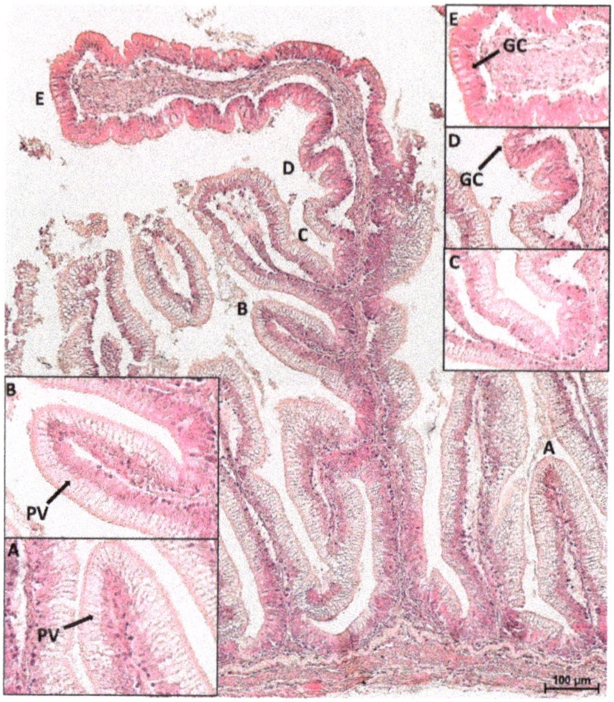

Figure 4. HE stained section of a complex fold in the distal intestine: parietal villi (**A**), as well as villi emerging from the basal part of the fold (**B**,**C**), were both covered by pinocytotic vacuoles (PV). At one point, the epithelium morphology along the fold, drastically changed: pinocytotic vacuoles tended to disappear (**D**). The apex of the fold presented goblet cells (GC) and non-vacuolated enterocytes (**E**).

3.3. Quantitative Sterological Analysis

Volume estimation of each intestinal layer indicated that the mucosa was the predominant layer in both intestinal tracts. Moreover, the mucosa volume was significantly higher in the distal than in the proximal intestine at all ages and remained unchanged during development (Figure 5). The bigger mucosa volume of the posterior tract was not due to an alteration of the epithelium:lamina propria ratio since it remained constant in the different groups and indicated a healthy intestinal tract.

3.4. Goblet Cells

Goblet cell numbers and vacuole volume were not homogenously distributed among the different intestinal tracts. Goblet cells were significantly more abundant and their vacuoles significantly bigger in the proximal intestine and in the apical part of the posterior intestine complex folds. Villi of the distal intestine and of the basal part of the complex folds shared with the pyloric caeca a smaller goblet cell number and vacuole volume (Figure 6).

The accurate quantitative analysis of goblet cells was limited to the proximal intestine and the pyloric caeca because the number of PAS-positive goblet cells in the distal part was confounded by the presence of numerous PAS-positive pinocytotic cells (Figure 7).

PAS staining revealed that most goblet cells had a low affinity for Periodic acid-Shiff (Figure 8), but most of them strongly reacted with Alcian Blue pH 2.5 (Figure 9). Therefore, in order to have a more detailed qualitative analysis, we performed a combined PAS-Alcian Blue staining. This revealed a range of blue stain intensity in all investigated tracts indicating a heterogenous production of complex carbohydrates. A few cells produced only neutral mucous (PAS positive-magenta cells), some produced acid mucous (AB positive-light blue cells), while the majority secreted a mixed combination of acid and neutral mucous (PAS-AB positive-dark blue cells) (Figure 10).

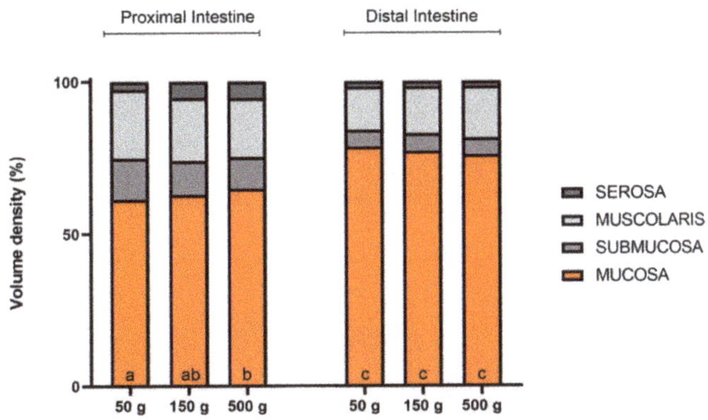

Figure 5. Volume density (Vv) estimation of rainbow trout's intestinal layers in the proximal and distal intestine during growth. Data are expressed as a percentage of the whole intestinal wall. [a–c] Different superscripts in the same layer indicate significant differences ($p < 0.05$) determined by two-way ANOVA (animal weight and intestinal tract: independent variables).

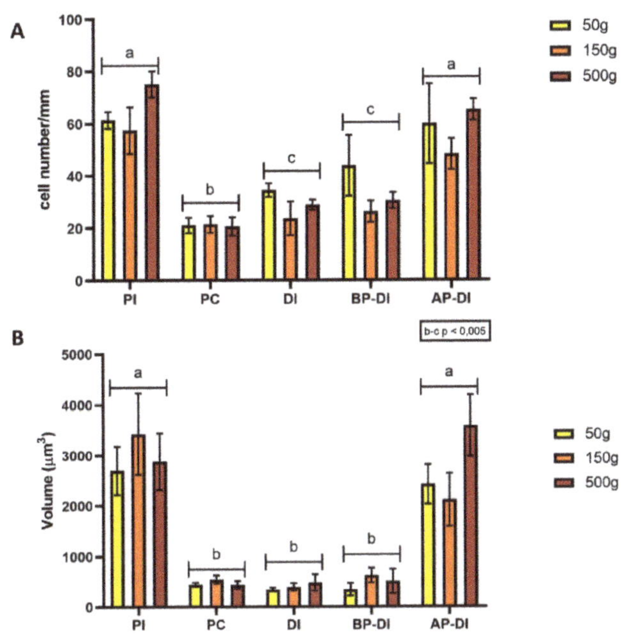

Figure 6. Goblet cells number (**A**) and volume (**B**) within five regions of the intestinal tract: proximal intestine (PI), pyloric caeca (PC), distal intestine (DI), basal (BP-DI) and apical part (AP-DI) of the complex folds of the distal intestine. Values are measured at three different stages of development (50, 150 and 500 g). [a–c] Different superscripts within the same histogram indicate significant differences ($p < 0.05$ or $p < 0.005$) measured by two-way ANOVA (animal weight and intestinal tract: independent variables).

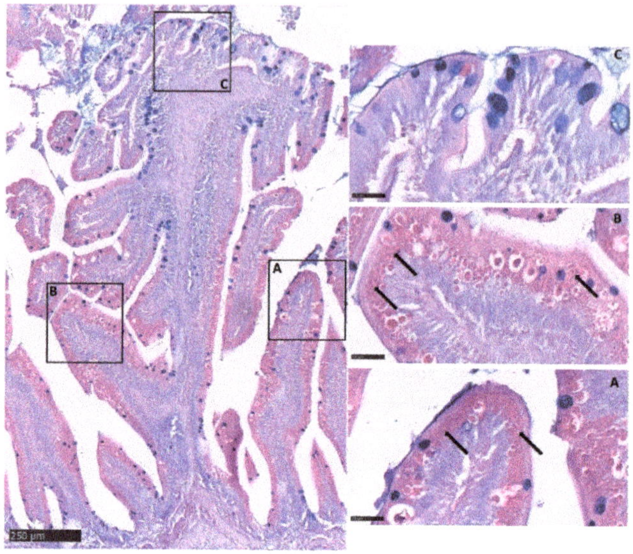

Figure 7. Alcian Blue–Periodic acid–Schiff (AB-PAS) stained section showing the presence of numerous PAS-positive pinocytotic vacuoles (arrows) on the villi of the distal intestine (**A**), as well as, on the basal part of the complex folds (**B**); whereas, they were completely absent at the fold apex (**C**).

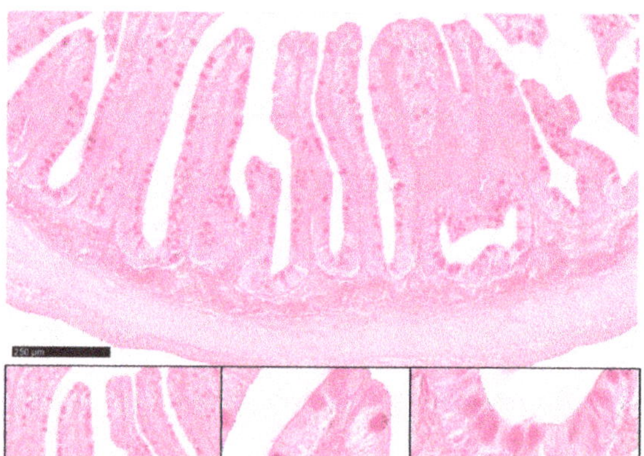

Figure 8. Representative figure of PAS stained section of the anterior intestine of rainbow trout.

In the proximal intestine, the distribution of the different types of goblet cell was the same in short and long villi. In both types of villi, however, the number of goblet cells producing acid mucus significantly increased in 500 g individuals in parallel with the onset of villi branching (Figure 11). We did not observe the same trend in the pyloric caeca where the quality of mucus secretion remained constant along development (Figure 12).

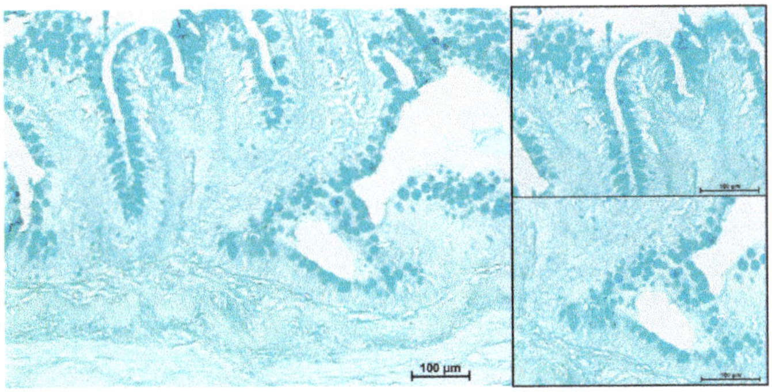

Figure 9. Representative figure of AB pH 2.5 stained section of the anterior intestine of rainbow trout.

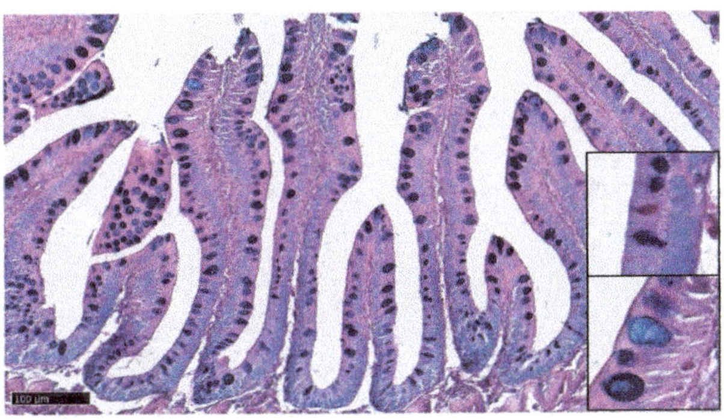

Figure 10. AB-PAS stained section showing the presence of a heterogenous populations of goblet cells. Most mucus-secreting cells produced a mixed secretion (PAS-AB positive-dark blue cells), some secreted acid mucins (AB positive-light blue cells), and a smaller amount produced neutral mucus (PAS positive-magenta cells).

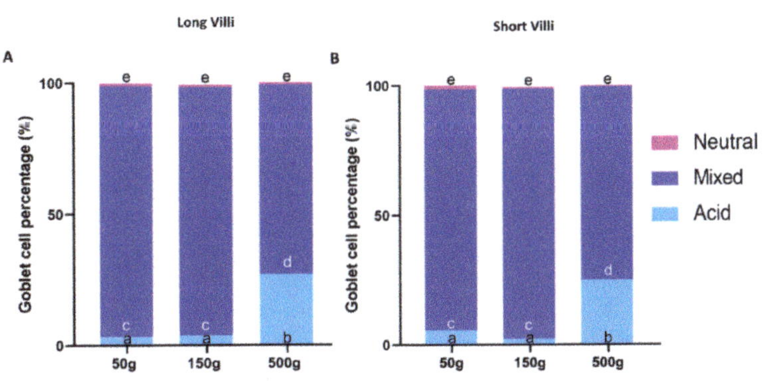

Figure 11. Estimation of the percentage of goblet cells secreting acid mucins (AB positive), mixed mucins (AB/PAS positive) and neutral mucins (PAS positive) in the proximal intestine of both short (**A**) and long (**B**) villi along rainbow trout development. $^{a–e}$ Different superscripts within the same histogram indicate significant differences ($p < 0.05$) measured by one-way ANOVA (animal weight: independent variable).

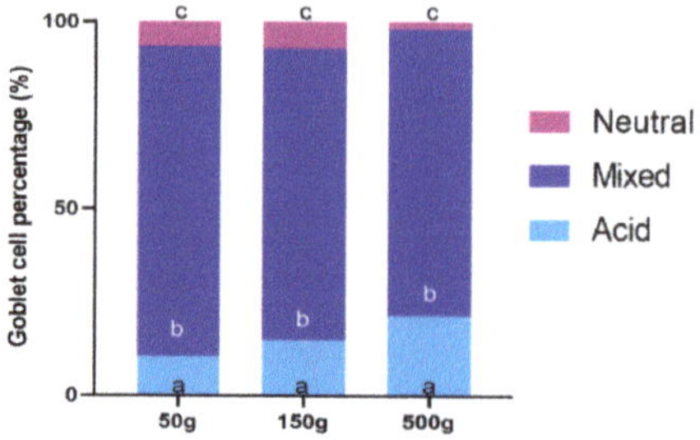

Figure 12. Estimation of the percentage of goblet cells secreting acid mucins (AB positive), mixed mucins (AB/PAS positive) and neutral mucins (PAS positive) in the pyloric caeca along rainbow trout development. In this region, the quality of mucus secretion remained constant along development. The same superscripts $^{a–c}$ within the same histogram indicate no significant differences ($p > 0.05$) measured by one-way ANOVA (animal weight: independent variable).

3.5. Intestinal Eosinophilic Granule Cells

Round cells positive to the Phloxine tartrazine staining were observed in the upper and lower granulosum layers of the submucosa along the whole intestinal length. At high magnification, these cells showed the presence of cytoplasmatic granules enabling their classification as eosinophilic granule cells (EGCs). Their shape and position do not correspond to the typical Paneth cell. These cells, however, were negative for both alkaline phosphatase and peroxidase, suggesting that EGCs do not possess the morphological characteristics of Mast cells (Figure 13).

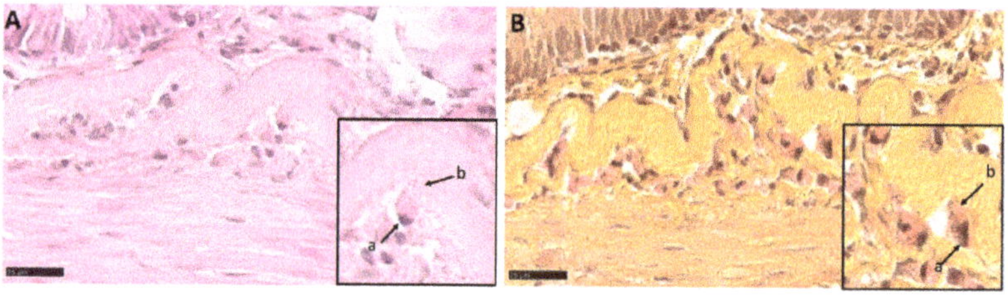

Figure 13. HE (**A**) and phloxine tartrazine (**B**) stained sections showing intestinal eosinophilic granule cells within the granulosum layers of submucosa (a—nucleus; b—eosinophilic granular cytoplasm).

3.6. Proliferation, Differentiation and Apoptosis

A strong PCNA signal was detected within the intestinal folds at the basis of the villi, confirming that the stem/progenitor cell zone is located here at all ages, in all investigated regions. In the proximal intestine, PCNA was not limited to the folds but was also expressed in correspondence of villi branching points. Mature, fully differentiated epithelial cells were identified by the histochemical detection of the alkaline phosphatase (AP) activity. As expected, the signal was localized along the villi length and at their apex.

As for the goblet cells also, the AP/PCNA staining pattern was not homogeneous along the intestine and we could identify two patterns. One characterized by a restricted proliferating area accompanied by an expanded differentiated zone present in the proximal intestine and in the apical region of the complex folds of the distal intestine. The other characterized by an extended proliferating area accompanied by a restricted differentiated zone present in pyloric caeca and distal intestine villi and in the basal part of the complex folds (Figures 14 and 15).

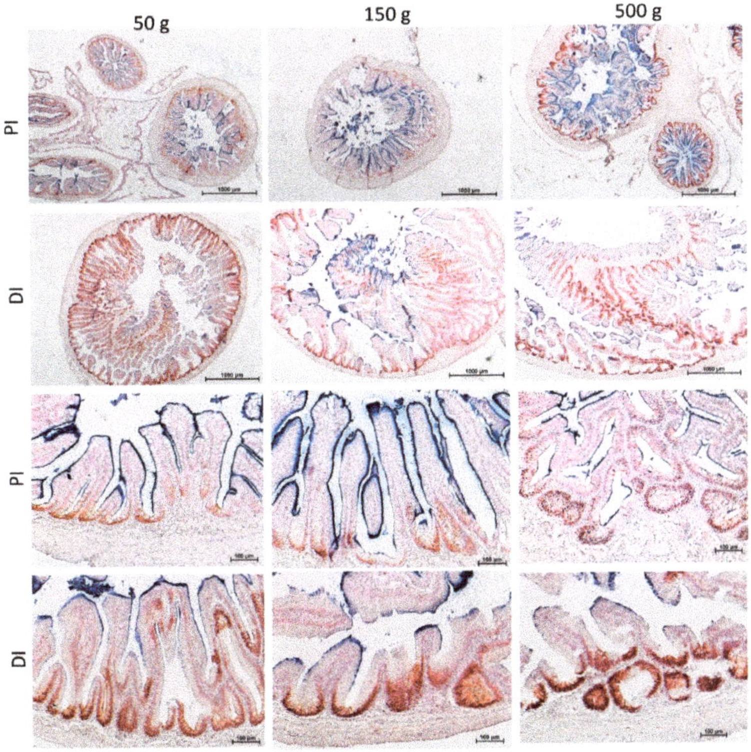

Figure 14. Proliferating cell nuclear antigen (PCNA) immunolocalization was observed within the intestinal folds at the basis of the villi in both proximal (PI) and distal (DI) intestine in all investigated stages. PCNA^{+} cells were also found in correspondence of villi branching points in the proximal intestine. Cell proliferation was more intense and widespread in the distal compared to the proximal intestine in all investigated stages. Fully differentiated enterocytes expressing alkaline phosphatase (AP) activity were observed at the villi apex, but conversely, its expression was stronger and more widespread in the proximal compared to the distal intestine in all ages.

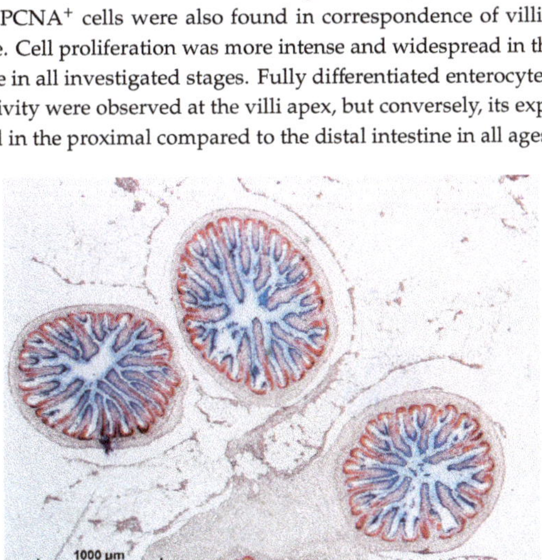

Figure 15. Representative figure of PCNA combined AP stained section pf pyloric caeca showing a proliferating pattern similar to that observed in the distal intestine and in the basal part of the complex folds.

Cells undergoing apoptosis were identified using the TUNEL assay and showed different patterns along the intestine. Only a few cells undergoing apoptosis were visible in the proximal intestine and on the apical part of the complex folds, whereas, many more were detected in the pyloric caeca, in the distal intestine villi and on the basal part of the folds (Figures 16 and 17). Apoptotic cells presented two different morphology: round and slender. Round cells were present along the entire intestinal mucosa and could be detected exfoliating into the intestinal lumen (Figure 18). Slender cells were found exclusively in the pyloric caeca, in the villi and in the basal part of the complex folds of the distal intestine. The same cells displayed a dark eosinophilic cytoplasm if stained with eosin, periodic acid-Schiff or phloxine.

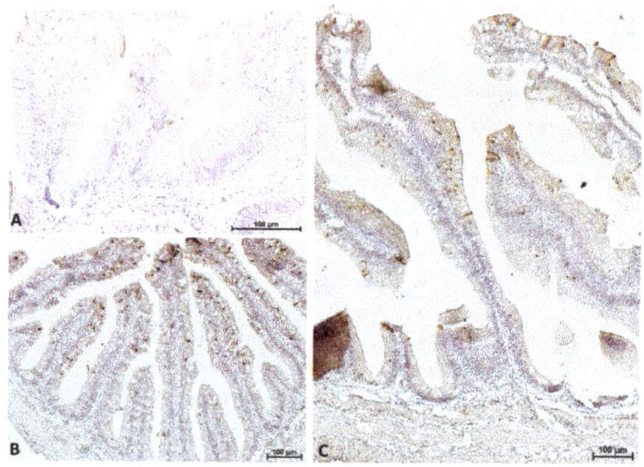

Figure 16. Representative figure of cells undergoing apoptosis showing only a few round apoptotic cells in the anterior tract (**A**) and many more apoptotic cells, both round and slender in the pyloric caeca (**B**) and in the posterior intestine (**C**).

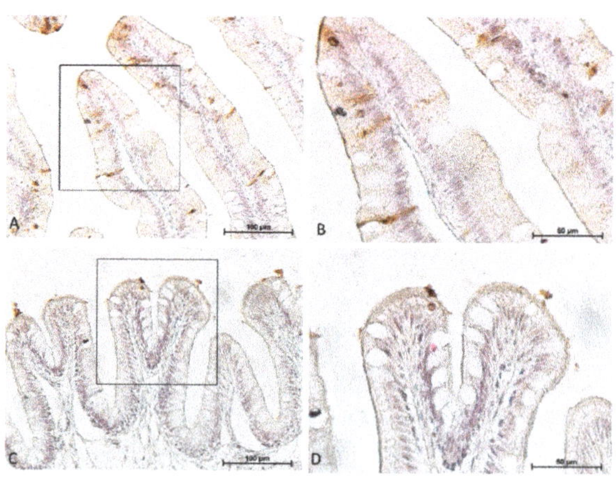

Figure 17. Representative figure of cells undergoing apoptosis showing numerous, both round and slender apoptotic cells in the basal part of the complex folds (**A**,**B**) and only a few round apoptotic cells in the apical part of the complex folds (**C**,**D**).

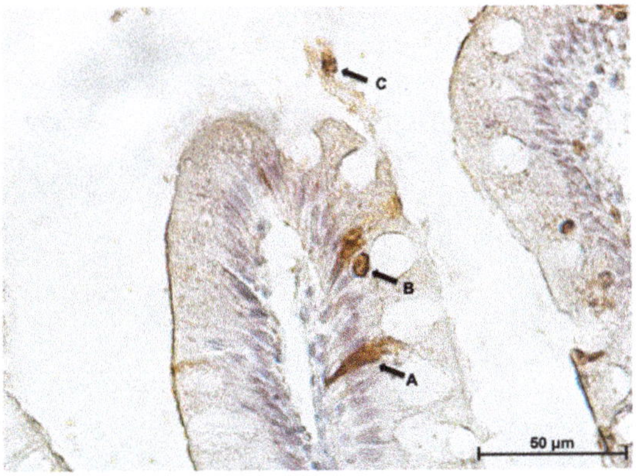

Figure 18. In situ detection of cells undergoing apoptosis showing three different apoptotic stages: A: early apoptotic stage; B: apoptotic body: C: cells exfoliating into the intestinal lumen.

Overall, proliferation, differentiation and apoptotic rate were not homogenously distributed along the intestinal epithelium. The pyloric caeca, the distal intestine wall as well as the basal part of the complex folds were characterized by high proliferation and high apoptotic rate but low alkaline phosphatase expression, which also appeared to be fragmented. In contrast, the proximal intestine and the apical part of the complex folds were characterized by a low PCNA expression, few round apoptotic cells and a strong alkaline phosphatase signal (Figure 19).

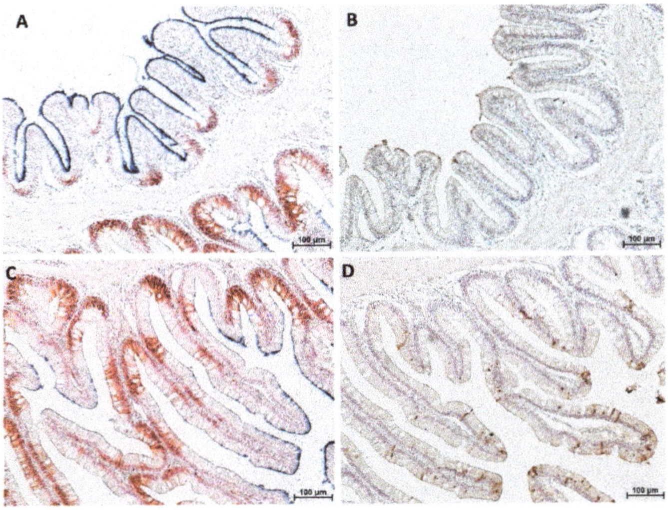

Figure 19. Representative figure of PCNA immunolocalization (red), histochemical detection of differentiated cells (blue) and in situ detection of apoptotic cells (brown). The apical part of the complex folds was characterized by a low PCNA expression, few round apoptotic cells and a strong alkaline phosphatase signal (**A**,**B**). In contrast, high proliferation and high apoptotic rate associated with low alkaline phosphatase expression in the basal part of the complex folds were observed (**C**,**D**).

4. Discussion

Here, we performed a detailed characterization of the intestinal epithelial cells lining the intestinal tract in rainbow trout along the first year of development.

Macroscopically, the morphology of rainbow trout intestine corresponds to the general description of this organ in teleost fish [17] with the peculiar presence of complex folds protruding from the distal intestinal wall towards the lumen. These structures have been previously described in close species like *Salmo trutta and Salmo salar* [9,16] as well as in more distant ones like sharks [18] but are not common to all teleost species [17,19,20]. At present, their detailed morphology remains poorly known, and their function is even less understood [16]. It is interesting to note that a circular tunica muscularis was present at the apex of these folds, while the tunica submucosa was reduced to a thin layer of connective tissue with neither the compactum nor granulosum layers. This arrangement suggests that the complex folds can contract independently from the rest of the distal intestine wall, possibly separating two functional compartments [16].

The wide variation of villi length observed in the proximal intestine prompted us to divide the villi into two populations based on their length. This enabled us to notice that, at 500 gr, only the long villi increased their average length, while the number of short villi diminished. At the same time, villi fusion and branching became evident. Some authors sustain that villi fusion represents a typical inflammation sign [9,21], whereas others support the hypothesis that mucosa folds simply get more complex in parallel with growth [16]. Based on our observations, we hypothesize that branching may be accompanied by or due to the short villi fusing into the long ones suggesting possible functional differences between the two villi types. Consistent with previous observations in Salmo salar [15], we did not observe villus branching at any age in the pyloric caeca, but in this tract, supranuclear pinocytotic vacuolization became visible in enterocytes of 500 gr individuals that could be related to lipid accumulation [9]. Similar vacuolization was present in the enterocytes lining the villi and the basal part of the complex folds of the distal intestine, but not in the proximal intestine nor in the apical part of the complex folds. In the adult goldfish, Carassius auratus these vesicles are known to be specialized in the uptake of large intact molecules [22]

Goblet cells followed two different patterns: numerous, swollen, actively secreting cells, in the proximal intestine and in the apical part of the distal intestine complex folds; scarce and inactive in the pyloric caeca and in the rest of the distal intestine. Our observation that the number of goblet cells was significantly more abundant in the proximal than in the distal intestine is in contrast with other data present in the literature [8,19]. However, a recent study suggests that the greater abundance of goblet cells within the posterior tract could be a misinterpretation due to the massive presence of pinocytotic vacuoles in the apical part of the enterocyte's cytoplasm [23]. Moreover, goblet cell density is strongly affected by several external factors comprising nutrition, physiology and immunology [24,25], and this must be considered as a possible source of variability. On the contrary, our finding of goblet cells producing different mucous substances agrees with previous results in several teleost fish, including rainbow trout [4,8,10]. Interestingly, the significant increase of acid glyconjugates observed in the anterior intestine, the increase of villus branching and the appearance of vacuolized enterocytes in the pyloric caeca all occurred in correspondence with the increase of lipid concentration in the diet. Consequently, since qualitative changes in mucins composition are associated with gastrointestinal disorders, our data suggest the hypothesis that the last diet change may have caused a mild stress [24].

We identified eosinophilic granule cells (EGCs) within the submucosa granular layers. in agreement with previous observations in the *Salmo salar* [26]. EGC are considered the functional equivalent of Paneth cells and indeed, EGCs found in Rainbow trout were positive to Phloxine-tartrazine staining that is specific for Paneth cells. They play a pivotal role in intestinal mucosa defense by secreting antimicrobic peptides and in intestinal stem cells regulation [27]. However, their position outside the epithelium is not compatible with that of *bona fidae* Paneth cells. Recent data report that the stratum granulosum hosts numerous eosinophilic mast cells involved in the inflammatory process [28], so that some authors simply consider eosinophilic granule cells to be mast cells [29]. Our data do not support this conclusion because rainbow trout ECG did not contain alkaline phosphatase nor peroxidase as opposed to mast cells described in fish.

The localization of proliferative cells at the base of the villi in the pyloric caeca and in both anterior and posterior intestine confirmed that the stem/progenitor cells zone is located within the intestinal folds in all ages, in agreement with previous reports in *Salmo salar* [16] and *Salmo trutta* [30]. However, cell proliferation was not homogeneous in the different intestinal districts but followed the dual pattern described for goblet cells: in the proximal intestine and apical part of the distal intestine complex folds, proliferation was restricted to the bottom of intestinal folds, whereas in the pyloric caeca, in the rest of the distal intestinal the signal was more spread along the villus length. Interestingly, spots of cell proliferation were observed also along the villi length but only in the proximal intestine of 500 gr individuals. We hypothesize that these spots correspond to newly formed branches.

Alkaline Phosphatase (AP) is considered a marker for mature enterocytes [31]; however, we observed a clear signal on the apical part of all differentiated cells. This is in agreement with previous observations in mouse small intestine [32]. As expected, the extension of the cell proliferating compartment was inversely proportional to that of fully differentiated cells. However, it was interesting to note that, once again, this distribution also followed a double pattern as described for the other aspects: the anterior intestine and the apical part of the distal intestine complex folds showing low proliferation and extensive differentiation while the opposite occurred in the pyloric caeca and in the rest of the distal intestine. Moreover, low proliferation corresponded to low frequency of apoptotic cells, while high proliferation corresponded to high apoptotic rate. These observations suggest that the proximal intestine and the apical part of the complex folds are subjected to a lower renewal rate compared to the pyloric caeca and the rest of the distal intestine. Apoptotic cells had two different morphologies: round and slender. Round cells correspond to the classical morphological features of apoptotic bodies already described in mouse small intestine [33]. They were localized along the villi length, at their tip as well as exfoliating into the intestinal lumen. Slender cells, with a strong eosinophilic cytoplasm, were observed only in the distal intestine, on the apical part of the complex folds and in the pyloric caeca. These features correspond to the early apoptotic stages in which dying cells assume the typical funnel-like structure with crescent nuclei and display a strongly acidophilic cytoplasm due to chromatin condensation [34].

5. Conclusions

Our data indicate that pyloric caeca, the villi and the basal portion of the complex folds in the distal intestine have a number of common morphological characteristics that set them apart from the proximal intestine and from the apical portion of the distal intestine complex folds. This indicates that rainbow trout intestine is characterized by the discontinuous distribution of two different morphological and functional compartments. Therefore, our results suggest that rainbow trout intestine is physiologically arranged to mingle digestive and absorptive functions along its length.

Author Contributions: Conceptualization, F.G. and T.A.L.B.; methodology, N.V., R.P., M.S. and G.S.; data collection and validation, N.V. and R.P.; formal statistical analysis, N.V. and R.P.; writing—original draft preparation, F.G. and N.V.; writing—review and editing, F.G.; supervision, F.G. and T.A.L.B.; project administration, F.G. and T.A.L.B.; funding acquisition, F.G. and T.A.L.B. All authors have read and agreed to the published version of the manuscript.

Funding: This research has received funding from the European Union's Horizon 2020 research and innovation programme under grant agreement No 828835 and from Skretting ARC, Norway.

Acknowledgments: The authors are grateful to Rebecca Heavyside for helpful suggestions and discussion, and Daphne Siciliani, Federica Camin and Lodovico Sterzi for their help with sampling and morphological analysis. N.V., R.P., T.A.L.B. and F.G. are members of COST Action 16119.

Conflicts of Interest: The authors declare that the research was conducted in the absence of any commercial or financial relationships that could be construed as a potential conflict of interest.

References

1. Crawford, S.S.; Muir, A.M. Global introductions of salmon and trout in the genus oncorhynchus: 1870–2007. *Rev. Fish Biol. Fish.* **2008**, *18*, 313–344. [CrossRef]
2. Caimi, C.; Gasco, L.; Biasato, I.; Malfatto, V.; Varello, K.; Prearo, M.; Pastorino, P.; Bona, M.C.; Francese, D.R.; Schiavone, A.; et al. Could dietary black soldier fly meal inclusion affect the liver and intestinal histological traits and the oxidative stress biomarkers of Siberian sturgeon (*Acipenser baerii*) juveniles? *Animals* **2020**, *10*, 155. [CrossRef] [PubMed]
3. Barker, N.; van Oudenaarden, A.; Clevers, H. Identifying the stem cell of the intestinal crypt: Strategies and pitfalls. *Cell Stem Cells* **2012**, *11*, 452–460. [CrossRef] [PubMed]
4. Elia, A.C.; Capucchio, M.T.; Caldaroni, B.; Magara, G.; Dörr, A.J.M.; Biasato, I.; Biasibetti, E.; Righetti, M.; Pastorino, P.; Prearo, M.; et al. Influence of hermetia illucens meal dietary inclusion on the histological traits, gut mucin composition and the oxidative stress biomarkers in rainbow trout (*Oncorhynchus mykiss*). *Aquaculture* **2018**, *496*, 50–57. [CrossRef]
5. Verdile, N.; Mirmahmoudi, R.; Brevini, T.A.L.; Gandolfi, F. Evolution of pig intestinal stem cells from birth to weaning. *Animal* **2019**, *3*, 1319. [CrossRef] [PubMed]
6. Fagundes, K.R.C.; Rotundo, M.M.; Mari, R.B. Morphological and histochemical characterization of the digestive tract of the puffer fish Sphoeroides testudineus (Linnaeus 1758) (*Tetraodontiformes: Tetraodontidae*). *An. Acad. Bras. Ciênc.* **2016**, *88*, 1615–1624. [CrossRef] [PubMed]
7. Ray, A.K.; Ringø, E. The gastrointestinal tract of fish. *Aquac. Nutr.* **2014**, *41*, 1–13. [CrossRef]
8. Khojasteh, S.M.B.; Sheikhzadeh, F.; Mohammadnejad, D.; Azami, A. Histological, histochemical and ultrastructural study of the intestine of rainbow trout (*Oncorhynchus mykiss*). *World Appl. Sci. J.* **2009**, *6*, 1525–1531.
9. Li, Y.; Kortner, T.M.; Chikwati, E.M.; Munang'andu, H.M.; Lock, E.J.; Krogdahl, Å. Gut health and vaccination response in pre-smolt Atlantic salmon (*Salmo salar*) fed black soldier fly (*Hermetia illucens*) larvae meal. *Fish Shellfish Immunol.* **2019**, *86*, 1106–1113. [CrossRef]
10. Hamidian, G.; Zirak, K.; Sheikhzadeh, N.; Oushani, K.A.; Shabanzadeh, S.; Divband, B. Intestinal histology and stereology in rainbow trout (*Oncorhynchus mykiss*) administrated with nanochitosan/zeolite and chitosan/zeolite composites. *Aquac. Res.* **2018**, *49*, 1803–1815. [CrossRef]
11. Dall'aglio, C.; Mercati, F.; Faeti, V.; Acuti, G.; Trabalza-Marinucci, M.; de Felice, E.; Tardella, F.M.; Franciosini, M.P.; Casagrande Proietti, P.; Catorci, D.; et al. Immuno-and glyco-histochemistry as a tool to evaluate the oregano supplemented feed effects in pig gut. *Eur. J. Histochem.* **2020**, *64*, 3110. [CrossRef]
12. Mercati, F.; Dall'Aglio, C.; Acuti, G.; Faeti, V.; Tardella, F.M.; Pirino, C.; de Felice, E.; Scocco, P. Oregano feed supplementation affects glycoconjugates production in swine gut. *Animals* **2020**, *10*, 149. [CrossRef] [PubMed]
13. Albl, B.; Haesner, S.; Braun-Reichhart, C.; Streckel, E.; Renner, S.; Seeliger, F.; Wolf, E.; Wanke, R.; Blutke, A. Tissue sampling guides for porcine biomedical models. *Toxicol. Pathol.* **2016**, *44*, 414–420. [CrossRef]
14. Nyengaard, J.R. Stereologic methods and their application in kidney research. *J. Am. Soc. Nephrol.* **1999**, *10*, 1100–1123.
15. Burnstock, G. The morphology of the gut of the brown trout (*Salmo trutta*). *J. Cell Sci.* **1959**, *3*, 183–198.
16. Lkka, G.; Austb, L.; Falk, K.; Bjerkås, I.; Koppang, E.O. Intestinal morphology of the wild atlantic salmon (*Salmo salar*). *J. Morphol.* **2013**, *274*, 859–876. [CrossRef] [PubMed]
17. Hassanpour, M.; Joss, J. Anatomy and histology of the spiral valve intestine in juvenile australian lungfish, neoceratodus forsteri. *Open Zool. J.* **2009**, *2*, 62–85. [CrossRef]
18. Henningsen, A.D.; Whitaker, B.R.; Walker, I.D. Protrusion of the valvular intestine in captive smalltooth sawfish and comments on pristid gastrointestinal anatomy and intestinal valve types. *J. Aquat. Anim. Health* **2005**, *17*, 289–295. [CrossRef]
19. Çinar, K.; Şenol, N. Histological and histochemical characterization of the mucosa of the digestive tract in flower fish (*Pseudophoxinus antalyae*). *J. Vet. Med. Ser. C Anatom. Histol. Embryol.* **2006**, *35*, 147–151. [CrossRef] [PubMed]
20. Argyriou, T.; Clauss, M.; Maxwell, E.E.; Furrer, H.; Sánchez-Villagra, M.R. Exceptional preservation reveals gastrointestinal anatomy and evolution in early actinopterygian fishes. *Sci. Rep.* **2016**, *6*, 18758. [CrossRef]

21. Glover, C.N.; Petri, D.; Tollefsen, K.E.; Jørum, N.; Handy, R.D.; Berntssen, M.H.G. Assessing the sensitivity of Atlantic salmon (*Salmo salar*) to dietary endosulfan exposure using tissue biochemistry and histology. *Aquat. Toxicol.* **2007**, *84*, 346–355. [CrossRef] [PubMed]
22. Gauthier, G.F.; Landis, S.C. The relationship of ultrastructural and cytochemical features to absorptive activity in the goldfish intestine. *Anat. Rec.* **1972**, *172*, 675–701. [CrossRef]
23. Arman, S.; Üçüncü, S.I. Histochemical characterization of convict cichlid (*Amatitlania nigrofasciata*) intestinal goblet cells. *Pak. J. Zool.* **2017**, *49*, 445–453. [CrossRef]
24. Bosi, G.; Arrighi, S.; di Giancamillo, A.; Domeneghini, C. Histochemistry of glycoconjugates in mucous cells of Salmo trutta uninfected and naturally parasitized with intestinal helminths. *Dis. Aquat. Organ.* **2005**, *64*, 45–51. [CrossRef]
25. Kim, J.J.; Khan, W.I. Goblet cells and mucins: Role in innate defense in enteric infections. *Pathogens* **2013**, *2*, 55–70. [CrossRef] [PubMed]
26. Sveinbjornsson, B.; Olsen, R.; Paulsen, S. Immunocytochemical localization of lysozyme in intestinal eosinophilic granule cells (EGCs) of Atlantic salmon, *Salmo salar* L. *J. Fish Dis.* **1996**, *19*, 349–355. [CrossRef]
27. Clevers, H.C.; Bevins, C.L. Paneth cells: Maestros of the small intestinal crypts. *Annu. Rev. Physiol.* **2013**, *75*, 289–311. [CrossRef]
28. Krystel-Whittemore, M.; Dileepan, K.N.; Wood, J.G. Mast cell: A multi-functional master cell. *Front. Immunol.* **2016**, *6*, 620. [CrossRef]
29. Reite, O.B.; Evensen, Ø. Inflammatory cells of teleostean fish: A review focusing on mast cells/eosinophilic granule cells and rodlet cells. *Fish Shellfish Immunol.* **2006**, *20*, 192–208. [CrossRef]
30. Dezfuli, B.S.; Giari, L.; Lui, A.; Squerzanti, S.; Castaldelli, G.; Shinn, A.P.; Manera, M.; Lorenzoni, M. Proliferative cell nuclear antigen (PCNA) expression in the intestine of Salmo trutta trutta naturally infected with an acanthocephalan. *Parasites Vectors* **2012**, *5*, 198. [CrossRef]
31. Lallès, J.P. Intestinal alkaline phosphatase: Novel functions and protective effects. *Nutr. Rev.* **2014**, *72*, 82–94. [CrossRef] [PubMed]
32. Shin, J.; Carr, A.; Corner, G.A.; Tögel, L.; Dávaos-Salas, M.; Tran, H.; Chueh, A.C.; Al-Obaidi, S.; Chionh, F.; Naseem, A.; et al. The intestinal epithelial cell differentiation marker intestinal alkaline phosphatase (ALPi) is selectively induced by histone deacetylase inhibitors (HDACi) in colon cancer cells in a kruppel-like factor 5 (KLF5)-defendent manner. *J. Biol. Chem.* **2014**, *289*, 25306–25316. [CrossRef] [PubMed]
33. Moss, S.F.; Attia, L.; Scholes, J.V.; Walters, J.R.F.; Holt, P.R. Increased small intestinal apoptosis in coeliac disease. *Gut* **1996**, *39*, 811–817. [CrossRef] [PubMed]
34. Wang, Y.; George, S.P.; Roy, S.; Pham, E.; Esmaeilniakooshkghazi, A.; Khurana, S. Both the anti-and pro-apoptotic functions of villin regulate cell turnover and intestinal homeostasis. *Sci. Rep.* **2016**, *6*, 35491. [CrossRef] [PubMed]

animals

MDPI

Article

Immunohistochemical Analysis of Intestinal and Central Nervous System Morphology in an Obese Animal Model (*Danio rerio*) Treated with 3,5-T2: A Possible Farm Management Practice?

Roberta Imperatore [1,2], Lea Tunisi [2,3], Isabella Mavaro [3,4], Livia D'Angelo [3], Chiara Attanasio [3,4], Omid Safari [5], Hamidreza Ahmadniaye Motlagh [5], Paolo De Girolamo [3], Luigia Cristino [2], Ettore Varricchio [1] and Marina Paolucci [1,6,*]

1 Department of Science and Technology (DST), University of Sannio, 82100 Benevento, Italy; rimperatore@unisannio.it (R.I.); etvarric@unisannio.it (E.V.)
2 Endocannabinoid Research Group, Institute of Biomolecular Chemistry-National Research Council (ICB-CNR), 80078 Pozzuoli, Italy; lea.tunisi@unina.it (L.T.); luigia.cristino@icb.cnr.it (L.C.)
3 Department of Veterinary Medicine and Animal Productions, University of Naples Federico II, 80137 Naples, Italy; isabella.mavaro@unina.it (I.M.); livia.dangelo@unina.it (L.D.); chiara.attanasio@unina.it (C.A.); degirola@unina.it (P.D.G.)
4 Center for Advanced Biomaterials for Healthcare, Istituto Italiano di Tecnologia, 80125 Naples, Italy
5 Department of Fisheries, Faculty of Natural Resources and Environment, Ferdowsi University of Mashhad, Mashhad 9177948974, Iran; omid_safary@yahoo.com (O.S.); ahmadnia@um.ac.ir (H.A.M.)
6 Institute of Food Science, National Research Council (ISA-CNR), 83100 Avellino, Italy
* Correspondence: paolucci@unisannio.it

Received: 6 June 2020; Accepted: 1 July 2020; Published: 3 July 2020

Simple Summary: The obesity induced by overconsumption of nutrients leads to systemic inflammation and alters metabolic homeostasis by acting on central nervous system and peripheral tissues such as intestine. The 3,5-diiodo-L-thyronine (3,5-T2) is well-known for its positive role on fat mass and lipid metabolism, and at date, it is widely used as a drug for the treatment of obesity. However, the safe and effective dose as well as the possible adverse effects of this molecule have not been sufficiently explored. In this study, we analyzed the role of 3,5-T2 in regulating central and peripheral inflammation in diet-induced obese (D.I.O.) model of zebrafish. We found that 3,5-T2 sustained the intestinal alteration caused by D.I.O., as indicated by the high levels of pro-inflammatory cytokines, accompanied by a significant effect of 3,5-T2 on body weight and central inflammation in D.I.O. zebrafish. Therefore, the suggested potential use of 3,5-T2 to contrast obesity should be viewed with caution. We conclude that the zebrafish model can help to better understand the fundamental beneficial and side effects of 3,5-T2, which is of great importance to define the possible use of this metabolite of thyroid hormones as a drug in different diseases including obesity.

Abstract: The 3,5-diiodo-L-thyronine (3,5-T2) is an endogenous metabolite of thyroid hormones, whose administration to rodents fed high-fat diet (HFD) prevents body weight increase and reverts the expression pattern of pro-inflammatory factors associated to HFD. The diet-induced obese (D.I.O.) zebrafish (Danio rerio) has been recently used as an experimental model to investigate fundamental processes underlying central and peripheral obesity-driven inflammation. Herein, we aim to understand the role of 3,5-T2 in regulating central and peripheral inflammation in D.I.O. model of zebrafish. 3,5-T2 (10 nM and 100 nM) was administered with the obesity-inducing diet (D.I.O. with 3,5-T2) or after 4 weeks of obesity-inducing diet (D.I.O. flw 3,5-T2). 3,5-T2 significantly increased the body weight and serum triglyceride levels in D.I.O. zebrafish in both conditions. Moreover, 3,5-T2 sustained or increased inflammation in the anterior (AI) and mid (MI) intestine when administered with the obesity-inducing diet, as indicated by the immunoexpression of the inflammatory markers tumor-necrosis factor-α (TNFα), cyclooxygenase 2 (COX2), calnexin, caspase 3, and proliferating cell nuclear antigen (PCNA). On the contrary, when 3,5-T2 was administered after the obesity-inducing diet, partly reverted the intestinal alteration induced by D.I.O. In addition, brain inflammation, as indicated by the increase in the activation of microglia, was detected in D.I.O. zebrafish and D.I.O. treated with 3,5-T2. These findings reveal that the effects of 3,5-T2 on fish intestine and brain can deviate from those shown in obese mammals, opening new avenues to the investigation of the potential impact of this thyroid metabolite in different diseases including obesity.

Keywords: 3,5-diiodo-L-thyronine; zebrafish; diet-induced obesity; intestinal inflammation; brain inflammation

---

1. Introduction

Abnormal or excessive fat accumulation are generally accompanied by low chronic central and peripheral [1–3]. One of the main factors contributing to the development of obesity is the consumption of a high-fat diet (HFD) [1,4], characterized by high leptin levels in humans [5], mice and zebrafish (Danio rerio) [6–8] and increase in tumor necrosis factor α (TNF-α) and interleukin 6 (IL-6) levels [9]. Such pro-inflammatory factors are released into the blood stream contributing to the development of a chronic low-grade peripheral and central inflammatory status [10,11]. Many of the pathological conditions related to obesity derive from the impairment of mitochondrial respiration [12]. Since obesity causes serious pathologies, such as diabetes, hypertension, and dyslipidemia [11], its treatment is essential to prevent their development. Effective anti-obesity drugs able to counteract the excess of body adiposity without undesirable side effects are still needed. The use of thyroid hormones has been recently suggested to reduce weight and lipid accumulation. The 3,5-diiodo-L-thyronine (3,5-T2), an endogenous metabolite of thyroid hormones, has been identified as a biologically active iodothyronine [13–15], which acts as an allosteric regulator of cytochrome oxidase (COX) activity, and regulates mitochondrial activity and respiration [16], leading to a rapid increase in mitochondrial oxygen consumption [15,17]. Through these actions, 3,5-T2 seems to be able to revert the proinflammatory pattern activated in rats by HFD [18], playing important metabolic activities in HFD-fed rodents [19]. Moreover, chronic administration of 3,5-T2 (250 μg/100 g BW for 14 or 28 days i.p.) in diet-induced obese mice has beneficial effects on adiposity, serum leptin, and energy expenditure [20]. However, high levels of 3,5-T2 decrease body weight and blood glucose in obese mice, but induce thyrotoxicosis [21]. 3,5-T2 hypolipidemic effects have been studied in several animal models and it is reported that the effects of 3,5-T2 on metabolic efficiency are conserved across species [19,22]. However, little is known about the potential role of 3,5-T2 in the central and peripheral inflammation. Many studies are in progress to better understand whether 3,5-T2 can be a potential anti-obesity agent, but also to define its time- and dose-dependent activity, as well as the possible occurrence of undesirable side effects [16,19].

Zebrafish has been recognized as an excellent animal model to investigate the fundamental processes underlying human metabolic and inflammatory diseases, because of its similarity to mammals in terms of gut and brain functions as well as immunity related genes [23]. In zebrafish, as in mammals, the consumption of HFD induces general inflammation [3]. Diet-induced obese (D.I.O.) zebrafish model shows common pathophysiological pathways with obese mammalians, suggesting that zebrafish can be used as obesity model. Studies performed both in rodents and zebrafish have shown that the prolonged consumption of HFD leads to intestinal inflammation [1–3]. As in humans and rodents, also in zebrafish obesity-induced inflammation involves the activation of inflammatory mediators such as chemokines and cytokines (TNFα and interleukins) which are also highly conserved between zebrafish and mammals [3].

Despite the anatomical differences, the thyroid system is highly conserved in teleosts [24,25]. Indeed, several studies reported an important role of 3,5-T2 in regulating the development, growth, and metabolism in fish. Recently, Olvera et al. [26] showed that 3,5-T2 regulates genes involved in cell signaling and transcriptional pathways in the brain of tilapia (Oreochromis niloticus), while Little [27] and Navarrete-Ramirez [28] revealed that 3,5-T2 regulates thermal acclimation in zebrafish and growth in tilapia, respectively.

Considering that HFD is a major cause of systemic inflammation, D.I.O. zebrafish has been confirmed as a reliable animal model to identify putative pharmacological targets for the treatment of obesity and 3,5-T2 has been proposed to reduce obesity-induced inflammation, in this work we investigate the effect on peripheral and central inflammation of supraphysiological 3,5-T2 administration in the D.I.O. zebrafish model. In particular, to study the inflammation pathway we analyzed the gut morphology and the expression of the main inflammatory markers together with the related brain microglia activation. In addition, we describe the side effects of 3,5-T2 administration, providing useful information concerning the potential pharmacological use of this molecule.

2. Materials and Methods

2.1. Animals

Animals were housed under standard conditions of photoperiod (14:10 Light/Dark; ZT0, 9 Anti Meridian) and temperature (28 °C) [29]. The current study was carried out in the Aquaculture Lab of the Ferdowsi University of Mashhad (FUM). This project was approved by FUM animal ethics committee. Fish used in this study were treated in accordance with the European Commission recommendation 2007/526/EC and 2010/63/UE on revised guidelines for the accommodation and care of animals used for scientific purposes. All efforts were made to minimize fish suffering. Zebrafish did not receive medical treatment prior or during the experience. No deaths occurred before the experiment endpoint.

2.2. Zebrafish Feeding and Treatment

In feeding experiments, adult male zebrafish (0.35 ± 0.05 g) were divided into four dietary groups (n = 10 fish per group). (1) Control group: zebrafish fed once a day for 4 weeks with peeled Artemia salina cysts (22% fat, 44% proteins, 16% carbohydrates; Aqua Schwarz) in a weight-maintaining amount (5 mg Artemia per fish) (ctrl). (2) D.I.O. zebrafish group (diet-induced obesity): zebrafish made obese by feeding for 4 weeks with the same amount of Artemia salina of the control group, but three times a day (D.I.O.). (3) D.I.O. zebrafish followed by 3,5-T2 group: zebrafish made obese by 4 weeks of diet-inducing obesity and then fed with the same amount of Artemia salina of the control group adding 3,5-diiodothyronine (3,5-T2) (Sigma Aldrich) for 4 weeks (D.I.O. flw 3,5-T2). 3,5-T2 was added to the water at two final concentrations (10 nM and 100 nM). (4) D.I.O. zebrafish fed with 3,5-T2 group: zebrafish fed for 4 weeks with diet inducing obesity and at the same time treated with 3,5-T2 added to the water at two final concentrations (10 nM and 100 nM) (D.I.O. with 3,5-T2). Supraphysiological doses of 3,5-T2 were used according to Garcia et al. [30,31]. The addition of 3,5-T2 to the water was performed considering that the use of hydrophobic drugs by immersion provide an efficient, non-invasive, and minimally stressful mean of chronic administration in aquatic vertebrates [31]. The water was changed every day at 9:00 a.m. and 3,5-T2 was added. The efficacy of the administration of iodothyronine by immersion has been reported by Garcia and colleagues [30,31]. The D.I.O. feeding protocol used in this study was adapted from a previous study by Mania et al. [32]. After the trial, the zebrafish were fasted overnight and then suppressed. For every treatment, both the brain and the intestine from five fish were collected for histological analysis and immunohistochemistry. The 100-nM 3,5-T2 concentration showed toxicity with high mortality, therefore its administration was suspended.

2.3. Body Weight Measurement

Zebrafish body weight was determined by a precision analytical scale. Body weight (n = 10 animals/diet) was recorded at the beginning (week 0) and at the end (week 4 or 8) of the experiment. Specimens were fasted for 24 h and then suppressed.

2.4. Serum Lipid Parameters

At the end of the feeding experiments, blood was collected from 24-h-fasted zebrafish of all dietary groups (n = 10 animals/diet). For analyses of serum triglycerides, blood was collected by tail ablation. Serum was obtained by centrifugation at 2500× *g* for 10 min at 4 °C. Serum of three zebrafish was pooled for detection. Total triglycerides were measured using enzymatic colorimetric assays (BioAssay Systems, San Jose, CA, USA) following the manufacturer's instructions.

2.5. Morphological Analysis

Intestine samples were promptly fixed in 4% formalin in 0.01 M phosphate-buffered saline (PBS) pH 7.4 for no longer than 24 h at 4 °C, dehydrated in a graded series of ethanol, cleared with xylol, and embedded in paraffin. Samples were cut in 7 μm sections using a microtome [33]. For morphological analysis, anatomic comparable sections of intestinal bulb, herein indicated as anterior- (AI) and mid- (MI) intestine were deparaffinized with xylol and stained with hematoxylin-eosin (H&E). Histological sections were examined under light microscopy with a Leica DMI6000 equipped with Leica DFC340 cooled digital CCD camera (Leica Microsystems, Buccinasco MI, Italy) to obtain microscopic images at 20× magnification. To count the number of the goblet cells, anatomically comparable sections of AI and MI were stained with Alcian blue (1 g of Alcian blue, pH 2.5, 3 mL/L of acetic acid, and 97 mL of distilled water) for 1 h. Thereafter, the slides were rinsed in tap water for 10 min, oxidized in periodic acid (5 g/L) for 5 min, rinsed in lukewarm tap water for 10 min, and dehydrated in alcohol and clarified in xylol. Histological sections were examined by using the same microscope mentioned before at 10× magnification. In order to estimate the number of goblet cells, sections (n = 5 animal per treatment; n = 3 pairs of sections/animal, each section selected at a 50 μm distance to avoid counting the same cells) were analyzed by two independent operators blinded to the experimental protocol. For each section AI and MI were divided into three regions and goblet cells were counted in each of them.

2.6. Single Immunohistochemistry

For immunohistochemistry, the deparaffinized anatomic comparable sections of AI and MI were stained by the avidin-biotin immunohistochemical technique. The following antibodies were used: monoclonal antibodies raised in mouse against tumor-necrosis factor-α (TNFα) (code ab1793, Abcam, Cambridge, UK), polyclonal antibodies raised in rabbit against cyclooxygenase 2 (COX2) (code 69720, NovaTeinBio, Woburn, MA, USA), polyclonal antibodies raised in rabbit against calnexin (code NB100-1965, Novus Biologicals, Centennial, CO, USA), polyclonal antibodies raised in rabbit against caspase 3 (code ab13847, Abcam, Cambridge, UK), and monoclonal antibodies raised in mouse against proliferating cell nuclear antigen (PCNA) (code ab29, Abcam, Cambridge, UK). The sections were incubated for 5 min in 0.1% H_2O_2 to inactivate the endogenous peroxidase activity and then incubated for 30 min with 10% normal goat serum (NGS) (Vector Laboratories, Burlingame, CA, USA) in 0.1 M Tris-buffered saline, pH 7.6, containing 0.3% Triton X-100. Thereafter, slides were incubated overnight at 4 °C with primary antibodies diluted 1:200 in NGS. After several rinses, the sections were incubated at room temperature for 2 h with biotinylated goat anti-mouse or goat anti-rabbit immunoglobulin at the appropriate dilution (Vector Laboratories, Burlingame, CA, USA), followed by 1 h incubation in the avidin-biotin complex (ABC Kit; Vectastain, Vector Laboratories, Burlingame, CA, USA) diluted in tris-buffered saline according to the manufacturer's instructions and then in 0.05% of 3′-diaminobenzidine (DAB) for 10 min (DAB Sigma Fast, Merck Life Science S.r.l., Milano, Italy). Histological sections were examined under light microscopy with a Leica DMI6000 equipped with Leica DFC340 cooled digital CCD camera (Leica Microsystems, Wetzlar, Germany) and the images were acquired at 10× and 20× magnification [33]. To quantify the density of TNFα-, COX2-, and calnexin-positive signal, each section of AI and MI was divided into three regions (n = 5 animals per treatment; n = 3 pairs of sections/animal, each section selected at 50 μm distance). Digital images were acquired under constant light illumination and magnification, using a digital camera working on gray levels (JCV FC 340FX, Leica Microsystems, Buccinasco MI, Italy). Densitometric analysis of TNFα, COX2, and calnexin peroxidase-based immunostaining was performed by measuring optical density using the image analysis software Image Pro Plus® version 6.0 (MediaCybernetics, Rockville, MD, USA) working on a logarithmic scale of absorbance. In each region, the optical density zero value was assigned to the background (i.e., a tissue portion devoid of stained cells) [33]. Quantitative analysis of mean dead and proliferating cells was performed by counting the caspase 3- or PCNA-positive cells with the nucleus on the focal plane within a box measuring 2 × 104 mm^2 in the AI or MI (n = 5 animals per treatment; n = 3 pairs of sections/animal, each section selected at 50 μm distance to avoid to count same cells) [34]. All histological analyses were performed by two independent operators blinded to the type of treatment.

2.7. Immunofluorescence

For single immunofluorescence, the polyclonal antibody raised in rabbit against ionized calcium-binding adapter molecule 1 (Iba1) (1:200; code 019-19741, WAKO Chemicals, Neuss, Germany) was used. After incubation with the primary antibody, the brain sections were washed several times in PBS and incubated for 2 hours at room temperature with donkey anti rabbit Alexa 488-conjugated secondary antibodies (1:100; Invitrogen, ThermoFisher Scientific, Waltham, Massachusetts, U.S.). Tissue sections were washed in PBS and all slides were coverslipped with Aquatex mounting medium (Merck, Darmstadt, Germany). The immunostained sections were observed with a confocal microscopy Nikon Eclipse Ti2 (Nikon, Florence, Italy) equipped with x-y-z motorized stage, a digital camera DS-Qi2 (Nikon, Florence, Italy), and the acquisition and Image analysis software NIS-Elements C (Nikon, Florence, Italy). Digital images were acquired using the 20–40× objectives. We collected serial Z-stacks of images throughout the area of interest (6–10 planes with an increment varying 0.5–1 μm). Images were deconvolved using the imaging deconvolution software by application of ten iterations. Serial Z plane images were collapsed into a single maximum projection image. Micrographs were saved in TIFF format and adjusted for light and contrast before being assembled on plates using Adobe Photoshop 6.01 (Adobe Systems, San Jose, CA, USA) [33]. The number of cells positive for Iba-1 was determined within a box measuring 2 × 104 μm^2 that was placed in the lateral, central, and medial areas of hypothalamic nucleus (n = 5 animals per treatment; n = 3 pairs of sections/animal, each section selected at 50 μm distance to avoid to count the same cells). To avoid cell overcounting, only cells with the nucleus on the focal plane were considered [35]. Iba-1-positive cells were identified as resting (with small somata bearing long, thin, and ramified processes) and activated microglia (with hypertrophy together with retraction of processes to a length shorter than the diameter of the somata) or dystrophic microglia. Dystrophic microglia was recognized by debris consisting of several cells displaying fragmented processes and an irregularly shaped cell body as previously demonstrated in humans [35].

2.8. Controls

All the antibodies used are reported as reactive for zebrafish. Their specificity was validated with control samples, including: (1) Omission of primary or secondary antibody staining; (2) pre-absorptions of each primary antibody with an excess of the relative peptide (TNFα, and caspase 3) (Figure S1).

2.9. Statistical Analysis

The data are expressed as mean ± SEM and analyzed with GraphPad Prism 6 software, version 6.05 (GraphPad, Inc., San Diego, CA, USA). One-way analysis of variance with Bonferroni's post-hoc test was adopted for the analysis of normally distributed data. For the experiment, including densitometric analysis Kruskal–Wallis Anova nonparametric test followed by Dunn's post hoc test was adopted for the analysis of non-normally distributed data. A p-value < 0.05 was considered as significant.

3. Results

3.1. 3,5-T2 Increased Body Weight and Serum Triglyceride Levels in D.I.O. Zebrafish

At the end of the experiment significant differences in the body weight were observed among groups. The body weight significantly increased in D.I.O. with respect to the control (ctrl: 0.40 ± 0.013 gr vs. D.I.O.: 0.78 ± 0.021 gr, $p < 0.0001$). The treatment with 3,5-T2 at a concentration of 10 nM caused a significant increase in the body weight both in D.I.O. flw 3,5-T2 (ctrl: 0.40 ± 0.013 gr vs. D.I.O. flw 3,5-T2: 0.97 ± 0.032 gr, $p < 0.0001$) and D.I.O. with 3,5-T2 (ctrl: 0.40 ± 0.013 gr vs. D.I.O. with 3,5-T2: 0.93 ± 0.033 gr, $p < 0.0001$) compared to the control. Moreover, the D.I.O. flw 3,5-T2 and D.I.O. with 3,5-T2 showed an increase in the body weight significantly higher than D.I.O. (D.I.O.: 0.78 ± 0.021 gr vs. D.I.O. flw 3,5-T2: 0.97 ± 0.032 gr, $p < 0.0001$; D.I.O.: 0.78 ± 0.021 gr vs. D.I.O. with 3,5-T2: 0.93 ± 0.033 gr, $p < 0.001$) (Figure 1).

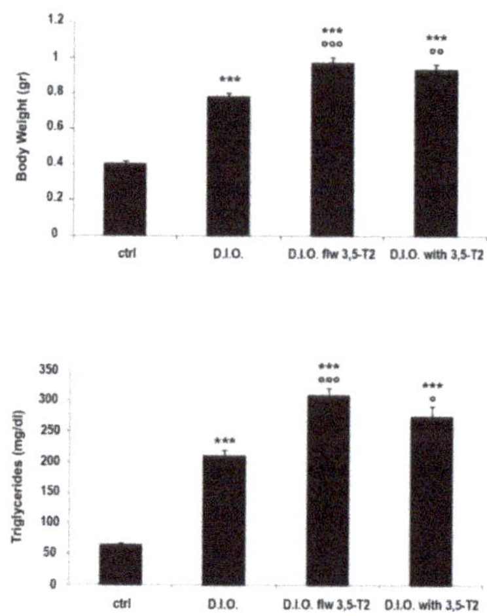

Figure 1. 3,5-T2 effect on the body weight and triglyceride levels. ctrl (control zebrafish), D.I.O. (diet-induced obesity zebrafish), D.I.O. flw 3,5–T2 (D.I.O. zebrafish followed by 3,5–T2), D.I.O. with 3,5–T2 (D.I.O. zebrafish treated with 3,5–T2). Data are expressed as mean ± SE. *** $p < 0.0001$ compared to the control group; °°° $p < 0.0001$, °° $p < 0.0001$, and ° $p < 0.05$ compared to D.I.O.

The significant increase in the body weight was accompanied by the increase in serum triglycerides. In particular, D.I.O. zebrafish presented significantly elevated plasma triglyceride levels with respect to the control (ctrl: 65.33 ± 1.84 mg/dL vs. D.I.O.: 209.67 ± 9.22 mg/dL, $p < 0.0001$). The treatment with 3,5-T2 at a concentration of 10 nM caused a significant increase in the triglycerides both in D.I.O. flw 3,5-T2 (ctrl: 65.33 ± 1.84 mg/dL vs. D.I.O. flw 3,5-T2: 307.67 ± 13.04 mg/dL, $p < 0.0001$) and D.I.O. with 3,5-T2 (ctrl: 65.33 ± 1.84 mg/dL vs. D.I.O. with 3,5-T2: 274.11 ± 17.34 mg/dL, $p < 0.0001$), compared to the control. Moreover, the D.I.O. flw 3,5-T2 and D.I.O. with 3,5-T2 showed an increase in triglycerides significantly higher than D.I.O. (D.I.O.: 209.67 ± 9.22 mg/dL vs. D.I.O. flw 3,5-T2: 307.67 ± 13.04 mg/dL, $p < 0.0001$; D.I.O.: 209.67 ± 9.22 mg/dL vs. D.I.O. with 3,5-T2: 274.11 ± 17.34 mg/dL, $p < 0.05$) (Figure 1).

3.2. 3,5-T2 Sustained or Increased Intestinal Morphological Alterations Induced by D.I.O.

Histological analysis of the intestine is presented in Figure 2. Control zebrafish showed a normal structure of both AI and MI (Figure 2A and B respectively). On the contrary, D.I.O. showed alteration of the intestinal folds (villi) which appeared ragged and irregular both in AI and MI. Debris from ragged villi were detected in the intestinal lumen. In D.I.O., villi alteration was accompanied by a significant increase in the number of goblet cells in AI and MI compared to the control (AI: ctrl: 44 ± 1 vs. D.I.O.: 73 ± 1, $p < 0.05$; MI: ctrl: 31 ± 1 vs. D.I.O.: 55 ± 1, $p < 0.05$). 3,5-T2 treatment led to thinner villi with the consequent enlargement of the gut lumen, together with a significant increase in goblet cell number both in D.I.O. flw 3,5-T2 and D.I.O. with 3,5-T2 with respect to the control zebrafish. More specifically, the quantitative analysis of histological slides stained with Alcian blue, showed a statistically significant increase in the number of goblet cells in the MI of D.I.O. flw 3,5-T2 and D.I.O. with 3,5-T2 (ctrl: 31 ± 1 vs. D.I.O. flw 3,5-T2: 63 ± 3, $p < 0.0001$; ctrl: 31 ± 1 vs. D.I.O. with 3,5-T2: 51 ± 3, $p < 0.05$) and in the AI of D.I.O. flw 3,5-T2 and D.I.O. with 3,5-T2 in comparison to the control (ctrl: 44 ± 0.8 vs. D.I.O. flw 3,5-T2: 100 ± 1, $p < 0.0001$; ctrl: 44 ± 0.8 vs. D.I.O. with 3,5-T2: 98 ± 1, $p < 0.0001$). Moreover, a significant increase in the number of goblet cells was found in the AI of D.I.O. flw 3,5-T2 with respect to the D.I.O. (D.I.O.: 73 ± 1 vs. D.I.O. flw 3,5-T2: 100 ± 1, $p < 0.05$). Although no significant differences were found in the goblet cell number in the AI of D.I.O. with 3,5-T2 and in MI of D.I.O. with 3,5-T2 and D.I.O. flw 3,5-T2 with respect to the D.I.O. zebrafish, they show an increase of morphological alterations with thinner villi and basal membrane disconnection (Figure 2 and Figure S2).

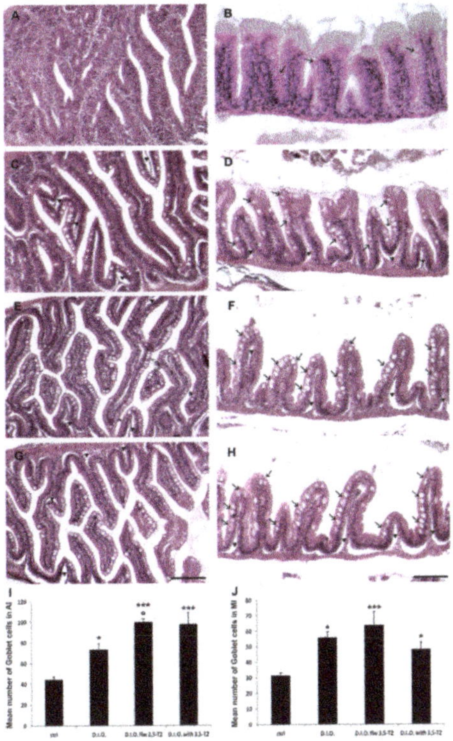

Figure 2. Hematoxylin and eosin (H&E) staining of anterior (AI) and mid (MI) intestine of ctrl (control zebrafish), D.I.O. (diet-induced obesity zebrafish), D.I.O. flw 3,5–T2 (D.I.O. zebrafish followed by 3,5–T2), D.I.O. with 3,5–T2 (D.I.O. zebrafish treated with 3,5–T2). (**A**) AI and (**B**) MI of control zebrafish. (**C**) AI and (**D**) MI of D.I.O. (**E**) AI and (**F**) MI of D.I.O. flw 3,5–T2. (**G**) AI and (**H**) MI of D.I.O. with 3,5–T2. Arrows indicate goblet cells, arrowheads indicate ragged villi. (**I**,**J**) Bar graphs showing the number of goblet cells in the (**I**) AI and (**J**) MI of ctrl, D.I.O., D.I.O. flw 3,5–T2 and D.I.O. with 3,5–T2. Goblet cells were counted based on Alcian Blue staining. Data are expressed as mean ± SE. *** $p < 0.0001$, * $p < 0.05$ compared to the control group. ° $p < 0.05$ compared to D.I.O. Scale bar: 100 μm.

3.3. The Cotreatement with 3,5-T2 Sustained Intestinal Inflammation Induced by D.I.O.

The morphological alterations of zebrafish intestine induced by D.I.O. and sustained by 3,5-T2 treatment were accompanied by the increased expression of pro-inflammatory markers. In particular, TNFα and COX2 immunoreactivity increased in the enteroendocrine and goblet cells of D.I.O. This increase in TNFα and COX2 immunoexpression in the intestinal cells was generally maintained in D.I.O. with 3,5-T2, showing positivity also in M-like vacuolated cells, while it was partly reduced in D.I.O. flw 3,5-T2 (Figures 3 and 4).

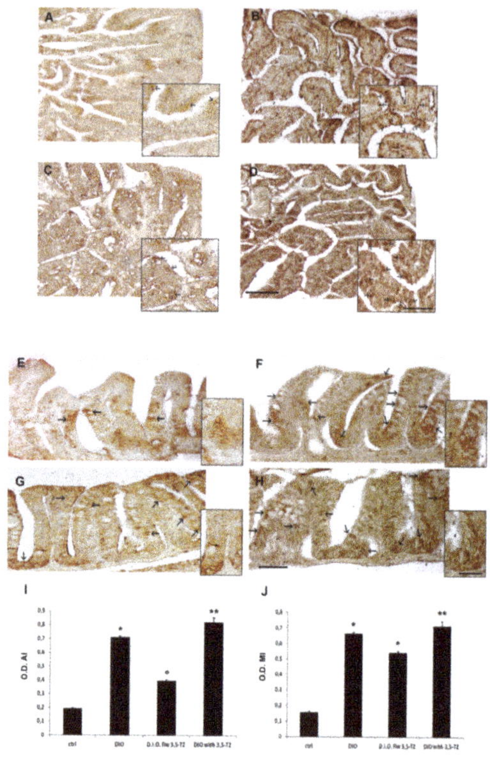

Figure 3. TNFα immunostaining in the anterior (AI) and mid (MI) intestine of ctrl (control zebrafish), D.I.O. (diet-induced obesity zebrafish), D.I.O. flw 3,5–T2 (D.I.O. zebrafish followed by 3,5–T2), D.I.O. with 3,5–T2 (D.I.O. zebrafish treated with 3,5–T2). (**A**) AI of ctrl zebrafish (**B**) AI of D.I.O. (**C**) AI of D.I.O. flw 3,5–T2. (**D**) AI of D.I.O. with 3,5–T2. (**E**) MI of ctrl zebrafish. (**F**) MI of D.I.O. (**G**) MI of D.I.O. flw 3,5–T2. and (**H**) MI of D.I.O. with 3,5–T2. The arrows indicate TNFα immunoexpression in the enteroendocrine and goblet cells. (**I**,**J**) Bar graphs showing TNFα optical density (O.D.) in the (**I**) AI and (**J**) MI of ctrl, D.I.O., D.I.O. flw 3,5–T2 and D.I.O. with 3,5–T2. Data are expressed as mean ± SE. ** $p < 0.001$, * $p < 0.05$ compared to the control group. ° $p < 0.05$ compared to D.I.O. Scale bar: 100 μm in the low magnification and 50 μm in the higher magnification in the boxes of A–D. 50 μm in the low magnification and 25 μm in the higher magnification in the boxes of E–H.

To confirm the increase of TNFα and COX2 immunoexpression in the epithelial cells a densitometric analysis was performed. The densitometric analysis showed a significant increase of TNFα optical density in the epithelium of both AI and MI of D.I.O. with respect to the ctrl (AI: ctrl 0.19 ± 0.01 vs. D.I.O. 0.71 ± 0.009, $p < 0.05$; MI: ctrl 0.15 ± 0.01 vs. D.I.O. 0.66 ± 0.006, $p < 0.05$). A significant increase of TNFα optical density was found also in the MI of D.I.O. flw 3,5-T2 (ctrl 0.15 ± 0.01 vs. D.I.O. flw 3,5-T2 0.54 ± 0.006, $p < 0.05$) and in the AI and MI of D.I.O. with 3,5-T2 (AI: ctrl 0.19 ± 0.01 vs. D.I.O.

with 3,5-T2 0.82 ± 0.05, $p < 0.001$; MI: ctrl 0.15 ± 0.01 vs. D.I.O. with 3,5-T2 0.71 ± 0.05, $p < 0.001$) compared to the control. On the contrary, a significant decrease of TNFα optical density was found in the AI of D.I.O. flw 3,5-T2 with respect to D.I.O. zebrafish (D.I.O. 0.71 ± 0.009 vs. D.I.O. flw 3,5-T2 0.39 ± 0.01, $p < 0.05$) (Figure 3).

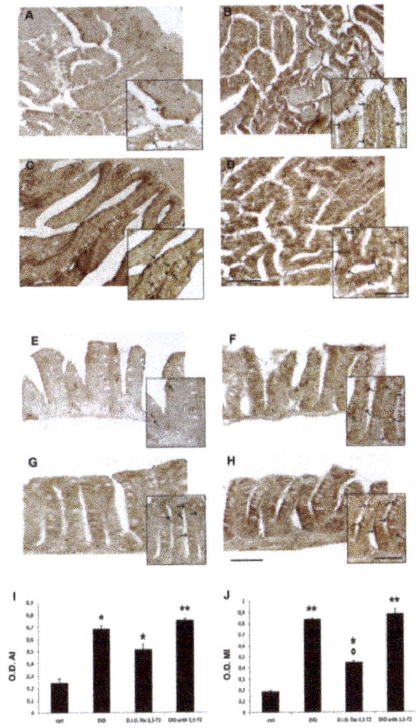

Figure 4. COX2 immunostaining in the anterior (AI) and mid (MI) intestine of ctrl (control zebrafish), D.I.O. (diet-induced obesity zebrafish), D.I.O. flw 3,5–T2 (D.I.O. zebrafish followed by 3,5–T2), D.I.O. with 3,5–T2 (D.I.O. zebrafish treated with 3,5–T2). (**A**) AI of ctrl zebrafish (**B**) AI of D.I.O. (**C**) AI of D.I.O. flw 3,5–T2. (**D**) AI of D.I.O. with 3,5–T2. (**E**) MI of ctrl zebrafish. (**F**) MI of D.I.O. (**G**) MI of D.I.O. flw 3,5 T2. and (**H**) MI of D.I.O. with 3,5–T2. The arrows indicate COX2 immunoexpression in the enteroendocrine and goblet cells. (**I**,**J**) Bar graphs showing COX2 optical density (O.D.) in the (**I**) AI and (**J**) MI of ctrl, D.I.O., D.I.O. flw 3,5–T2 and D.I.O. with 3,5–T2. Data are expressed as mean ± SE. ** $p < 0.001$, * $p < 0.05$ compared to the control group. ° $p < 0.05$ compared to D.I.O. Scale bar: 100 μm in the low magnification and 50 μm in the higher magnification in the boxes.

The inflammatory status was confirmed by COX2 optical density with significant increase in D.I.O. with respect to the control (AI: ctrl 0.24 ± 0.06 vs. D.I.O. 0.68 ± 0.05, $p < 0.05$; MI: ctrl 0.18 ± 0.01 vs. D.I.O. 0.83 ± 0.01, $p < 0.001$). COX2 optical density significantly increased in the AI and MI of D.I.O. flw 3,5-T2 (AI: ctrl 0.24 ± 0.06 vs. D.I.O. flw 3,5-T2 0.51 ± 0.08, $p < 0.05$; MI: ctrl 0.18 ± 0.01 vs. D.I.O. flw 3,5-T2 0.44 ± 0.01, $p < 0.05$) and AI and MI of D.I.O. with 3,5-T2 (AI: ctrl 0.24 ± 0.06 vs. D.I.O. with 3,5-T2 0.75 ± 0.02, $p < 0.001$; MI: ctrl 0.17 ± 0.01 vs. D.I.O. with 3,5-T2 0.88 ± 0.05, $p < 0.001$) with respect to the ctrl. On the contrary, a significant decrease of COX2 optical density was found in the MI of D.I.O. flw 3,5-T2 with respect to the D.I.O. zebrafish (D.I.O. 0.83 ± 0.01 vs. D.I.O. flw 3,5-T2 0.44 ± 0.01, $p < 0.05$) (Figure 4).

3.4. The 3,5-T2 Sustained Inflammation Induced by D.I.O. Was Accompanied by ER-Stress in the Anterior and Mid Intestine

The inflammation induced by D.I.O. was accompanied by ER-stress, as indicated by calnexin immunoexpression that increased in the enteroendocrine and goblet cells of D.I.O. This ER-stress was sustained by the cotreatment with 3,5-T2, while the 3,5-T2 post-treatment partly reverted this condition both in the AI and MI. As shown by the densitometric analysis, a significant increase in calnexin optical density was found in the AI and MI of D.I.O. (AI: ctrl 0.24 ± 0.03 vs. D.I.O. 0.72 ± 0.02, $p < 0.05$; MI: ctrl 0.22 ± 0.05 vs. D.I.O. 0.77 ± 0.03, $p < 0.001$), AI and MI of D.I.O. flw 3,5-T2 (AI: ctrl 0.24 ± 0.03 vs. D.I.O. flw 3,5-T2 0.48 ± 0.02, $p < 0.05$; MI: ctrl 0.22 ± 0.05 vs. D.I.O. flw 3,5-T2 0.51 ± 0.02, $p < 0.05$) and in the AI and MI of D.I.O. with 3,5-T2 (AI: ctrl 0.24 ± 0.03 vs. D.I.O. with 3,5-T2 0.80 ± 0.03, $p < 0.001$; MI: ctrl 0.22 ± 0.05 vs. D.I.O. with 3,5-T2 0.69 ± 0.03, $p < 0.05$) with respect to the control. Interestingly, a significant decrease in calnexin optical density was found in the AI and MI of D.I.O. flw 3,5-T2 with respect to D.I.O. (AI: D.I.O. 0.72 ± 0.02 vs. D.I.O. flw 3,5-T2 0.48 ± 0.02, $p < 0.05$; MI: D.I.O. 0.77 ± 0.03 vs. D.I.O. flw 3,5-T2 0.51 ± 0.02, $p < 0.001$) (Figure 5).

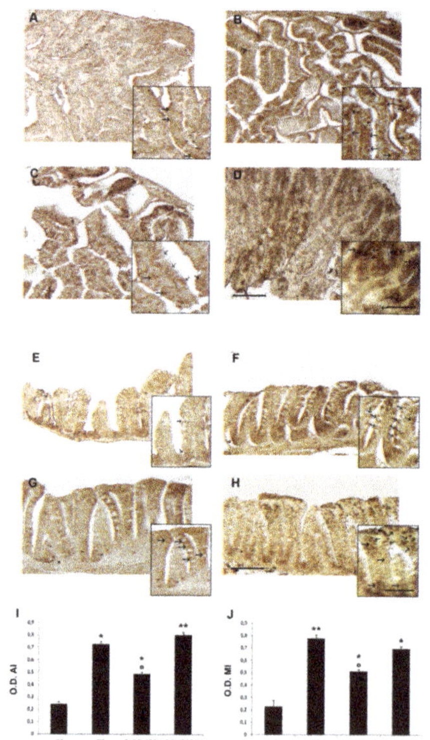

Figure 5. Calnexin immunostaining in the anterior (AI) and mid (MI) intestine of ctrl (control zebrafish), D.I.O. (diet-induced obesity zebrafish), D.I.O. flw 3,5–T2 (D.I.O. zebrafish followed by 3,5–T2), D.I.O. with 3,5–T2 (D.I.O. zebrafish treated with 3,5–T2). (**A**) AI of ctrl zebrafish (**B**) AI of D.I.O. (**C**) AI of D.I.O. flw 3,5–T2. (**D**) AI of D.I.O. with 3,5–T2. (**E**) MI of ctrl zebrafish. (**F**) MI of D.I.O. (**G**) MI of D.I.O. flw 3,5–T2. and (**H**) MI of D.I.O. with 3,5–T2. The arrows indicate calnexin immunoexpression in the enteroendocrine and goblet cells. (**I**,**J**) Bar graphs showing calnexin optical density (O.D.) in the (**I**) AI and (**J**) MI of ctrl, D.I.O., D.I.O. flw 3,5–T2 and D.I.O. with 3,5–T2. Data are expressed as mean ± SE. ** $p < 0.001$, * $p < 0.05$ compared to the control group. ° $p < 0.05$ compared to D.I.O. Scale bar: 100 µm in the low magnification and 50 µm in the higher magnification in the boxes.

3.5. The 3,5-T2 Sustained Inflammation Induced by D.I.O. Was Accompanied by Alteration of the Cell Turnover in the Anterior and Mid Intestine

Inflammation and ER-stress induced by D.I.O. and sustained by 3,5-T2 were accompanied by intestinal epithelial apoptosis, as indicated by the increase in caspase-3 immuno-positive cells. In particular, an increase of caspase-3 positive epithelial cells was found in D.I.O., D.I.O. flw 3,5-T2, and D.I.O. with 3,5-T2 respect to the control. An increase of caspase 3 positive cells was observed at the base of the villi, mainly in the AI of D.I.O. and in the MI of D.I.O. with 3,5-T2.

The cell count showed a significant increase of caspase 3 positive cell number in D.I.O. (AI: ctrl 28 ± 1 vs. D.I.O. 119 ± 3, $p < 0.001$; MI: ctrl 13 ± 1 vs. D.I.O. 42 ± 1, $p < 0.05$), in the MI of D.I.O. flw 3,5-T2 (ctrl 13 ± 1 vs. D.I.O. flw 3,5-T2 78 ± 1.5, $p < 0.05$), and in the AI and MI of D.I.O. with 3,5-T2 zebrafish (AI: ctrl 28 ± 1 vs. D.I.O. with 3,5-T2 82 ± 1, $p < 0.05$; MI: ctrl 13 ± 1 vs. D.I.O. with 3,5-T2 88 ± 2, $p < 0.001$) compared to the control. Interestingly, a significant decrease in caspase 3 positive cell number was found in the AI of D.I.O. flw 3,5-T2 with respect to D.I.O. (D.I.O. 119 ± 3 vs. D.I.O. flw 3,5-T2 63 ± 1, $p < 0.05$), while a significant increase of caspase 3 positive cell number was found in the MI of D.I.O. with 3,5-T2 compared with D.I.O. (D.I.O. 42 ± 1 vs. D.I.O. with 3,5-T2 88 ± 2, $p < 0.05$) (Figure 6).

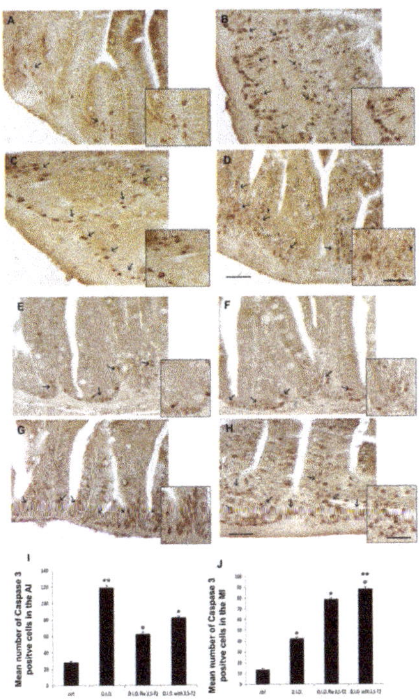

Figure 6. Caspase 3 immunostaining in the anterior (AI) and mid (MI) intestine of ctrl (control zebrafish), D.I.O. (diet-induced obesity zebrafish), D.I.O. flw 3,5–T2 (D.I.O. zebrafish followed by 3,5–T2), D.I.O. with 3,5–T2 (D.I.O. zebrafish treated with 3,5–T2). (**A**) AI of ctrl zebrafish (**B**) AI of D.I.O. (**C**) AI of D.I.O. flw 3,5–T2. (**D**) AI of D.I.O. with 3,5–T2. (**E**) MI of ctrl zebrafish. (**F**) MI of D.I.O. (**G**) MI of D.I.O. flw 3,5–T2. and (**H**) MI of D.I.O. with 3,5–T2. The arrows indicate caspase 3 positive cells in the folds. (**I**,**J**) Bar graphs showing the mean number of caspase 3 positive cells in the (**I**) AI and (**J**) MI of ctrl, D.I.O. flw 3,5–T2, D.I.O. and D.I.O. with 3,5–T2. Data are expressed as mean ± SE. ** $p < 0.001$, * $p < 0.05$ compared to the control group. ° $p < 0.05$ compared to D.I.O. Scale bar: 50 μm in the low magnification and 25 μm in the higher magnification in the boxes.

Since the intestine is characterized by a continuous cell turnover, we analyzed also the expression of the proliferative cells. The increase of apoptotic cells was paralleled by the reduction in proliferating cells, as indicated by the decrease in PCNA immuno-positive cells (Figure 7). Specifically, the number of PCNA positive cells was significantly reduced in the MI of D.I.O. with respect to the control (ctrl 113 ± 5 vs. D.I.O. 62 ± 3, $p < 0.05$). The reduction was found also in the AI of D.I.O. flw 3,5-T2 (ctrl 139 ± 2 vs. D.I.O. flw 3,5-T2 79 ± 3, $p < 0.05$), and in the AI and MI of D.I.O. with 3,5-T2 (AI: ctrl 139 ± 2 vs. D.I.O. with 3,5-T2 57 ± 2, $p < 0.001$; MI: ctrl 113 ± 5 vs. D.I.O. with 3,5-T2 55 ± 4, $p < 0.001$) compared to the control. Moreover, a significant decrease in PCNA positive cell number was found in the AI of D.I.O. with 3,5-T2 with respect to the D.I.O. (D.I.O. 105 ± 1 vs. D.I.O. with 3,5-T2 57 ± 2, $p < 0.05$) (Figure 7).

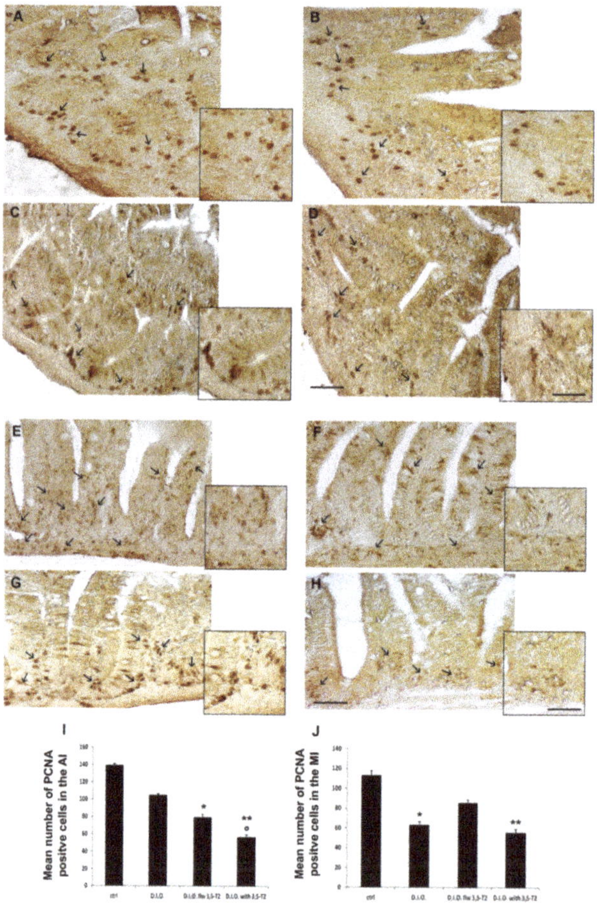

Figure 7. PCNA immunostaining in the anterior (AI) and mid (MI) intestine of ctrl (control zebrafish), D.I.O. (diet-induced obesity zebrafish), D.I.O. flw 3,5–T2 (D.I.O. zebrafish followed by 3,5–T2), D.I.O. with 3,5–T2 (D.I.O. zebrafish treated with 3,5–T2). (**A**) AI of ctrl zebrafish (**B**) AI of D.I.O. (**C**) AI of D.I.O. flw 3,5–T2. (**D**) AI of D.I.O. with 3,5–T2. (**E**) MI of ctrl zebrafish. (**F**) MI of D.I.O. (**G**) MI of D.I.O. flw 3,5–T2. and (**H**) MI of D.I.O. with 3,5–T2. The arrows indicate PCNA positive cells in the folds. (**I**,**J**) Bar graphs showing PCNA optical density in the (**I**) AI and (**J**) MI of ctrl, D.I.O., D.I.O. flw 3,5–T2 and D.I.O. with 3,5–T2. Data are expressed as mean ± SE. ** $p < 0.001$, * $p < 0.05$ compared to the control group. ° $p < 0.05$ compared to D.I.O. Scale bar: 50 μm in the low magnification and 25 μm in the higher magnification in the boxes.

3.6. 3,5-T2 Sustained or Increased Brain Inflammation Induced by D.I.O.

The peripheral morphological alteration and the activation of the inflammation induced by D.I.O. and sustained by 3,5-T2 were paralleled by the inflammation in the central nervous system. By single immunohistochemistry, we found an increase in Iba1 immunoreactivity coupled with the typical morphology of activated microglia. Microglia, the primary immune cells of the central nervous system, is normally present in the resting state, but it becomes activated under pathological conditions or inflammation state. Iba-1-positive cells were identified as resting (with small somata bearing long, thin, and ramified processes) in the control group, and activated (with hypertrophy together with retraction of processes that appeared shorter than the diameter of the somata) or dystrophic microglia (with fragmented processes and an irregularly shaped cell body) in D.I.O., D.I.O. flw 3,5-T2 and D.I.O. with 3,5-T2 (Figure 8).

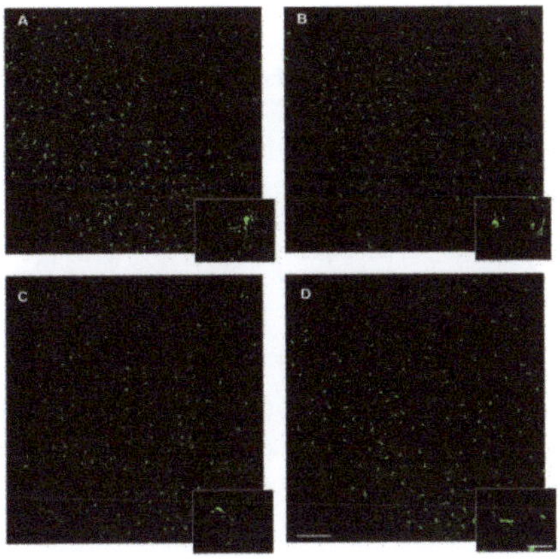

Figure 8. Iba1 immunostaining in the brain of zebrafish. (**A**) Representative image of the hypothalamus of control zebrafish showing resting microglia depicted in the box with high magnification. (**B**) Representative image of D.I.O. (diet-induced obesity zebrafish) hypothalamus showing activated or dystrophic microglia depicted in the box with high magnification. (**C**) Representative image of the hypothalamus of D.I.O. followed by 3,5-T2 showing activated or dystrophic microglia depicted in the box with high magnification. (**D**) Representative image of the hypothalamus of D.I.O. treated with 3,5-T2 showing activated or dystrophic microglia depicted in the box with high magnification. Scale bar: 100 μm in the low magnification and 25 μm in the higher magnification in the boxes.

In particular, the quantitative analysis of resting, activated, or dystrophic microglia in the hypothalamus of D.I.O. zebrafish showed a significant increase of activated microglia with respect to the control (ctrl: 1.78 ± 0.4 vs. D.I.O.: 7.44 ± 0.9, $p < 0.0001$), with a significant increase of dystrophic microglia (ctrl: 0.5 ± 0.2 vs. D.I.O.: 2.2 ± 0.17, $p < 0.0001$). Moreover, the D.I.O. flw 3,5-T2 and D.I.O. with 3,5-T2 also have showed a significant increase of activated microglia (ctrl: 1.78 ± 0.4 vs. D.I.O. flw 3,5-T2: 5.67 ± 0.5, $p < 0.0001$; ctrl: 1.78 ± 0.4 vs. D.I.O. with 3,5-T2: 8 ± 0.67, $p < 0.0001$) and dystrophic microglia (ctrl: 0.5 ± 0.2 vs. D.I.O. flw 3,5-T2: 2.33 ± 0.13, $p < 0.0001$; ctrl: 0.5 ± 0.2 vs. D.I.O. with 3,5-T2: 3.67 ± 0.19, $p < 0.0001$) compared to the control. More interestingly, the D.I.O. with 3,5-T2 showed a significant increase in the number of dystrophic microglia with respect to the D.I.O. (D.I.O.: 2.2 ± 0.17 vs. D.I.O. with 3,5-T2: 3.67 ± 0.19, $p < 0.0001$) (Figure 9)

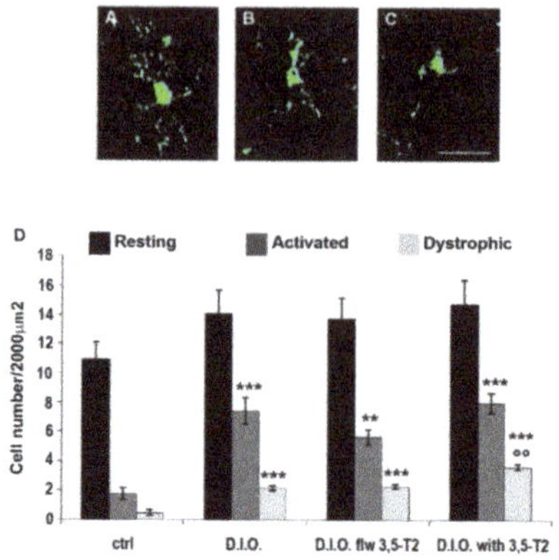

Figure 9. Morphological analysis of microglia in the brain of zebrafish. (**A–C**) Representative images of different microglial morphology. (**A**) resting microglia, (**B**) activated microglia, and (**C**) dystrophic microglia. Scale bar: 25 µm. (**D**) Quantitative analysis of the resting, activated, and dystrophic-like cells in the hypothalamus of ctrl (control zebrafish), D.I.O. (diet-induced obesity zebrafish), D.I.O. flw 3,5–T2 (D.I.O. zebrafish followed by 3,5–T2), D.I.O. with 3,5–T2 (D.I.O. zebrafish treated with 3,5–T2). Data are expressed as mean ± SE. *** $p < 0.0001$, ** $p < 0.001$ compared to the control group. °° $p < 0.05$ compared to D.I.O.

4. Discussion

3,5-T2 is well-known for its role in promoting metabolic rate and oxygen consumption, leading to beneficial metabolic effects on HFD rodents [20,36]. Goglia and colleagues have shown the positive role of 3,5-T2 on mitochondrial functions, fat mass, and lipid metabolism, supporting the use of 3,5-T2 as a drug for the treatment of obesity [14,37,38]. Although the ability of 3,5-T2 to increase metabolism has been reported [39], the safe and effective dose as well as the possible adverse effects of this molecule have not been sufficiently explored. Moreover, a lack of studies exists on the possible effects of 3,5-T2 on the intestine and the brain.

Given the highly conserved gut function and immunity-related genes between mammals and zebrafish, the latter has become an interesting animal model to investigate the essential processes underlying intestinal inflammation and injury [40]. Recent studies have revealed a well-conserved organization of the intestinal system and brain between zebrafish and mammals [29,41,42]. Moreover, strong evidence supports that the overconsumption of nutrients leads to systemic inflammation and alters metabolic homeostasis acting on central and peripheral systems [3,43]. Previous studies have shown the increase in intestinal inflammation in D.I.O. zebrafish [1,3,6], but no data on the role of 3,5-T2 are available.

Here we show the 3,5-T2 functional role in peripheral and central inflammation in D.I.O. zebrafish model. We detected that the dose of 3,5-T2 can affect not only the health status, but also the survival of zebrafish. In particular, our results indicate that a high dose of 3,5-T2 (100 nM) caused death, while a lower dose (10 nM) did not affect survival, although increased the morphological alterations in the intestine caused by D.I.O, and sustained the peripheral and central inflammation. The high mortality induced by the 100 nM dose could be explained by the reported thyrotoxic effect induced by chronic administration of high doses of 3,5-T2 in mice [21].

The study presented here demonstrates that both the post- and contemporary- treatment with 3,5-T2 (10 nM) of D.I.O. zebrafish, strongly affected weight and triglyceride levels. This first result is in contrast with previous studies which report a positive or null effect of 3,5-T2 on weight and fat mass in high fat fed rodents [39,44].

Studies in humans and animal models, such as rodents and zebrafish, report cases of intestinal inflammation induced by a HFD [45–47]. Such alterations are induced and modulated by pro-inflammatory cytokines, that are well-known to play a key role in metabolic diseases [10]. 3,5-T2 is associated with many modern lifestyle-induced conditions, and zebrafish has emerged as an important model for human endocrine diseases including obesity, diabetes, and metabolic syndrome [48]. In this study, 3,5-T2 treatment led to a worsening of the intestinal morphological alterations caused by a high-fat diet. In particular, in the D.I.O. zebrafish, the mucosal architecture was damaged, as revealed by (i) the cell fusion at the apical border of the villi, (ii) the thinning of the villi themselves and iii) the significant increase of goblet cell number. The post- or contemporary-treatment with 3,5-T2 of D.I.O. zebrafish sustained these morphological alteration and in particular increased the number of goblet cells in the anterior intestine, which is the first gut segment providing the nutrient absorption.

However, the post- and contemporary- treatment with 3,5-T2 (10 nM) of D.I.O. zebrafish showed different effect on the activation of pro-inflammatory markers, with the contemporary treatment sustaining the effect of D.I.O. and the post-treatment partially reverting the effect of D.I.O. Mainly, our results are in disagreement with the beneficial role of 3,5-T2 reported in many studies [14,19,36], since the comparative analysis of D.I.O. vs. 3,5-T2-treated D.I.O. zebrafish indicates a role for 3,5-T2 in the upregulation of pro-inflammatory markers. In this study, 3,5-T2 co-treatment of D.I.O. zebrafish generally sustained the increase of pro-inflammatory markers, such as TNFα and COX2, two major mediators of inflammation. On the contrary, the post-treatment with 3,5-T2 of D.I.O. zebrafish induced a decrease in the expression of these inflammatory markers. The mitochondrium is one of the cellular targets of 3,5-T2, where it stimulates the respiration via a direct action on the cytochrome C oxidase (COX), with a consequent reduction of the respiratory efficiency [13–16]. 3,5-T2 leads to a rapid increase in mitochondrial oxygen consumption, which is then reflected at the whole animal level [17]. The increase of COX expression is one of the major markers of inflammation and, interestingly, in our study the contemporary treatment of D.I.O. zebrafish with 3,5-T2 sustained the enhancement of COX2 expression induced by D.I.O. This effect could be due to the alteration of mitochondrial number and oxidative function induced by chronic HFD and which can change with type of fat contained in the diet inducing obesity [49,50].

This inflammatory pathway activated by D.I.O. and maintained by 3,5-T2 induced cellular stress, as confirmed by the significant increased expression of calnexin, a marker of alteration of the endoplasmic reticulum activity (ER-stress) in intestinal enteroendocrine and goblet cells. On the contrary, the post-treatment with 3,5-T2 reduced the activation of this inflammatory pathway, as confirmed by the decrease of calnexin expression.

Often inflammation and ER-stress are accompanied by apoptotic events. In agreement with the inflammation status, 3,5-T2 co-treatment of D.I.O. zebrafish caused the increase of caspase 3-immunoexpression in the mid intestine. Since it is well-known that in the zebrafish intestine a constant cell turnover is at the basis of the correct morphology and activity of the villi [40], we analyzed also the number of proliferating cells. As expected, the enhancement of apoptotic cells was paralleled by the 3,5-T2 sustained decrease in proliferating cells, which can explain the reduction of villi size with the consequent enlargement of the gut lumen.

Therefore, the contemporary treatment with 3,5-T2 of D.I.O. zebrafish did not ameliorate the inflammation status induced by diet, on the contrary, it seemed to sustain the high expression of markers that play a substantial role in the modulation of inflammation. However, the post-treatment with 3,5-T2 of D.I.O. zebrafish partially reverted the inflammation status induced by the high-fat diet, maybe because of the association of 3,5-T2 treatment with a change of diet. Thus, the use of 3,5-T2 as pharmacological agent and diet supplement should be considered with caution, taking into consideration that it is a metabolic derivate of thyroid hormones, in turn, able to exert side effects on target organs, such as intestine, liver, kidney, and heart. Moreover, while most of the studies reporting beneficial effects of 3,5-T2 have been performed in rodent models of obesity with chronic hypothyroidism [51–54], data on the effects of 3,5-T2 on metabolism in HFD euthyroid animal models or obese models brought back to a normal diet are scarce [55]. In a recently published study, the treatment with 3,5-T2 performed in diet-induced obese euthyroid mice, resulted in the reduction of body fat mass, but with an undesired negative feedback inhibition of the hypothalamus–pituitary–thyroid (HPT) axis accompanied by increased heart weight [20].

In this study, the 3,5-T2-sustained inflammation seemed to occur in both AI and MI, when 3,5-T2 treatment was given at the same time of the obesity-inducing diet, while when given after D.I.O. and in association with a normal diet it partly reverted the inflammation status. The intestine is one of the primary organs responsible for the absorption of nutrients and can be modified by the diet. Through the well-known bi-directional gut–brain axis, the intestine can induce changes in the central nervous system and consequently affect the brain functions [56,57]. We show here that the contemporary treatment with 3,5-T2 negatively modulated the expression of pro-inflammatory markers supporting the alteration of the intestine morphology and the activation of inflammation in the gut induced by D.I.O. Such pro-inflammatory effect was accompanied by an increase of inflammation in the central nervous system. Indeed, we detected a strong microglia activation in the brain of D.I.O. zebrafish treated with 3,5-T2, with an increase of dystrophic microglia in numerous brain regions, such as the hypothalamus, which is the main feeding center of the central nervous system. Therefore, we can hypothesize that the co-treatment with 3,5-T2 during obesity inducing diet, led to a reinforcement of the pro-inflammatory pathways at peripheral level and consequent intestine morphological alteration, able to induce an increase of triglyceride levels and to sustain inflammation at central nervous system level through the gut-brain axis. These side effects probably affect nutrient absorption and the central regulation of food intake, leading to the increase in the body weight and triglycerides.

To the best of our knowledge, this is the first immunohistochemistry analysis of intestine and brain tissues from D.I.O. zebrafish treated with 3,5-T2. Densitometric and quantitative approaches showed strong effects of 3,5-T2 on body weight and triglycerides accompanied by a sustained peripheral and central inflammation in D.I.O. zebrafish. These findings are surprising since they are in contrast with previous studies in HFD rodents reporting beneficial effects of 3,5-T2 on the metabolic rate [14,58]. However, our results are in agreement with the conclusions of recent studies on 3,5-T2 action in rat hepatic nuclei [28,59], where 3,5-T2 showed toxic effects in the liver [60]. We should take into account that the effects of 3,5-T2 can be influenced by the model, the dietary conditions, the diet fat composition, and by 3,5-T2 concentration, uptake, metabolism, and elimination. Overall, administration of 3,5-T2 might exert or sustain peripheral and central adverse effects induced by diet and additional studies are required to further delineate the mechanisms whereby 3,5-T2 acts and the specific contexts in which it can be efficacious or toxic. Therefore, the potential use of 3,5-T2 to fight obesity, as it is suggested in several publications or commentaries, should be viewed with caution, posing special attention to the role played by the diet associated with 3,5-T2 treatment.

5. Conclusions

Since it has been reported that the thyromimetic effects of 3,5-T2 treatment may affect many organs, resulting in long-term undesirable effects, a reliable model to study the side effects of 3,5-T2 on important organs targeted by thyroid hormones (e.g., pituitary, brain, gut, heart, bone, muscle) is clearly useful to evaluate any potential risk. Zebrafish has been recognized as an excellent animal model for studying human metabolic and inflammatory diseases. We suggest that the adult zebrafish is a reproducible and adequate model to test the effects of 3,5-T2 on the gut–brain axis, with particular attention to the systemic and central inflammation induced by the diet. This work underlies how the zebrafish model can help to better understand the fundamental beneficial and side effects of 3,5-T2, which is of great importance to define the possible use of this metabolite of thyroid hormones as a drug in different diseases including obesity.

Supplementary Materials: The following are available online at http://www.mdpi.com/2076-2615/10/7/1131/s1, Figure S1: Antibody specificity control. Figure S2: Alcian blue staining of the zebrafish intestine.

Author Contributions: R.I. designed the experiment, carried out tissue processing, immunohistochemistry, and analyzed the data. O.S. and H.A.M. designed the experiment and took care of the animal husbandry. L.C., L.D., P.D.G., C.A. analyzed the data and wrote the manuscript. L.T., I.M., and E.V. carried out tissue processing, immunohistochemistry and quantitative analysis. M.P. designed the experiment, analyzed the data and wrote the manuscript. All authors have read and agreed to the published version of the manuscript.

Funding: This study was supported by Fondi Ricerca di Ateneo (FRA), University of Sannio.

Acknowledgments: The authors wish to acknowledge Elena Silvestri for her assistance during the zebrafish treatment.

Conflicts of Interest: The authors declare that the research was carried out in the absence of any commercial or financial relationships that could be considered as a potential conflict of interest.

References

1. Araujo, J.R.; Tomas, J.; Brenner, C.; Sansonetti, P.J. Impact of high-fat diet on the intestinal microbiota and small intestinal physiology before and after the onset of obesity. *Biochimie* **2017**, *141*, 97–106. [CrossRef] [PubMed]
2. Saltiel, A.R.; Olefsky, J.M. Inflammatory mechanisms linking obesity and metabolic disease. *J. Clin. Invest.* **2017**, *127*, 1–4. [CrossRef] [PubMed]
3. Arias-Jayo, N.; Abecia, L.; Alonso-Sáez, L.; Ramirez-Garcia, A.; Rodriguez, A.; Pardo, M.A. High-Fat diet consumption induces microbiota dysbiosis and intestinal inflammation in zebrafish. *Microb. Ecol.* **2018**, *76*, 1089–1101. [CrossRef] [PubMed]
4. Oka, T.; Nishimura, Y.; Zang, L.; Hirano, M.; Shimada, Y.; Wang, Z.; Umemoto, N.; Kuroyanagi, J.; Nishimura, N.; Tanaka, T. Diet-induced obesity in zebrafish shares common pathophysiological pathways with mammalian obesity. *BMC Physiol.* **2010**, *10*, 21. [CrossRef]
5. Kieffer, T.J.; Habener, J.F. The adipoinsular axis: Effects of leptin on pancreatic β-cells. *Am. J. Physiol. Endocrinol. Metab.* **2000**, *278*, E1–E14. [CrossRef]
6. Bessesen, D.H. Update on obesity. *J. Clin. Endocrinol. Metab.* **2008**, *93*, 2027–2034. [CrossRef]
7. Casper, R.C.; Sullivan, E.L.; Tecott, L. Relevance of animal models to human eating disorders and obesity. *Psychopharmacology* **2008**, *199*, 313–329. [CrossRef]
8. Carnovali, M.; Luzi, L.; Terruzzi, I.; Banfi, G.; Mariotti, M. Metabolic and bone effects of high-fat diet in adult zebrafish. *Endocrine* **2018**, *61*, 317–326. [CrossRef]
9. Wannamethee, S.G.; Tchernova, J.; Whincup, P.; Lowe, G.D.; Kelly, A.; Rumley, A.; Wallace, A.M.; Sattar, N. Plasma leptin: Associations with metabolic, inflammatory and haemostatic risk factors for cardiovascular disease. *Atherosclerosis* **2007**, *19*, 418–426. [CrossRef]
10. Rainone, V.; Schneider, L.; Saulle, I.; Ricci, C.; Biasin, M.; Al-Daghri, N.M.; Giani, E.; Zuccotti, G.V.; Clerici, M.; Trabattoni, D. Upregulation of inflammasome activity and increased gut permeability are associated with obesity in children and adolescents. *Int. J. Obes.* **2016**, *40*, 1026–1033. [CrossRef]
11. Ellulu, M.S.; Patimah, I.; Khaza'ai, H.; Rahmat, A.; Abed, Y. Obesity and inflammation: The linking mechanism and the complications. *Arch. Med. Sci.* **2017**, *13*, 851–863. [CrossRef] [PubMed]
12. Bournat, J.C.; Brown, C.W. Mitochondrial dysfunction in obesity. *Curr. Opin. Endocrinol. Diabetes Obes.* **2010**, *17*, 446–452. [CrossRef] [PubMed]
13. Cioffi, F.; Senese, R.; Lanni, A.; Goglia, F. Thyroid hormones and mitochondria: With a brief look at derivatives and analogues. *Mol. Cell. Endocrinol.* **2013**, *379*, 51–61. [CrossRef] [PubMed]
14. Goglia, F. The effects of 3,5-diiodothyronine on energy balance. *Front. Physiol.* **2014**, *5*, 528. [CrossRef] [PubMed]
15. Lanni, A.; Moreno, M.; Goglia, F. Mitochondrial actions of thyroid hormone. *Compr. Physiol.* **2016**, *6*, 1591–1607. [CrossRef]
16. Senese, R.; de Lange, P.; Petito, G.; Moreno, M.; Goglia, F.; Lanni, A. 3,5-Diiodothyronine: A Novel Thyroid Hormone Metabolite and Potent Modulator of Energy Metabolism. *Front. Endocrinol.* **2018**, *9*, 427. [CrossRef]
17. Lanni, A.; Moreno, M.; Lombardi, A.; Goglia, F. Calorigenic effect of diiodothyronines in the rat. *J. Physiol.* **1996**, *494 (Pt 3)*, 831–837. [CrossRef]

18. Silvestri, E.; Cioffi, F.; De Matteis, R.; Senese, R.; de Lange, P.; Coppola, M.; Salzano, A.A.; Scaloni, A.; Ceccarelli, M.; Goglia, F.; et al. 3,5-Diiodo-L-thyronine affects structural and metabolic features of skeletal muscle mitochondria in high-fat-diet fed rats producing a co-adaptation to the glycolytic fiber phenotype. *Front. Physiol.* **2018**, *9*, 194. [CrossRef]
19. Coppola, M.; Cioffi, F.; Moreno, M.; Goglia, F.; Silvestri, E. 3,5-diiodo-L-thyronine: A Possible Pharmacological Agent? *Curr. Drug. Deliv.* **2016**, *13*, 330–338. [CrossRef]
20. Jonas, W.; Lietzow, J.; Wohlgemuth, F.; Hoefig, C.S.; Wiedmer, P.; Schweizer, U.; Köhrle, J.; Schürmann, A. 3,5-Diiodo-L-Thyronine (3,5-T_2) Exerts Thyromimetic Effects on Hypothalamus-Pituitary-Thyroid Axis, Body Composition, and Energy Metabolism in Male Diet-Induced Obese Mice. *Endocrinology* **2015**, *156*, 389–399. [CrossRef]
21. Da Silva Teixeira, S.; Filgueira, C.; Sieglaff, D.H.; Benod, C.; Villagomez, R.; Minze, L.J.; Zhang, A.; Webb, P.; Nunes, M.T. 3,5-diiodothyronine (3,5-T2) reduces blood glucose independently of insulin sensitization in obese mice. *Acta Physiol.* **2017**, *220*, 238–250. [CrossRef] [PubMed]
22. Coppola, M.; Glinni, D.; Moreno, M.; Cioffi, F.; Silvestri, E.; Goglia, F. Thyroid hormone analogues and derivatives: Actions in fatty liver. *World J. Hepatol.* **2014**, *6*, 114–129. [CrossRef] [PubMed]
23. D'Angelo, L.; Lossi, L.; Merighi, A.; de Girolamo, P. Anatomical features for the adequate choice of experimental animal models in biomedicine: I. Fishes. *Ann. Anat.* **2016**, *205*, 75–84. [CrossRef]
24. Pappalardo, A.; Porreca, I.; Caputi, L.; De Felice, E.; Schulte-Merker, S.; Zannini, M.; Sordino, P. Thyroid development in zebrafish lacking Taz. *Mech. Dev.* **2015**, *3*, 268–278. [CrossRef]
25. Porreca, I.; De Felice, E.; Fagman, H.; Di Lauro, R.; Sordino, P. Zebrafish bcl2l is a survival factor in thyroid development. *Dev. Biol.* **2012**, *15*, 142–152. [CrossRef] [PubMed]
26. Olvera, A.; Martyniuk, C.J.; Buisine, N.; Jimenez-Jacinto, V.; Sanchez-Flores, A.; Sachs, L.M.; Orozco, A. Differential transcriptome regulation by 3,5-T2 and 3′,3,5-T3 in brain and liver uncovers novel roles for thyroid hormones in tilapia. *Sci. Rep.* **2017**, *7*, 15043. [CrossRef]
27. Little, A.G.; Seebacher, F. Thyroid hormone regulates cardiac performance during cold acclimation in zebrafish (Danio rerio). *J. Exp. Biol.* **2014**, *217 (Pt 5)*, 718–725. [CrossRef]
28. Navarrete-Ramirez, P.; Luna, M.; Valverde, R.C.; Orozco, A. 3,5-di-iodothyronine stimulates tilapia growth through an alternate isoform of thyroid hormone receptor beta1. *J. Mol. Endocrinol.* **2014**, *52*, 1–9. [CrossRef]
29. Imperatore, R.; D'Angelo, L.; Safari, O.; Motlagh, H.A.; Piscitelli, F.; de Girolamo, P.; Cristino, L.; Varricchio, E.; Di Marzo, V.; Paolucci, M. Overlapping Distribution of Orexin and Endocannabinoid Receptors and Their Functional Interaction in the Brain of Adult Zebrafish. *Front. Neuroanat.* **2018**, *30*, 12–62. [CrossRef]
30. García, G.C.; Jeziorski, M.C.; Valverde, R.C.; Orozco, A. Effects of iodothyronines on the hepatic outer-ring deiodinating pathway in killifish. *Gen. Comp. Endocrinol.* **2004**, *135*, 201–209. [CrossRef]
31. García, G.C.; López-Bojorquez, L.; Nuñez, J.; Valverde, R.C.; Orozco, A. 3,5-Diiodothyronine in vivo maintains euthyroidal expression of type 2 iodothyronine deiodinase, growth hormone, and thyroid hormone receptor beta1 in the killifish. *Am. J. Physiol. Regul. Integr. Comp. Physiol.* **2007**, *293*, R877–R883. [CrossRef] [PubMed]
32. Mania, M.; Maruccio, L.; Russo, F.; Abbate, F.; Castaldo, L.; D'Angelo, L.; de Girolamo, P.; Guerrera, M.C.; Lucini, C.; Madrigrano, L.; et al. Expression and distribution of leptin and its receptors in the digestive tract of DIO (diet-induced obese) zebrafish. *Ann. Anat.* **2017**, *212*, 37–47. [CrossRef] [PubMed]
33. Imperatore, R.; D'Angelo, L.; De Girolamo, P.; Cristino, L.; Paolucci, M. Identification of Orexin and Endocannabinoid Receptors in Adult Zebrafish Using Immunoperoxidase and Immunofluorescence Methods. *J. Vis. Exp.* **2019**, *148*. [CrossRef] [PubMed]
34. Varricchio, E.; Russo, F.; Coccia, E.; Turchini, G.; Francis, D.; De Girolamo, P.; Paolucci, M. Immunohistochemical and immunological detection of ghrelin and leptin in rainbow trout oncorhynchus mykiss and murray cod maccullochella peelii peelii as affected by different dietary fatty acids. *Microsc. Res. Tech.* **2012**, *75*, 771–780. [CrossRef]
35. Streit, W.J.; Xue, Q.S. Life and death of microglia. *J. Neuroimmun. Pharmacol.* **2009**, *4*, 371–379. [CrossRef]
36. Antonelli, A.; Fallahi, P.; Ferrari, S.M.; Di Domenicantonio, A.; Moreno, M.; Lanni, A.; Goglia, F. 3,5-diiodo-L-thyronine increases resting metabolic rate and reduces body weight without undesirable side effects. *J. Biol. Regul. Homeost. Agents* **2011**, *25*, 655–660.
37. Moreno, M.; de Lange, P.; Lombardi, A.; Silvestri, E.; Lanni, A.; Goglia, F. Metabolic effects of thyroid hormone derivatives. *Thyroid* **2008**, *18*, 239–253. [CrossRef]

38. Cioffi, F.; Lanni, A.; Goglia, F. Thyroid hormones, mitochondrial bioenergetics and lipid handling. *Curr. Opin. Endocrinol. Diabetes Obes.* **2010**, *17*, 402–407. [CrossRef]
39. Hernandez, A. 3,5-Diiodo-L-Thyronine (T2) in Dietary Supplements: What Are the Physiological Effects? *Endocrinology* **2015**, *156*, 5–7. [CrossRef]
40. Brugman, S. The zebrafish as a model to study intestinal inflammation. *Dev. Comp. Immunol.* **2016**, *64*, 82–92. [CrossRef]
41. Wang, Z.; Du, J.; Lam, S.H.; Mathavan, S.; Matsudaira, P.; Gong, Z. Morphological and molecular evidence for functional organization along the rostrocaudal axis of the adult zebrafish intestine. *BMC Genomics* **2010**, *11*, 392. [CrossRef] [PubMed]
42. Goldsmith, J.R.; Jobin, C. Think small: Zebrafish as a model system of human pathology. *J. Biomed. Biotechnol.* **2012**, *2012*, 817341. [CrossRef] [PubMed]
43. Patterson, E.; Ryan, P.M.; Cryan, J.F.; Dinan, T.G.; Ross, R.P.; Fitzgerald, G.F.; Stanton, C. Gut microbiota, obesity and diabetes. *Postgrad. Med. J.* **2016**, *92*, 286–300. [CrossRef] [PubMed]
44. Vatner, D.F.; Snikeris, J.; Popov, V.; Perry, R.J.; Rahimi, Y.; Samuel, V.T. 3 3,5 Diiodo-L-Thyronine (T2) does not prevent hepatic steatosis or insulin resistance in fat-fed sprague dawley rats. *PLoS ONE* **2015**, *20*, e0140837. [CrossRef] [PubMed]
45. Cani, P.D.; Bibiloni, R.; Knauf, C.; Waget, A.; Neyrinck, A.M.; Delzenne, N.M.; Burcelin, R. Changes in gut microbiota control metabolic endotoxemia-induced inflammation in high-fat diet-induced obesity and diabetes in mice. *Diabetes* **2008**, *57*(6), 1470–1481. [CrossRef] [PubMed]
46. Landgraf, K.; Schuster, S.; Meusel, A.; Garten, A.; Riemer, T.; Schleinitz, D.; Kiess, W.; Körner, A. Short-term overfeeding of zebrafish with normal or high-fat diet as a model for the development of metabolically healthy versus unhealthy obesity. *BMC Physiol.* **2017**, *17*, 4. [CrossRef]
47. Wong, S.; Stephens, W.Z.; Burns, A.R.; Stagaman, K.; David, L.A.; Bohannan, B.J.; Guillemin, K.; Rawls, J.F. Ontogenetic differences in dietary fat influence microbiota assembly in the zebrafish gut. *MBio* **2015**, *6*, e00687-15. [CrossRef]
48. Zang, L.; Maddison, L.A.; Chen, W. Zebrafish as a Model for Obesity and Diabetes. *Front. Cell Dev. Biol.* **2018**, *6*, 91. [CrossRef]
49. Cole, M.A.; Murray, A.J.; Cochlin, L.E.; Heather, L.C.; McAleese, S.; Knight, N.S.; Sutton, E.; Jamil, A.A.; Parassol, N.; Clarke, K. A high fat diet increases mitochondrial fatty acid oxidation and uncoupling to decrease efficiency in rat heart. *Basic Res. Cardiol.* **2011**, *106*, 447–457. [CrossRef]
50. Ørgensen, W.; Rud, K.A.; Mortensen, O.H.; Frandsen, L.; Grunnet, N.; Quistorff, B. Your mitochondria are what you eat: A high-fat or a high-sucrose diet eliminates metabolic flexibility in isolated mitochondria from rat skeletal muscle. *Physiol. Rep.* **2017**, *5*, e13207. [CrossRef]
51. De Lange, P.; Cioffi, F.; Senese, R.; Moreno, M.; Lombardi, A.; Silvestri, E.; De Matteis, R.; Lionetti, L.; Mollica, M.P.; Goglia, F.; et al. Nonthyrotoxic prevention of diet-induced insulin resistance by 3,5-diiodo-L-thyronine in rats. *Diabetes* **2011**, *60*, 2730–2739. [CrossRef] [PubMed]
52. Grasselli, E.; Voci, A.; Demori, I.; Canesi, L.; Goglia, F.; Lanni, A.; Gallo, G.; Vergani, L. 3,5-Diiodo-L-thyronine modulates the expression of genes of lipid metabolism in a rat model of fatty liver. *J. Endocrinol.* **2012**, *212*, 149–158. [CrossRef] [PubMed]
53. Lombardi, A.; Senese, R.; Busiello, R.A.; Cioffi, F.; Goglia, F.; Lanni, A. 3,5-diiodo-L-thyronine activates brown adipose tissue thermogenesis in hypothyroid rats. *PLoS ONE* **2015**, *10*(2), e0116498. [CrossRef] [PubMed]
54. Cavallo, A.; Priore, P.; Gnoni, G.V.; Papa, S.; Zanotti, F.; Gnoni, A. 3,5-Diiodo-L-thyronine administration to hypothyroid rats rapidly enhances fatty acid oxidation rate and bioenergetic parameters in liver cells. *PLoS ONE* **2013**, *8*, e52328. [CrossRef]
55. Goldberg, I.J.; Huang, L.S.; Huggins, L.A.; Yu, S.; Nagareddy, P.R.; Scanlan, T.S.; Ehrenkranz, J.R. Thyroid hormone reduces cholesterol via a non-LDL receptor-mediated pathway. *Endocrinology* **2012**, *153*, 5143–5149. [CrossRef]
56. Al-Asmakh, M.; Anuar, F.; Zadjali, F.; Rafter, J.; Pettersson, S. Gut microbial communities modulating brain development and function. *Gut Microbes* **2012**, *3*, 366–373. [CrossRef]
57. Cryan, J.F.; Dinan, T.G. Mind-altering microorganisms: The impact of the gut microbiota on brain and behaviour. *Nat. Rev. Neurosci.* **2012**, *13*, 701–712. [CrossRef]
58. Goglia, F. Biological effects of 3,5-diiodothyronine (T(2)). *Biochemistry* **2005**, *70*, 164–172. [CrossRef]

59. Koerner, D.; Schwartz, H.L.; Surks, M.I.; Oppenheimer, J.H. Binding of selected iodothyronine analogues to receptor sites of isolated rat hepatic nuclei. High correlation between structural requirements for nuclear binding and biological activity. *J. Biol. Chem.* **1975**, *250*, 6417–6423.
60. Lietzow, J.; Golchert, J.; Homuth, G.; Völker, U.; Jonas, W.; Köhrle, J. 3,5-T2 alters murine genes relevant for xenobiotic, steroid, and thyroid hormone metabolism. *J. Mol. Endocrinol.* **2016**, *56*, 311–323. [CrossRef]

MDPI
St. Alban-Anlage 66
4052 Basel
Switzerland
Tel. +41 61 683 77 34
Fax +41 61 302 89 18
www.mdpi.com

Animals Editorial Office
E-mail: animals@mdpi.com
www.mdpi.com/journal/animals

www.ingramcontent.com/pod-product-compliance
Lightning Source LLC
Chambersburg PA
CBHW061756210926
51850CB00006B/574
9783036568829